Insights into Speciality Inorganic Chemicals

Insights into Speciality Inorganic Chemicals

Edited by

David Thompson

Consulting Chemist, Reading, UK

Cover diagram

Top left: Cutaway of an automobile catalytic converter, showing the ceramic monolith inside.* Chapter 7 describes how these work (page 145).

Bottom left: An electrical microcircuit.* The use of inorganic chemicals and metals in the electronics industry is described in Chapter 9 (page 197).

Right: Structure of a cuprate superconductor (see Figure 12, page 288). A discussion of high temperature superconductivity is included in Chapter 11 (page 275).

*Courtesy of Johnson Matthey plc

A catalogue record for this book is available from the British Library.

ISBN 0-85404-504-X

Published by The Royal Society of Chemistry,
Thomas Graham House, Science Park, Milton Road, Cambridge CB4 4WF, UK

Typeset by Vision Typesetting, Manchester, UK
Printed and bound by Athenaeum Press Ltd, Gateshead, Tyne and Wear, UK

Foreword

Technology provides the means whereby a society strives to achieve its goals. Whether competitive societies and cultures can agree on common goals and global laws to regulate their activities so as to achieve a stable equilibrium with one another and with nature still awaits the judgments of history. Meanwhile, a rapidly evolving technology continues to diffuse throughout the whole. Advanced materials are the enablers of the new technologies, and our ability to control the physical and chemical properties of materials will prove critical to our future capacity to manage our human destiny. However, ultimately it is how we choose to use this technology in the twenty-first century that will determine the future of humanity.

Over the past century, industrial empiricism has increasingly given way to the molecular design of materials, including their defects, to perform a specific engineering function. The ability to design materials has, in turn, been built on three pedestals that are being continuously enhanced through a positive feedback from the activities that are built upon them. These three pedestals are the development of improved instrumentation with which to probe materials at the molecular level; an ever deepening understanding of the factors and mechanisms that control molecular architecture and processes; and the extension of our range of synthetic techniques to include 'soft' chemistry and high-pressure synthesis as well as film growth under vacuum or a rapid quenching from the melt that prevents crystallization. It is now possible to control on an industrial scale the growth of individual layers of atoms of exquisite chemical purity deposited epitaxially on single-crystal substrates, and we can look forward to an ability to arrange molecular structures and perform different functions in ordered sequences as nature does on biological membranes.

This extraordinary development has come about through a dynamic interplay between engineering and science: a particular new material or insight into a special material property suggests an engineering opportunity; a perceived engineering need drives the design/search/improved processing and properties of a material to provide the desired property. With the increase in our ability for the molecular design of both intrinsic and extrinsic material properties, the case for a dynamic interplay between engineering practice, instrumentation development, and scientific inquiry becomes ever more compelling. However, the mounting costs of the scientific/engineering enterprise are forcing cutbacks in the industrial investment in longer-term technical development; and a changing world order is causing governments to rethink priorities with respect to the training of scientists and the support of research. This juxtaposition of opportunity on the one hand and, on the other, the loss of confidence in high places in how to direct the scientific/engineering enterprise cost effectively, makes particularly timely the appearance of the unassuming exposition contained in this book.

Dr Thompson's extensive experience of working with academia from a base in industrial management has impressed upon him the reality of the opportunity and the importance to its realization of a more symbiotic relationship between the academic and industrial cultures; it is this insight that has motivated this book. As one who has built a career attempting to bridge the worlds of physics, chemistry, and engineering of materials, I share the authors' hopes that the examples of use-motivated scientific advances described herein will stimulate the present and future actors in the field of materials science and engineering.

John B. Goodenough
University of Texas at Austin, USA

March 1995

Preface

Speciality Inorganic Chemicals are defined as high added value inorganic substances used in commercial products or processes. This book covers a wide diversity of topics in inorganic chemistry, catalysis, and materials science. Throughout, the various chapters demonstrate how good basic research, carried out in university or industry, can lead to important applications and innovations relevant to community needs.

The topics covered include refining, pharmaceuticals, and both homogeneous and heterogeneous catalysts for chemical processing. Catalysts for stereospecific synthesis are also described; and catalytic systems for pollution control considered, with special emphasis on those for motor vehicles. There are four chapters on materials with special electrical and magnetic properties, *i.e.* fast ion conductors, materials for the electronics industry, magnetic materials, and superconducting materials. Then there are three chapters on inorganic colours; ceramics, glasses, and hard metals; and advanced cementitious materials. The chemistry of corrosion inhibitors is then described and this is followed by a chapter on inorganic chemicals for water purification. Finally there are chapters on inorganic chemistry in the nuclear industry, and catalysts and photocatalysts for solar energy conversion.

The forward looking 'trends' sections included at the end of each chapter should stimulate constructive thoughts on directions for future research which will lead to further applications for speciality inorganic chemicals and the concepts surrounding their use.

The book is principally written for final year undergraduates and newly qualified graduates and PhDs who are starting their careers in industry, but experienced researchers in universities and industry will also find the book very useful. Chapters have been contributed by both industrial and academic researchers and both specific and general references are given at the end of each chapter to stimulate further reading.

David Thompson

March 1995

About the Editor

David Thompson's research career has included participation in and management of university and industrial research in the fields of organic and inorganic chemistry, catalysis, and materials science, in UK and USA universities (Imperial College, London; and University of California, Los Angeles), ICI, and Johnson Matthey. He has more than 70 publications (papers and reviews) covering topics in all these areas, including books entitled 'Reactions of Transition Metal Complexes', published in 1968, and 'Universities and Industrial Research', published in 1995. The emphasis throughout his career has been on how best to progress relevant ideas from basic research into the development of useful new products and processes. Dr Thompson is currently employed as a consulting chemist advising Industrial Organizations in the UK, USA, France, and Japan on research projects in the catalyst, chemicals, and material science areas, often carried out in collaboration with universities.

Contents

Contributors

Christopher F. J. Barnard, *Johnson Matthey Technology Centre, Blount's Court, Sonning Common, Reading RG4 9NH, UK*

Peter Boden, *Department of Materials Engineering and Materials Design, University of Nottingham, University Park, Nottingham NG7 2RD, UK*

Geoffrey C. Bond, *Department of Chemistry, Brunel University, Uxbridge, Middlesex UB8 3PH, UK*

Michael Bowker, *Department of Chemistry, University of Reading, PO Box 224, Whiteknights Park, Reading RG6 2AD, UK*

John M. Brown, *Dyson Perrins Laboratory, South Parks Road, Oxford OX1 3QY, UK*

Christopher R. S. Dean, *James M. Brown Limited, Napier Street, Fenton, Stoke-on-Trent, Staffordshire ST4 4NX, UK*

Simon P. Fricker, *Johnson Matthey Technology Centre, Blount's Court, Sonning Common, Reading RG4 9NH, UK*

Fred P. Glasser, *Department of Chemistry, University of Aberdeen, Meston Walk, Old Aberdeen AB9 2UE, UK*

Philip D. Gurney, *Johnson Matthey Technology Centre, Blount's Court, Sonning Common, Reading RG4 9NH, UK*

Alexander Harper, *AEA Technology, Harwell, Didcot, Oxfordshire OX11 0RA, UK*

Andrew Harrison, *Department of Chemistry, University of Edinburgh, The King's Buildings, West Mains Road, Edinburgh EH9 3JJ, UK*

Carole C. Harrison, *Department of Chemistry and Physics, The Nottingham Trent University, Clifton Lane, Nottingham NG11 8NS, UK*

Philip G. Harrison, *Department of Chemistry, University of Nottingham, University Park, Nottingham NG7 2RD, UK*

John Hill, *formerly employed by Johannesburg Consolidated Investment Company, PO Box 590, Johannesburg 2000, South Africa; current address: Grove House, Farthingdown, Holywell Lake, Wellington, Somerset TA21 0EH, UK*

Edward W. Hooper, *AEA Technology, Harwell, Didcot, Oxfordshire OX11 0RA, UK*

John T. S. Irvine, *School of Chemistry, The University, St. Andrews, Fife KY16 9ST, UK*

Richard W. Joyner, *Research Office, Newton Building, Nottingham Trent University, Burton Street, Nottingham NG1 4BU, UK*

Don Kingerley, *Department of Materials Engineering and Materials Design, University of Nottingham, University Park, Nottingham NG7 2RD, UK*

Andrew Mills, *Department of Chemistry, University of Wales, Swansea, Singleton Park, Swansea SA2 8PP, UK*

Adrian W. Parkins, *Department of Chemistry, King's College, Strand, London WC2R 2LS, UK*

Richard J. Seymour, *Johnson Matthey Technology Centre, Blount's Court, Sonning Common, Reading RG4 9NH, UK*

Robert C. T. Slade, *Department of Chemistry, University of Exeter, Stocker Road, Exeter EX4 4QD, UK*

David T. Thompson, *'Newlands', The Village, Whitchurch Hill, Reading RG8 7PN, UK*

Owen J. Vaughan, *Johnson Matthey Technology Centre, Blount's Court, Sonning Common, Reading RG4 9NH, UK*

CHAPTER 1

Introduction

DAVID THOMPSON

We define **Speciality Inorganic Chemicals** as high added value inorganic elements, compounds, and materials used in commercial products or processes. It is the principal objective of this book to demonstrate how high quality basic or 'upstream'[1] research, followed by effective applied R&D, can lead to the development of **useful products** for the market place. Effective collaboration between university and industrial research personnel has frequently formed a basis for successful ventures of this type.

The book addresses the needs of final year undergraduate students and newly qualified graduates and postgraduates who are moving into industry at the start of their careers. The various chapters demonstrate how **careful study** of the properties of inorganic chemicals and gaining a mechanistic understanding of a topic in the inorganic chemistry and materials fields can lead to significant advances in technological areas with **new applications** identified. This kind of scientific advance can be made in both industrial and academic research laboratories; but only when we keep in touch with the needs of society, partly via contact with the commercial world, will their full potential be realized.

Enlightened interaction between members of the **academic** and **industrial** communities has frequently been an effective catalyst for identifying the relevance of new technology and evaluating its commercial potential. It is because universities and industrial companies have different cultures and objectives that such contacts are mutually stimulating and the resulting collaborations provide helpful contexts for fruitful projects.[1]

The field covered by the book is very wide and includes an indication of **refining** processes which provide the chemical elements of interest in this book, and the use of inorganic chemicals in such diverse and effect areas as **pharmaceuticals** and building materials, but the theme running throughout is that careful study of the properties of inorganic substances can lead to the identification of significant applications for these materials.

Inorganic **catalysts** have major applications in chemical processing and pollution control, and a significant part of the book is devoted to examining the many ways in which inorganic chemistry plays a part in the various aspects of this technology. The field covers some of the largest industrial processes such as petroleum reforming and fat hardening but also includes sophisticated stereospecific synthesis of pharmaceuticals. Some of the reactions take place homogeneously in solution, and others heterogeneously in the gas or liquid phase on high surface area inorganic materials.

Sophisticated **materials chemistry** has been influential in shaping the development of applications in the electronics, electrochemical, and related industries. Progress in the fields of fast ion conductors, and superconducting and magnetic materials is reviewed in detail. The chemical nature of ceramics, glasses, and hard metals is considered and the wide scope for their use in both small and large scale applications indicated. Cementitious materials and colours are two aspects of materials science which have been chosen to illustrate the fact that scientific approaches to studies in these areas can lead to significant new applications for industries which traditionally built their technological base largely on experience and empirical methods.

The use of inorganic chemicals in various aspects of **water purification** is described, and methods for **corrosion prevention** considered. The significant contributions of inorganic chemistry in **nuclear energy** generation technology are discussed, and the potential of catalytic processes in **solar energy** conversion evaluated.

It is noticeable that many of the areas where applications have emerged are **multidisciplinary**. Magnetism has inputs from chemists, physicists, materials scientists, and engineers; heterogeneous catalysis from physical, organic, and inorganic chemists; electronics from physicists, materials scientists, and chemists; pharmaceuticals from inorganic chemists, biology, and medicine; and so on. The 'breakthrough' discovery that soluble platinum compounds can be used to successfully treat cancers has led pharmaceutical firms to ask whether inorganics will be found to have other new uses in medicine. Organic compounds no longer have the same pre-eminence amongst types of compounds considered for biological screening.

Effects required in the marketplace are often approached in a number of different ways, all of which may involve inorganic chemicals. For example, acetaldehyde can be made using Pd/Cu systems in both homogeneous and heterogeneous media, fast ion conductors or MOSFET technology can be used as the basis of chemical sensors, and nuclear or solar energy could eventually contribute the most effective non-fossil fuel energy source.

The field of **pollution control** is becoming increasingly important as the need for maintaining a good natural environment becomes stronger. A number of chapters describe technology which is contributing to this field. In addition to the mainstream themes of purification of gaseous and aqueous effluents, to which whole chapters of this book are devoted, many of the topics considered in other chapters also provide technology which will contribute to the solution of these problems – for example, hydrogen/oxygen fuel cells have virtually no effluent other than water.

Attached to each chapter there is a selection of references for general reading, as well as more specific references which relate directly to the text and a core list of journals where original papers on the topic are most likely to be found. During the course of the book we indicate the usefulness of good patents and emphasize the importance of reading relevant literature.

It is hoped that the forward looking **'trends'** sections at the end of each chapter will stimulate constructive thoughts on directions for future research which will lead to yet more applications for this rich field of endeavour. In these sections we have tried to identify needs and challenges which if met would provide new avenues for progress in the area being considered.*

REFERENCE

1. E. Konecny, C. Quinn, K. Sachs, and D. Thompson, 'Universities and Industrial Research', The Royal Society of Chemistry, Cambridge, UK, 1995.

*** Editorial notes:**

[1] Bold-face type is used throughout the text to highlight keywords.

[2] In general, modern IUPAC nomenclature is used for numbering Groups of the Periodic Table; thus transition metals scandium to zinc are Groups 3 to 12, and boron to neon Groups 13 to 18. An exception is made in Chapter 9, where the widely used notation for III–V and II–VI semiconductors is retained.

CHAPTER 2

Refining Processes Relevant to the Production of Speciality Inorganic Chemicals

JOHN HILL

1 INTRODUCTION: SCOPE OF THE CHAPTER

This chapter examines the techniques used to separate the constituents of mined ores. The business of mining and separating the contents of ore bodies is the **supply of metals** to the world's economy and to the activities to be described in further chapters of this book. As in all commercial activity this business is driven predominantly by the profit motive with the result that the choice of recovery and refining methods is heavily **driven by economics**. Indeed the decision to start mining a particular ore body will be essentially an economic one. Furthermore, many ore bodies are complex, containing more than one payable metal, and we will see how the ratio of one payable metal against the other will determine the choice of process.

Speciality inorganic chemicals are frequently compounds of those elements which we think of as precious or rare. The **rare earths, gold,** and the **platinum group metals (PGM)** are the obvious candidates for this category, but we will also examine the refining of **nickel, cobalt, titanium,** and **tungsten**. This is partly because compounds of these elements will feature in later chapters, but also since the recovery of the platinum group metals is inescapably linked with their occurrence in ore bodies containing nickel and cobalt. However, we have consciously omitted the refining of copper and silver since this is normally covered adequately in elementary courses.

The **separation** of the elements in a refinery feed is the exploitation of differences of **physical** and **chemical** properties. This will be the main focus of the refining chapter. In particular we have selected the platinum group metals for special attention as their refining provides an excellent example of complex metal extraction which is particularly appropriate in a text on speciality inorganic chemicals. We will also examine some of the important unit operations which have wide application in refining. To get an overall picture we need to understand the importance of some separation operations that employ principles not normally regarded as the province of a chemist.

Having opened this introduction by emphasizing the importance of economics, it is appropriate that we examine more closely how the cost elements of capital cost, operating cost, plant inventory, and product value intermesh and affect the way a refining operation is designed and operated.

After investigating the general principles of the refining activity we will then examine in more detail the flowsheets of some of the refineries producing the elements of interest to us in this book.

However, the subject is vast; indeed whole books have been written about the refining of a single element. Accordingly the approach will be a general description followed by a more detailed examination of particular refining steps. This will enable us to understand how relatively complex chemistry can be managed in an industrial environment, how the requirements of ultimate product purity often demand a specific impurity removal step and how aspects such as the health and safety of the operators, security of high value materials, and environmental concerns will influence the overall operation.

2 UNIT OPERATIONS IN REFINING

Refining to pure metals relies on various types of separations. Chemists will be familiar with operations such as precipitation and filtration, distillation, and evaporation and crystallization. Less familiar will be physical separations; high temperature operations; and large scale applications of electrolysis, ion exchange, and solvent extraction.

In this section we consider the basics of **separation processes** that fall into three broad categories:

- **Mineral Dressing**: the treatment of ores, usually carried out adjacent to the mine site, to produce concentrates of the valuable mineral while discarding the host material or gangue[1]
- **Pyrometallurgy**: separation procedures or thermal treatments carried out at temperatures above, say, 500°C
- **Hydrometallurgy**: separation processes carried out in aqueous solutions

2.1 Crushing and Milling

The **comminution** of run-of-mine ore is a necessary step before further treatment. Bearing in mind that the ore grades are usually below 10% of valuable mineral (and for gold and platinum group metals often below 0.0005% resulting in upwards of 200 tonnes of ore being required to produce 1 kg of metal), these first steps are operated at very large scale.

Plants able to treat many hundreds of tonnes of ore per day are common. Crushing of run-of-mine ore employs jaw[2] or gyratory[3] crushers, usually in two stages to provide a product with a size range of 1–6 mm. Milling is carried out in ball mills[4] or rod mills,[5] operated in a slurry with water, to produce ultimate particle sizes of 200–300 microns.

2.2 Froth Flotation[6]

Flotation is the most commonly used operation in mineral dressing. Briefly, arrangements are made so that when air is blown through an agitated suspension, certain minerals attach themselves to the bubbles and are floated on a froth, leaving the gangue particles as an underflow. The surfaces of specific minerals are treated with chemicals called collectors which render the particles hydrophobic so that they attach to air bubbles. 'Frothers' are added to the pulp to promote a stable froth which flows over the weir of the flotation machine.

2.3 Magnetic Separation and Gravity Concentration

These are further separation processes that are applied to minerals suspended in aqueous pulps. Magnetic separation[7] has particular application in the collection of platinum group metals from nickel/copper mattes, while gravity separation,[8] originally used for the collection of gold by 'panning', continues to have application in the recovery of high specific gravity (SG) minerals of PGM.

2.4 Smelting[9] and Converting[10]

The smelting of sulfide minerals to form a **'matte'** has particular application to the production of nickel, cobalt, and copper. The concentrate produced by flotation contains the sulfides from the ore, but is also contaminated with gangue materials which are usually siliceous. The concentrates are melted in electric arc furnaces to yield a **sulfide matte** and an **oxide slag**. Mattes are denser than slags (the SG of mattes is about 5, against 3 for slag) and the two phases are immiscible, so the process gives a good separation of valuable metal from the discard.

Iron is always present in sulfide concentrates and will collect in the matte unless it has been roasted to the oxide before the smelting step.

In a typical process the matte is transferred in molten form to a converter. This is a refractory lined vessel fitted with tuyeres to enable air to be blown through the molten matte.

The iron is oxidized and 'converted' to a ferrosilicate slag by reaction with quartz:

$$2\,FeS + 3\,O_2 \rightarrow 2\,FeO + 2\,SO_2 \qquad (1)$$

$$2\,FeO + SiO_2 \rightarrow 2\,FeO.SiO_2 \qquad (2)$$

This process is capable of producing a matte containing less than 1% iron.

2.5 Solvent Extraction[11]

Solvent extraction has become a well established operation in hydrometallurgy. The ability of a water-immiscible organic reagent to extract elements selectively has been used in chemical analysis for decades. Application to industrial hydrometallurgy started in the nuclear industry (see page 446), and by the late 1960s commercial plants extracting copper from dilute copper sulfate liquors were being built, the liquors being obtained by heap-leaching of low grade copper ores. Further development has produced systems for refining nickel, cobalt, the rare earths, and the platinum group metals.

Figure 1 shows a typical solvent extraction flow sheet. In most cases satisfactory mixing and separating characteristics are obtained by using a mixture of the **active reagent** and an **inert diluent**. The mixed organic phase must satisfy a number of criteria:

- It must be stable and resistant to degradation in the extraction, scrubbing, and stripping stages.
- The **distribution coefficient** (D, the ratio of concentration in the organic phase versus the aqueous phase) must be high (>100) in the extraction step and low (<0.01) in the stripping step.
- The **selectivity of extraction** of the element of interest compared to others must be high. (Selectivity is defined by the **'separation factor'**, the ratio of the distribution coefficients of the two metals being separated).
- The **strip solution** must be amenable to further treatment to give a saleable product.

A variety of equipment systems are employed, **mixer-settlers** being the most common although column contactors are also used.

2.6 Ion Exchange[12,13]

Ion exchange (IX) resins are insoluble solid acids or bases that have the property of exchanging ions from solutions. IX resins are three dimensional macromolecules and are usually manufactured from spheres of styrene–divinyl benzene copolymers. **Cation exchange resins** are commonly based on sulfonic acids and **anion resins** on amines, although many other types are available.

IX resins are commonly used for water treatment by collecting small concentrations of impurities. However in refining there are two major applications:

- **Concentration of ions.** A valuable ion is collected from a dilute solution and subsequently eluted as a more concentrated solution. The recovery of PGM is an example of this procedure.
- **Ion exchange chromatography** can be used to separate ions with similar but sufficiently different affinities for the resin. Rare earth separation is an example.

2.7 Electrolysis[14]

Electrolysis is used both for **extraction** and for **refining** of metals. The process produces a pure metal by cathodic deposition from solution. In **electrowinning**, the metal is brought to the cell by a continuous flow of solution, while in

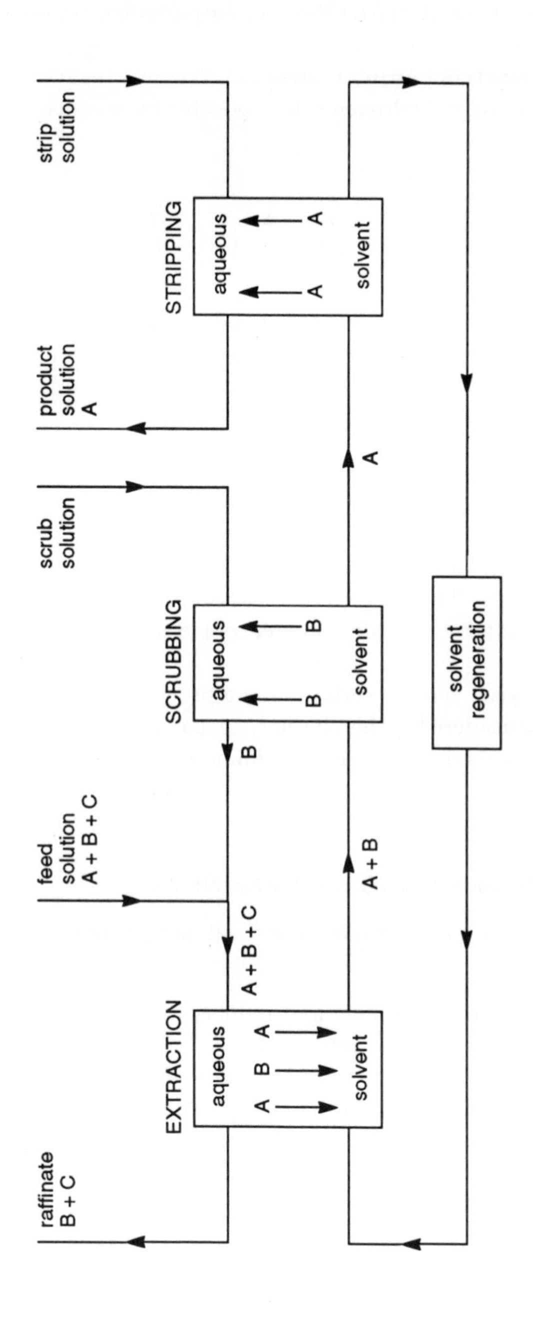

A = flow of extracted species
B = flow of impurity species (partially extracted)
C = flow of impurity species (not extracted)

Figure 1 *Typical solvent extraction flow sheet*

electrorefining the supply of metal is essentially by dissolution of an impure anode. The cathode is frequently a 'starter sheet' of the pure metal or occasionally a blank of another metal.

Consideration of the **electrochemical series** of elements suggests that only deposition of elements positive to hydrogen will be possible from aqueous solution. For example:

	Au	*Pt*	*Ag*	*Cu*	*H*
$E^{\ominus}$	+1.68	+1.4	+0.8	+0.34	0.0

However, the presence of a **hydrogen overpotential** on the surface of particular elements enables the deposition of coherent metal from solution. Examples of these are:

	Ni	*Co*	*Cd*	*Fe*	*Cr*
$E^{\ominus}$	−0.25	−0.28	−0.40	−0.44	−0.56

The production of a pure cathode relies on excluding more electropositive impurities from the catholyte. Thus in copper deposition concentrations of nickel and iron will not contaminate the deposit; however copper must not be present in a nickel catholyte.

In practice, electrolysis is a very powerful refining procedure and is employed for several of the metals considered in this chapter, *e.g.* cobalt, gold, and nickel, as well as being the major method employed for refining copper, zinc, aluminium, and silver and also for the production of lead and manganese.

3 THE ECONOMICS OF MINING AND REFINING

In order to bring pure metal to the market, costs will be incurred in four areas:

- the **capital cost** of building the mine and refinery.*
- the **operating cost** of mining and refining.
- the **inventory cost** resulting from the need to retain metal within the refining operation. This is variously called the **working capital** or **pipeline cost**.
- the **recovery cost**. This is a measure of the cost of metal lost in the refining process, in recognition that part of the expenditure incurred in the mine is lost by non-recovery in the refinery.

We now consider how the importance of these four elements of cost will vary as a result of the value of the product, the grade of the ore body in the mine, and the presence of more than one product in a complex ore body.

* In this section we will use 'refinery' to include all those operations which process run-of-mine ore through to pure product metal.

3.1 Effect of Ore Grade

The cost of bringing a quantity of metal to the refining operations is the combination of the cost of mining and the ore grade; *i.e.*

$$C_m = \frac{C_o}{\text{ore grade (kg tonne}^{-1})} \tag{3}$$

where C_m is the cost of mining 1 kg of contained metal and C_o is the cost of mining 1 tonne of ore.

Consider, on the one hand, a gold operation with an ore grade of 8 g $tonne^{-1}$, against a nickel operation with a grade of 0.4%. The cost of mining 1 kg of metal is 125 C_o for gold and 0.25 C_o for nickel.

Assuming that the ore mining costs are the same for each case, the cost per kg is 500 times greater for gold than for nickel. The justification for incurring the higher cost is the higher value of the gold (the **cost ratio** of gold to nickel is typically 2500).

The effect of these factors, namely higher mining cost and higher value of product, on refining is to place more emphasis on the recovery, which will justify extra operating expenditure in order to minimize losses.

3.2 Complex Ore Bodies

Complex ore bodies contain more than one payable metal, and the contributions made by the individual metals to the total revenue will influence the design and operation of the refining steps.

As an example of this, in Table 1 a comparison is drawn between typical complex ore bodies containing nickel and PGM at Sudbury, Canada, and from the Merensky Reef in South Africa. The Canadian deposit yields 71% of its revenue from nickel output while for the South African deposit 66% comes from platinum (and 87% from the total precious metals).

It follows that the Canadian operation will emphasize the maintenance and improvement of the nickel operations both with respect to recovery and operating cost, whereas the focus in the South African plants will be PGM. For example in the South African plants the installation of extra recovery steps to minimize PGM losses will be economic. It is noteworthy that a 1% change in the PGM recovery will affect the overall revenue by 0.9% in the South African operation against only 0.1% in the Canadian operation while the reverse effect will apply for nickel.

Whilst the first objective of a refining operation must be to produce metal to the defined purity specification, economic considerations will be a major motivation in the plant design.

4 OCCURRENCE OF ELEMENTS COVERED BY THIS CHAPTER

4.1 The Rare Earths[15]

Fifteen elements, atomic numbers 57–71 listed in Table 2, form the rare earth (RE) series. RE is something of a misnomer since, as a group, they occur very

Table 1 *Effect of metal ratios in a complex ore body*

	Typical ore from Sudbury, Canada		*Typical ore from Merensky Reef, South Africa*	
Element	*Ore Grade*	*Value of 1 tonne of ore (US$)*	*Ore Grade*	*Value of 1 tonne of ore (US$)*
Pt	0.34 g $tonne^{-1}$	3.92	4.78 g $tonne^{-1}$	55.17
Pd	0.36 g $tonne^{-1}$	1.39	2.03 g $tonne^{-1}$	7.83
Rh	0.03 g $tonne^{-1}$	0.91	0.24 g $tonne^{-1}$	7.25
Au	0.12 g $tonne^{-1}$	1.35	0.26 g $tonne^{-1}$	2.93
Total precious metals		7.57		73.18
Ni	1.2%	52.80	0.19%	8.36
Cu	0.7%	13.58	0.11%	2.13
Total Ni+Cu		66.38		10.49
TOTAL		$73.95		$83.67

Sources: G. G. Robson, 'Platinum 1985', Johnson Matthey plc, 1985; Metal prices at mid-September 1993

Table 2 *The rare earth elements*

Atomic Number	*Element*	*Symbol*	*Atomic Mass*
57	Lanthanum	La	138.9
58	Cerium	Ce	140.1
59	Praseodymium	Pr	140.9
60	Neodymium	Nd	144.2
61	Promethium	Pm	(147)
62	Samarium	Sm	150.4
63	Europium	Eu	152.0
64	Gadolinium	Gd	157.2
65	Terbium	Tb	158.9
66	Dysprosium	Dy	162.5
67	Holmium	Ho	164.9
68	Erbium	Er	167.3
69	Thulium	Tm	168.9
70	Ytterbium	Yb	173.0
71	Lutetium	Lu	175.0

widely, rank fifteenth in abundance, and together are somewhat more plentiful than zinc. RE minerals are normally classified as sources for light (lanthanum through to gadolinium) or heavy (terbium through to lutetium) elements although, in practice, most minerals contain some amount of all RE. **Monazite sand** contains all the RE, and is widely spread, with India and Brazil being the main sources. The primary source for the light RE is Bastnasite, a fluorocarbonate RE FCO_3 from California; while a by-product of uranium mining in Ontario, Canada produces most of the heavy RE.

4.2 Titanium[16]

Rutile, TiO_2, and **ilmenite**, $FeTiO_3$, are the only titanium minerals of commercial importance. Consumption of pure titanium dioxide as a white pigment in paint, paper, and plastics accounts for about 90% of the world titanium demand, with metal sponge production accounting for most of the rest. Beach sands in Australia are the major source, although operations in Canada, South Africa, and Russia are also significant.

4.3 Nickel[17]

Economic nickel deposits are either the **laterites** (oxides) or **sulfides**. About 60% of the world's land-based reserves in Western countries are found in laterite deposits, notably in New Caledonia, the Philippines, Australia, and Indonesia. Lateritic deposits occur predominantly in tropical climates as they are formed by the weathering of rocks such as peridotite and serpentine to produce heterogeneous mixtures of magnesium silicate and iron oxides, both containing small levels of nickel and cobalt.

Sulfide deposits occur primarily in Canada, Russia, and Southern Africa. The massive deposits of the Sudbury basin in Ontario came into production in 1885 and for the next 60 years supplied the Western world with the majority of its needs. The deposits at Norilsk in Siberia are known to be very substantial and Russia is presently the world's largest nickel producer. By comparison with Canada and Russia, South Africa is a small producer, but we shall see that the main driving force for mining there is platinum production with nickel (and cobalt) as by-products.

Nickel sulfides occur in a variety of minerals, the most common being **pyrrhotite**, $(Ni, Fe)_7S_8$, and **pentlandite** $(Ni, Fe)_9S_8$. Typically magnetite, Fe_3O_4, and chalcopyrite, $CuFeS_2$, are also present.

In general, high grade nickel (99% and better) is obtained from sulfide ores while laterites are processed to ferronickel, used for stainless steel production.

4.4 Cobalt[18]

Cobalt invariably is associated in small amounts with other metals. The major occurrence is in the copper mines of Zambia and Zaire, producing two thirds of Western world cobalt. However all nickel deposits contain low concentrations of cobalt, supplying the remaining third. Sulfides, arsenides, and carbonates such as **cobaltite** (CoAsS), **linnaeite** (Co_3S_4), and **sphaerocobaltite** ($CoCO_3$) are typical.

4.5 Gold

Gold is found chiefly as the **free metal** scattered in gravel or distributed in quartz veins. Small quantities are also found in lead and copper ores. South Africa, the United States, and Russia are predominant producers with Australia and Canada also important. However the high value of the metal results in gold being produced in small quantities by very many countries. Typical ore grades are 5–15 g $tonne^{-1}$.

4.6 Tungsten[19]

Tungsten occurs as **wolframite**, which is a series of chemical variations of (Fe, Mn)WO_4 or as **scheelite**, $CaWO_4$. The minerals are widely spread, the most important deposits being in China, which produces more than 50% of the world's requirements.

4.7 The Platinum Group Metals (PGM)[20]

The platinum group metals (ruthenium, rhodium, palladium, osmium, iridium, and platinum) almost always occur together and generally in association with nickel, cobalt, and copper. The South African deposits of the bushveld igneous complex in the Transvaal and, on a much smaller scale, the Stillwater mine in Montana, USA, are the only deposits mined primarily for PGM with nickel, copper and cobalt as by-products.

The Norilsk nickel deposits in Russia contain significant PGM values, but the proportion of the lower value palladium is high; hence the PGM are regarded as by-products of the nickel production.

The Sudbury nickel mines in Ontario, Canada, contain low levels of PGM which are produced as a by-product. Here the ratio of platinum to palladium is approximately 1:1.

PGM concentrations are extremely low and vary substantially. The South African and Stillwater deposits are typically 10 g $tonne^{-1}$ PGM, while those at Sudbury are an order of magnitude lower, at about 0.9 g $tonne^{-1}$.

There are very many individual PGM minerals present in the ore bodies, and most of these minerals are present to some degree in the PGM sources being exploited worldwide.

The sulfide minerals, usually associated with nickel minerals, are cooperite, PtS, vysotskite, PdS, braggite, (Pt,Pd)NiS, and laurite, Ru(Os,Ir)S_2. A series of alloys with bismuth and tellurium provides the minerals moncheite, PtBiTe, merenskyite, (Pt,Pd)BiTe, and kotulskite, PdBiTe. Iso-ferroplatinum, Pt_3Fe, and sperrylite, $PtAs_2$, are also widely spread. Rhodium occurs in solid solution in pentlandite, $(Ni,Fe)_9S_8$, together with significant levels of palladium, although when nickel sulfidation is absent an unnamed mineral complex, approximately (Pt,Cu,Ir,Rh)S, is seen.[21]

5 REFINING OF ELEMENTS COVERED BY THIS CHAPTER

5.1 The Rare Earths[15]

Monazite sand contains about 50% **rare earth (RE) oxide** and is normally treated with sulfuric acid to produce a mixed solution of RE. **Separation** of RE is achieved by selective absorption and elution from **ion exchange resins**. The elution product is collected in successive portions and reloaded in the same order onto a clean ion exchange column.

This process is repeated many times, ultimately to produce separated RE of sufficient purity.

Bastnasite is concentrated by flotation to produce a concentrate containing 70% RE. In one process, the chlorides of the mixed RE are produced by direct chlorination in the presence of carbon at 1200°C; at the same time a number of impurities with volatile chlorides are removed. Leaching and evaporation produces a mixed RE $Cl_3.6H_2O$ residue. Individual RE are then produced by ion exchange or **solvent extraction** using either tributyl phosphate (TBP) or the sodium salt of di(2-ethylhexyl)phosphoric acid (D2EHPA). The solvent extraction process, although complex (multiple systems containing typically 26 extraction, 4 scrubbing, and 8 stripping stages) is more efficient and less time-consuming than ion exchange, and has largely taken over from ion exchange methods. McGill has written a comprehensive description of RE occurrence and refining.[15]

5.2 Titanium[22]

We have noted that production of TiO_2 for use as a white pigment accounts for about 90% of the world titanium demand. Refining to titanium metal commences with rutile of about 95% TiO_2 content. The transformation to the metal is achieved in two steps:

$$TiO_2 + 2\,C + 2\,Cl_2 \rightarrow TiCl_4 + 2\,CO \quad (4)$$

$$TiCl_4 + 2\,Mg \rightarrow 2\,MgCl_2 + Ti \quad (5)$$

The first reaction is carried out in a **continuous chlorinator**, and the product vapours are passed through dust separation before condensing as impure $TiCl_4$. The impurities, such as $SiCl_4$, $CrCl_3$, *etc.* are separated by **fractional distillation**. The pure $TiCl_4$ is reduced with magnesium in a sealed reactor at 850°C. The molten $MgCl_2$ is discharged from the bottom of the reactor, and after cooling the titanium metal is removed and crushed before leaching Mg and $MgCl_2$ with acid. In alternative processes, sodium is used as the reductant and vacuum distillation used to remove the Mg and $MgCl_2$ before cooling the reactor.

The Ti sponge is compressed and vacuum melted to form ingots which can be rolled or drawn to sheet or tube.

5.3 Nickel

Nickel is produced from laterite (oxide) ores and from sulfide ores. The laterites are treated to produce a ferronickel alloy, or, alternatively, a sulfide concentrate which is further processed by the methods described below for sulfide ores. Sulfide ores are usually taken through the pure nickel metal by one of several processes.

5.3.1 Refining Laterite Ores.[23] Laterite ores occur in two principle types and are treated differently. The magnesium- and silica-rich **garnieritic ores** are rich in nickel, up to 3%, and it is economic to employ **pyrometallurgical processes** to produce ferronickel. The ore is intially mixed with coal or coke and heated in a rotary kiln to remove structurally bound water, and to carry out some reduction of the oxides. It is then smelted in an electric furnace. In the totality of the two operations nickel is reduced at high efficiency but only about 60% of the iron is reduced. The product **ferronickel** containing about 25% nickel is further refined in a converter to remove impurities such as silicon, phosphorus, carbon, and chromium. The product is cast into **ingots** for sale, primarily to stainless steel manufacturers.

A variant of the above process is to add sulfur to the reduced ore in the kiln, so that the product from the electric furnace is a mixed iron/nickel sulfide matte. This matte is then blown in a converter to oxidize the iron and remove it as a ferrosilicate slag (see page 7, Equations 1 and 2). The result is a nickel matte containing low levels of iron (<1%).

The **limonitic ores** are of lower nickel grade and contain more iron. The pyrometallurgical route would require excessive energy and be uneconomic, and would produce low grades of ferronickel. These ores are therefore treated by **hydrometallurgy**. There are two processes: pressure dissolution with sulfuric acid followed by precipitation with hydrogen sulfide to yield a nickel sulfide concentrate; alternatively a reduction roast to covert the nickel oxides to the metal followed by leaching with ammonia. The nickel dissolves as an ammine complex while iron is oxidized to ferric hydroxide which is discharged with the leach residues. Nickel is removed from the filtrate by stripping the ammonia so that basic nickel carbonate is precipitated. This carbonate is roasted to give a nickel oxide product.

5.3.2 Refining Sulfide Ores.[24] Sulfide ores are treated by one of two methods:

- **Smelting** to produce a matte which is then processed by hydrometallurgical routes.
- **Leaching** of the flotation concentrate with ammonia under pressure.

The second method is unable to recover platinum group metals (PGM) with the result that economic considerations restrict its application to ore bodies essentially free of PGM.

5.3.3 The Matte Smelting Process.[25] As we have seen, nickel sulfide ores NiS, Ni_3S_2, $(Ni,Fe)_9S_8$, *etc.* invariably contain cobalt, iron, and copper and almost always contain some level of PGM. After the mineral dressing processes of crushing, grinding, and flotation the concentrate will still contain substantial quantities of gangue materials such as silica and occasionally magnesia and alumina. The matte smelting process must therefore separate the silica, and reject the iron to produce a high grade matte containing nickel, copper, cobalt, and PGM. The process generally consists of three steps: roasting, smelting, converting.

The **roasting** step is designed to adjust the sulfur level in the concentrate by partial oxidation of the sulfide to sulfur dioxide by heating in air to about 600 °C.

The **smelting** step, carried out at about 1250 °C, is designed to reject the gangue materials and any iron oxidized in the roasting step as a slag, while producing a molten iron/nickel/copper/cobalt sulfide or matte. The **converting** step is designed to reject the iron by oxidation of the molten matte in the presence of added silica to produce a **ferrosilicate slag** (see page 7, Equations 1 and 2). The slag floats on the surface of the matte and is separated to leave a **high grade nickel/copper matte** containing cobalt and PGM. It will be appreciated that the nickel and copper content of the matte will vary according to the composition in the ore body, but after conversion the iron level will normally be reduced to less than 1%, while a typical sulfur level is 20–22%. However, as will be seen below, the conditions of converting may be varied to suit the subsequent refining process.

5.3.4 Separation of Copper and Nickel Sulfides.[26] Examination of the nickel–copper–sulfur ternary phase diagram shows that there are **three separate phases** present. These are a nickel sulfide phase Ni_3S_2, a copper sulfide phase Cu_2S, and, as a result of a deficiency of sulfur in normal converter mattes, a metallic phase rich in nickel which also contains copper and iron. Application of a **slow cooling procedure**, where mattes are cooled in insulated moulds to increase the grain size, enables the three phases to be separated after crushing and grinding. The metallic phase is recovered by **magnetic separation**, while the copper and nickel sulfides are separated by **froth flotation**.

5.3.5 Hydrometallurgy of Nickel Sulfides.[27] There are four main processes for producing pure nickel from nickel mattes:

- The Outokumpu process
- The Sherrit Gordon process
- The Falconbridge process
- The INCO Matte Electrolysis process[28]

These processes are those employed by the major producers of pure nickel in the Western world. Russia is, of course, a major producer, its total output probably being greater than the west, but as little is known of the processes used, it is not possible to describe them in detail.

The **Outokumpu Nickel process** was developed by the Finnish company and is operated at their refinery at Harjavalta, Finland, as well as under licence in Zimbabwe.

The process is a **multistage atmospheric leach** carried out in a sulfate media to yield a pure nickel solution for **electrowinning** against lead anodes. The leach solutions are purified for copper by cementation in countercurrent with the incoming matte and for cobalt by oxidative hydrolysis using the process to be described in Section 5.4.4. Electrowinning demands a highly purified feed solution to ensure a high quality cathode product. This purity is greatly assisted since the cobalt removal process also removes impurities such as iron, manganese, and arsenic.

The **Sherritt Gordon process** was developed by the Canadian company to process ores from their Lynn Lake, Manitoba mine. The process has been adapted to treat other ores, and is licenced to several companies outside Canada.

The process in its original form employed **ammonia leaching** of the froth flotation concentrates from the mine, followed by purification to give a solution containing nickel and cobalt ammines. This solution is **selectively precipitated** by hydrogen under pressure at approximately 180 °C to provide a pure nickel powder. The cobalt, together with a small amount of nickel, is discharged from the autoclave and treated with hydrogen sulfide to give a feed to the cobalt plant.

The ammonia leach process does not recover any PGM, as they are retained in the leach residue. Thus to treat inputs containing PGM an acid leach of high grade nickel matte produced in a smelting operation is used. In this variant the acid leach is carried out under pressure leading to a solution that is converted to the ammine complex in a later step. Provided the redox potential of the acid leach is carefully controlled the PGM are left in the acid leach residue, which is now at a suitable concentration for further processing.

The **Falconbridge process** is operated partly in Canada and partly in Norway. Ore mined in the Sudbury district of Ontario is processed to a nickel–copper matte which is shipped to Kristiansand, Norway, for refining.

At Kristiansand the incoming matte is **ground** and **separated** to size fractions above and below 45 microns. The smaller size material is **leached** with hydrochloric acid at 75 °C. The copper and PGM remain with the leach residue. The off-gas contains most of the sulfur from the matte as hydrogen sulfide and this is converted to elemental sulfur. Iron and cobalt are removed by **solvent extraction**. The purified solution is crystallized and the nickel chloride converted to nickel oxide by **high temperature oxidation** with air in a fluidized bed. Finally the nickel oxide is reduced to nickel by hydrogen at 600 °C.

The 45 micron oversize material is **leached** in a separate operation using **chlorine** gas as an oxidant. After purification the solution is **electrolysed** to give nickel cathodes using titanium anodes coated with a noble metal oxide. The chlorine evolved at the anode is collected and recycled to the leach stage. The residue from the countercurrent leaching contains essentially all the copper and PGM. The copper is recovered in an electrowinning tankhouse after the residue is roasted and leached with the sulfuric acid in the tankhouse spent electrolyte. The residue from the copper leach is further processed to provide a feed for the PGM refinery.

The **INCO Matte Electrolysis process.**[28] INCO produce pure nickel cathode at their Thompson, Manitoba refinery in Canada, by the **electrorefining** of nickel/copper matte anodes. The matte produced in the smelter is cast into anodes and electrolysed in a mixed nickel chloride/sulfate electrolyte. As described in Section 2.7, nickel electrolysis is carried out in a divided cell, and the feed to the cathode compartment must be of a high purity. At Thompson this is achieved by processing the anolyte with hydrogen sulfide to remove copper and arsenic, and with chlorine to remove cobalt. The sulfur in the matte does not dissolve under these conditions, and the spent anodes are removed from the cells when approximately two thirds of the nickel has dissolved. The anode sludge is separated and the elemental sulfur recovered by melting.

In this section on the hydrometallurgy of nickel sulfides we have summarized the principal features of the four major processes. More detailed descriptions are available in the references.

5.4 Cobalt

We have seen that cobalt is always produced as a by-product of either nickel or of copper. Thus the refining description will focus on separations from the host metal.

5.4.1 The Separation of Cobalt and Copper.[29] The **cobalt-bearing copper concentrates** from the mines of Zaire and Zambia are generally treated initially by a roast–leach–electrowinning flowsheet which produces copper cathode. The cobalt reports in the spent electrolytes from the tankhouse and a bleed from that stream provides the input to the cobalt operation. The solution at this point typically contains 30 $g\,l^{-1}$ copper as well as concentrations of iron, aluminium, and magnesium which vary from operation to operation. Copper is normally lowered to about 0.5 $g\,l^{-1}$ by further electrolysis and cementation using cobalt granules.

Other impurities are removed by hydrolysis with lime, and filtration yields a solution of sufficient purity for **cobalt electrowinning**. This demands that elements which would contaminate the electrodeposit (copper, iron, zinc, *etc.*) are reduced to below 5 $mg\,l^{-1}$, however others such as magnesium and manganese can be tolerated at 5–10 $g\,l^{-1}$. The cobalt is deposited onto stainless steel cathodes in a 4–5 day cycle. The deposit is brittle and can be readily removed from the stainless steel cathodes. Final treatment consists of **vacuum degassing** at 850 °C. Purity is typically 99.95% with nickel the major impurity.

5.4.2 The Separation of Cobalt and Nickel.[30] The separation of cobalt from nickel in aqueous solution is a challenging problem in **hydrometallurgy** as their adjacent position as transition metals in the Periodic Table results in similar chemical behaviour. The challenge is to identify and exploit those differences which do exist. There are four major techniques employed commercially:

- sulfide precipitation
- precipitations exploiting oxidation/reduction processes
- ion exchange resin processes
- solvent extraction processes

5.4.3 Sulfide Precipitation. Since the solubility of cobalt sulfide is lower than nickel sulfide, a separation by this precipitation is feasible. However, as the ratio of cobalt to nickel in plant solutions is typically low, the technique produces cobalt filter cakes contaminated with as much as four parts of nickel for one of cobalt, and further processing is required.

5.4.4 Precipitations Exploiting Oxidation/Reduction Processes. The thermodynamic equilibrium relationships between the +2 and +3 oxidation states in solution and as precipitated solids are very similar for nickel and cobalt. However practical separations are not only controlled by equilibria; activity effects and kinetics also play a considerable part. Exploitation of these effects is used in a number of ways. For example, nickel can be **selectively reduced** at high purity from **mixed sulfate** solutions by hydrogen under pressure. Moreover if **mixed sulfides** are **oxidized** with air under pressure in the presence of ammonia the cobalt is oxidized to the cobalt(III) Co^{3+} state with the nickel remaining as nickel(II) Ni^{2+}. This enables a separation by precipitating the sparingly soluble nickel(II) ammonium sulfate.

Other processes rely on the fact that the formation of Ni^{3+} is more difficult than the equivalent cobalt ion, and when formed is much less stable. Strong oxidants, such as chlorine, Caro's acid (peroxomonosulfuric acid, H_2SO_5), or ammonium persulfate are required. Alternatively nickel(III) hydroxide is produced by anodic oxidation of a nickel(II) hydroxide slurry and then used to remove cobalt in an exchange reaction.

The choice between these oxidation methods varies from plant to plant depending on the particular constraints of the operation. For example, in plants employing the electrolysis of nickel metal or nickel matte anodes in a mixed chloride/sulfate electrolyte, chlorine gas is suitable. In plants employing electrowinning of nickel against inert lead anodes in a sulfate media, anode corrosion will occur at electrolyte chloride concentrations above 10 $mg\,l^{-1}$, and nickel(III) hydroxide is an appropriate oxidant.

Caro's acid does remove cobalt very efficiently, producing a granular, easily filtered precipitate, but since it is only stable in the presence of large excesses of sulfuric acid, its use is restricted to those processes which can accommodate such an excess.

5.4.5 Ion Exchange Processes. Ion exchange separations employing both cation and anion processes are possible, but have only limited application in commercial refining. It is instructive to examine the reasons for this situation.

In chloride solutions cobalt forms complex chloro anions such as $[CoCl_3]^-$ and $[CoCl_4]^{2-}$. There are no nickel equivalents. However the cobalt anions are only stable in high chloride concentrations when common impurity elements such as zinc, copper, iron, and lead form anions and are also retained on the resin. Thus, although the primary objective of separation from nickel is achieved, the disadvantage of impurity contamination is such that alternative methods are preferred.

Cation exchange in sulfate solutions is used for scavenging nickel from plant intermediate solutions containing suitably high cobalt to nickel ratios. However, a process that removes the predominant element from the minor constituent is unattractive, and bearing in mind the typical low cobalt : nickel ratio in the common ore bodies, such processes have no commercial application for the primary separation.

5.4.6 Solvent Extraction Processes. As with ion exchange, cationic and anionic processes are possible, and here both are used commercially. **Anionic** solvent extraction processes are operated in refineries employing chloride media, while **cationic** processes are used in plants with sulfate media.

Extraction by chelating reagents such as hydroxyoximes or substituted 8-hydroxyquinolines has been proposed, but since they extract nickel away from cobalt the disadvantages seen in cationic ion exchange are again present. In commercial practice it is the **alkyl phosphorus acids**, which selectively extract cobalt from nickel, that are commonly used. The **separation factors** obtained are a complex function of acid type, cobalt concentration, and temperature. The separation factor increases in the series phosphoric $(RO)_2P(O)OH$ < phosphonic $R(RO)P(O)OH$ < phospinic $R_2P(O)OH$. Specific examples of these three reagents are di(2-ethylhexyl)phosphoric acid; (2-ethylhexyl)phosphonic acid 2-ethylhexyl ester; and di(2,4,4-trimethylpentyl)phosphinic acid.[31,32,33]

At the same time an increase of temperature and cobalt concentration will also have a substantial effect on the separation factor. This remarkable change in separation factor is caused by a transition in the cobalt complex present in the organic phase. At low temperature and cobalt concentration the pink hydrated octahedral complex is present. At higher temperatures and concentrations a blue anhydrous tetrahedral complex is formed. Nickel does not exhibit this behaviour and the separation factor for the alkyl phosphoric acid system can change from about 10 at 25 °C to about 80 at 50 °C.

The use of these solvent extraction systems enables the production of very high quality cobalt metal (by electrolysis) or alternatively high purity cobalt sulfate hexahydrate crystal, while achieving high levels of recovery to product.

5.5 Gold

Gold recovery from ores in modern plants relies on dissolution with cyanide in an aerated pulp of milled ore:

$$Au + 2\,CN^- \rightarrow [Au(CN)_2]^- + e^- \tag{6}$$

$$O_2 + 2\,H_2O + 4\,e^- \rightarrow 4\,OH^- \tag{7}$$

The run-of-mine ore is crushed and milled to ensure that all the gold particles are available for the leaching process. The **leach process** is maintained at pH 10.5–11.0 by addition of lime.

5.5.1 Recovery of Gold from the Cyanide Leach Solution. Having produced pulp of waste ore in a dilute solution (typically 5 $mg\,l^{-1}$ Au) of the $[Au(CN)_2]^0$ complex there are three techniques applied to recover the gold:

– Filtration followed by **cementation** with zinc:

$$[Au(CN)_2]^- + e^- \rightarrow Au + 2\,CN^- \tag{8}$$

$$Zn + 4\,CN^- \rightarrow [Zn(CN)_4]^{2-} + 2\,e^- \tag{9}$$

- **Adsorption** onto active carbon, directly from the pulp without filtration (the CIP or carbon-in-pulp process)
- An equivalent process using an **ion exchange** resin (the RIP or resin-in-pulp process).

Before applying the **zinc cementation** process the solutions are de-aerated under vacuum to avoid the loss of zinc by the reaction:

$$2\,Zn + 8\,CN^- + O_2 + 2\,H_2O \rightarrow 2\,[Zn(CN)_4]^{2-} + 4\,OH^- \qquad (10)$$

A typical cementation product will contain 30–40% gold, 20–30% zinc and, from South African mines, 4% silver, the balance being other base metals and sulfur. After leaching the bulk of the zinc with sulfuric acid, the product is smelted to an impure bullion.

The **carbon-in-pulp** (CIP) processes[34] rely on the adsorption of $[Au(CN)_2]^-$ onto active carbon. The carbon is relatively coarse, ranging from 1–3 mm, so that it can be separated from the milled ore by passing over screens. The gold is recovered from the carbon by a three stage process:

- Acid washing to remove calcium
- Soaking in 30 $g\,l^{-1}$ NaCN solution
- Eluting with water

Gold in the eluate is recovered by either zinc cementation or by electrolysis.

Resin-in-pulp (RIP) processes operate on very similar principles to CIP while employing an anion exchange resin. RIP processes are much favoured in Russian operations, while not finding much application elsewhere. The advantage of RIP is that it is more selective than CIP, and much less susceptible to poisoning by organic contamination (flotation reagents, engineering lubricants, *etc.*). However recovery of gold from resin is more difficult, particularly if strong-base resins are employed.

The chemistry of cyanide leaching of gold ores and recovery of the gold from solution has been described by Nicol, Fleming, and Paul.[35]

5.5.2 Refining to Pure Gold. The impure gold produced by zinc cementation is refined further by the **Miller Chlorine process**.[36] The gold is melted and chlorine injected through a graphite tube. Iron, lead, and zinc are rejected as their volatile chlorides, while copper and silver chlorides float on the surface of the melt and can be skimmed off. Normally a purity of 99.6% Au, suitable for sale as investment bars, is achieved.

To produce gold of 99.99% quality, electrolytic refining[37] is necessary. 99.6% Au is cast into anodes and electrorefined at 60 °C in an electrolyte containing 100 $g\,l^{-1}$ Au and 100 $g\,l^{-1}$ HCl. The cathodes are alternatively titanium blanks or rolled gold sheets. The deposit is a porous gold sponge which is thoroughly washed before melting to 99.99% quality bars.

5.6 Tungsten[38]

The production of pure tungsten relies on the **extraction** of tungsten from the ore, preparation of pure ammonium paratungstate or tungstic oxide, **reduction** to metal powder, and **conversion** of the powder to massive metal by pressing, sintering, and rolling at elevated temperatures.

Wolframite ores, $(Fe,Mn)WO_4$, are digested with sodium or potassium hydroxide under pressure. The alkali metal tungstate is purified by recrystallization and then converted to tungstic acid, H_2WO_4, by acidification with hydrochloric acid.

Scheelite, $CaWO_4$, is decomposed to an insoluble form of tungstic acid by treatment with hydrochloric acid. Leaching with sodium hydroxide follows.

Purification of the impure sodium tungstate employs techniques which vary from producer to producer, but typically arsenic and phosphorus are removed by the addition of magnesium chloride to precipitate arsenates or phosphates. Conversion of sodium tungstate to ammonium tungstate is normally carried out by **liquid ion exchange**. Crystallization to ammonium paratungstate, $3(NH_4)_20.7WO_3.11H_2O$, followed by calcination at 400 °C produces an oxide, generally assigned the formula W_40_{11}. Reduction with hydrogen produces the metal powder.

5.7 The Platinum Group

5.7.1 Introduction. The Platinum Group is a family of six elements (Groups 8, 9, and 10 of the Periodic Table) consisting of ruthenium, rhodium, palladium, osmium, iridium, and platinum. As we have seen in Section 4.7 these elements almost always occur together, although the ratio one with another may vary widely. The platinum group metals (PGM) exhibit an immensely complex chemistry. Since they invariably occur with lesser quantities of gold and silver the **refining process** must achieve **at least eight distinct separations**, as well as the elimination of unwanted impurities.

In the **'classical'** or 'conventional' processes the separations rely almost entirely on **precipitation** and **filtration**. These processes, which are still operated with minor variations in a number of major PGM refineries in the world, are not too different to those used in the original isolation of the elements, and are a useful introduction to the complicated chemistry of the PGM.[39]

There are significant disadvantages in the classical processes, and as a result **improved techniques**, mostly based on **solvent extraction** and **ion exchange**, have been developed in the last two decades and brought into production.

This section on PGM refining examines basic aspects of the chemistry, the classical processes, the benefits of solvent extraction, and reviews one of the solvent extraction processes.

5.7.2 PGM Process Chemistry.[40] The separation of PGM by hydrometallurgy must start by dissolving the feed material in an aqueous medium. In practice,

Table 3 *PGM oxidation states and chloro complexes*

Metal	*Oxidation state*	*Structure*	*Major species*	*Comments*
Platinum	+2	Square planar	$[PtCl_4]^{2-}$	Redox potential
	+4	Octahedral	$[PtCl_6]^{2-}$	controlled
Palladium	+2	Square planar	$[PdCl_4]^{2-}$	Pd^{4+} is only stable at
	+4	Octahedral	$[PdCl_6]^{2-}$	high redox potential
Rhodium	+3	Octahedral	$[RhCl_6]^{3-}$	Level of aquation
	+3	Octahedral	$[Rh(H_2O)_nCl_{6-n}]^{n-3}$	determined by chloride ion concentration
Iridium	+3	Octahedral	$[IrCl_6]^{3-}$	Iridium will form
	+4	Octahedral	$[IrCl_6]^{2-}$	aquated species, determined by chloride ion concentration
Ruthenium	+3	Octahedral	$[RuCl_6]^{3-}$	Complex equilibria,
	+3	Octahedral	$[Ru(H_2O)Cl_5]^{2-}$	depending on redox
	+4	Octahedral	$[RuCl_6]^{2-}$	potential and chloride ion concentration
Osmium	+3	Octahedral	$[OsCl_6]^{3-}$	Os^{4+} is the more
	+4	Octahedral	$[OsCl_6]^{2-}$	stable
Gold	+3	Square Planar	$[AuCl_4]^-$	

chloride is the only effective system and a combination of **kinetic** and **thermodynamic** factors are involved in the dissolution process:

- the high electrochemical potential of the reaction taking the metal through to its lowest stable oxidation state
- slow kinetics of the dissolution reaction
- the tendency of the metal to form inert oxide films when exposed to an oxidative attack, which slows the dissolution

The last two effects restrict the rate of solubility more strongly for the other platinum metals (OPM* – ruthenium, rhodium, osmium, and iridium) than for platinum and palladium. This is especially true if the mixed metal feed concentrate has been subjected to roasting in air. However, at best, the separation of OPM from platinum and palladium is only partial.

The formation of a large number of **anionic complexes** is the characteristic of PGM which distinguishes them from other elements. Some of the chloro complexes

* The OPM are variously described as the 'insoluble metals' or 'secondary PGM'. However these metals *are* soluble under the right conditions and 'secondary PGM' is also used to categorize PGM which are being recovered from scrap and industrial residues (as opposed to production from primary ore bodies). Therefore in this text we will use the acronym OPM to collectively represent ruthenium, rhodium, osmium, and iridium.

which feature in refining sequences are shown in Table 3. The manipulation of redox potential and chloride ion concentration is a major feature in the construction of refining sequences by both classical and solvent extraction techniques.

Kinetics also play a major part in refining PGM. Differences in ligand exchange rates are considerable and can be employed to effect a separation between species. In general, rates follow broad comparisons as:

Pd^{2+} Au^{3+}	Very labile
Ru^{3+} Rh^{3+} Ir^{3+} Os^{3+}	Moderately labile
Pt^{2+} Ru^{4+} Pd^{4+}	Moderately inert
Pt^{4+} Ir^{4+} Os^{4+}	Very inert

These kinetic differences are often the mechanism for precipitations which give good selectivity for a specific element. Examples are the selective reduction of gold from a mixed PGM solution and the precipitation of palladium by dimethyl glyoxime. Gold reduction by reagents such as sulfur dioxide or hydrazine is controlled by fast kinetics in addition to the higher potential of Au^{3+} compared to the PGM. The Pd^{2+} dimethyl glyoxime complex will be rapidly formed and act as a separation from a Pt^{4+} chloro complex.

Application of these thermodynamic and kinetic factors leads to a number of general processes which feature in almost all PGM refining processes:

- **Silver** is precipated as the chloride, essentially totally insoluble in solutions with low chloride ion concentration.
- The **separation of platinum** from base metals and from the rest of the PGM most frequently relies on the formation on the very stable $[PtCl_6]^{2-}$ ion. This ion can be separated by liquid ion exchange in a solvent extraction process or by precipitation as the ammonium salt $(NH_4)_2PtCl_6$.
- **Ruthenium** and **osmium** have very stable tetroxides which can be distilled away from other elements, both PGM and base metals.
- **Iridium**, like platinum, will form a very stable chloroanion in the +4 oxidation state.
- **Palladium** is usually separated by processes that rely on fast ligand exchange kinetics, such as precipitation of the diammine $Pd(NH_3)_2Cl_2$.

Utilization of these and many other principles leads to a number of possible separation sequences. We will examine a typical classical process and then consider modern solvent extraction processes.

5.7.3 The Classical Process.[41] The classical processes rely on separations utilizing **precipitation** and **filtration** for the majority of their operations. A typical set of processes is shown in Figures 2 and 3. There are variations operated by the various refineries, but the basic principles remain similar.

The incoming PGM **concentrate** (Figure 2) is partially dissolved by attack with aqua regia or alternatively by passing chlorine gas into a suspension in hydrochloric acid. The objective is to maximize the separation between platinum, palladium, and gold on the one hand and the OPM on the other. The incoming concentrate may be subjected to an oxidizing roast to restrict the solubility of the OPM.

Gold is the first metal to be removed. This is achieved by reduction using sulfur dioxide, iron(II) sulfate, or hydrazine and produces a reasonably pure product.

Platinum is precipitated as the ammonium chloroplatinate $(NH_4)_2PtCl_6$ by addition of ammonium chloride. The precipitate is filtered, washed, and calcined to an impure sponge. This sponge will be dissolved and reprecipitated at least once to produce a product of 99.95% purity or better.

The filtrate from the platinum precipitation is treated with ammonia to form palladium tetrammine chloride $Pd(NH_3)_4Cl_2$ in solution. Acidification with hydrochloric acid produces the insoluble diammine $Pd(NH_3)_2Cl_2$. This salt is further purified after filtration by redissolving in ammonia, and the clarified solution reprecipitated with acid. Calcination then produces a pure **palladium** sponge, typically of 99.98% purity or better.

In the classical process the production of the OPM is difficult and time consuming (Figure 3). The residue from the initial dissolving of platinum and palladium will contain the OPM and also impurities such as silver, base metals, and siliceous slag materials. The first task is to smelt the residue with lead and then to 'part' with nitric acid. Silver, lead, and the other base metals will dissolve. The OPM are rendered refractory by the **lead smelt** and are solublized by a **fusion with sodium peroxide**. The peroxide fusion product is dissolved in hydrochloric acid and the ruthenium and osmium distilled as their tetroxides by air sparging the boiling solution while adding sodium chlorate or sodium bromate. The ruthenium and osmium tetroxides are collected in a hydrochloric acid scrub solution, which converts the **ruthenium** to the chloro complex. However, **osmium** remains as the tetroxide and can be redistilled provided the redox potential is maintained with hydrogen peroxide or nitric acid.

After the ruthenium/osmium distillation the excess oxidants are destroyed by an extended boil and iridium is precipitated as the chloroiridate by addition of ammonium chloride. Since the solution is in a fully oxidized state after the ruthenium oxidation, it is the **iridium(IV)** salt that is formed, $(NH_4)_2IrCl_6$.

The **rhodium** remaining in solution is then precipitated as the pentaamminechlorodichloride $[Rh(NH_3)_5Cl]Cl_2$.

All of these precipitates are impure and require further refining before achieving a saleable purity. It can be seen that the refining process is complex. The separations are imperfect so that substantial recycles are necessary to achieve the required combination of pure products and high recovery. In addition, the effluent solutions initially contain PGM and before discharge must be treated by cementation, firstly with iron then with zinc, to recover the PGM values.

5.7.4 Disadvantages of Classical Processes.[42] A fundamental disadvantage of a chemical process employing precipitation techniques is the inability to effect

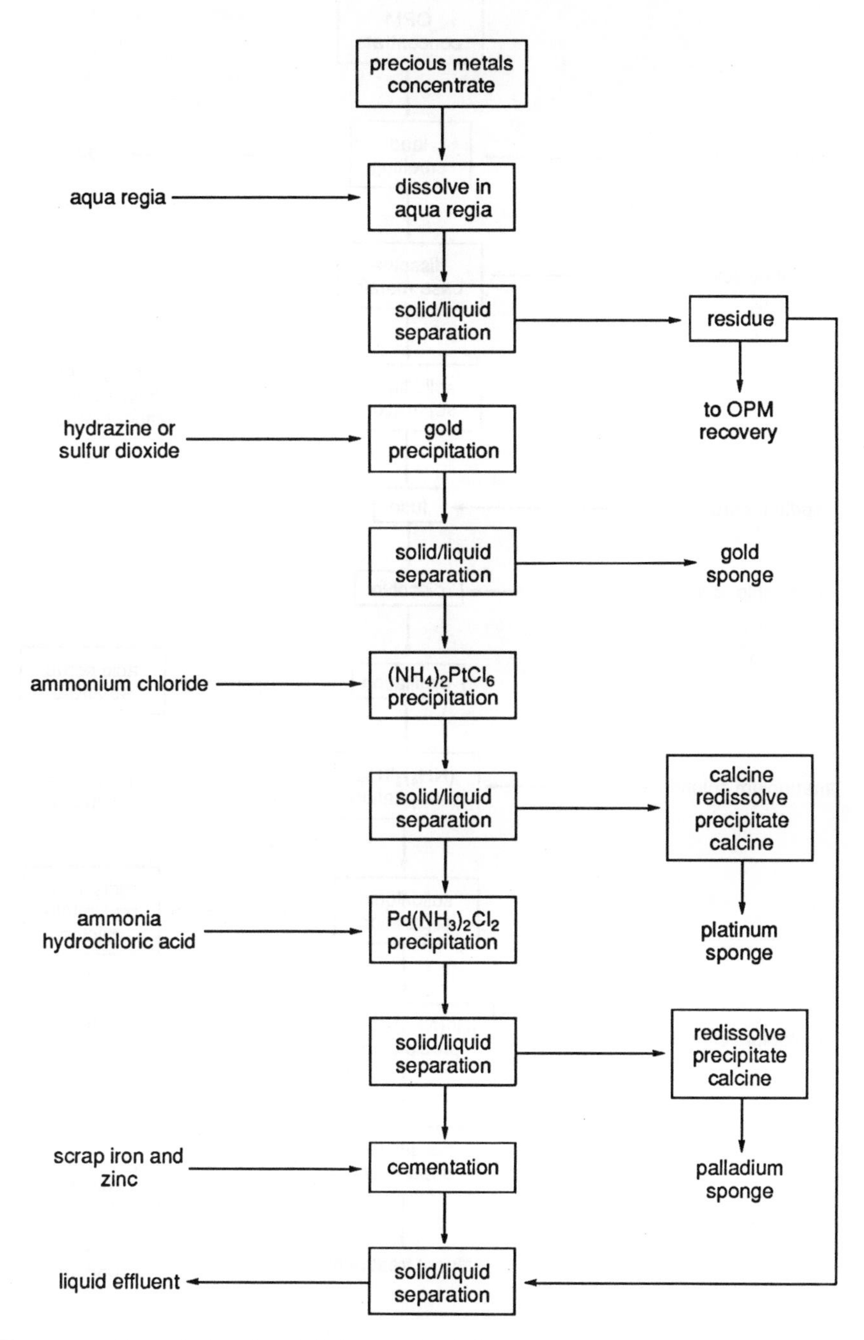

Figure 2 *Classical separation of platinum and palladium*

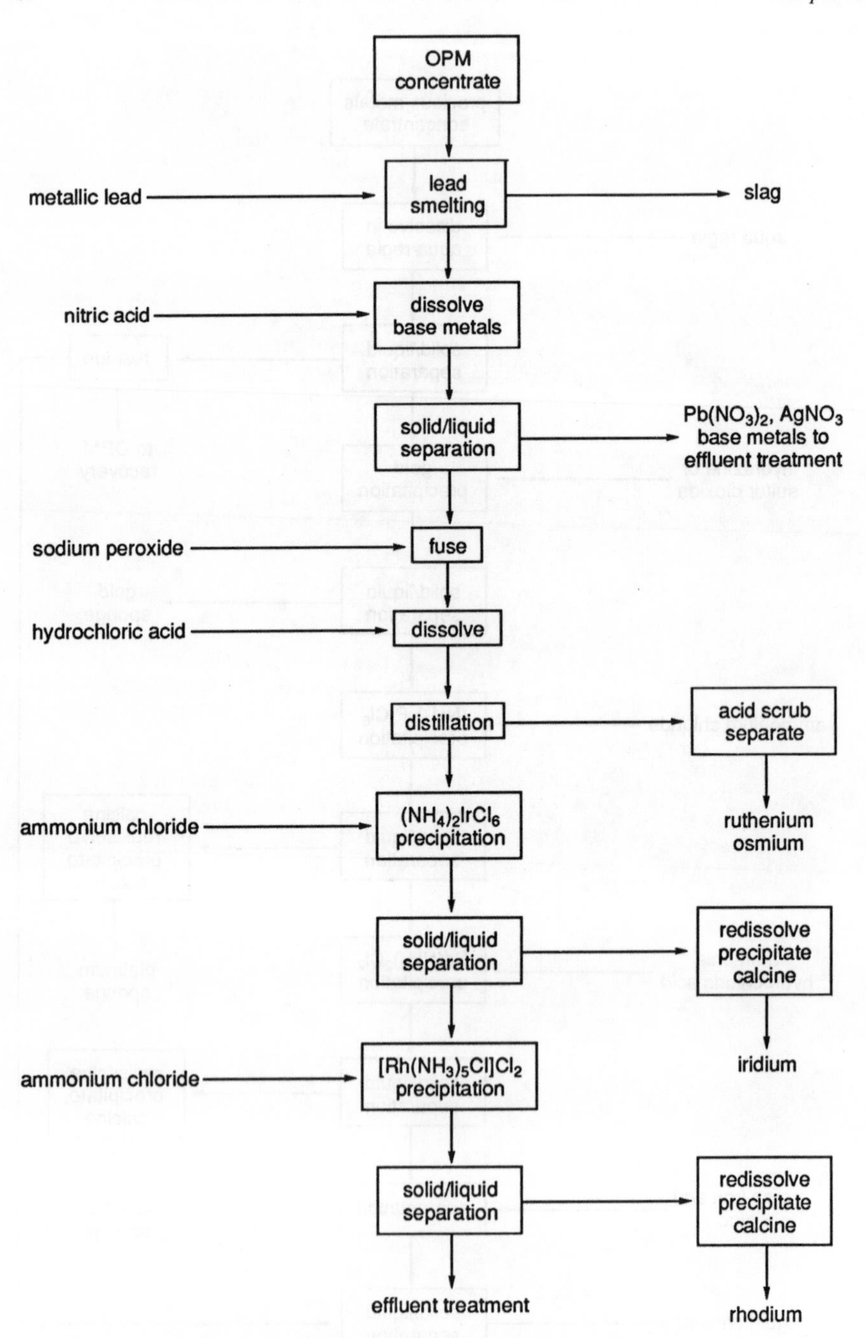

Figure 3 *Classical separation of OPM (ruthenium, osmium, iridium, and rhodium)*

efficient separations of the precipitated component. Even if conditions are favourable to the complete removal of the desired elements from solution, the precipitate is contaminated by mother liquor containing undesired elements. These may be removed by thorough washing, but this operation is rarely successful as undesired species may adhere to the precipitate, or even coprecipitate. In a complex multi-element system the problems of separation are considerable, and extensive chemical knowledge and expertise is required to separate and purify the individual elements. Recycling, retreatment and re-refining operations are required, even in the most efficient process. These are costly both in materials and labour, and also in processing time which, because of their high intrinsic value, is a crucial issue in PGM refining.

5.7.5 Solvent Extraction Processes.[43] Mechanisms for the solvent extraction separation of PGM fall into three classes:

Solvation. These processes involve the extraction of chloro complexes with neutral reagents:

$$[MCl_m]^{n-} + y\overline{S} \rightarrow \overline{[MCl_m]^{n-}.yS} \qquad (11)$$

(The bar indicates the organic phase). Examples of this mechanism are the extraction of gold (and impurity elements such as iron, tellurium, selenium, and antimony) by ketones or ethers.

Compound Formation. This involves a direct bonding of the metal to the extractant. If this is to occur, the chloro complex must be amenable to ligand exchange. In practice only bivalent palladium is sufficiently labile for this mechanism. Hydroxyoximes with the general formula Ar.C(NOH)R and organic sulfides R_2S are suitable extractants:

$$[PdCl_4]^{2-} + \overline{2\,R_2S} \rightarrow \overline{PdCl_2.2R_2S} + 2\,Cl^- \qquad (12)$$

Ion Exchange. Since PGM exist as anionic chloro complexes in the solution of interest it is extraction by organic bases that is effective.

Protonated organic amines of the general formula $[NR_nH_{4-n}]^+$ are suitable extractants. The strength of the ion pair bond will follow the series primary < second < tertiary < quaternary amine.

This results in primary amines providing weak extraction and easy stripping, with quaternaries offering the reverse. In practice a satisfactory balance is obtained with the secondary and tertiary amines. The chloro anion $[PtCl_6]^{2-}$, as an example, is extracted from solutions of low acidity (0–1 M free hydrochloric acid) and stripped using 10 M hydrochloric acid.

Organic amides offer a different approach to ion pair extraction. Compounds of the type R.C(O)NHR will only protonate at high acidity. Reaction 13 proceeds to the right in 6 M hydrochloric acid:

$$R.CONHR + H^+ \rightleftharpoons (R.CONH_2R)^+ \qquad (13)$$

Thus the organic amides offer a system where the chloro anion is extracted from a high acidity feed and the stripping is carried out with water.

With the great variety of commercially available extractants and the complexity of the chemistry of the solutions involved it is not surprising that the various PGM producers using solvent extraction have evolved different flowsheets. Some general conditions and constraints will however apply to all systems:

- Amphoteric elements, and base metals which form anionic complexes in chloride media must be removed to a low level before attempting a PGM separation.
- **Gold** is extracted by all reagents, so must be extracted early in the refining sequence.
- **Rhodium** is not extractable by any commercially available reagent, so will report to the final raffinate.
- **Ruthenium** in the form of its very stable nitrosyl complex $[Ru(NO)Cl_5]^{2-}$ is extractable by organic bases. However, the classical distillation of the tetroxide RuO_4 is also available.
- **Iridium(II)** is moderately labile, and in low acidity solutions will partially hydrolyse to weakly acidic complexes such as $[IrCl_4(H_2O)_2]^-$ or the neutral $[IrCl_3(H_2O)_3]$ complex. These are not extracted by the amines which readily extract the anions $[PtCl_6]^{2-}$, $[PdCl_4]^{2-}$, or $[IrCl_6]^{2-}$.
- Although **palladium** is sufficiently labile to enable extraction by ligand exchange using organic sulfides or hydroxyoximes, the kinetics are nevertheless slow.
- All solvent extraction refining systems commence with the complete dissolution of all the PGM and gold in the concentrate, in contrast to the imperfect separation of OPM from platinum, palladium, and gold in the classical process.

As a result of these factors there is a degree of commonality in all the refineries employing solvent extraction. The basic sequence is shown in Figure 4. Beyond this we need to describe the alternatives for ruthenium and osmium separation and to consider more fully the extraction processes for the key elements.

Osmium is invariably removed by distillation, usually during or immediately after the dissolution step 1, although in one process the separation is carried out by distillation after platinum solvent extraction, step 5.

Ruthenium is commonly separated by distillation of the tetroxide either after concentrate dissolution or after platinum extraction. However one process uses solvent extraction of the nitrosyl complex $[Ru(NO)Cl_5]^{2-}$ with tributyl phosphate.

In some refineries **gold** is removed from the solvent extraction feed by reduction with iron(II) sulfate or sulfur dioxide. However extraction by methyl isobutyl ketone (MIBK) or dibutylcarbitol (butex) is more common. MIBK has the additional benefit that impurities such an iron and tellurium are also extracted, thus giving a cleaner solution for subsequent operations.

Palladium extraction is characterized by the slow kinetics associated with the ligand exchange mechanism. There are two categories of extractants employed: sulfides and hydroxyoximes.

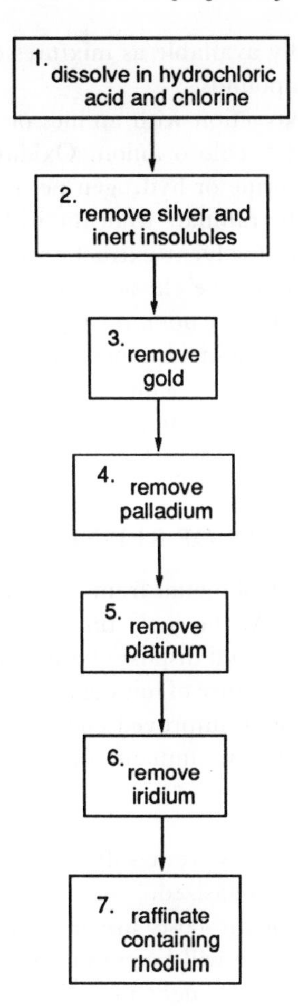

Figure 4 *Basic refining sequence in a PGM solvent extraction refinery. (PGM = platinum group metals)*

Dihexyl and dioctyl sulfides are employed in batch stirred reactors, it taking some hours to reach equilibrium. Oximes with the general structure **(I)** or **(II)** exhibit somewhat faster kinetics than the sulfides so that continuous operations in mixer–settler equipments are practical. The extraction kinetics may be enhanced by the addition of phase transfer agents such as organic amines.

(I) **(II)**

Platinum is recovered by extraction with tributyl phosphate $(C_4H_9O)_3PO$ or with secondary or tertiary amines, *e.g.* R_2NH or R_3N where R is typically C_8–C_{12}.

Such amines are commercially available as mixtures of isomers, there being no advantage in using pure compounds.

Iridium is recovered by extraction with amines or amides, having first been oxidized to form the $[IrCl_6]^{2-}$ chloro anion. Oxidation may be achieved by boiling in the presence of chlorine or hydrogen peroxide.

Rhodium will remain in the raffinate from the iridium solvent extraction.

It must be noted that whilst solvent extraction techniques provide a much cleaner separation of PGM than the classical process, nevertheless, with some exceptions, the solutions obtained are not pure enough for direct conversion to the product metal. Thus the strip liquor obtained from a typical platinum solvent extraction process will contain about 1% impurities, and further refining steps are required. These are normally precipitations similar to those used in the classical process.

6 THE FUTURE DIRECTION OF REFINING

We have seen that the recovery of metals from ore bodies is usually a large scale and thus high cost operation. At the same time, by its very nature, a refining operation is required to separate and dispose of impurities. Bearing in mind these points, it is self-evident that the future of mining and refining will be the quest for lower costs and for processes with improved environmental standards.

The twin pressures of cost containment and environmental improvement suggest possible future directions:

- Methods for recovering metals from sulfide ores while minimizing sulfur dioxide emissions will be emphasized.
- Since a high proportion of the costs are incurred in the mine, improved recovery from run-of-mine ore will be emphasized.
- Alternatives to cyanide will be sought for leaching gold.
- Recognizing that rhodium is the most valuable PGM and also the last element to be produced in the refining sequence, efforts will continue to devise a solvent extraction method which is selective for this over the other PGM.

7 REFERENCES

7.1 Specific

1. 'SME Mineral Processing Handbook', ed. N. L. Weiss, Society of Mining Engineers, New York, 1985.
2. N. L. Weiss, in reference 1, p. 3B-2.
3. S. C. Westerfeld, in reference 1, p. 3B-28.
4. C. A. Rowland, in reference 1, p. 3C-26.
5. C. A. Rowland, in reference 1, p. 3C-44.
6. N. Arbiter, in reference 1, p. 5-1.
7. D. M. Hopstock, in reference 1, p. 6-11.
8. P. H. O'Neill and C. M. Romanowitz, in reference 1, p. 18-26.

9. B. J. DiSanto, in reference 1, p. 12-1.
10. J. N. Anderson, in reference 1, p. 12-39.
11. W. C. Hazen, in reference 1, p. 13-38.
12. R. R. Porter, in reference 1, p. 13-33.
13. A. C. Reents, in 'Chemical and Process Technology Encyclopedia', ed. D. M. Considine, McGraw-Hill, New York, 1974, p. 618.
14. F. H. Chapman, in reference 1, p. 13-59.
15. I. McGill, in 'Ullman's Encyclopedia of Industrial Chemistry', VCH Publishers, Weinheim, 5th edn, 1993, p. 607.
16. F. F. Aplan, in reference 1, p. 27-14.
17. B. Terry, A. J. Monhemius, and A. R. Burkin, in 'Extractive Metallurgy of Nickel', ed. A. R. Burkin, Wiley, Chichester, 1987, p. 1.
18. L. F. Theys, in reference 1, p. 17-5.
19. F. F. Aplan, in reference 1, p. 27-17.
20. G. G. Robson, 'Platinum 1985', Johnson Matthey plc, London, 1985, p. 19.
21. E. D. Kinloch, *Economic Geology*, 1982, **77**, 1328.
22. 'Encyclopedia of Chemical Technology', ed. M. Grayson, Wiley, New York, 3rd edn, 1983, vol. 23, p. 114.
23. A. J. Monhemius, in reference 17, p. 51.
24. B. Terry, in reference 17, p. 7.
25. B. Terry, in reference 17, p. 17.
26. K. Sproule, G. A. Harcourt, and L. S. Renzoni, in 'Extractive Metallurgy of Nickel, Copper and Cobalt', American Institute of Mechanical Engineers, New York, 1961, p. 33.
27. A. R. Burkin, in reference 17, p. 98.
28. Various Authors, *Trans. Can. Inst. Min. Metall.*, 1964, **67**, 223.
29. J. Aird, R. S. Celmer, and A. V. May, *Mining Magazine*, 1980, 320.
30. D. S. Flett, in reference 17, p. 76.
31. D. S. Flett, United Kingdom Patent 1598561, 1981.*
32. T. Ogata, S. Namihisa, and T. Fujii, US Patent 4246240, 1981.*
33. W. A. Rickelton, A. J. Robertson, and D. R. Burley, US Patent 4353883, 1982.*
34. G. J. McDougall and C. A. Fleming, in 'Ion Exchange and Sorption Processes in Hydrometallurgy', ed. M. Streat and D. Naden, Wiley, Chichester, 1987, p. 56.
35. M. J. Nicol, C. A. Fleming, and R. L. Paul, in 'The Extractive Metallurgy of Gold in South Africa', ed. G. G. Stanley, South African Institute of Mining and Metallurgy, Johannesburg, 1987, p. 831.

* **Patents** are a very useful source of technical information, but effort is required to make use of them. Patent literature is presented in a language that is to some extent strange and unattractive to the layman, resulting from patents being legal documents and from the need for words which have a precise and unambiguous meaning. Nevertheless, the effort required to understand will be rewarded, as patents frequently provide a more complete description of technology than elsewhere and sometimes represent the only source of information on a subject. Furthermore, to place the patented invention in context, the author usually provides a description of the position before the invention, 'the state of the art', and these descriptions are frequently an excellent review of the relevant technical literature. The above patents are good examples of these points.[31–33]

36. K. G. Fisher, in reference 35, p. 621.
37. K. G. Fisher, in reference 35, p. 626.
38. 'Encyclopedia of Chemical Technology', ed. M. Grayson, Wiley, New York, 3rd edn, 1983, vol. 23, p. 417.
39. D. McDonald and L. B. Hunt, 'A History of Platinum and its Allied Metals', Johnson Matthey plc, London, 1982.
40. R. I. Edwards, The Metallurgical Society of the American Institute of Mechanical Engineers paper selection A75-59, 1975.
41. G. B. Harris, in 'Precious Metals 1993', ed. R. K. Misra, International Precious Metals Institute, Allentown, 1993, p. 352.
42. G. B. Harris, in reference 41, p. 357.
43. D. S. Flett, 'Solvent Extraction in Precious Metals Refining'; paper given at Seminar of International Precious Metals Institute entitled 'Sampling, Assaying and Refining of Precious Metals', London, 10–11 October 1982.

7.2 General

'SME Mineral Processing Handbook', ed. N. L. Weiss, Society of Mining Engineers, New York, 1985.

D. McDonald and L. B. Hunt, 'A History of Platinum and its Allied Metals', Johnson Matthey plc, London, 1982.

7.3 Core Journals

Transactions of the Institute of Mining and Metallurgy – Section C – Minerals Processing and Extractive Metallurgy
Minerals Engineering
Metallurgical Transactions B – Process Metallurgy
Journal of the South African Institute of Mining & Metallurgy
International Journal of Mineral Processing
Canadian Mining Journal
Journal of Metals
Journal of Chemical Technology and Biotechnology
Hydrometallurgy
Physics and Chemistry of Minerals

N.B. When separation methods are required for refining operations, 'Separation Technology' and similar journals should be consulted. Analytical chemistry journals are also sources of relevant ideas.

CHAPTER 3

Medical Applications of Inorganic Chemicals

CHRISTOPHER F.J. BARNARD, SIMON P. FRICKER, AND OWEN J. VAUGHAN

1 INTRODUCTION

1.1 The Biological Role of Metals

Living organisms are complex, self-contained, highly organized **chemical reactors** where molecules are continually synthesized and broken down. A casual glance would suggest that the majority of the chemistry taking place in this reactor is organic chemistry as the elements C, H, N, O, P, and S account for 99.3% of the atoms in a human body. The remaining 0.7% are inorganic elements and play a vital role in the biochemistry and physiology of biological systems both as structural and functional components.[1]

The most obvious structural role for metals is that of **calcium in bone**. Bone is composed of an organic matrix, collagen, onto which hydroxyapatite, $[3Ca_3(PO_4)_2.Ca(OH)_2]$, is deposited. Metals are also important in maintaining the structural integrity of many proteins. One of the reasons why proteins can perform very specific biological functions is that they can fold into a precise three-dimensional structure. In many proteins this is maintained by metal ions coordinating to specific amino acid residues within the protein molecule thus stabilizing its three-dimensional structure. One example is the **zinc finger proteins,** so called because the protein chain is looped into a structure that resembles fingers by the coordination of Zn to the sulfur atoms on cysteine residues on opposite sides of the finger loops. Their shape allows these proteins to bind to specific sites on nucleic acids in a precise spatial arrangement.

Metals also participate in a number of important physiological **'signalling'** functions. Transmembrane concentration gradients of Na^+ and K^+ create a potential difference across nerve cell membranes. Changes in this membrane potential, caused by changes in ion flux, are responsible for the transmission of nerve impulses. Hormone-induced changes in intracellular calcium concentrations are important in regulating cell function, often by activating enzymes. Calcium is also involved in the mechanism of muscle contraction.

The ability of metals to coordinate with a variety of ligands makes them important components of many **enzyme-catalysed reactions** where they are able to form, or stabilize, intermediates in the reactions. Metals with a catalytic role are often bound in strained or distorted coordination geometries in the enzyme. Examples of this are the zinc-containing enzymes carbonic anhydrase (which catalyses the hydration of carbon dioxide to bicarbonate), carboxypeptidase A (which catalyses the hydrolysis of the carboxyl-terminal peptide bond of polypeptides), and liver alcohol dehydrogenase (which catalyses the interconversion of alcohols and aldehydes). Metals can act as Lewis acids at the enzyme active site in pH ranges where proton catalysis is ineffective.

The **redox properties** of transition metals have also been exploited by living organisms. Metal cofactors acting as redox catalysts can operate over a wide range of potentials, for example the haem (iron porphyrin) group of the cytochrome P450 enzymes operates between −300 and +800 mV. These enzymes hydroxylate a variety of substrates as part of the detoxification of foreign chemicals. Transition metals participate in energy metabolism as components of the redox units of the mitochondrial electron transport chain. In mitochondria the stepwise transfer of electrons down a potential gradient is linked to a chemical reaction, the phosphorylation of adenosine diphosphate (ADP) to the 'energy-rich' triphosphate (ATP). Inorganic chemistry is therefore an essential part of the biochemical and physiological processes of the human body.

1.2 Inorganic Medicines

Inorganic compounds were first used as medicines in the early days of civilization. The ancient Egyptians used **copper** to **sterilize water**, the Greek physician Hippocrates was using mercury as a drug in 400BC, and later Paracelsus used **mercurous chloride** as a **diuretic**. Inorganic chemistry's involvement in modern medicine began at the beginning of the twentieth century with Erlich's discovery in 1910 of salvarsan, an **arsenic** organometallic compound (see Figure 1), for the effective **treatment of syphilis**. **Antimony** compounds (for example, see Figure 1) were developed for the treatment of the parasitic disease leishmaniasis in 1912, in the 1920s gold cyanide was used to treat tuberculosis (based on the earlier observations by Robert Koch in 1890), and in 1929 gold compounds were introduced for the treatment of arthritis.

A major event in the history of inorganic medicinal chemistry was the discovery of the **anti-tumour properties** of **platinum** compounds by Barnett Rosenberg. Cisplatin is one of the most successful anti-cancer drugs developed in recent years. This discovery provided the stimulus for inorganic chemists to embark upon a continuing search for novel metal-based cancer drugs.

There are now many metal compounds used in medicine, including the platinum anti-cancer drugs and **gold** drugs for the treatment of **arthritis**.[2] Their use is varied and diverse. **Magnesium** and **aluminium oxides** are widely used as **antacids** both as over-the-counter and prescription medicines. Colloidal **bismuth subcitrate** is used to treat **peptic ulcers**. This is based upon its anti-bacterial activity against *Helicobacter pylori*, a gram negative bacterium

Salvarsan

p-Aminophenylstibonic acid

Figure 1 *Organometallic anti-parasitic drugs*

associated with chronic gastritis. The anti-microbial properties of **ionic silver** are utilized in silver sulfadiazine, marketed as Flamazine^{TM}, for topical treatment of **burn wounds**. **Lithium** salts are used in **psychiatry** for the treatment of manic depression. Recent research has shown that lithium inhibits one of the enzymes of inositol phosphate metabolism (inositol phosphates are known to be important molecules for the regulation of cellular function). **Sodium nitroprusside**, $Na_2[Fe(NO)(CN)_5]$, is used in emergency situations to treat **hypertension**. This compound acts by releasing nitric oxide, which has been shown to be the Endothelium-Derived Relaxing Factor (EDRF), a hitherto unidentified molecule responsible for causing vasodilation of the arteries and a resultant decrease in blood pressure.

1.3 'Chance' versus 'Design'

Practically all of the medical uses for inorganic molecules discussed in the previous section were the result of chance discoveries. However, rational drug design does have an increasingly important place in modern inorganic medicinal chemistry. In many instances rational design has exploited the earlier chance discoveries. This is particularly apparent in the development of 'second generation' drugs. The aim of this chapter is to describe specific examples where innovative medicinal inorganic chemistry has contributed to drug design and development.

The role of chemical modification in designing drugs with reduced toxicity or an improved method of administration will be described, with examples from platinum and gold chemistry for use in cancer treatment and arthritis respectively. A description of the use of metal diagnostic agents, such as Gd for NMR imaging and radionuclides for diagnosis and therapy will be given with a discussion on the design of ligands to maximize their utility. The application of coordination chemistry in the design of chelating agents to remove excess toxic metal ions will be discussed from the point of view of iron overload. The chapter will conclude with a brief description of potential future developments in medicinal inorganic chemistry.

2 PLATINUM ANTI-CANCER DRUGS

2.1 Introduction

The anti-tumour activity of platinum compounds was first recognized by Rosenberg and coworkers in the 1960s. They observed that under the influence of

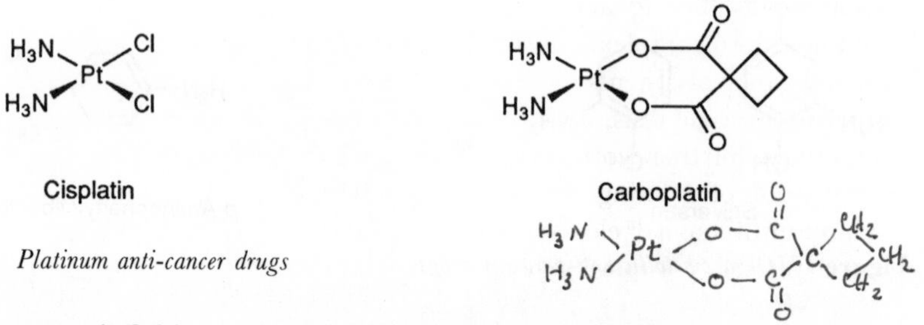

Cisplatin

Carboplatin

Figure 2 *Platinum anti-cancer drugs*

an electromagnetic field generated using platinum electrodes, *E. coli* bacteria grew as long filaments, many times longer than normal cells. Investigation showed the effect to be due to low concentrations of platinum compounds formed in the culture medium. This evidence of a much greater inhibition of cell division than cell growth suggested that the compounds would have anti-tumour properties and this was subsequently confirmed by testing at the National Cancer Institute (NCI), USA.

On publication of these results (and prior to evidence of clinical utility) work began in a number of centres to develop an understanding of the mechanism of action of platinum compounds and to correlate the structure of the compounds with their activity.

The **clinical development** of the first platinum anti-cancer drug **cisplatin** (*cis*-$[PtCl_2(NH_3)_2]$, see Figure 2) was hampered by the side effects of the treatment. Initially, kidney toxicity limited the doses so that only occasional responses were obtained. However, improved methods of administration were developed using hydration and diuretics which allowed higher doses to be given. Significant activity was then achieved, particularly in genitourinary disease. Approval for the marketing of cisplatin for use in cancer treatment was given in a number of countries in the late 1970s and it is now widely used around the world.

During this period of clinical development for cisplatin, alternative compounds were being sought which might offer greater activity or reduced toxicity. The marketing of cisplatin gave added impetus to these programmes. The mechanism of action and structure/activity studies had a major influence on the selection of the 'second generation' candidates including **carboplatin**, $[Pt(CBDCA)(NH_3)_2]$ (CBDCA = 1,1-cyclobutanedicarboxylate, see Figure 2), the only one to have gained widespread acceptance to date.

2.2 Structure/Activity Studies

A number of metal compounds had been evaluated for anti-tumour effects prior to the discovery of the activity of platinum compounds but none had shown sufficient activity to maintain interest.[3] However, after the publication of Rosenberg's work the field of metal-based agents was re-examined, with NCI playing a major role in testing a wide variety of compounds. By 1980 over 10 000 compounds had been screened including 1100 containing platinum. In particular, early studies examined compounds of the other platinum group metals since there are many similarities in the chemistry of these elements. The species initially identified by

Rosenberg as having activity were *cis*-$[PtCl_2(NH_3)_2]$ and *cis*-$[PtCl_4(NH_3)_2]$. Ammine chloro complexes of rhodium and ruthenium showed filament-inducing properties in experiments with *E. coli* and were found to have slight anti-tumour properties in some tests, but this activity was less marked than for the platinum compounds. Further work therefore concentrated on platinum compounds with the intention of defining the optimum ligand set.

Platinum(II) has a d^8 electron configuration and forms primarily four-coordinate, square planar complexes. Systematic study of the series of compounds $[PtCl_x(NH_3)_y]$ ($x=1$ to $4, y=4-x$) has indicated that the **neutral *cis* complex** is the **most active**.[4] Slight activity was noted for $[PtCl_3(NH_3)]^-$ but not for any of the other compounds. Accumulation studies suggested that this was not because the charged compounds were prevented from entering cells, although a greater barrier to entry is to be expected, but rather that the compounds lacked the appropriate reactivity to achieve an anti-tumour effect.

A broader study of the two isomers of compounds $[PtCl_2L_2]$ (L=amine) confirmed anti-tumour activity for many *cis* compounds but no activity for the *trans* forms.[3] The amine ligands in such complexes are inert to substitution while the chloride ligands can both be displaced suggesting a requirement for a pair of *cis* leaving groups for activity. Examination of a wider range of neutral ligands L, *e.g.* sulfur- and phosphorus-based ligands, also revealed a marked drop in activity. For amine ligands the general order of activity primary > secondary ≫ tertiary is seen, a series which is believed to relate to the hydrogen bonding ability of these ligands (see below).

In aqueous solution cisplatin undergoes hydrolysis and subsequent deprotonation reactions:

$$cis\text{-}[PtCl_2(NH_3)_2] + H_2O \rightleftharpoons cis\text{-}[PtCl(OH_2)(NH_3)_2]^+ + Cl^-$$

$$cis\text{-}[PtCl(OH_2)(NH_3)_2]^+ \rightleftharpoons cis\text{-}[PtCl(OH)(NH_3)_2] + H^+ \qquad pK_a = 6.85$$

$$cis\text{-}[PtCl(OH_2)(NH_3)_2]^+ + H_2O \rightleftharpoons cis\text{-}[Pt(OH_2)_2(NH_3)_2]^{2+} + Cl^-$$

$$cis\text{-}[Pt(OH_2)_2(NH_3)_2]^{2+} \rightleftharpoons cis\text{-}[Pt(OH)(OH_2)(NH_3)_2]^+ + H^+ \qquad pK_a = 5.93$$

$$cis\text{-}[Pt(OH)(OH_2)(NH_3)_2]^+ \rightleftharpoons cis\text{-}[Pt(OH)_2(NH_3)_2] + H^+ \qquad pK_a = 7.87$$

In blood the high chloride level (*ca.* 0.1 M) maintains *cis*-$[PtCl_2(NH_3)_2]$ as the major species, but inside the cell (with only 0.004 M Cl^-) *cis*-$[PtCl(H_2O)(NH_3)_2]^+$, *cis*-$[Pt(OH_2)_2(NH_3)_2]^{2+}$, and their hydroxo-forms predominate. Substitution occurs much more readily for the water ligands than for chloride or hydroxide and so the aqua complexes are the most likely reaction intermediates. As the pK_a values for these compounds are close to 7 a significant proportion of the platinum from cisplatin will be present as these reactive species. By varying the anionic groups on the complex the rates of these reactions can be substantially altered, modifying the anti-tumour activity and toxicity of the compounds.

The ease of substitution of anionic groups in platinum(II) complexes has been determined using the series of compounds $[PtX(dien)]^+$ (X=anionic ligand). Readily substituted ligands include nitrate, sulfate, and water; while complexes containing ligands such as thiocyanate, nitrite, and cyanide are unreactive. Halide complexes have intermediate reactivity. The rates of reaction across the series vary by a factor of 10^6.

The stability of complexes containing anionic oxygen donor ligands can be increased by using chelating groups, *e.g.* malonate derivatives. Such compounds are very stable in water, because, while ring opening of the chelate does occur to yield a monodentate carboxylate group, the reverse reaction occurs readily keeping the concentration of this species very low. Solutions of these complexes thus show no degradation even after weeks of storage (in the absence of light, which promotes loss of the amine ligands). However, malonate complexes do react with nitrogen donor ligands sufficiently rapidly to have biological activity.[5] In this case, once the nitrogen donor is bound the strength of the platinum–nitrogen bond is sufficient to prevent the reverse reaction of displacement by the free carboxylate group of the monodentate malonate. The second substitution step displacing the malonate ligand can then occur readily as for other simple carboxylate compounds.

In vivo evaluation of series of complexes containing alternative anionic groups showed that where these ligands are readily displaced (*e.g.* nitrate) the complexes are highly toxic, reacting indiscriminately at many biological sites. The toxicity relative to the therapeutic dose decreases as the stability of the complexes increases over the series sulfate, chloride, malonate with the substitutionally inert thiocyanate compounds being inactive. The potency of the compounds also decreases as the compounds become more stable, with much of the dose of malonate complexes being excreted as unreacted compound. The neutral ligands can also influence the reactivity of the complex through steric and hydrogen bonding interactions so the optimum choice of reactivity for the anionic ligands differs slightly for different amine complexes, thus offering a range of possibilities for clinical candidates.

Octahedral complexes of the higher oxidation state **platinum(IV)** are also active anti-tumour agents. These compounds, with the platinum having a stable d^6 electron configuration, are very inert to substitution and it is believed that they must be reduced prior to interaction with biological ligands. This theory is supported by evidence that platinum(IV) complexes do not react with isolated DNA (deoxyribonucleic acid) and that platinum(II) metabolites are formed following the administration of platinum(IV) drugs. Variation in the ease of reduction of different complexes offers further scope for modifying the toxicity/activity balance, and the solubility and lipophilicity of the compounds can be significantly different from the platinum(II) products due to the additional pair of ligands, resulting in different pharmacokinetic behaviour.

2.3 Mechanism of Action

The original observation in *E. coli* bacteria that **cisplatin inhibited cell division** more strongly than cell growth suggested that the mechanism of action involved reaction with DNA rather than with other cellular macromolecules such as RNA (ribonucleic acid) or proteins. This was soon supported by studies of cisplatin's effect on macromolecular biosynthesis using tritium-labelled precursors. DNA synthesis was inhibited selectively and at cisplatin concentrations which intrastrand crosslink　　interstrand crosslink

Figure 3 *Platinum–DNA crosslinks*

were only minimally cytotoxic. Determination of the intracellular disposition of platinum in relation to the relative concentration and molecular weights of DNA, RNA, and proteins also suggested that DNA was the most likely site of action.[6]

Analysis of platinated DNA has shown that platinum is concentrated in guanine(G)–cytosine(C) rich regions.[7] The kinetically preferred sites of reaction between cisplatin and nucleic acid bases were found to be N-7 of guanine and adenine, with **binding to guanine** being dominant. This is also the case for synthetic oligonucleotides.

There are four possible types of bifunctional lesions which can be formed following an initial binding to DNA; chelation to a single base (G), an intrastrand **crosslink**, an interstrand crosslink, and a DNA–protein crosslink. Of these, there is no evidence for the first, despite a guanine N-7/O-6 chelate having been proposed some time ago. Of the other possible lesions the 1,2-intrastrand crosslink (*i.e.* binding to neighbouring bases on the same DNA strand, see Figure 3) has been viewed as the most important as this cannot be formed by the *trans* isomer. The structural restrictions imposed by the *trans* geometry limit this complex to 1,3- and 1,4-intrastrand links. The absence of this lesion may thus be used to explain the striking difference in cytotoxicity and anti-tumour activity of the two isomers.

As for the first reaction with DNA, the preferred second binding site is G producing a GG intrastrand crosslink. In reactions with isolated salmon sperm DNA this lesion accounted for approximately 60% of the platinum bound to DNA. This is a higher value than would be expected on statistical grounds after initial binding to guanine suggesting some preferrence for GG over single G in DNA.

Considerable effort has been expended to characterize the structure of the GG crosslink, primarily using solution state NMR experiments and solid state X-ray crystallography. The crosslink leads to both an unwinding (*ca.* 13°) and kinking (*ca.* 33°) of the DNA double helix. This distortion weakens but does not prevent base pairing with the complementary strand as determined by NMR experiments. The structural studies have also indicated that hydrogen bonding involving one ammine ligand and the 5′-phosphate group plays an important part in stabilizing the final structure and thus provides a possible explanation for the poor activity of tertiary amine compounds which lack this capability. The phosphate group has also been found to enhance the rate of reaction of aquated cisplatin with G suggesting that this interaction (both electrostatic and through hydrogen bonding) may also play a significant role in stabilizing the transition state complex.

2.4 Selection of Second Generation Drug Candidates

During the early stages of clinical trials with cisplatin it was apparent that the serious side-effects of the drug would limit its usefulness. Improvements in administration methods achieved some control of kidney toxicity in particular, but it was clear that an alternative compound with a better toxicity profile, while retaining the anti-tumour activity of cisplatin, was desirable. This therefore became the major goal of the selection process for second generation drugs.[8]

In general, for suitability as drugs with intravenous infusion as the route of administration, candidate compounds should be stable in solution and have sufficient aqueous solubility for formulation. These criteria act against compounds containing weakly bound ligands such as sulfate (poor stability) and against many of the simple amine chloro complexes (low solubility). The substituted malonate ligands provide stable complexes with a wide variation in solubility, largely controlled by the extent of intramolecular hydrogen bonding in the solid state, and so examples of this type of compound featured strongly in the work of many groups. Nevertheless, a very wide range of complexes were still available as potential candidates.

At the Institute of Cancer Research, a group of compounds showing similar activity to cisplatin in tumour screens and covering a variety of chemical structures were studied for haematological and major organ toxicities. This work identified JM-8 (carboplatin, see Figure 2) and JM-9 (iproplatin, see Figure 4) as cisplatin analogues with **little or no kidney toxicity** but of similar haematological toxicity. Carboplatin was preferred on the evidence of a human tumour xenograft model which suggested lower toxicity for an equivalent therapeutic dose. Subsequently, both of these derivatives received clinical evaluation with the results confirming that while both drugs were active in treating cancers, particularly ovarian cancer, fewer side-effects were observed with carboplatin. The latter is now widely used in the treatment of ovarian cancer, having received its first marketing approvals in 1986.

Other groups concentrated on identifying compounds of low toxicity which showed a lack of cross-resistance with cisplatin in tumour screens. It was hoped that such compounds could be used to treat patients relapsing after cisplatin therapy or whose tumours were resistant to cisplatin on initial treatment. Several of these incorporated the 1,2-diaminocyclohexane ligand in the complex (*e.g.* tetraplatin, see Figure 4). To date, none of these compounds have proved to be superior to cisplatin or carboplatin.

2.5 Conclusion

Platinum drugs are now widely used in **cancer chemotherapy**. Particular successes include the treatment of testicular and ovarian tumours and these agents are also used in combination with organic drugs for the treatment of many other forms of cancer. The introduction of carboplatin, with its narrow toxicity profile, has allowed out-patient treatment in many cases but still greater convenience for the patient could be achieved with an orally administered drug. The requirement

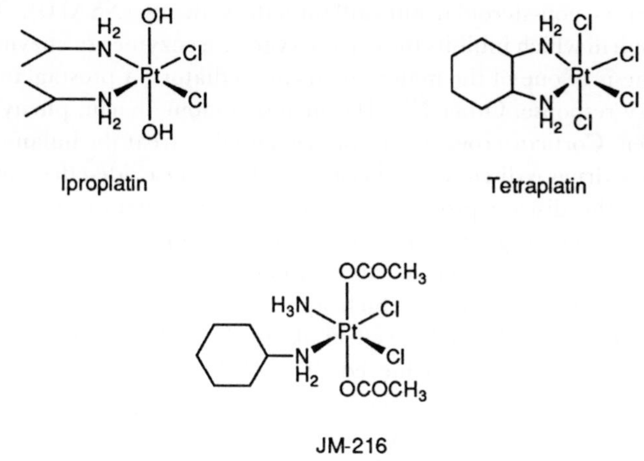

Iproplatin

Tetraplatin

JM-216

Figure 4 *Platinum(IV) compounds tested as anti-cancer drugs*

for effective oral absorption has lead to the development of a class of Pt(IV) carboxylate complexes, with JM-216 (see Figure 4) entering clinical trials in 1992.

After over 20 years of clinical application platinum anti-cancer drugs remain the focus of strong research activity. There is much still to be understood regarding the molecular basis of their mechanism of action and efforts are continuing in the search for new compounds for wider clinical application.

3 GOLD ANTI-ARTHRITIS DRUGS

3.1 Introduction

Rheumatoid arthritis is an inflammatory disease characterized by a progressive erosion of the joints resulting in deformities, immobility, and a great deal of pain. It belongs to the category known as autoimmune diseases in which the body's immune system (its defence mechanism against invasion by organisms such as bacteria and viruses) turns against itself. The immune system is a complex, multi-component system controlled by an intricate, interactive mechanism. If this control breaks down the immune system can mount a response by producing autoantibodies against 'self' components known as autoantigens. In rheumatoid arthritis these autoantibodies are immunoglobulin proteins called rheumatoid factors. The autoimmune response results in a malign growth of the synovial cells (the cells lining the joint) called a pannus and an infiltration of the joint space by cells of the immune system, primarily macrophages. The synovial and macrophage cells release degradative enzymes called proteinases which destroy the surrounding collagen and bone. This progressive inflammatory response is mediated by an overproduction of chemical mediators such as prostaglandins, leukotrienes, and cytokines.[9]

At present there is no effective cure for this painful condition. Initially, the therapeutic approach adopted by doctors is to treat the symptoms by the use of drugs known as non-steroidal anti-inflammatory drugs (NSAID). The classic NSAID is aspirin which inhibits the cyclo-oxygenase enzyme, an enzyme required for the synthesis of one of the major chemical mediators, a prostaglandin, of the inflammatory response. Other NSAIDs include indomethacin, phenylbutazone, and ibuprofen. Corticosteroids have also been used to treat the inflammation but the use of these drugs is discouraged because of the adverse side-effects of their long term use. As the disease progresses more aggressive therapy is required and immunosuppressive drugs such as azathioprine and methotrexate may be used. There is also a chemically diverse category of drugs known as disease modifying anti-arthritic drugs (DMARD) which appear to suppress the disease process. They do not produce an immediate effect like the NSAIDs but require up to 4 to 6 months before a response is achieved. The DMARDs include penicillamine, chloroquine, and a number of gold compounds. All are associated with possible side-effects (*e.g.* kidney toxicity and skin rash for penicillamine) which influences the selection of the most appropriate drug for each case.[10]

3.2 History of Medicinal Gold

The earliest medicinal uses of gold were a result of the mystical powers attributed to the lustrous element by ancient civilizations.[11] In medieval Europe alchemists had numerous recipes for an elixir known as *aurum potabile*, many of which we can say with certainty contained little gold. In the 17th century gold had gained entry into the new pharmacopoeias as a cordial for use in any ailment caused by a decrease in vital spirits, *e.g.* melancholy, fainting, fevers, and falling sickness. Its efficacy was summed up in the verse by Nicholas Culpepper:

'For gold is cordial
And thats the reason
Your raking misers
Live as long a season.'

In the 19th century gold formulations, usually a mixture of gold chloride and sodium chloride, were used, unsuccessfully, to treat syphilis. In 1890 the German bacteriologist Robert Koch discovered the **bacteriostatic** properties of **gold cyanide** against the tubercle bacillus. In the 1920s gold therapy was introduced for the treatment of tuberculosis. The suggestion that the tubercle bacillus was the cause of **rheumatoid arthritis** led to the use of gold therapy for this disease. Gold therapy soon proved to be ineffective against tuberculosis but, after 30 years of medical debate, a clinical study by the Empire Rheumatism Council in 1958 confirmed its effectiveness in the treatment of rheumatoid arthritis. The early gold drugs were gold(I) thiols such as **sodium aurothiomalate** (Myocrisin^{TM}) and **aurothioglucose** (Solganal^{TM}, see Figure 5). Myocrisin^{TM} is still to be found in the British National Formulary for use in the treatment of rheumatoid arthritis.

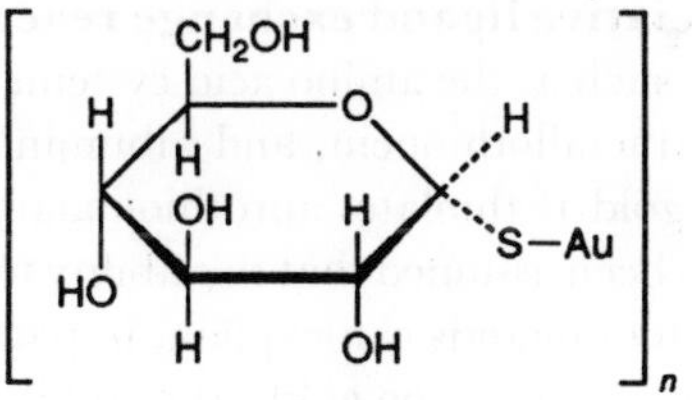

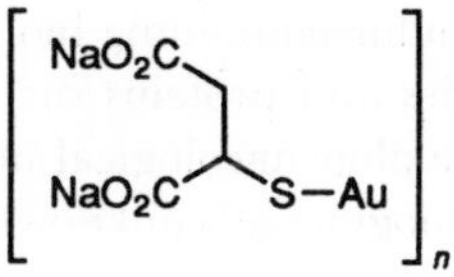

Aurothioglucose

Aurothiomalate

Figure 5 *Early gold anti-arthritis drugs*

3.3 Biological Chemistry of Gold

Gold can exist in a number of oxidation states: −I, 0, I, II, III, IV, and V. However only gold 0, I, and III are stable in aqueous, and therefore biological, milieu.[12] Both gold(I) and gold(III) are unstable with respect to gold(0) and are readily reduced by mild reducing agents with positive standard potentials of +1.68 and +1.42 volts respectively:

$$Au^{1+} + 1\,e^- \rightarrow Au(0) \qquad E = +1.68\ V$$
$$Au^{3+} + 3\,e^- \rightarrow Au(0) \qquad E = +1.42\ V$$

Gold(III) is readily reduced to Au(I) and this oxidizing power generally means that Au(III) compounds are toxic.

Au(I) commonly forms two-coordinate, linear complexes and can readily disproportionate to give Au(III) and Au(0). The gold compounds used in the treatment of arthritis are all **Au(I) thiolates**, that is they contain the AuSR unit where R is a suitable organic functional group. The SR ligand is the sole ligand in the drugs Myocrisin™ and Solganal™ (SR = thiomalate and thioglucose respectively). These compounds form polymeric structures both in solution and in the solid state, as has been demonstrated by X-ray techniques such as WAXS and EXAFS. This type of structure requires an 'extra' –SR group to cap the 'bare' gold atom at the end of the chain and, in fact, a slight excess of thiomalate is found in conventional preparations of Myocrisin™.

The gold(I) thiolate structure possesses a number of useful properties for a gold compound with a potential pharmacological use. As mentioned earlier, Au(III) compounds are potentially toxic due to their oxidizing ability and therefore the Au(I) oxidation state is to be preferred. Conversely, as Au(I) can disproportionate to Au(III) and Au(0) it must be stabilized by suitable ligands which can form strong covalent bonds with the metal. The ligand donor atom must be soft, *e.g.* sulfur or phosphorus. Harder atoms such as oxygen, nitrogen, and halogens yield unsuitable compounds. Whilst conferring stability upon the complex, these ligands should not be so strongly bound as to make the compound chemically unreactive. If this were the case the compound would be unable to participate in essential biotransformation reactions or react with the biological target molecules through which it would exert its pharmacological effect.

Gold thiol compounds can undergo **associative ligand exchange reactions** with sulfur-containing biological molecules, such as the amino acid cysteine, and peptides and proteins such as glutathione, metallothionein, and albumin. The precise pharmacological mechanism of the gold(I) thiolates aurothiomalate and aurothioglucose is unknown. However, it has been assumed that this ability to take part in exchange reactions with biological thiol ligands can explain, in part, the molecular target for these drugs. One proposal is that the gold drugs inhibit the action of the degradative enzymes responsible for the joint erosion in rheumatoid arthritis by binding to thiol groups on the protein.

3.4 Development of Auranofin

Despite the apparently favourable chemical properties of Myocrisin™ and Solganal™ as drugs, they both present problems to the rheumatologist. Both compounds are water soluble and have to be administered parenterally by a deep intramuscular injection at weekly intervals. This requires frequent visits by the patient to the rheumatologist, an inconvenience for both parties and painful for at least one. An initial dose of 10 mg is administered, followed by increased doses of up to 50 mg until the onset of remission when the dosing frequency can be reduced.

The drug is rapidly absorbed into the blood stream when administered in this way and the dose given results in an intial serum level of 4–10 $\mu g\ ml^{-1}$ gold. Some of the gold is then rapidly cleared from the bloodstream so that after 2 hours the level is 2–3 $\mu g\ ml^{-1}$ gold, which is then maintained. The gold is distributed to various tissues throughout the body, particularly the kidney where it eventually accumulates after repeated doses. This results in one of the major toxic side effects of the simple gold thiolates, damage to the kidney. Other adverse reactions include mouth ulcers, skin reactions, blood disorders, and occasionally liver toxicity. Despite these problems there are clear indicators of the therapeutic usefulness of gold therapy in rheumatoid arthritis and this provided the impetus for the search for a new gold drug with a better pharmacological and toxicological profile.

This challenge was taken up by the medicinal chemists at the then Smith, Kline and French laboratories.[13] Their key requirement for an improved gold drug was a compound that could be administered in small, daily doses giving stable serum levels of active drug at a therapeutic concentration. Ideally, this would result in a better therapeutic response and the avoidance of high serum gold levels would give fewer toxic side effects, particularly reduced nephrotoxicity. Daily intramuscular injections of the hypothetical 'ideal' drug were not a practical proposition from either the patient's or the clinician's point of view and therefore the new compound would have to have a different, more convenient, route of administration. The obvious choice, and the most frequently used route for medicines, is oral administration. So the aim of the medicinal chemists was to design an orally bioavailable gold drug incorporating the useful and necessary chemical features of the then available gold drugs.

Coordination of gold(I) with phosphine ligands stabilizes that oxidation state. Complexes of this type are generally nonionic and soluble in organic solvents. Being Au(I) they are two-coordinate, linear complexes. Importantly, they are also

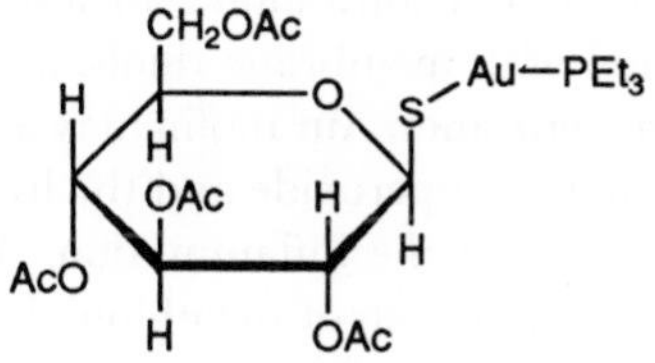

Figure 6 *Auranofin, used to treat rheumatoid arthritis*

structurally well-defined monomeric species. This is a key factor in simplifying the physical and chemical characterization required by today's stringent regulatory procedure. The polymeric gold(I) thiolates were approved for clinical use before such detail was required.

The medicinal chemists synthesized a series of compounds with the general formula R_3PAuSR^1 where R was an alkyl group and SR^1 a diverse range of ligands such as thioglucose, acetylated thioglucose, 2-mercaptoethanol and thiocyanate. These **trialkylphosphine Au(I) compounds** are all lipophilic and are readily absorbed after oral administration, satisfying one of the important criteria for the new drug. The compounds were evaluated in a pharmacological disease model of arthritis. Within a given series of related compounds the optimum pharmacological activity was seen with triethylphosphine complexes. Consideration of the requirements for formulation resulted in the selection of the compound containing tetra-acetylthioglucose as the thiolate ligand as a candidate for further development. This compound, [tetra-*O*-acetyl-*β*-D-(glucopyranosyl)thio](triethylphosphine)gold(I), is now available for the treatment of rheumatoid arthritis under the name of **auranofin** (Ridaura™, see Figure 6).

3.5 Pharmacology of Auranofin

Auronofin has a number of advantages over previous gold drugs.[14] Its oral bioavailability makes it easier to administer. Gold levels in the blood are lower and are maintained for a longer period of time and there is less retention of gold in the tissues. Approximately 25% of the administered gold is absorbed; the remaining 75% is excreted in the faeces. Auranofin is extensively metabolized. It is deacetylated in the intestine prior to absorption. The thioglucose ligand can be replaced by ligand exchange with biological thiol ligands. Ligand exchange with cell membrane-associated thiol groups has been postulated to be the mechanism of cellular uptake for auranofin. Once inside the cell the triethylphosphine is readily lost and oxidized to triethylphosphine oxide, a potentially toxic metabolite. This dissociation step controls the final cellular distribution and efflux of the gold. In this model, cellular uptake and distribution of gold is dependent upon local concentrations of thiol ligands in the biological milieu.

After metabolism the final active gold species is probably similar to the active species for the polymeric gold thiolates. Investigations into the effect of auranofin and aurothiomalate on various aspects of immune system function reveal a number of qualitative similarities but some minor quantitative differences. Other mechanistic studies suggest that auranofin can also stabilize cell membranes, presumably by interaction with membrane thiols, and inhibit cellular release of the degradative proteinase enzymes. Auranofin may also inhibit the production of reactive oxygen species such as superoxide and the hydroxyl radical, which have been implicated as mediators of the inflammation. Although there are enough differences between the cellular effects of aurothiomalate and auranofin to prompt suggestion of a unique mechanism of action of auranofin, many of these results were obtained *in vitro* using intact auranofin. Following metabolism *in vivo* the **mechanism of action of auranofin** is probably **similar to that of the earlier gold thiolate drugs**. Clinical studies have shown that the major advantage of auranofin over the other gold drugs is its **reduced toxicity**, resulting in fewer patients being withdrawn from therapy.[10,14] Renal toxicity is particularly reduced with auranofin. Its most common side-effect is diarrhoea.

3.6 Conclusion

The development of **auranofin** is a good example of the way medicinal chemistry can be applied to overcome problems of pharmacokinetics and toxicology associated with a class of drugs. In the final analysis it is clinical efficacy that counts. Rheumatoid arthritis is still a disease without a complete cure and even the disease-modifying drugs have limited effect on disease progression. In some quarters the utility of gold therapy is being questioned. A therapy that has been in existence for over sixty years may now be falling out of favour. Perhaps an **improved understanding of the molecular mechanism** of gold drugs may provide the impetus for further research to **improve gold therapy**.

4 MAGNETIC RESONANCE IMAGING

4.1 Introduction

One physical property of metal ions of great interest to medicine is **paramagnetism**. Magnetic resonance imaging (MRI) is an important diagnostic tool that exploits the differences in relaxation rates of water protons in different tissues. Paramagnetic metal complexes can shorten these proton relaxation times and, depending on their biodistribution, provide improved tissue contrast when administered *in vivo*. The metal ions currently favoured for this application are Gd(III), Fe(III), and Mn(II). Coordination complexes used as MRI contrast agents must have a high *in vivo* stability to prevent release of toxic metal ions, yet have open coordination sites for rapidly exchanging water ligands to enhance proton relaxation.

4.2 Gadolinium Agents

Gadolinium has found wide acceptance in the field of MRI.[15] Of all elements gadolinium has the strongest influence on proton relaxation time. This powerful effect can be attributed to a complex interplay of several factors including a strong

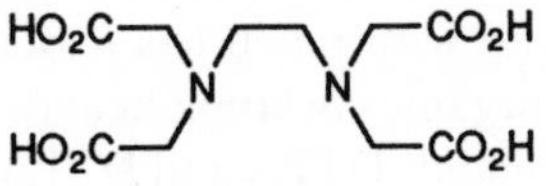

EDTA

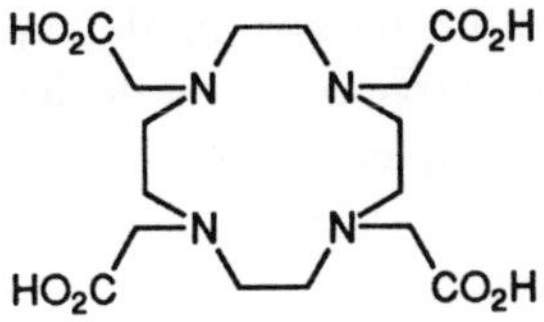

DOTA

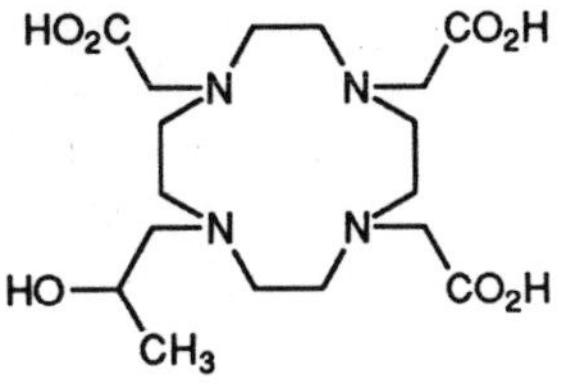

HP-DO3A

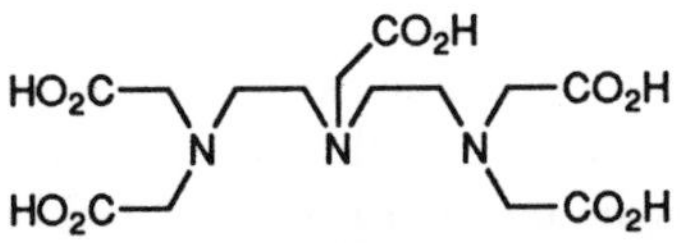

DTPA

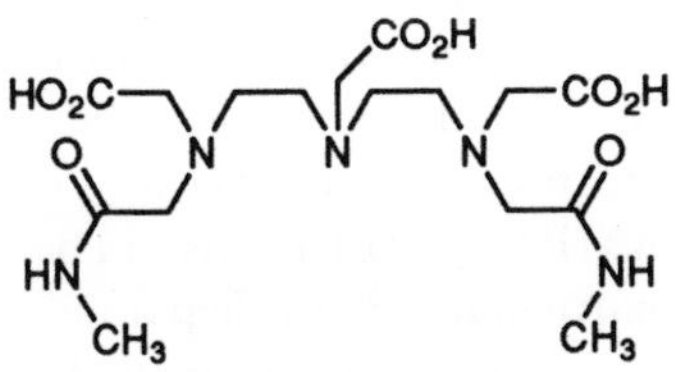

DTPA-BMA

Figure 7 *Ligands used for chelation of gadolinium for magnetic resonance imaging (MRI)*

magnetic moment, a long electron relaxation time, the configuration and mobility of molecules of hydration, and the proximity of hydrogen nuclei to the paramagnetic centre.

Chelating gadolinium to EDTA (ethylenediaminetetraacetic acid) and DTPA (diethylenetriaminepentaacetic acid, see Figure 7) reduces, but far from eliminates, gadolinium's strong influence on proton relaxation. The coordination number of Gd^{3+} is estimated to be 9 or 10. Thus, using DTPA with eight coordinating atoms as the chelating ligand leaves at least one or two metal binding sites free for water ligands. However, the complexation of gadolinium with EDTA produced little or no improvement in toxicity compared with gadolinium chloride. Whether the toxicity is caused by the intact complex or is due to dissociation of the gadolinium ion from the EDTA ligand in the body is not clear. Gadolinium chelation with DTPA produces a compound with much less toxicity.

The structures of several organic chelating ligands used to complex gadolinium are shown in Figure 7. The ligands DTPA and DTPA-BMA are based on the same linear triamine framework. For both ligands the bonds to gadolinium are formed from three nitrogen atoms and five oxygen atoms. For DTPA-BMA two of the charged carboxylate oxygen atoms are replaced with uncharged amide oxygen atoms. The effect of this modification is to reduce the charge count on the ligand from −5 in DTPA to −3 in DTPA-BMA, and, when complexed with Gd^{3+}, from overall −2 in $Gd(DTPA)^{2-}$ to overall 0 in Gd(DTPA-BMA).

1,4,7,10-tetracarboxymethyl-1,4,7,10-tetraazacyclododecane (DOTA) and 1,4,7-tricarboxymethyl-10-(2-hydroxypropyl)-1,4,7,10-tetraazacyclododecane (HP-DO3A) are based on a macrocyclic tetraamine framework. In this case, Gd(HP-DO3A) with three charged carboxylate oxygen atoms and one neutral hydroxyl oxygen atom is nonionic, whereas $Gd(DOTA)^{1-}$ with its four charged carboxylate oxygen atoms bears an overall charge of −1. The neutral hydroxyl group in Gd(HP-DO3A) is not acidic like carboxylates, and remains protonated in solution; it does remain bound to the gadolinium atom.

Thus, $Gd(DTPA)^{2-}$ is an ionic linear chelate, Gd(DTPA-BMA) is a nonionic linear chelate, $Gd(DOTA)^{1-}$ is an ionic macrocycle, and Gd(HP-DO3A) is a nonionic macrocycle. These physicochemical differences lead to greater formulation and dosing flexibility for the uncharged chelates, and greater kinetic inertia to ligand release for the Gd chelates built around a macrocyclic ligand framework. Two of these agents have been approved as MRI contrast agents in the clinic, Gd(DTPA) as Magnevist™ and Gd(HP-DO3A) as ProHance™. These compounds are injected intravenously and are confined to the extracellular space. Their major clinical application is in imaging the central nervous system.

Further development of MRI contrast agents will be aimed at introducing more **organ specific agents**, particularly for the hepatobiliary system. Some variation in the coordination behaviour of the paramagnetic ion can be achieved by using Fe(III) or Mn(II) instead of Gd(III) allowing different pharmacological properties to be obtained.

5 RADIOPHARMACEUTICALS

5.1 Introduction

Radiopharmaceuticals, drugs containing a radionuclide, have found use both as diagnostic and therapeutic agents. While the ideal pharmacological properties might best be provided by purely organic molecules, all suitable radionuclides are heavier elements with the majority being metals. Consequently, coordination compounds have found widespread applications in the area of nuclear medicine.

The **metal-based radiopharmaceutical** may be considered to be one of two types: (1) **metal essential**, whereby the properties of the coordination compound determine its biological distribution; or (2) **metal linked**, whereby the biological distribution of the coordination compound is determined by the properties of the carrier molecule, such as an antibody, with the metal or metal complex determining the ultimate geometry and stability of the radiopharmaceutical. The agent is administered into circulating blood which has a pH of *ca.* 7.4 and contains various proteins (*e.g.* transferrin, a protein which transports ferric ions in the bloodstream) and enzymes, any of which may compete with the carrier ligands and challenge the integrity of the complex of interest. Clearly, the radiopharmaceutical must be stable long enough to reach its target.

Diagnostic and therapeutic radiopharmaceuticals require radionuclides with different nuclear properties. For diagnostic imaging applications, the radiation must be able to penetrate the body and be detectable externally to the patient. Hence, the requirement is for a radionuclide able to emit photons of energy typically in the range 80–300 keV. Radionuclides decaying, by gamma ray (γ) or positron (β^+) emission without accompanying alpha (α) or beta (β^-) emission, to a relatively stable daughter nuclide in a relatively short time (a half-life of less than a day) are potentially suitable for incorporation into diagnostic radiopharmaceuticals. The half-life must be sufficiently long to allow synthesis of the radiopharmaceutical (usually carried out in the hospital), administration to the patient and imaging with the available instrumentation, but short enough to minimize exposure for the patient. For radiotherapeutic applications, which require cell destruction, the need is for radionuclides that decay by particle emission (α, β^-, Auger e^-) with a typical half-life of 1–10 days, with the energy of the particle emitted during decay and the radionuclide half-life being dependent on the particular radiotherapeutic application in question.

5.2 Technetium

In the field of **diagnostic nuclear medicine**, ^{99m}Tc has found widespread applications, so much so that in some chemical form or other, it is used in about 90% of radiodiagnostic scans performed in USA. The radioisotope ^{99m}Tc ($t_{\frac{1}{2}} = 6.02$ h, $E_\gamma = 140$ keV) is readily available at a relatively low cost. It has no β emissions and emits only low energy Auger electrons. Technetium occupies a central position in the second row transition metal series. As a consequence, many technetium complexes, of varying oxidation states from -1 to $+7$, with differing coordination geometries (4–9), are known. These features thus provide ample opportunity for the inorganic chemist to design ligand systems around the ^{99m}Tc. These differing ligand environments will largely dictate the chemical, physical, and biological properties, and hence the complex's specificity, for radioimaging applications. Several ^{99m}Tc radiopharmaceuticals have been approved throughout the world for assessing the status of organ function, or morphology, or for determining the pathological status of a patient. An excellent review by Jurisson *et al.* discusses the ^{99m}Tc radiopharmaceuticals currently approved.[16]

Ceretec™ [^{99m}TcO(*d,l*-HM-PAO); HM-PAO = hexamethyl propylene amine oxime; see Figure 8] has been approved as a **cerebral perfusion imaging agent** for the evaluation of stroke. It may also have uses in the evaluation of other cerebral disease states such as haematomas and Alzheimer's disease, but at present it has not been approved for these applications. The development of Ceretec™

TcO(PnAO) TcO(*d,l*-HM-PAO)

Figure 8 *Technetium complexes used as radiopharmaceuticals*

was based upon the compound TcO(PnAO) (PnAO = propylene amine oxime; 3,3,9,9-tetramethyl-4,8-diazaundecane-2,10-dionedioxime). It was shown in rat biodistribution studies that this neutral lipophilic coordination complex was able to cross the brain blood barrier. However, it also diffused back out of the brain with too short a half-life to permit imaging. As a consequence, scientists at Amersham International synthesized and characterized TcO(*d,l*-HM-PAO), which was not only taken up by the brain but was also retained sufficiently long to allow imaging. The HM-PAO ligand loses two amine protons and an oxime proton on coordination to the Tc(v) oxo core giving a neutral complex. In the brain Ceretec^{TM} undergoes a transformation to a more hydrophilic species that is unable to diffuse back out of the brain. Thus, brain retention is achieved with this complex. The ^{99m}TcO(*meso*-HM-PAO) analogue undergoes a similar transformation, but at a much slower rate. Consequently, it is able to diffuse out of the brain. Clearly, the configuration of the ligand has a profound effect on the biological properties of the coordination complex.

Technetium phosphate and phosphonate complexes have found use as bone imaging agents. The ability of ^{99m}Tc diphosphonates to accumulate in the skeletal metastases of cancer patients was established in the 1970s. These technetium coordination compounds presumably function by exploiting the affinity of the diphosphonate group for the hydroxyapatite structure of bone, with the preference for metastatic lesions reflecting a difference in relative metabolism. Currently, three different diphosphonate ligands complexed with ^{99m}Tc are used as **skeletal imaging agents** (Figure 9). The three ligands differ in the functional groups on the carbon atom between the two phosphorus atoms. The diphosphonate ligands can be considered as doubly bidentate or bidentate–tridentate systems, with the ability to complex two metals simultaneously. The compound ^{99m}Tc(MDP) (MDP = methylene diphosphonate, see Figure 9) is marketed by several companies as a skeletal imaging agent to delineate areas of altered osteogenesis. It is a Tc(IV) complex that has a high affinity for sites of actively growing bone, and is believed to be a mixture of oligomers and polymers. During the 24 hours after injection, any ^{99m}Tc(MDP) not retained by the skeleton is excreted.

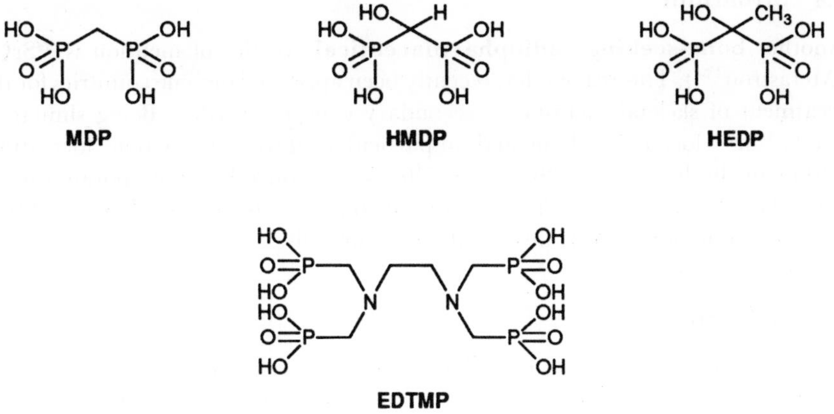

MDP HMDP HEDP

EDTMP

Figure 9 *Diphosphonate ligands used in technetium, rhenium, and samarium radiopharmaceuticals*

5.3 Rhenium and Samarium

If radionuclides with α or energetic β emissions can be targeted to disease sites, **radiotherapeutic agents** can be obtained. Agents for the palliation of bone cancer pain have been prepared by incorporating appropriate β-emitting radionuclides into diphosphonate complexes. The properties of a radionuclide considered appropriate for the palliative treatment of bone pain are a 1–5 day half-life with 0.1–1 MeV β^- energy. Suitable radionuclides are ^{186}Re and ^{153}Sm.

The chemistry of the ^{99m}Tc diphosphonate bone agents has been extended to rhenium, the third row congener of technetium in Group 7. Indeed, it has been shown that ^{186}Re(HEDP) (HEDP = hydroxyethylidene diphosphonate, Figure 9), like the ^{99m}Tc diphosphonates, is a complex mixture of species. However, although the biolocalization of the two compounds is very similar, there is one major difference. Re(HEDP) is more readily oxidized (to ReO_4^-) *in vivo* than is Tc(HEDP). This leads to an increase in abnormal to normal bone uptake with time for rhenium because ReO_4^- is washed off normal bone and excreted. This is beneficial from the radiotherapeutic standpoint because it reduces the radiation dose to normal tissue.

Another bone-seeking radiopharmaceutical ^{153}Sm(EDTMP) (EDTMP = ethylenediamine-*N*,*N*,*N'*,*N'*-tetrakis(methylenephosphonic acid), see Figure 9) also shows *in vivo* localization comparable to the ^{99m}Tc diphosphonate agents. Although the structure of ^{153}Sm(EDTMP) has not been determined, the predominant species in solution at physiological pH appears to be the fully deprotonated complex $Sm(EDTMP)^{5-}$ with a smaller percentage of the monoprotonated complex $Sm(HEDTMP)^{4-}$. The samarium is presumably coordinated to EDTMP through the two amine nitrogens and the oxygens from all four phosphonate groups.

Both ^{186}Re(HEDP) and ^{153}Sm(EDTMP) are currently in phase III clinical trials in USA and elsewhere for use in the palliative treatment of metastatic bone cancer pain.

5.4 Strontium

Another **bone-seeking radiopharmaceutical** worthy of mention is $^{89}SrCl_2$ (Metastron^{TM}). This reagent has recently been approved in some countries for the treatment of skeletal metastases (secondary cancer growths). Being similar to Ca^{2+}, $^{89}Sr^{2+}$ localizes in bone and, in particular, at sites of new bone formation. Although the half-life of ^{89}Sr is long (50.5 days), once it is incorporated in an osteoblast (bone-forming cell) it remains deposited at the metastatic site and thus delivers the majority of its dose to the tumour cells.

5.5 Gallium

The successful development of gallium-based agents has its roots in the 1950s, when ^{72}Ga was extensively studied in animals and shown to localize in normal bone tissue. During the 1960s, as ^{67}Ga became available, it was noted that the isotope, when administered as the citrate, accumulated in certain malignancies. The mechanism of **tumour accumulation** is still not well understood. However, it has been shown that when ^{67}Ga citrate is injected into the patient, essentially all the radioisotope is bound to transferrin. It is this labelled transferrin which is taken up by and accumulated in the tumour. The ^{67}Ga is supplied in a solution containing high levels of citrate in order to suppress hydrolysis to $Ga(OH)_3$ and to permit transchelation in serum to the circulating transferrin. Gallium(III) mimics Fe(III) in its binding to molecules in serum, due to the similar charge/size ratios of the two ions. However, Ga(III) is not reduced *in vivo* and does not become incorporated into haemoglobin.

Successful tumour images with ^{67}Ga citrate have been obtained in bronchogenic carcinomas, lymphomas, Hodgkin's disease, hepatomas, and melanomas. Results in the case of gynaecological, gastrointestinal, and genitourinary tract neoplasms have been less encouraging because of excessive accumulations of the radioisotope in normal abdominal tissues, such as the gastrointestinal tract, liver, and spleen, which tend to obscure small lesions. Another difficulty with ^{67}Ga citrate is its lack of specificity for tumour tissue, as it also tends to localize in inflammatory sites such as pneumonitis. Consequently, ^{67}Ga may find further applications for the detection of non-neoplasms such as abscesses, rather than for the detection of tumours.

The ready availability of the positron-emitting ^{68}Ga (half-life, 68 minutes), from a $^{68}Ge/^{68}Ga$ generator, permits the use of this radioisotope for **positron emission tomography** (PET). However, to avoid the transchelation of Ga(III) complexes *in vivo* by transferrin, kinetically inert complexes of Ga(III) need to be developed in order for the radiopharmaceutical potential of gallium to be fully realized. Several groups have demonstrated that N_3X_3 ligands such as TX-TACNH$_3$ (1,4,7-tris{3,5-dimethyl-2-hydroxybenzyl}-1,4,7-triazacyclononane) form Ga(III) complexes with very high *in vitro* and *in vivo* stability. These types of ligands essentially encapsulate the gallium ion and insulate it from competing ligands, such as transferrin.

5.6 Monoclonal Antibodies for Targeting

One of the differences between tumour and normal cells is the presence of different immunological markers on the cell surface. This offers the possibility of preparing antibodies directed against them, which, after radiolabelling, may be useful as carriers of a radioisotope to the tumour site. This involves providing a stable coordination site for the metal ion sufficiently removed from the antigen binding site of the antibody to prevent loss of the antibody's immunoreactivity. Monoclonal antibodies (MAb), or fragments of monoclonal antibodies (Fab), are used as carriers for radionuclides, as well as drugs or toxins, to target sites of cancer, cardiovascular, and other diseases for both diagnosis and treatment. (For a more detailed description of the role of antibodies see reference 9).

Recently, methods for the **attachment of** ^{99m}Tc, ^{123}I, ^{111}In, and other **radioisotopes to monoclonal antibodies** specific for tumour-associated antigens have been developed and, in this form, have been found to be useful for the detection and therapy of diverse tumour types.

Several radiolabelled MAbs and Fabs are in clinical trials. One compound of this type, Myocint^{TM}, ^{111}In-DTPA-antimyosin, has been approved for **imaging** myocardial necrosis. The DTPA molecule (diethylenetriaminepentaacetic acid, see Figure 7, page 49) is conveniently linked to a Fab fragment of the antimyosin antibody, and requires only the addition of $^{111}InCl_3$ to the lyophilized kit. The faster blood clearance of the Fab fragment compared with the whole antibody makes it more suitable for imaging. The ^{111}In presumably binds to the DTPA through the three amino and three or four of the carboxylate functional groups, the fifth carboxylate being unavailable for coordination to In as it is covalently linked to the Fab fragment of antimyosin.

Another radiolabelled monoclonal antibody conjugate of note is OncoScint CR/OV^{TM}, ^{111}In-DTPA-CYT-103, for the evaluation of colorectal and ovarian cancers. The DTPA is attached to the carbohydrate region of the antibody. The MAb targets a specific antigen (TAG-72) found on $>75\%$ of colorectal tumours.

Other radiolabelled coordination compounds, using radioisotopes such as ^{105}Rh, ^{109}Pd, and ^{199}Au, which have been studied *in vivo*, offer great potential as **radiotherapeutic agents**. However, the conditions required to synthesize Rh(III) and Pd(II) complexes require the metal ions to be complexed to the bifunctional chelate before conjugation to the MAb or Fab. This represents a departure from the conventional approach of adding the radioisotope to the previously linked bifunctional chelate–antibody. The aqueous chemistry of gold needs to be further developed such that the gold radiolabel is not lost as a consequence of *in vivo* reduction to the metal.

6 IRON CHELATION

6.1 Introduction

Iron overload is a potentially fatal disorder, damaging the heart, liver, and other organs. It occurs as a consequence of repeated blood transfusions or increased gastrointestinal absorption of iron, or both, as in the disease β-thalassaemia, which affects some 100 000 people annually throughout the world. Patients with iron overload conditions need treatment with iron chelation therapy. These patients can be maintained in good health by repeated transfusion and chelation from the first year of life. Many other conditions can be treated by chelation to prevent overload with iron, such as sickle-cell anaemia, aplastic anaemia, and chronic renal failure.[17,18]

Iron toxicity is believed to be related to its redox chemistry. The electron transfer associated with the oxidation of Fe(II) to Fe(III) produces free radicals (*e.g.* $OH^{\bullet}$ and $O_2^{\bullet -}$) which can react with the fatty acids of cell membranes, causing lipid peroxidation.[19] The cell membrane damage eventually manifests itself mainly in the form of cardiac, hepatic, and endocrine failure. Consequently, **iron has to be complexed effectively** if its toxic effects are to be avoided.

6.2 Potential Therapeutic Agents

Desferrioxamine (DFB, see Figure 10), a hexadentate trihydroxamic acid of microbial origin, has been the mainstay of iron (and aluminium) chelation for the past 20 years. Although DFB is generally effective, its cost, rapid metabolism, toxic side-effects and the cumbersome subcutaneous method of administration offer several targets for improvement.

A wide variety of other natural and synthetic hydroxamic acids have been evaluated. However, their use as iron chelators is limited by their hydrophilicity and lability in acid solution, which prevents oral application and their partition into cell membranes.

Several other ligand classes have been extensively studied and are worthy of mention.[20] Catechols, unlike hydroxamic acids, are stable in acid. They also have extremely high affinities for iron. However, catechol groups oxidize very easily and there is considerable hydrogen ion interference at pH 7.4, due to their high pK_a values. Consequently, relatively few catechol derivatives have been screened *in vivo*.

Hydroxypyridinone ligands have features in common with both hydroxamic acids and catechols. Like hydroxamic acids, such as DFB, the hydroxypyridinones are monoprotic acids which bind to metals via the oxyanion and oxo groups. The deprotonated ligands also have a zwitterionic aromatic resonance form which is isoelectronic with the catecholate dianion. **Charge delocalization in the ligand** contributes to the great stability of these Fe(III) complexes. As well as having a high affinity for iron, the hydroxypyridinone ligands, such as 1,2-dimethyl-3-hydroxypyrid-4-one (DMHP, see Figure 10), have a high selectivity for iron over other biologically important metals and stability in physiological conditions.

Aminocarboxylic acids, such as nitrilotriacetic acid (NTA, see Figure 10) and ethylenediaminetetraacetic acid (EDTA, see Figure 7, page 49), do not have a particularly high or selective affinities for iron. However, the situation for phenolic aminocarboxylic acids is somewhat different. Several compounds of this type, such as HBED (*N,N'*-bis(2-hydroxybenzyl)ethylenediamine-*N,N'*-diacetic acid) and HPED (*N,N'*-bis(2-hydroxyphenyl)ethylenediamine-*N,N'*-diacetic acid, see Figure 10) have been evaluated as iron chelators. The high stability of the Fe(III) chelates

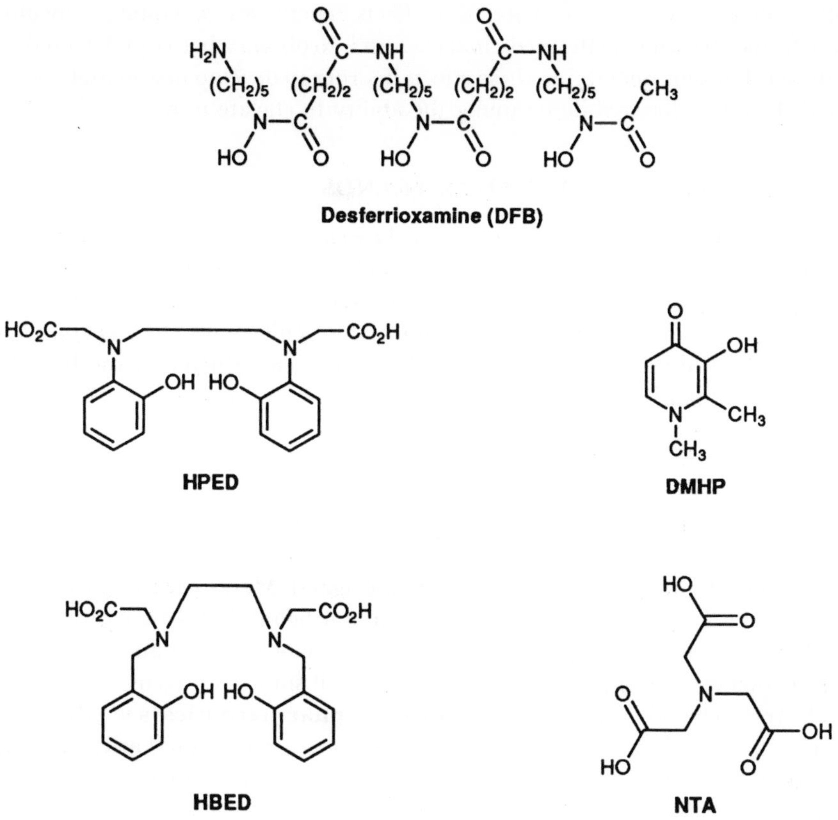

Desferrioxamine (DFB)

HPED

DMHP

HBED

NTA

Figure 10 *Chelating ligands for iron*

of these ligands is due to the high affinity of Fe(III) for the phenolate groups present in the ionized ligand, and to the orientation of these groups so as to permit their participation in chelate ring formation. In these Fe(III) complexes the ligands are hexadentate and the stability constants of the complexes are amongst the highest known, while the stability constants for complexes with divalent ions, which may interfere *in vivo*, are relatively low.

Unfortunately, despite the fact that many are capable of removing iron from both animals and humans as effectively as DFB, these ligands all display significant **toxicity**, such as impaired central nervous system activity and severe weight loss. In general, this toxicity appears to be associated with the free ligand rather than the iron chelate.

6.3 Conclusion

These studies indicate the variety of factors which influence the application of inorganic chemistry principles to biological systems. While iron may be chelated with high affinity and selectivity, side-effects remain an overriding difficulty. Clearly, the challenge to the medicinal chemist is to obtain a better understanding of the mechanisms underlying these effects in order to develop new ligands which avoid these problems while retaining the ability to chelate iron.

7 CONCLUSIONS AND FUTURE TRENDS

Inorganic drugs occupy an important niche in modern medicine. The chemical properties of transition metals (variable oxidation state, ability to coordinate ligands in a precise spatial arrangement and to undergo ligand exchange reactions) are well suited for the design of molecules with the capability of pharmacological interaction in biological systems. **The ability to modify the ligand** set around the metal enables the chemist to modulate the biological properties of the compound. Altering the ligands can result in a change in chemical reactivity which can in turn alter toxicity: for example, replacing the *cis* chlorides of cisplatin with the bidentate ligand CBDCA produces a less nephrotoxic compound. Drug delivery can be modified in a similar fashion. The oral bioavailability of the gold drug auranofin was made possible by the substitution of a thiolate ligand by a phosphine ligand. **Metal radionuclides** can be directed to specific biological targets by suitable ligands. ^{99m}Tc can be targeted to bone by the phosphonate ligand and monoclonal antibodies have been investigated as a means of targeting agents to cell surface antigens.

The research and development of **inorganic pharmaceuticals** is a dynamic, ongoing process. This is particularly evident in the search for novel **anti-tumour agents**. Recent work on platinum drugs has lead to the development of a class of orally bioavailable compounds. The specificity of cisplatin for the GG sequence on DNA has stimulated investigation into metal complexes with different sequence specificities. The DNA intercalator ethidium (see Figure 11) has been shown to alter the DNA binding of cisplatin which has prompted the synthesis of a range of platinum–ethidium complexes with unique DNA interactions.[21] DNA interactions have also been demonstrated for other metal complexes. Rhodium phenanthrenequinone diimine complexes interact with nucleic acids in a site specific fashion via intercalation into the major groove of the nucleic acid helix.[22] The specificity of binding can be modified by variations in the ancillary, non-intercalating ligands. Both of these examples suggest that it may be possible in the future to produce metal anti-tumour agents with novel mechanisms of action.

Exploitation of the inorganic chemistry within the body provides a biochemical rationale for the design of metal pharmaceuticals. At present this is a relatively unexplored area. One notable example, captopril (see Figure 12), is used in the treatment of hypertension. This drug is an inhibitor of angiotensin converting enzyme (ACE). This enzyme, which cleaves the peptide Angiotensin I to give the active hormone Angiotensin II, an important regulator of blood pressure, is a zinc-containing metalloenzyme. Captopril inhibits the enzyme by interaction of the thiol group with the zinc ion of the enzyme.

Superoxide dismutase (SOD) mimics offer another opportunity for **exploitation of bioinorganic chemistry**. Superoxide is a mediator of inflammation in a

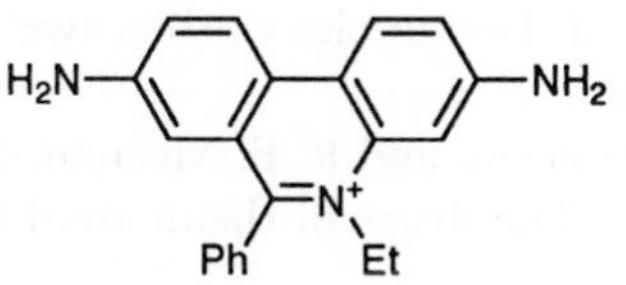

Figure 11 *The ethidium ion, a DNA intercalator*

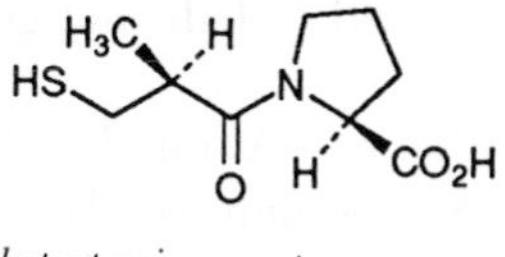

Figure 12 *Captopril, an anti-hypertensive agent*

number of disease states and is removed by SOD, a metalloenzyme with either copper or manganese at the active site. During the catalytic cycle the metal is reduced and superoxide is converted to hydrogen peroxide. Several **SOD mimics** have been described such as iron porphyrins, manganese desferrioxamine, and copper–amino acid complexes, and this a continuing focus of attention for a number of pharmaceutical companies.[19] The bioinorganic chemistry of the cell provides a **diversity of molecular targets** for the future development of **inorganic medicines**.

8 SELECTED BIBLIOGRAPHY/FURTHER READING

1. J. J. R. Frausto da Silva and R. J. P. Williams, 'The Biological Chemistry of the Elements. The Inorganic Chemistry of Life', Clarendon Press, Oxford, 1991.
2. P. J. Sadler, 'Inorganic chemistry and drug design', *Adv. Inorg. Chem.*, 1991, **36**, 1.
3. M. J. Cleare, 'Transition Metal Complexes in Cancer Chemotherapy', *Coord. Chem. Rev.*, 1974, **12**, 349.
4. J.-P. Macquet and J.-L. Butour, 'Platinum–Amine Compounds: Importance of the Labile and Inert Ligands for their Pharmacological Activities toward L1210 Leukemia Cells', *J. Natl. Cancer Inst.*, 1983, **70**, 899.
5. U. Frey, J. D. Ranford, and P. J. Sadler, 'Ring-opening Reactions of the Anti-cancer Drug Carboplatin: NMR Characterization of *cis*-[Pt$(NH_3)_2$(CBDCA-O)(5′-GMP-N7)]in Solution', *Inorg. Chem.*, 1993, **32**, 1333.
6. J. J. Roberts and A. J. Thomson, 'The Mechanism of Action of Antitumour Platinum Compounds', *Prog. Nucleic Acid Res. Mol. Biol.*, 1979, **22**, 71.
7. E. L. M. Lempers and J. Reedijk, 'Interactions of Platinum Amine Compounds with Sulfur-containing Biomolecules and DNA Fragments', *Adv. Inorg. Chem.*, 1991, **37**, 175.
8. K. R. Harrap, 'Platinum Analogues: Criteria for Selection' (Cancer Chemotherapy vol. 1), edited by F. M. Muggia, Martinus Nijhoff, Massachusetts, USA, 1983, p. 171.

9. I. M. Roitt, 'Essential Immunology', Blackwell Scientific Publications, Oxford, 6th edn, 1988.
10. D. T. Felson, J. J. Anderson, and R. F. Meenan, 'The comparative efficacy and toxicity of second-line drugs in rheumatoid arthritis', *Arthritis Rheum.*, 1990, **33**, 1449.
11. G. J. Higby, 'Gold in Medicine. A review of its use in the West before 1900', *Gold Bull.*, 1982, **15**, 130.
12. R. V. Parish and S. M. Cottrill, 'Medicinal gold compounds', *Gold Bull.*, 1987, **20**, 3.
13. 'Current Clinical Practice Series 7. Auranofin', Proceedings of a Smith Kline & French International Symposium, ed. H. A. Capell, D. S. Cole, K. K. Manghani, and R. W. Morris, Excerpta Medica, Amsterdam, 1983.
14. 'Ridaura^{TM} (auranofin): 1986 Smith Kline & French International Symposium', ed. T. G. Davis, *Scand. J. Rheumatol., Supplement 63,* 1986.
15. M. F. Tweedle, 'Physicochemical Properties of Gadoteridol and Other Magnetic Contrast Agents', *Invest. Radiol., Supplement 1,* 1992, **27**, S2.
16. S. Jurisson, D. Berning, W. Jia, and D. Ma, 'Coordination Compounds in Nuclear Medicine', *Chem. Rev.*, 1993, **93**, 1137.
17. C. G. Pitt, G. Gupta, W. E. Estes, H. Rosenkrantz, J. J. Metterville, A. L. Crumbliss, R. A. Palmer, K. W. Nordquest, K. A. Sprinkle-Hardy, D. R. Whitcomb, B. R. Byers, J. E. L. Arcenaux, C. G. Gaines, and C. V. Sciortino, 'The Selection and Evaluation of New Chelating Agents for the Treatment of Iron Overload', *J. Pharm. Exp. Therap.*, 1979, **208**, 12.
18. C. G. Pitt, Y. Boa, J. Thompson, M. C. Wani, H. Rosenkrantz, and J. Metterville, 'Esters and Lactones of Phenolic Amino Carboxylic Acids: Prodrugs for Iron Chelation', *J. Med. Chem.*, 1986, **29**, 1231.
19. B. Halliwell, 'Drug antioxidant effects. A basis for drug selection?', *Drugs*, 1991, **42**, 569.
20. G. J. Kontoghiorghes, 'Oral Iron Chelation is Here', *Brit. Med. J.*, 1992, **303**, 1279.
21. M. V. Keck and S. J. Lippard, 'Unwinding of Supercoiled DNA by Platinum–Ethidium and Related Complexes', *J. Am. Chem. Soc.*, 1992, **114**, 3386.
22. A. M. Pyle and J. K. Barton, 'Probing nucleic acids with transition metal complexes', in 'Progress in Inorganic Chemistry: Bioinorganic Chemistry', ed. S. J. Lippard, John Wiley and Sons, Inc., 1990, vol. 38, p. 413.

9 KEY JOURNALS

Inorganic Chemistry

Journal of the American Chemical Society
Inorganic Chemistry
Journal of the Chemical Society, Dalton Transactions
Journal of the Chemical Society, Chemical Communications

Biological Chemistry

Biochemistry
Journal of Medicinal Chemistry
Journal of Inorganic Biochemistry
Biometals

Medically-orientated

Cancer
Cancer Research
Antimicrobial Agents and Chemotherapy
Arthritis and Rheumatism
Journal of Rheumatology
Journal of Nuclear Medicine
Free Radical Biology and Medicine

CHAPTER 4

Inorganic Materials as Catalysts for Chemical Processing

GEOFFREY C. BOND

1 INTRODUCTION: SCOPE OF THE CHAPTER

The purpose of this chapter is to describe the use of inorganic materials to catalyse chemical processes. The phenomenon of catalysis is as old as life itself, for all living organisms depend on organic catalysts called enzymes for their functioning: they are large and complex molecules containing protein and sometimes a metal ion in a prosthetic group, for example magnesium in chlorophyll. The first intentional use of catalysis was undoubtedly in the fermentation of sugar to make alcoholic drinks (mead, wine, beer, *etc.*), but although catalysis was recognized and named by Berzelius in 1835 a number of decades were to pass before synthetic inorganic materials began to make a significant impact on the chemical industry. We cannot review the early work in any detail, but probably the first major triumph for catalysis was the development of an effective material for synthesizing ammonia from hydrogen and nitrogen (1911). This, together with the discovery soon afterwards of a means of oxidizing ammonia selectively to nitrogen monoxide, hence enabling the manufacture of nitric acid, paved the way for the **heavy inorganic chemical industry**. Catalysis had arrived.

Substances capable of acting catalytically are classified according to the phase in which they operate. **Homogeneous catalysts** function in the same phase as the reactants: thus reactions in gases and even in the solid phase can be catalysed by substances of the same phase, but most commonly it is the liquid phase that is involved. Very many species perform as homogeneous catalysts in liquids: acids, bases, protons, hydroxyl ions, and especially inorganic complexes of transition metals. Some of these applications are described in Chapter 5. **Enzymes** are a special type of homogeneous catalyst, because the molecules are so large that they often behave as colloids, and therefore represent an intermediate situation between homogeneous and **heterogeneous catalysts**. In the latter, there is a distinct phase boundary between catalyst and reactants, so that the possible combinations of phase are as follows: (a) catalyst, liquid; reactants, gaseous; (b) catalyst, solid; reactants, gaseous; (c) catalysts, solid; reactants, liquid (or liquid + gas). The last two categories are the more important.

Although we shall be chiefly concerned with the applications of catalysts in chemical processing, *i.e.* in the manufacture of useful chemical commodities, the basic principles of catalysis apply equally in other fields, such as the control of harmful gaseous effluents (see Chapter 7). The energy required for a catalytic process need not be thermal; radiation of the right sort allows **photocatalysis** to occur, and this effect is being exploited in the photolysis of water to give hydrogen and oxygen, and as a means of removing traces of organic substances from impure water (see Chapter 18). Electrical energy can substitute for thermal energy in **electrocatalysis**, as in the electrolysis of water or electrocatalytic oxidation of organic molecules. In the reverse mode, electrical energy can be generated directly from energy released in chemical reactions by **fuel cells**. In the broader sense, the principles of interfacial phenomena also allow us to understand something about **tribology** (the science of lubrication) and **electroplating**.

In the following sections we first outline the basic principles by which heterogeneous catalysis works, and the describe the types of inorganic substance which can act as a catalyst, how they are prepared, and what reactions they can assist. Although they have been in use in the chemical industry for less than a century, catalysts have facilitated a great many chemical operations: it is important to appreciate their power and their limitations, and that is what this chapter attempts to describe.

2 BASIC PRINCIPLES OF HETEROGENEOUS CATALYSIS

2.1 How Catalysis Works

A catalyst is a substance that increases the rate at which a chemical system approaches equilibrium, without being consumed in the process: catalysis is the phenomenon of a catalyst in action. The word 'catalysis' was selected by Berzelius to cover a number of cases where mere traces of some chemical species markedly accelerated the rate of reaction. It is derived from Greek words meaning 'to break down', and is related to 'analysis' (to break apart), 'photolysis' (to break with light), *etc.* We think Berzelius meant that in catalysis the normal rules of chemistry seem not to apply and the usual constraints which stop molecules reacting no longer operate.

However, the laws of thermodynamics are sacrosanct, and a catalyst can only speed up the rate of a reaction which is thermodynamically feasible, *i.e.* one which can proceed however slowly in its absence. It also follows that the position of equilibrium finally attained must be the same whether a catalyst is present or not. What stops molecules reacting is the potential energy barrier which has to be surmounted before the product state is reached. The height of this barrier is the activation energy or the enthalpy of activation: an increase in rate requires that the height of the barrier be lowered, and it is now generally accepted that **the principal function of a catalyst is to decrease the activation energy of the non-catalysed reaction** (Figure 1).

We can divide the inorganic solids which can act catalytically into (a) metals and (b) ionic compounds, the principal examples of the latter being (i)

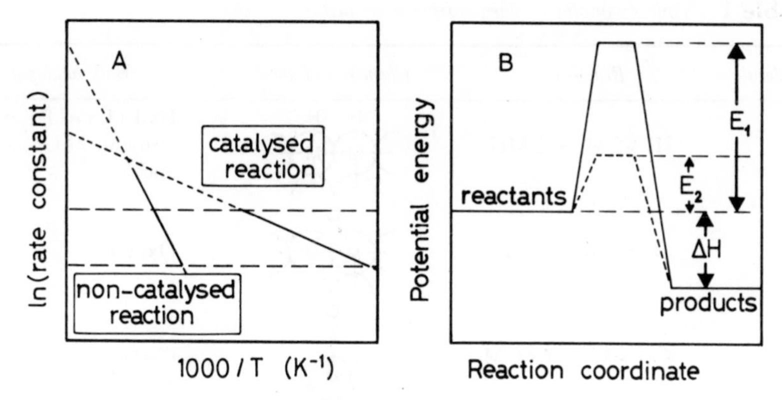

Figure 1 *Two ways of regarding the lowering of activation energy by a catalyst. A: The Arrhenius plot for the catalysed reaction has a lower slope and the measurable rate (between the horizontal broken lines) is in a lower temperature range than for the non-catalysed reaction B: The height of the potential energy barrier, which equates to the activation energy, is lowered from* E_1 *for the non-catalysed reaction to* E_2 *for the catalysed reaction because a more favourable reaction path has been found*

chalcogenides (oxides, sulfides) of the transition and post-transition elements and (ii) oxides of pre-transition elements. It is more informative to illustrate the basic principles in the context of catalysis by metals (Section 2.2). Ionic solids will be considered afterwards.

2.2 Metals as Catalysts

Metals are covalent solids which are in general hard because of the strength of the bonds between the atoms; when a crystal is cleaved, the bonds break homolytically, so that the atoms at the new surface possess one or more **free valencies**. This betokens extreme reactivity, and indeed the surfaces of all metals except the most noble rapidly acquire an oxide coating on exposure to the air, and silver tarnishes by forming a layer of sulfide.

The initial step in the reaction of a molecule with a clean metal surface is termed chemical adsorption or **chemisorption**, in the course of which the molecule if diatomic may dissociate into two atoms or if polyatomic may re-structure, re-hybridize or decompose in some way. Some of the processes involved in chemisorption are illustrated in Table 1. The driving force in chemisorption is the saturation of free valencies at the surface, as this lowers the surface free energy. New chemical compounds are formed, some of which have analogies in three-dimensional chemistry, others of which do not (see Table 1). We have rapidly reached the heart of the mystery, because **the act of chemisorption often accomplishes easily what in the absence of a catalytic surface would be prohibitively expensive in energy**. Consider the dissociation of the hydrogen molecule. In the gas phase this requires 435 kJ mol^{-1} and temperatures above 1300 K are needed to achieve a measurable concentration of hydrogen

Table 1 *Some examples of chemisorption on metals*

Molecule	*Reaction*	*Chemisorbed state*	*Bulk analogue*
H_2	$H_2+2\,M \rightarrow 2\,MH$	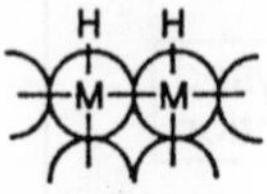 H H / M M	Hydrides of Pd and of elements of Groups 4 and 5
O_2	$O_2+2\,M \rightarrow OM_2$	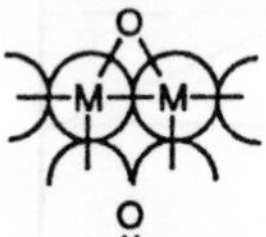 O / M M	Oxides
CO	$CO+M \rightarrow CO\text{–}M$	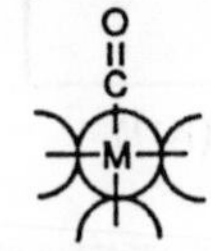 O=C / M	Carbonyls, $Ni(CO)_4$, $Ru_3(CO)_{12}$, *etc.*
	or		
	$CO+M_2 \rightarrow CO\text{–}M_2$	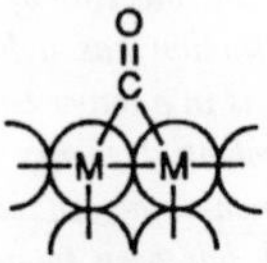 O=C / M M	
C_2H_4	$C_2H_4+2\,M \rightarrow C_2H_4\text{–}M_2$	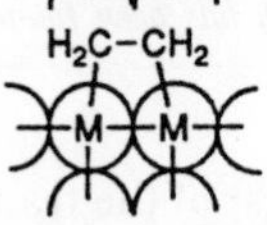 $H_2C\text{–}CH_2$ / M M	–

atoms; but some metal surfaces will chemisorb hydrogen at 20 K! This is because the surface acts to stabilize the atoms in a chemisorbed state.

There is one other type of adsorption that is important. This is physical adsorption or **physisorption**, where the force acting between the molecule and the surface is due to dipole (electrostatic) interactions or to van der Waals (dispersion) forces. These are the forces acting between molecules in liquids, and are negligible at temperatures more than 100 K above the boiling point. The chief value of physisorption is to provide a means of bringing a molecule close to the surface without expenditure of energy (Figure 2).

Not all molecule–metal surface reactions occur with equal facility. Just as in any area of chemical reactivity, there are rules governing whether chemisorption will occur, and to what extent: the essential rule is that **the system's free energy must decrease**, and this means (because of the loss of translational movement and the resulting decrease in entropy) the process must be exothermic. With few exceptions, the reactivity of some simple molecules towards any metal surface increases in the sequence:

$$N_2 < H_2 < CO < C_2H_4 < C_2H_2 < O_2$$

and metals can then be classified according to how many of these molecules are chemisorbed under specified conditions. A simplified version of this classification is shown in Table 2, where it is seen that the metals most active in chemisorption are those in the centre of the transition series. This has led to the idea that unpaired

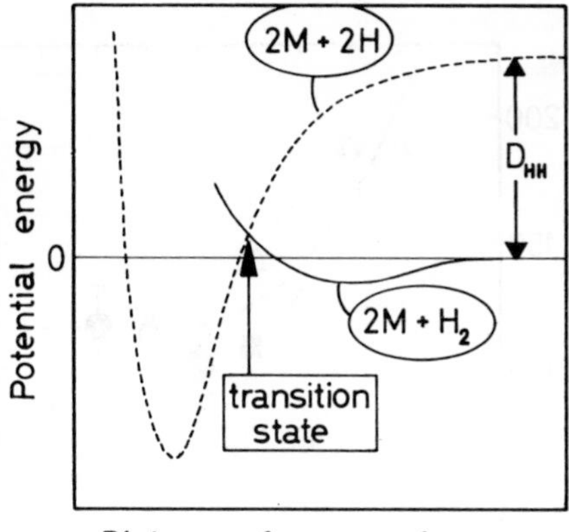

Figure 2 *Potential energy curves for the chemisorption (broken line) and physisorption (solid line) of H_2 on a metal. The activation energy is given by the height of the transition state above the potential energy zero and the depths of the wells give the respective enthalpies of adsorption*

Table 2 *A classification of metals according to their propensity for chemisorption*

Group	*Metals*	O_2	C_2H_2	C_2H_4	*CO*	H_2	N_2
A	Ti, Zr, Hf, V, Nb, Ta, Cr, Mo, W, Fe, Ru, Os	+	+	+	+	+	+
$B_{1,2}$	Co, Rh, Ir, Ni, Pd, Pt	+	+	+	+	+	−
B_3	Mn, Cu	+	+	+	+	±	−
C	Al, Au[a]	+	+	+	+	−	−
D	Li, Na, K	+	+	−	−	−	−
E	Mg, Ag, Zn, Cd, In, Si, Ge, Sn, Pb, As, Sb, Bi	+	−	−	−	−	−

+, strong chemisorption; ±, weak; −, unobservable
[a] Au does not chemisorb O_2

d-electrons are required in the formation of chemisorption bonds and so sp metals (Mg, Al, Cu, Au, *etc.*) are comparatively ineffective in chemisorption.

This qualitative ranking of metals can be improved on by evaluating the strength of the bond formed between the atom or molecule and the surface. There are a number of ways in which this can be estimated, but the most direct is through calorimetric measurement of the heat released when chemisorption occurs. In some cases the bond becomes weaker as the surface coverage increases, so it is usual to take the enthalpy of adsorption extrapolated to zero coverage as the measure of the bond strength. By way of illustration, Figure 3 shows, in the case of hydrogen, how this varies with position of the metal in the Periodic Classification.

In the cases of oxygen and nitrogen, there is a marked similarity between bonds formed at the surfaces of metals and those in corresponding three-dimensional compounds. Thus there is a good correlation between the enthalpies of oxygen chemisorption and of formation of the most stable bulk oxide (Figure 4). Metals which cannot form nitrides do not chemisorb nitrogen strongly. With hydrogen,

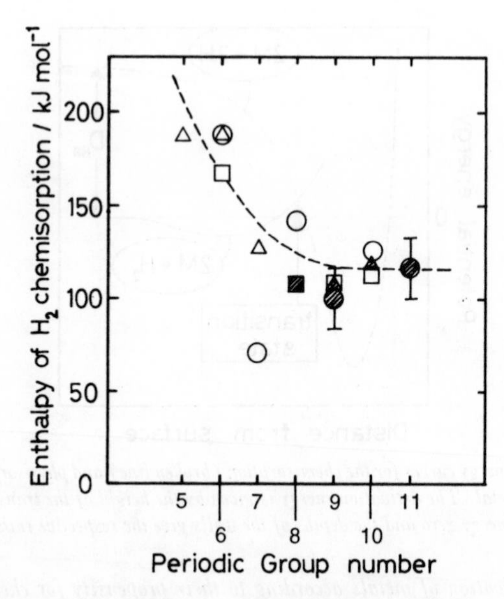

Figure 3 *Periodic variation of the initial enthalpy of chemisorption of H_2 on transition metals. Open points, evaporated metal films; hatched points, silica-supported metals. Circles, metals of the first long period; squares, those of the second long period; triangles, those of the third long period*

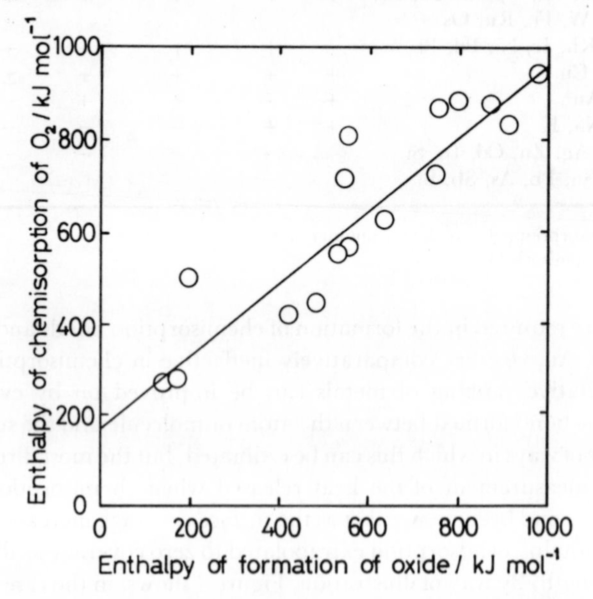

Figure 4 *Dependence of the enthalpy of chemisorption of O_2 on the enthalpy of formation of the most stable oxide at 298 K*

chemisorption enthalpies correlate well with those for other compounds, such as oxides. In chemisorption and catalysis we are concerned with **chemistry in two dimensions**: it is as fascinating and rich as the three-dimensional variety.

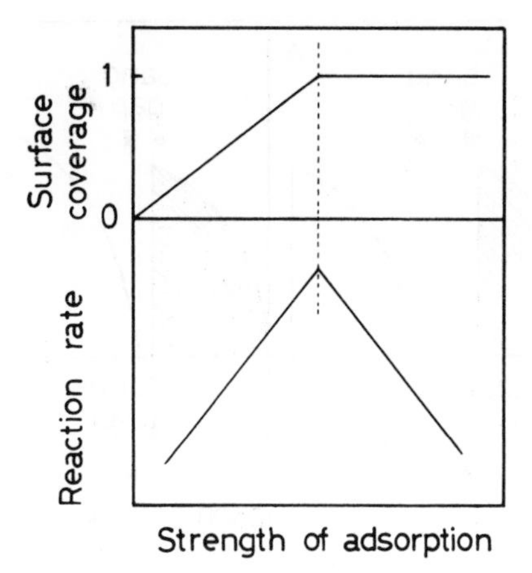

Figure 5 *Diagrammatic representation of how surface coverage and reaction rate vary with strength of adsorption: the Volcano Principle (see text)*

2.3 The Volcano Principle

These concepts concerning how molecules react with metal surfaces are only useful in the present context if they help us to understand the various catalytic activities which different metals show. Fortunately there are two simple and very general principles which apply: (1) **A catalysed reaction will only occur between two molecules if both can be chemisorbed**. Thus the information in Table 2 can be used, at least qualitatively, to predict which metals might be active as catalysts for a given reaction. For example, only the metals in Group A of Table 2 can catalyse ammonia synthesis, because the others do not chemisorb nitrogen. (2) The second general principle tries to say something more quantitative: **maximum catalytic activity results when the reactants are adsorbed with the minimum strength needed to achieve full surface coverage**. This is sometimes called the **Volcano Principle** (see Figure 5), because it predicts that a plot of rate for a given reaction on a number of different catalysts against the strength of adsorption will pass through a maximum: to the left, adsorption is too weak to give full surface coverage, while on the right the adsorbed species are too strongly held to be reactive.

In this way we can understand from Figure 3 why the noble metals of Groups 8, 9, and 10 are the most active in catalytic hydrogenation, and why the sp metals (except copper), which hardly chemisorb hydrogen at all, are virtually useless in this respect. This principle also rationalizes the high activity of iron, ruthenium, and osmium in ammonia synthesis, and as we shall see it applies to other classes of solids besides metals.

2.4 Oxides as Catalysts

We shall not be able to treat the other classes of catalytic solids in the same detail; however, both oxides and sulfides are important as catalysts, and some account of

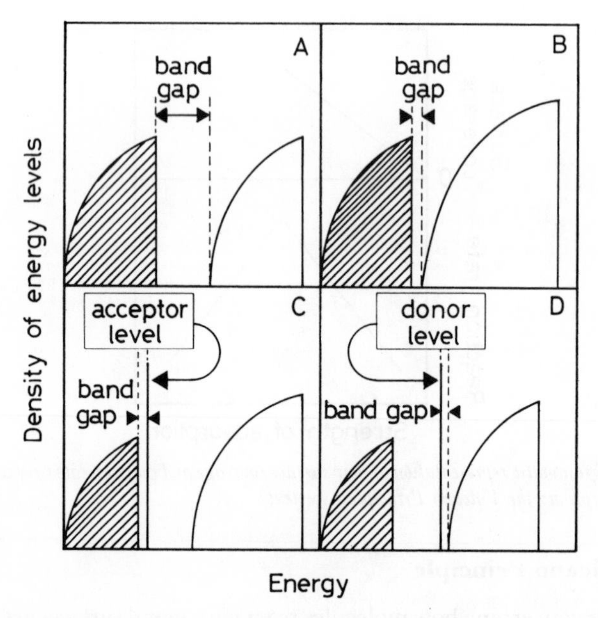

Figure 6 *Simplified band diagrams illustrating types of semiconductivity: shaded areas represent filled valence bands, unshaded areas the vacant conduction bands*
*A: insulators (*e.g. *MgO, SiO_2)*
*B: intrinsic semiconductors (*e.g. *Ge, Si, GaAs)*
*C: extrinsic p-type semiconductors (*e.g. *NiO)*
*D: extrinsic n-type semiconductors (*e.g. *ZnO)*

their properties must be given.

Oxides whose cations can change their oxidation state in one or other direction will behave as extrinsic or impurity semiconductors and are likely to show some catalytic activity in oxidation reactions. Oxides such as alumina and magnesia are **insulators** because there is a large band gap between the filled valence band representing the electrons in the closed outer shell and the vacant conduction band (see Figure 6A), and it is a basic tenet of the Band Theory of Solids that electrical conduction only occurs when a partially-filled band is present. In **intrinsic semiconductors** such as germanium and gallium arsenide, the band gap is narrow and some electrons can jump from the filled to the empty band under the influence of heat or light, resulting in a low level of conductivity (see Figure 6B).

In **extrinsic semiconductors**, small amounts of species in higher or lower oxidation states ('impurities') can provide means of inducing sufficient degrees of incomplete occupation for semiconductivity to be shown (see also Figures 6C and 6D). Where there is a vacant **acceptor level** close to the top of the filled valence band, some electrons can jump across the band gap, thus creating vacancies in the valence band. On the other hand, the presence of a filled **donor level** just below the vacant conduction band can enable some electrons to enter the latter, thus making the conduction band partially filled.

In nickel(II) oxide, some additional oxygen is taken up as oxide ion on heating in air, and a corresponding small number of Ni^{2+} ions are oxidized to Ni^{3+}, which form the acceptor level: it is therefore a **p-type semiconductor** (p for positive hole conduction) and oxygen is chemisorbed in a similar way:

$$Ni^{2+} + O_2 \rightarrow Ni^{3+}{-}O^- \tag{1}$$

With zinc oxide, on the other hand, there is no accessible higher oxidation state, and on heating in air it loses a little oxygen due to the formation of a few interstitial zinc atoms, which form a donor level. It is therefore an **n-type semiconductor** (n for negative charge carrier); it cannot chemisorb oxygen unless the surface is first partially reduced with the formation of surface anion vacancies, which can then be re-oxidized:

$$Zn^{2+}\,O^{2-} \rightarrow Zn^0\,\square_s \xrightarrow{O_2} Zn^{2+}\,O^{2-} \tag{2}$$

When oxides of this type act catalytically, the surface undergoes successive reductions and oxidations in a redox mechanism associated with the names of Mars and van Krevelen. With p-type semiconductors, the chemisorbed oxygen atoms are the oxidizing agents.

Although simple oxides such as these find some use for non-selective or deep oxidations (H_2 to H_2O, CO to CO_2, CH_4 to $CO_2 + H_2O$, *etc.*), of much greater interest and utility are **compound oxides** containing two (or sometimes more) different cations, each of which plays a distinct role in effecting the complex series of steps which constitute a **selective catalytic oxidation**. These oxides are chiefly based on molybdenum, vanadium, or antimony, and the active species are often stoichiometric compounds of the two component oxides. An early and important example of this type of catalysis is the use of 'bismuth molybdate' for oxidizing propene to propenal (acrolein) and thence by reaction with ammonia to form cyanoethene (acrylonitrile).

$$CH_2{=}CH{-}CH_3 \rightarrow CH_2{=}CH{-}CHO \xrightarrow[-H_2O]{+NH_3} CH_2{=}CH{-}C{\equiv}N \tag{3}$$

Other important classes of selective oxidations include dehydrogenation of alkanes (*e.g.* n-butane to 1,3-butadiene), oxidation of alkanes (*e.g.* n-butane to maleic anhydride), and oxidation of aromatics (e.g. *o*-xylene to phthalic anhydride). Other useful binary oxides include the Fe–Mo, Sn–Sb, and V–P systems.

2.5 Acidic Oxides as Catalysts

Oxides of metals and semi-metals having no d-electrons have cations that are unable to change their oxidation state except in extreme circumstances; oxides such as alumina, silica, beryllia, magnesia, and boria are thermally very stable and are electrical insulators; the first-named are especially used in the manufacture of ceramic ware. They are also suitable for use as **supports** for finely-divided metals (see page 77). Their structures terminate with hydroxyl groups which may show

$$O^{2-} \quad M^{2+}(OH^-) \quad OH^- \underset{+H_2O}{\overset{-H_2O}{\rightleftharpoons}} O^{2-} \quad M^{2+} \quad O^{2-}$$

$$O^{2-} \quad M^{2+} \quad O^{2-} + C_2H_5OH \longrightarrow O^{2-} \quad M^{2+}(OH^-) \quad OH^- + C_2H_4$$

Scheme 1 *Dehydration of ethanol on an oxide surface*

either acidic or basic character, depending on the electronegativity difference between the cation and the oxygen:

$$MO^- H^+ \rightleftharpoons M-OH \rightleftharpoons M^+ OH^- \quad (4)$$

Their surfaces are dehydrated on heating, and may be re-hydrated: they therefore act as **dehydration** catalysts, converting for example alcohols to alkenes (see Scheme 1).

The number of strong acid centres is much increased if two acidic oxides from this group, having cations of different charge, are combined. For example, silica–alumina ($\sim 10\%$ Al_2O_3) is non-stoichiometric and amorphous, and substitution of Si(IV) by Al(III) induces a charge imbalance that has to be rectified by a cation (*e.g.* H^+, Na^+, *etc.*). The dehydrated state of the acidic form contains **Lewis acid centres** (*i.e.* Al:) and the hydrated state protons which act as **Brønsted acid centres**. The latter in particular can initiate reactions requiring carbocationic intermediates (*e.g.* $C_2H_5^+$), and this material has been widely used as a catalyst for cracking long-chain hydrocarbons, polymerization, isomerization, alkylation, and reactions of the Friedel–Crafts type.

A particularly important class of acidic solids is the crystalline aluminosilicates, known as **zeolites**. These have the general formula $M_v(AlO_2)_x(SiO_2)_y.zH_2O$, and they exist in a great variety of structures, many of which are permeated by very small pores of uniform size, through which small molecules can diffuse but into which large molecules cannot enter. They therefore behave as **molecular sieves**. Some occur naturally (faujasite, erionite) and many others have been synthesized. They are widely used in petroleum reforming processes, especially in **catalytic cracking**, and they also catalyse many reactions of interest to the petrochemical industry. They are highly active, indeed so active that they are often used diluted with the less active amorphous material, and they owe their properties to the acid strength of their protons, the high electric field gradients which exist in the cavities formed by the intersection of the pores, and to their shape selectivity. Recently there has also been much interest in analogues such as the aluminophosphates ($AlPO_4$) which can also crystallize in forms having porous networks. Indeed silica itself can be prepared as a zeolite (silicalite), and many heteroatoms have been incorporated into the Si–O–Al or Al–O–P frameworks to induce additional catalytic effects; examples include iron, vanadium, and titanium.

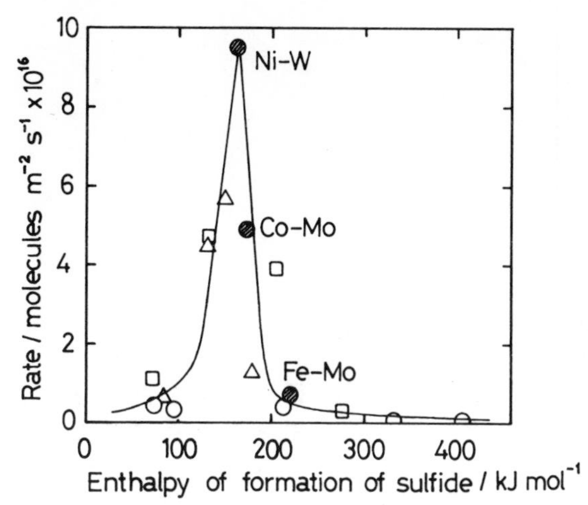

Figure 7 *Volcano curve for the hydrodesulfurization of dibenzothiophene: specific rate at 673 K vs enthalpy of formation of sulfide. Filled points are for binary sulfides, the activities of which are plotted against the average enthalpy of formation; for the other symbols, see the legend to Figure 3*

2.6 Sulfides as catalysts

We conclude this section with a few brief remarks concerning sulfides as catalysts. Their principal use is in **hydrodesulfurization** (HDS), that is, removal of sulfur from organic sulfur compounds, especially heterocyclic molecules, by reaction with hydrogen; hydrogen sulfide is formed. For this reaction, which is widely used to produce sulfur-free fuels, the principal catalyst comprises cobalt and molybdenum on an alumina support. The active elements start as oxides, but are converted into sulfides during use. The mechanism involves adsorption of the sulfur atom in the reactant at an anion vacancy, after which the molecule decomposes, leaving the sulfur atom behind as S^{2-}. Hydrogen then recreates the anion vacancy, forming hydrogen sulfide in the process. Catalytic activity depends on the strength with which sulfide ions are bound to the surface, and this in turn is related to the enthalpy of formation of the bulk sulfide. Study of the behaviour of a number of single and binary metal sulfides for the hydrodesulfurization of dibenzothiophene has produced a very nice volcano curve (Figure 7), showing that optimal activity is given by the nickel–tungsten sulfide system, which has an average enthalpy of formation of about 160 kJ mol^{-1}.

3 A CLASSICATION OF HETEROGENEOUSLY CATALYSED REACTIONS

3.1 Classes of Catalytic Materials and the Reactions They Catalyse

In this short survey of the principles underlying heterogeneous catalysis, it has been expedient to refer to a few examples of the kinds of reaction each class of

Table 3 *A classification of heterogeneously-catalysed reactions*

Class of material	*Type of reaction*	*Examples*	*Types of catalyst*
Metals, alloys	Hydrogenation	$3\,H_2 + N_2 \rightarrow 2\,NH_3$	Fe, Ru, Os
		$CO + 2\,H_2 \rightarrow CH_3OH$	Cu
		$CO + H_2 \rightarrow CH_4$, higher alkanes and oxygenates (methanation, Fischer–Tropsch synthesis)	Fe, Co, Ni, Ru
		H_2 + unsaturated acids (fat-hardening)	Ni
		H_2 + alkenes, alkynes, aromatics, –CHO, $-NO_2$, *etc.*	Pd, Pt, Rh
	Dehydrogenation	cyclo-$C_6H_{12} \rightarrow C_6H_6$	Pt, Ni
		$CH_3OH \rightarrow HCHO$	Ag
	Dehydrocyclization	n-$C_6H_{14} \rightarrow C_6H_6$	Pt
	Hydrogenolysis	$C_2H_6 \rightarrow 2\,CH_4$	Ni, Ru
		$C_6H_5Cl \rightarrow C_6H_6 + HCl$	Pd
	Oxidation	$NH_3 \rightarrow NO$	Pt–Rh
		H_2, CO, CH_4, *etc.* $\rightarrow CO_2 + H_2O$	Pt, Pd, Rh
	Redox	$CO + NO \rightarrow CO_2 + N_2$	Pt–Rh
Semiconducting and compound oxides	Oxidation	$CH_4 \rightarrow C_2H_4$, C_2H_6	MgO
		H_2, CO, CH_4, *etc.* $\rightarrow CO_2 + H_2O$	CuO, NiO, MnO_2
		$C_3H_6 \rightarrow CH_2{=}CH{-}CHO$	CuO, Sn–Sb oxide, Bi–Mo oxide
		n-$C_4H_{10} \rightarrow$ maleic anhydride	V–P oxide
		o-xylene $\rightarrow$ phthalic anhydride	V–Ti oxide
		n-$C_4H_{10} \rightarrow C_4H_6$	Fe–Mo oxide
		$SO_2 \rightarrow SO_3$	$Na_2O{-}V_2O_5$

Semiconducting and compound oxides (contd.)	Redox	$NO + NH_3 \rightarrow N_2 + H_2O$	V–Mo–Ti oxide
		$CO_2 + H_2 \rightleftharpoons CO + H_2$ (water–gas shift)	Fe_3O_4–Cr_2O_3, $Cu_xZn_{1-x}O/Al_2O_3$
	Reduction/hydrogenation	$C_2H_4 + H_2 \rightarrow C_2H_6$	ZnO, Cr_2O_3
		$CO + H_2 \rightarrow CH_3OH$	$ZnCr_2O_4$, $Cu_xZn_{1-x}O/Al_2O_3$
		$CH_3COOH \rightarrow C_2H_5OH$	$CuCr_2O_4$
Sulfides	Hydrodesulfurization	thiophene $\rightarrow C_4H_6 + H_2S$	Co–Mo/Al_2O_3
	Hydrogenation	$CH_3COOH \rightarrow C_2H_5OH$	Re_2S_7
Insulator/acidic oxides/zeolites	Hydrolysis, hydration	Oxirane $\rightarrow HOCH_2CH_2OH$	H_3PO_4/clay
		$C_2H_4 \rightarrow C_2H_5OH$	H_3PO_4/clay
	Dehydration	$C_2H_5OH \rightarrow (C_2H_5)_2O + C_2H_4$	Al_2O_3
	Polymerization	$C_3H_6 \rightarrow (C_3H_6)_n$	SiO_2–Al_2O_3
	Isomerization	n-$C_4H_{10} \rightarrow$ iso-C_4H_{10}	SiO_2–Al_2O_3
	Cracking	$(CH_2)_n \rightarrow$ alkanes, *etc.*	RE-exchanged Y zeolites
	Alkylation	$C_6H_6 + C_3H_6 \rightarrow C_6H_5.C_3H_7$	$AlCl_3$

Note: the reactions quoted are merely illustrative examples, but they include some of the more industrially important processes; the types of catalyst listed are the more active or more widely used.

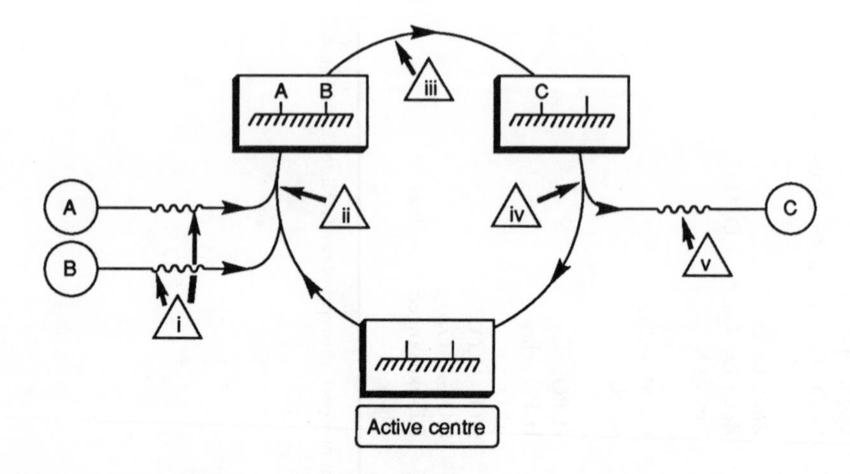

Scheme 2 *Catalytic cycle for the reaction of* $A + B \rightarrow C$

material can assist, but, especially in the section concerning metals, only a very few cases were quoted. We must now attempt a more comprehensive overview.

There are two important points to bear in mind. The first, which emerges from the last section, is that in each class of material the reactants (and where appropriate the products also) must show a chemical affinity for the surface in order for chemisorption to occur. The second principle is almost a statement of the obvious: the catalyst must be chemically stable under the conditions of the reaction, and not be converted by the reactants into another phase. As with all generalizations, there will be occasional exceptions, but these concepts are very useful guides. So, for example, base metals cannot act as oxidation catalysts because they are too easily oxidized, and noble metal oxides cannot be used for hydrogenation because they are too readily reduced. Only the most noble metals can act as oxidation catalysts at high temperature, and only very stable oxides have hydrogenation properties (Cr_2O_3, ZnO).

Table 3 provides a fuller list of the types of reaction catalysed by each class of material, with some illustrative examples.

3.2 Elementary Steps in a Catalysed Reaction

So far we have thought of a catalytic reaction as a chemical change occurring on the surface of a solid, but this is only one of several processes that need to operate satisfactorily if all is to go well. The situation is explained in Scheme 2 where the five crucial steps are identified by small Roman numerals. First of all, the reactants have to be brought to the neighbourhood of the surface (step i). This is a process of **mass-transport** and, if it is slower than the steps that follow, it becomes rate-limiting. This is usually undesirable, as the catalyst is then under-utilized. Secondly, the reactants must be chemisorbed in the right form and with the most suitable strength if efficient catalysis is to result (step ii). In many cases not all parts of a surface are equally suited, due to variations in local arrangements of atoms or ions (*e.g.* different crystal faces): those atoms, ions or groups of surface entities that are appropriate are termed **active centres** or **active sites**, and they sometimes account for only a small fraction of the total surface. Then follows the catalytic act (step iii), and the desorption of the product (step iv), and its transport away from the neighbourhood of the surface (step v), in consequence of which the original active centre is regenerated and the cycle can be repeated, in theory indefinitely.

By means of this figure we can also identify the points at which trouble may occur. In addition to mass-transport limitation, noted above, the existence of impurities amongst the reactants may cause difficulty; if they chemisorb strongly they will block the active centres and the catalyst will be **poisoned**. The effect is similar to that of carbon monoxide on haemoglobin. Equally if the reaction produces by-products that do not desorb easily, activity will be lost through **self-poisoning**. This often occurs in reactions of hydrocarbons on metals, when dehydrogenated species, known generically as 'coke', became firmly fixed to the surface. Great skill is needed to design catalytic materials that avoid all these pitfalls, and in the next section we shall examine some aspects of how this is done.

4 PREPARATION AND CHARACTERIZATION OF CATALYSTS

4.1 Introduction

The science of catalyst preparation encompasses the making of a few grams for laboratory use or for fundamental research to the manufacture of catalysts for industrial use on a scale of tens or even hundreds of tons. Small amounts of catalyst for use in organic synthesis are readily purchased, and many bottles on the laboratory shelf contain compounds having some catalytic properties. It is fortunately quite simple to prepare small amounts of highly effective catalysts, and in this section we outline some of the principles and procedures that should be followed for each class of material. Large scale manufacture is altogether more complex and expensive; we indicate some of the unit operations that are involved, and some of the problems encountered.

4.2 Preparation of Metal Catalysts

Metals are used as catalysts in one of three forms: macroscopic, promoted, or supported. Macroscopic forms include wire, gauze, sheet, evaporated metal films (all polycrystalline), and single crystals which are used in fundamental research. Promoted metals have the active metal as the main component, but they contain small but vitally important amounts of **promoters** which stabilize the active phase and enhance its activity.

In **supported metal catalysts** the active phase is in the form of extremely small particles which are stabilized by being attached to a material which is usually inert, of high surface area, microporous, and typically an oxide such as silica or alumina, or activated carbon; this is termed the support. Since catalysis only occurs on atoms at the surface, expensive metals such as platinum generally need to be used in a form that maximizes the fraction of surface atoms, that is, the

degree of dispersion. It is easily calculated that to obtain 50% dispersion, one has to use particles only about 2 mm in size, and containing *ca.* 200 atoms. Fortunately it is surprisingly easy to make such small groups of atoms on the surface of a support, using one of the following methods:

(1) **Impregnation**: a solution of a metal salt is absorbed into the pores of the support, and the solvent evaporated. After calcination (optional), reduction by hydrogen or another reducing agent produces metal particles in the desired size range.
(2) **Ion exchange**: in this method, also called **equilibrium adsorption**, exchangeable protons or hydroxyl groups on the support are replaced by cations or anions containing the catalytic element. The solvent is evaporated and the preparation continued as in (1).
(3) **Deposition–precipitation**: here the support is suspended in a solution of a metal salt, from which the insoluble hydroxide is precipitated onto the support by adding hydroxide ions. A good way to do this is to generate the hydroxide ions homogeneously by the decomposition of urea:

$$CO(NH_2)_2 + H_2O \rightarrow CO_2 + 2\,NH_3 \tag{5}$$

$$NH_3 + H_2O \rightleftharpoons NH_4{}^+\,OH^- \tag{6}$$

(4) **Co-precipitation**: in this procedure the support and the precursor to the active metal are formed at the same time, for example, by precipitation. Thus, adding alkali to a solution of aluminium and nickel nitrates gives a mixed hydroxide, which after calcination and reduction leads to a nickel/alumina catalyst.

Unsupported metals in finely divided form find some limited use for catalysing reactions at or near room temperature. **Metal blacks** are prepared by chemical reduction of a solution of a salt, using formic acid, hydrazine, *etc.*; use of tetrahydridoborates or -aluminates (*e.g.* $LiAlH_4$, $NaBH_4$) can however lead to incorporation of the Group 13 element in their final product. Perhaps the most widely used catalyst of this type is **Raney nickel**, made by dissolving aluminium from a nickel–aluminium alloy by means of an alkali.

There are very many variations and elaborations of these methods, and a number of other specialized procedures of limited use. The deposition of complexes of metals in the zero-valent state (carbonyls, acetylacetonates, *etc.*) either from the vapour phase or from solution can give metal particles containing only 10–15 atoms, *i.e.* 100% dispersed. Some of the above procedures can be used to make supported alloy particles containing two metals, but it is not always easy to ensure complete homogeneity.

While the support is usually in the form of either fine powder or larger pieces of regular or irregular shape, depending on the reactor configuration in which it is to be used (Sections 4.5 and 5.1), the support itself is sometimes supported on inert, non-porous bodies to facilitate its use. A particularly interesting and useful material for this purpose is the **ceramic honeycomb** or **monolith**, the simplest form of which resembles a wax honeycomb, but made from α-alumina or mullite. The high-area support, which is applied as a thin layer by a process known as **wash-coating**, is then impregnated with a salt of the catalytic metal, and this is then processed as before. The low resistance to gas flow that these bodies exhibit make them particularly suitable for use where high space velocities have to be used, for example, in vehicle exhaust treatment (see Chapter 7).

4.3 Preparation of Oxides and Sulfides

These are frequently used as unsupported bulk compounds, or as compounds physically mounted on larger particles of inert materials (ceramic spheres, for example) that give the catalyst the physical structure its method of use requires. Binary oxides may be made by fusion of the components or by a carefully controlled precipitation. Sulfides often begin life as oxides and become sulfided during use.

There has been a growing interest in and use of **supported oxides** in which a single monolayer of a reactive oxide is attached or **grafted** onto a supporting oxide. The V_2O_5/TiO_2 catalyst used for oxidation of *o*-xylene to phthalic anhydride (Table 3, page 74) is of this type, and the vanadium species so created are unlike those existing at the surface of V_2O_5 crystals. In the Co–Mo/Al_2O_3 hydrodesulfurization (HDS) catalyst (see also Table 3) the alumina surface is partially covered by monolayer patches of MoO_3 (MoS_2 after sulfidation), the edges of which are decorated with cobalt ions, which participate in the active centres.

4.4 Preparation of Zeolites

Although many zeolitic minerals occur naturally, some of the most interesting and useful are only available by synthesis. The general procedure is to heat solutions or suspensions of the components in the desired proportions; a few (Zeolite A, for example) precipitate on heating below 373 K, but for most it is necessary to use hydrothermal conditions (for example, 423 K) with the reactants in an autoclave. For some less common forms, it is also necessary to include a **templating ion**, typically a tetraalkylammonium ion. Zeolites are metastable, and the time allowed for the synthesis, which ranges from hours to days, is critical, because after the optimum time the product changes towards more stable forms of lower surface area.

The role played by the templating ion is not altogether clear, and indeed the mechanism by which zeolitic networks are constructed is somewhat obscure. The template cannot simply act to form silicate and aluminate ions into the right shape, since the same template can lead to different structures if other conditions are changed, and different templates can lead to the same structure. No comprehensive theory to account for these observations appears to be available, but a clue is provided by the fact that there has been no successful preparation of a zeolite in an organic medium. The presence of water seems to be essential, and it has been suggested that the observations may be explained by supposing that a **water clathrate** is formed around the template, and that it is around a shell of water molecules that the silicate and aluminate ions arrange themselves. Such complex hydrates are in fact known; tetra-*tert*-butylammonium hydroxide crystallizes with an amazing 32 waters of crystallization, and in the simplest water clathrate structure, the water molecules form a pentagonal dodecahedron; this is also precisely the disposition of water molecules in the large cavities of fully-hydrated zeolite A. This identity of structures provides a concept on which it ultimately may be possible to build a more rational theory of zeolite synthesis.

4.5 Large Scale Manufacture of Catalysts

Unless some very special conditions are needed, as for example with the synthesis of zeolites, the **unit operations** performed in manufacturing catalysts or their precursors on a large scale are the classical ones of dissolution, precipitation, filtration, washing, drying, and calcination. Reduction may be needed for some metal catalysts, although *in situ* reduction in the reactor is often preferred. The final physical form adopted for the catalyst will depend on the type of reactor in which it is to be used (Section 5.1). Fluidized-bed reactors need small spherical particles of hard material, while a variety of forms are employed in fixed-bed reactors: granules, tablets, pellets, rings, and extrudates of various shapes and sizes are available to optimize contact of reactants with the catalyst at minimum cost in terms of pressure drop through the bed. In liquid or gas–liquid systems, the catalyst is normally a fine powder, although granular material can be used in **trickle-column** or **spinning-basket reactors**.

Although product quality is of paramount importance, every effort has to be made to minimize wastage and to keep production costs low. In the manufacture of noble-metal-containing catalysts, very great care has to be taken to account for every trace of these expensive commodities. However, catalysts for many major industrial processes are required to function, with periodic *in situ* reactivation as necessary, for several years, since the value of the production lost when a catalyst charge is changed can be very great. In such cases the original cost of making the catalyst may represent only a small fraction of the operating costs of the production process.

Performing a unit operation on a large scale is often more difficult than doing it in the laboratory. To take a simple example, 100 ml of water can be heated to boiling over a bunsen burner in a minute or two, but there is no way to boil 100 l of water in the same time. In general, the larger the mass of material, the longer the operation will take, even although the equipment used is proportionately larger. One operation that can cause difficulty is drying: with precursors to supported metal catalysts, the location of the active phase within a porous support granule may depend critically on the rate at which water is removed and on the temperature of the bed. Total control over all stages in the large scale manufacture of catalysts calls for the combined skills of chemists and chemical engineers.

4.6 Physical Characterization of Catalysts

The acid test of any catalyst preparation, whether the scale be large or small, is 'does it work in its intended application?' However, for purposes of quality control in the factory, it is often easier to measure one or more convenient physical properties in order to give at least a preliminary verdict. For basic research designed to achieve a better understanding of the correlation between physical structure, chemical composition, and catalytic behaviour, a full and accurate description of a catalyst's physical character is also an absolute necessity.

Many catalysts and supports have a high surface area, arising from the fact that either the basic particles are small or they are porous. It is important to know the surface area and the porous character of the total catalyst, as this will determine its activity and the ease with which mass-transport to the active centres will occur. Pores are classified as macro-, meso-, or micropores, according to their size. For cylindrical non-intersecting pores, the internal surface areas (that is, the area due to the pores alone) is given by

$$S = \frac{2V}{r}$$

where V is the total pore volume and r the mean pore radius. Surface areas are most usually measured by determining a nitrogen physisorption isotherm at liquid nitrogen temperature (78 K) and then applying the BET (Brunauer–Emmett–Teller) equation to estimate the monolayer capacity, and hence the surface area. The pore size distribution can be found by employing mercury porosimetry; the force required to persuade liquid mercury into a pore is a function of its radius, and by following the change in volume of mercury over a sample as a function of the applied pressure, the pore size distribution can be derived with the help of the Kelvin equation.

It is then necessary to establish what chemical phases are present. This is conveniently done using X-ray diffraction, the diffraction lines providing a fingerprint with which, by comparison with those of known substances, the phase or phases present (if in significant amount) can be identified. The extent of broadening of the X-ray lines also provides a measure of particle size in the range of 5–50 nm. Infrared and Raman spectroscopy are also helpful in recognizing what bulk phases are present, and thermal analytical methods (thermogravimetric and differential thermal analysis, TGA and DTA, and temperature-programmed reduction, TPR) are valuable adjuncts.

At an early stage in the investigation of a catalyst, it becomes necessary to know the overall chemical composition. Sometimes it is sufficient to analyse only for the major or active component; at other times, the presence or absence of trace impurities may be decisive. Classical wet chemical analysis is now little used, the important methods being **atomic absorption spectroscopy, ion chromatography**, and **inductively-coupled plasma** (ICP) **spectroscopy**. **Secondary-ion mass spectrometry** (SIMS), in which ions knocked off the sample by energetic incident radiation are mass-analysed, is also of use, but the equipment is expensive (as indeed are some of the other items). The golden rule is: only measure what it is essential to know.

It is, however, the surface composition which is of the utmost importance in determining catalytic activity, for it is only the surface that the reactants see, and it is there that impurities may congregate. Techniques for analysing the outermost layers of a catalyst are therefore of great importance. Probably the most useful is

X-ray photoelectron spectroscopy (XPS), also known as **ESCA** (electron spectroscopy for chemical analysis). In this technique the incident X-ray ejects a photoelectron, the kinetic energy of which equals that of the X-ray minus that needed to remove the electron from its energy level to infinity. Measurement of the kinetic energy of the ejected electron then allows its **binding energy** to be estimated, and for an electron of a given type (*e.g.* 1 s) this value is characteristic of a particular element. Since the depth below the surface from which electrons can escape is usually only a few nm, the method is almost surface sensitive. **Static SIMS**, which uses comparatively low energy incident radiation, is also surface-sensitive and can detect very small amounts of impurities.

More detailed information on the morphology of the individual building-blocks of a catalyst is available by using **scanning electron microscopy** (SEM); this provides images of particles larger than about 10 nm, and, in addition, energy analysis of emitted secondary X-rays (EDX) permits elemental analysis over quite small regions of the sample. Even closer examination is possible with **transmission electron microscopy** (TEM), the resolution increasing with electron energy; instruments using 100 or 200 keV electrons are commonly available, and a few use energies as high as 10^6 eV. Lattice planes in crystalline materials can be separated, and detailed images of zeolite structures obtained. The electron diffraction mode also enables structural analysis to be carried out on very small areas. This technique is of particular utility in examining supported metal catalysts, as it is usually easy to observe metal particles larger than 1 nm in size on supports containing only nuclei of fairly low mass (C, Al_2O_3 or SiO_2), for then the contrast is high. Size distributions can be obtained by sizing a sufficiently large number of particles (500–1000) and with high resolution instruments the shapes of particles can be established.

A widely-used method for determining the *average* size of metal particles in such catalysts is the **selective chemisorption** procedure. After careful cleaning and outgassing the sample so that the surface of the metal particles is free of adsorbed molecules, an isotherm is measured using a gas which chemisorbs only on the metal and not on the support: hydrogen, carbon monoxide, and oxygen are the most frequently employed. From the monolayer volume and with an assumed value for the surface stoichiometry (*i.e.* the adsorbed species to surface metal atom ratio) the number of superficial metal atoms is derived. This together with the total number of metal atoms present, as determined by analysis, gives the dispersion: a further assumption concerning particle shape then allows the mean size to be estimated. This procedure is of course subject to a number of caveats, and cross-checking with at least one other method is desirable; it does however require only comparatively inexpensive apparatus, and despite its limitations enjoys wide popularity.

There are other recently-developed techniques for getting detailed information on catalysts and their surfaces. With **Extended X-ray Absorption Fine Structure** (EXAFS), the diffracted X-ray suffers interference with electrons on neighbouring atoms, and this gives rise to 'fine structure' immediately above the X-ray absorption edge, from which structural information can be extracted. In **scanning tunnelling microscopy** (STM), electron flow due to a large applied potential difference between the surface and a very fine point close to it can yield images of the surface on an atomic scale.

Much basic research on surfaces is carried out with single crystals of metals, which can be cleaned and studied under **ultra high vacuum** (UHV) conditions ($\sim 10^{-10}$ Torr). Two of the most important methods for studying such surfaces, and molecules and atoms adsorbed on them, are **low-energy electron diffraction** (LEEDS) and **high-resolution electron energy loss spectroscopy** (HREELS). In the former a crystalline surface acts as a diffraction grating, and the image due to diffracted electrons reveals the form of the surface lattice (cubic, hexagonal, *etc.*). A surprising observation has been the frequency with which surfaces *reconstruct* under vacuum to form single layers having lower surface energy: the effect is somewhat akin to that of surface tension in liquids. Ordered arrays of chemisorbed atoms, radicals, or molecules will also contribute to the observed diffraction pattern, and the lattice they form and some structural information can also be deduced. In HREELS, reflected electrons lose part of their energy by exciting vibrations in adsorbed molecules, and the resulting spectrum, at somewhat low resolution, reveals adsorption bands related to those seen in infrared spectroscopy.

This finally leads us to mention methods which reveal information on the structure of molecules chemisorbed on highly dispersed metals and on oxides. Quite the most useful and widely employed techniques are based on **infrared spectroscopy**: they include Fourier-Transform IR (FTIR) spectroscopy, Reflection-Absorption IR spectroscopy (RAIRS) and Diffuse-Reflectance IR-FT spectroscopy (DRIFTS). The chemisorbed carbon monoxide molecule acts as a useful probe for surface structure and composition, as the C–O vibration absorbs strongly and responds sensitively to the nature of the C-surface bond. The difference in frequency between the bridged and linear structures (see Table 1, page 66) is typically ~ 200 cm^{-1}; the IR spectra of carbonyl complexes such as $Ni(CO)_4$ and $Fe_2(CO)_9$ have helped in assigning the observed bands. The type of acid centres existing at the surfaces of oxides and zeolites is also revealed through the IR spectrum shown by adsorbed pyridine; on Lewis centres it is adsorbed virtually unchanged, while on Brønsted centres it forms a pyridinium ion, the spectrum of which is somewhat different.

Raman spectroscopy is complementary to IR in that it operates through different selection rules (Raman scattering occurs when the vibration produces a change in polarizability), so that for example species in aqueous solution can be recognized. Especially in the recently available 'FT Raman' mode, information on structures of adsorbed species becomes available.

5 HETEROGENEOUS CATALYSTS IN CHEMICAL PROCESSING

5.1 Methods of Performing Heterogeneously-catalysed Reactions

Some passing reference has already been made to types of reactor used in chemical processing (Section 4.5) in introducing the subject of large scale manufacture of catalysts. We must now put this information into a more ordered form and provide some additional information.

There are really three basic types of catalytic reactor: (i) the **static** or **batch reactor**; (ii) the **continuous-flow reactor**; and (iii) the **continuous-stirred**

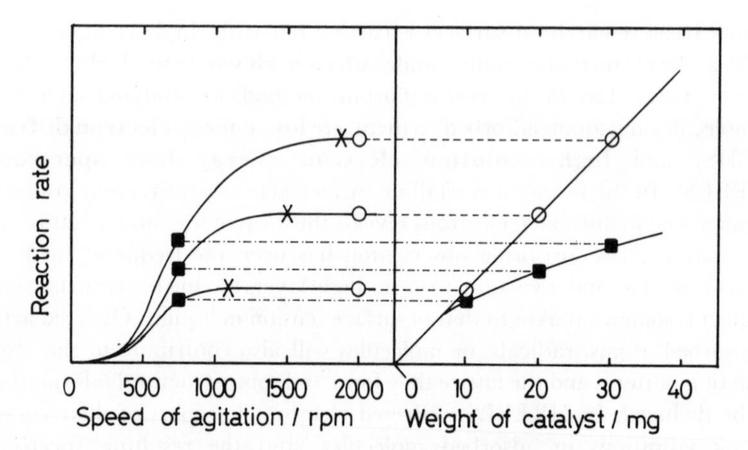

Figure 8 *Rate of reaction in a three-phase system as a function of speed of agitation at various catalyst weights, and as a function of catalyst weight at high and low agitation speeds. The speed needed to overcome diffusion control (X) increases with weight of catalyst*

tank reactor (CSTR). The first type is chiefly used where one of the reactants is liquid: reactants and catalyst are charged into a reactor, which may be pressurized (*i.e.* an autoclave), and the reaction is allowed to proceed to completion, after which the products are separated from the catalyst, which is if possible re-used. Fat-hardening (see below) is carried out in this way. In three-phase systems (*i.e.* solid + liquid + gas), it is usually important to try to avoid mass-transport limitation (see Section 3.2) and this requires effective agitation to minimize the distance that the dissolved gas has to diffuse through the liquid to the solid surface. Stirring, with baffles set into the walls of the reactor, or shaking if the reactor is small, will help to achieve this objective; alternatively the catalyst itself can be rotated, as with the spinning-basket reactor. Mass-transport limitation is recognized (Figure 8) by (i) the rate being dependent on catalyst weight raised to a power less than one, (i) the rate varying with conditions of agitation, and (iii) a small temperature coefficient.

The CSTR is a variant of the batch reactor, but here the reactants flow in, and the products out, continuously; the rates of flow are adjusted so that the **mean residence time** allows the desired degree of conversion. The catalyst must be in a granular form, so that it is not carried from the reactor, and the spinning-basket configuration can be used in a CSTR.

Major gas-phase processes are of the continuous-flow type, and there are two distinct methods of operation: (i) fixed-bed reactor, (ii) fluidized-bed reactor (see Section 4.5). In the former, the reactor tube is packed with pieces of catalyst typically between 1 mm and 1 cm in size, chosen to give the best compromise between too difficult access to the active centres and too high a pressure drop to be economically acceptable. A frequent problem with such reactors is how to remove the heat of reaction which otherwise would cause the bed temperature to rise, with possible harmful consequences to the catalyst. Two solutions have been found. In the first, the reactor diameter is reduced to about 1 cm, and a very large number of such tubes ($\sim$1000) are fixed into a cannister of diameter $\sim$50 cm through which a heat-transfer fluid is circulated. This device is termed a **multitubular reactor**. In a second design, the catalyst is placed in shallow beds in a reactor of larger diameter, and between each section there is a coil through which the heat-transfer fluid flows, thus lowering the temperature of the gas before it reaches the next section of the bed.

In a fluidized-bed reactor, the catalyst is in the form of small, roughly spherical particles, so that when the gas flow rate is increased past a critical value the whole bed immediately expands and becomes agitated, and resembles a boiling liquid. Continuous movement of the catalyst minimizes the formation of hot-spots which often plague fixed-bed systems, ensures a more or less uniform temperature throughout the bed and encourages heat transfer to the reactor walls.

The three major classes of reactor have other manifestations and subdivisions. The batch reactor is sometimes used to examine gas-phase reactions in the laboratory, especially where reactants are scarce or expensive. However to avoid mass-transport limitation, the reactants can be made to circulate around a closed loop and through the catalyst bed by means of a magnetically-operated pump. Such a reactor is termed a **closed-circuit** or **racetrack reactor**.

Finally it may be useful to make a few remarks about deactivation of catalysts and its consequences. As noted in Section 3.2, all catalysts are liable to lose activity during use. There are many reasons for this, the most important being (i) mechanical instability (breaking, cracking, loss of material from the outer surface through attrition), (ii) sintering, that is, loss of surface area of the active phase due to particle growth, especially troublesome with supported metals, (iii) poisoning due to impurities in the feedstock, and (iv) poisoning due to strongly retained by-products of the reaction. In this last category, formation of carbon deposits or 'coke' in reactions of hydrocarbons in the various stages of petroleum reforming is an outstanding problem.

Mechanical weakness can only be repaired by replacing the catalyst with a better one. Some adventitious poisons are 'reversible' in the sense that, if their supply is stopped, the toxic molecules will slowly desorb and the catalyst will reactivate. Sintering and self-poisoning often occur together, as is the case with petroleum-reforming catalysts (see below); much effort has been devoted to finding means to reactivate catalysts *in situ*, that is, without removing them from the reactor, as this is less expensive in down-time and in manufacturing costs than replacing the catalyst with a new charge. With Pt/Al_2O_3 catalysts, coke can be removed by burning, but in order to redisperse sintered metal particles it is necessary to treat with a source of chlorine and an oxidizing agent to reconstitute the original chloroplatinate ion $[PtCl_6]^{2-}$, which can then be again reduced to well-dispersed metal.

5.2 Reactions of Synthesis Gas

In the distant past, many basic organic chemicals such as methanol, ethanol, and acetic acid were made either by fermentation or by destructive distillation of natural materials including wood and coal. Synthesis gas, or **syngas** for short, was introduced many years ago as a source of a great variety of useful organic molecules; it consists of hydrogen and carbon monoxide, and was originally made via the reaction of steam with red-hot coke:

$$C + H_2O \rightarrow H_2 + CO \tag{7}$$

$$C + 2\,H_2O \rightarrow 2\,H_2 + CO_2 \tag{8}$$

These reactions are endothermic, so the bed of coke had periodically to be re-heated by the following exothermic reactions:

$$C + O_2 \rightarrow CO_2 \tag{9}$$

$$2\,C + O_2 \rightarrow 2\,CO \tag{10}$$

More recently, carbon sources rich in hydrogen (*e.g.* methane, naphtha) have replaced coke, and syngas is now made by a **steam reforming** process such as

$$CH_4 + H_2O \rightarrow 3\,H_2 + CO \tag{11}$$

Finally the composition can be fine-tuned by means of the **water gas shift equilibrium**

$$CO + H_2O \rightleftharpoons CO_2 + H_2 \tag{12}$$

Processes (11) and (12) require catalysts: steam reforming uses promoted Ni/Al_2O_3, while water gas shift uses either iron–chromium oxides (Fe_3O_4/Cr_2O_3) at 670 to 770 K or $Cu–ZnO/Al_2O_3$ at 460 to 530 K.

Formation of an amazing range of organic molecules from syngas is thermodynamically feasible, and many of the possible processes can be realized in practice by the appropriate choice of catalyst and reaction conditions. Products of principal interest can be grouped as follows: (i) methane, (ii) higher alkanes, (iii) alkenes, (iv) methanol, and (v) higher alcohols. Aldehydes, ketones, and carboxylic acids are included amongst other possible products. Formation of methane (*i.e.* methanation) is not of great importance, except for removing traces of carbon monoxide when very pure hydrogen is needed, as in ammonia synthesis (see below). Both nickel and ruthenium are very active for methanation (see Tables 2 and 3, pages 67 and 74) and methane is always the favoured product at low pressures and high temperatures.

Methanol is now manufactured from syngas on a very large scale with the help of a $Cu–ZnO/Al_2O_3$; formerly $ZnO–Cr_2O_3$ was used. The milder hydrogenation properties of copper ensures that the C–O bond is not broken, an essential prerequisite if an oxygenated product is to result.

The conversion of syngas into products containing more than one carbon atom is called the **Fischer–Tropsch synthesis**. Alkanes, alkenes, and alcohols can be made; there is a vast literature, both open and patent, describing how the choice of catalyst, particularly the choice of promoters, and reaction conditions can affect product yields. It is only possible here to mention a few salient features of the results. The yield decreases logarithmically with increasing number of carbon

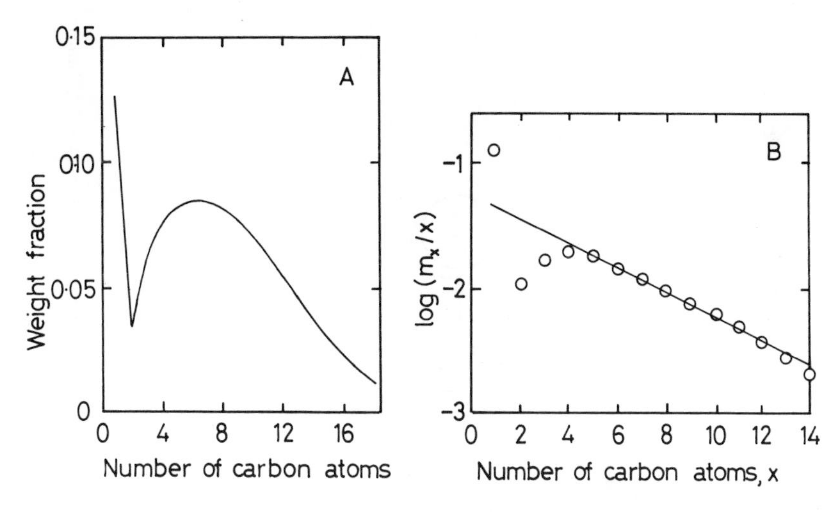

Figure 9 *(a) Product distribution for alkanes formed over a cobalt–thoria catalyst. (b) The same distribution as an Anderson–Schulz–Flory plot ($\alpha = 0.81$)*

atoms, giving what is referred to as the **Anderson–Schulz–Flory distribution**. The chain growth parameter, α, is derived from the plot of log(yield) versus carbon number (Figure 9); its value depends on temperature and pressure, long-chain compounds being most abundant at low temperature and high pressure.

For example, ruthenium catalysts can produce very high molecular weight hydrocarbons (*e.g.* greases and waxes) under these conditions, but at high temperature and low pressure hydrogenolysis is a competing reaction, and as noted above, methane is then the chief product. Under the appropriate conditions, promoted iron and cobalt catalysts produce higher oxygenates, while nickel catalysts produce mainly alkanes. Much interest has been shown in attempts to develop processes for selectively making lower alkenes (ethane, propene) for use as feedstock for petrochemical processing; they are thought to be initial products, subsequently hydrogenated to alkanes. Copper catalysts can be modified to make a certain proportion of higher alcohols, and both rhodium and iridium catalysts that give good yields of ethanol can be prepared. Some of the elementary steps thought to be involved in chain growth are illustrated in Scheme 3A, and an outline of those responsible for formation of oxygenated products is shown in Scheme 3B.

The very great interest shown in Fischer–Tropsch chemistry over the past 15 years no doubt originated in the political uncertainties associated with the major oil-producing states, especially those of the Middle and Near East. However, the need to replace oil-derived products with synthetic fuels and to develop alternative sources of raw materials for the petrochemical industry on a large scale has not really arisen as yet, and it is only in South Africa, devoid as it is of indigenous crude oil, where coal-based fuels have been economically produced. The exigencies of World War II however caused the Fischer–Tropsch process to be operated on a moderate scale in both Germany and Japan.

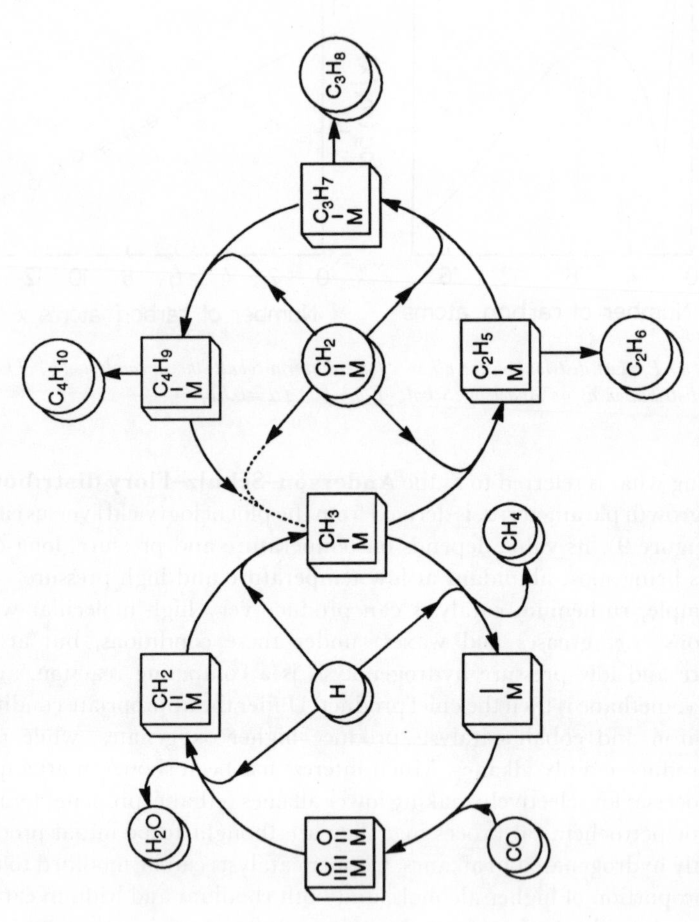

Scheme 3A *Mechanism of synthesis of methane and higher alkanes from 'synthesis gas'. The left-hand cycle shows the route to methane, mediated by a pool of H atoms; the right-hand cycle shows the route to higher alkanes formed by methylene insertion into alkyl–metal bonds*

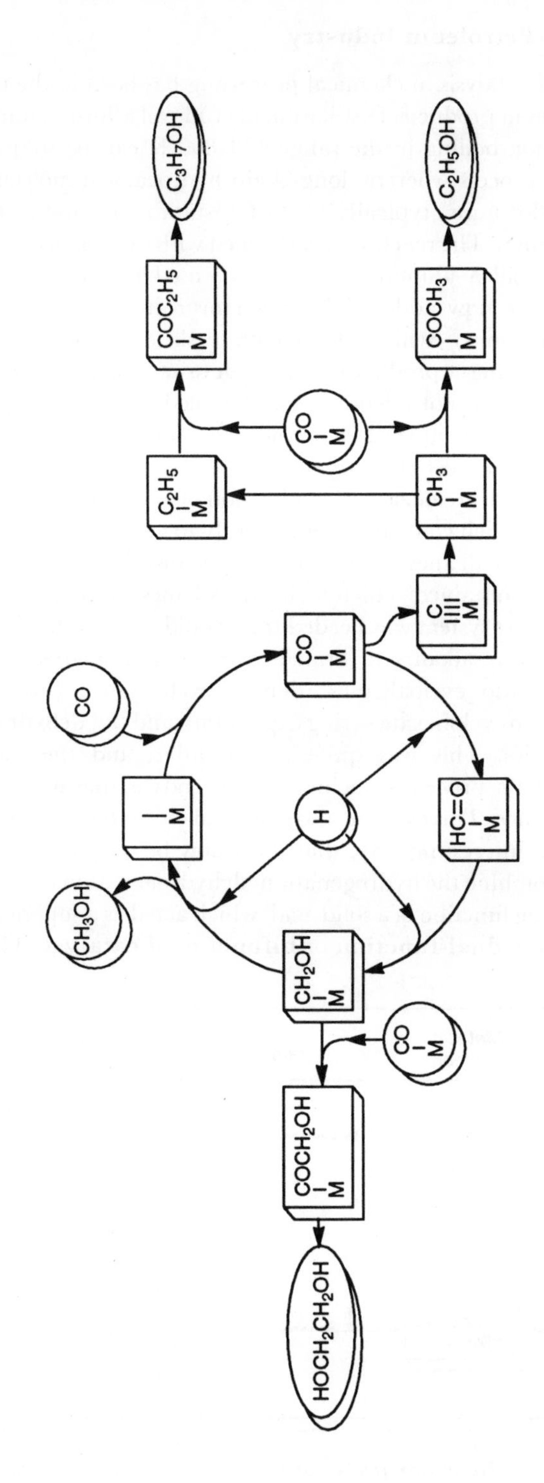

Scheme 3B *Mechanism of synthesis of oxygenated products from synthesis gas. The processes depicted in the right-hand part of the scheme proceed by reduction and methylene insertion as shown in Scheme 3A.*

5.3 Catalysis in the Petroleum Industry

The largest single use of catalysis in chemical processing has been in the refining and reforming of petroleum products. Distillation of crude oil affords a number of products, and the fraction boiling in the range 490–800 K can be subjected to **catalytic cracking**, a process whereby long-chain hydrocarbon molecules are broken down into smaller units, typically C_5 to C_8, suitable for use as fuel for internal combustion engines. The reaction is performed with hydrogen over a solid acid catalyst; amorphous silica–alumina was originally used, but zeolites (Section 2.5) of higher activity are now added. The mechanism involves formation of carbocations (carbonium ions), which also facilitate skeletal isomerization to branched alkanes, thus giving a product of higher **octane rating**. Even so, the values achieved (85–90) were not adequate for the needs of high-compression engines, where octane ratings of at least 98 are needed. Additional processes had therefore to be developed to meet this target.

Octane rating gives a measure of ease of combustion of a hydrocarbon vapour, and classes of compound which contribute significantly to good performance as a fuel include branched-chain alkanes and aromatics. The distillation cut available for further processing contained chiefly linear alkanes and cycloalkanes ('naphthenes'), so a catalyst system was needed that would perform the following functions: (i) convert linear alkanes to branched alkanes (isomerization); (ii) convert linear alkanes into cycloalkanes (dehydrocyclization); (iii) convert cyclopentane derivatives to cyclohexanes (ring expansion); and (iv) dehydrogenate cycloalkanes to aromatics. This was quite a tall order, and the successful development of a catalytic process to achieve these ends is one of the major triumphs of catalytic science. This exercise also extended our understanding of the theory of catalysis in an interesting way, that must now be outlined.

The solution was to combine the hydrogenation–dehydrogenation function of a metal with the isomerizing function of a solid acid, which acted as a support for the metal, in what is termed a **dual-function** or **bifunctional catalyst**. The steps

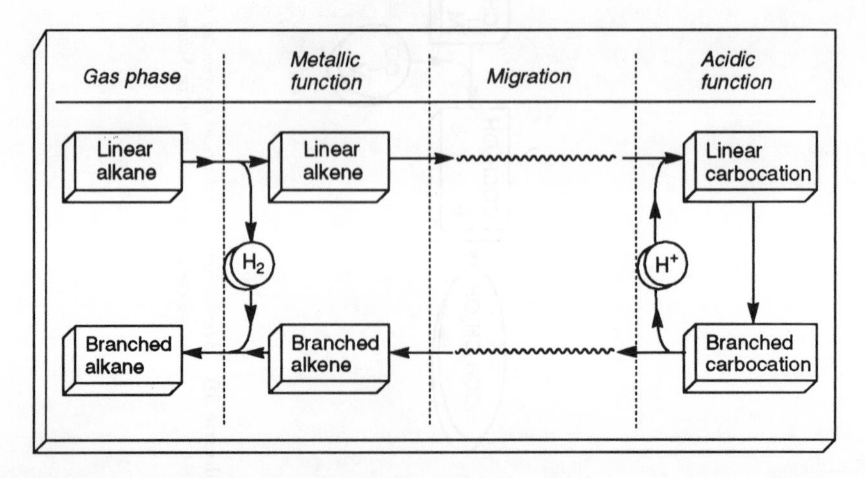

Scheme 4 *Mechanism of the skeletal isomerization of an alkane on a bifunctional catalyst*

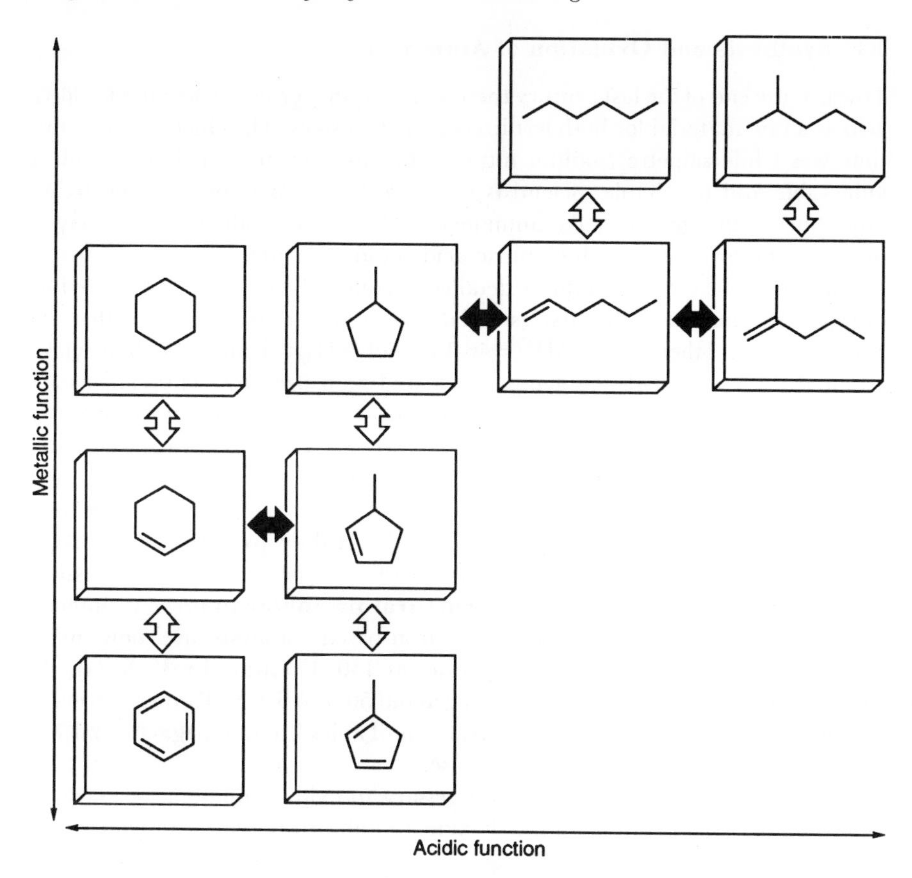

Scheme 5 *Pathways for the reaction of n-hexane on a bifunctional catalyst*

that can then occur with a linear alkane are shown in Scheme 4. It is first dehydrogenated at the metal, then migrates through the gas phase to an acidic site where it becomes a carbocation, isomerizes, reverts to an alkene, and migrates back to the metal where it is rehydrogenated to a branched alkane. In forming aromatics from a linear alkane, it is believed that C_6 intermediates are required (Scheme 5); thus methylcyclopentene can isomerize via its carbocationic form into cyclohexene, which then rapid dehydrogenates to benzene. The total process is known as **catalytic reforming**.

Nickel was originally used as the metal, but this was soon replaced by platinum; now bimetallic systems are most often used, platinum in combination with rhenium, iridium, or tin being the most common. These catalysts are less liable to deactivation through formation of toxic 'carbonaceous deposits' or 'coke' than is pure platinum. Amorphous silica–alumina has been used as the acidic support, but now it is chlorided alumina which is generally employed. With periodic reactivation in the reactor, involving oxidation, redispersal and re-reduction of the metal (see Section 4.2), a single charge of catalyst can be expected to last from three to five years.

5.4 Synthesis and Oxidation of Ammonia

Towards the end of the last century there was a rapidly growing demand for nitric acid as a raw material for both fertilizers and explosives. The chief source at that time was Chile salpetre (sodium nitrate), but its extreme remoteness made it vulnerable and unreliable. Scientists therefore began to study ways of 'fixing' atmospheric nitrogen to make ammonia, and then to oxidize it selectively to nitrogen monoxide, from which nitric acid could be made.

Limitations of space prohibit a detailed account of early attempts to reduce nitrogen to ammonia; the first essential was however to appreciate that the reaction was exothermic [$-\Delta H^{\ominus} = 46$ kJ (mol NH_3)$^{-1}$] and that high yields required the use of both low temperature and high pressure. This called for a catalyst, but a rather special one which would be stable and unaffected by traces of impurities in the reactants. The fortuitous discovery that an iron catalyst made from iron ore obtained from Galliväre in Sweden met all the requirements enabled the BASF (Badische Anilin und Soda Fabrik) company to set up the first synthetic ammonia plant, using what is often called the **Haber process**. The 'magic' ingredients which made all the difference were a few percent of both potassium and alumina, the latter acting as a **structural promoter** to prevent sintering during use. More complex promoted iron-based catalysts are now made synthetically. The process usually operates at 150–350 atm (15–35 MPa), the equilibrium conversion to ammonia being about 50% at 670 K. Ruthenium is also extremely active as a catalyst for this reaction (Tables 2 and 3, pages 67 and 74), and has found some limited industrial use.

In order to make nitric acid, it is necessary to oxidize ammonia to nitrogen monoxide, then oxidize this to the dioxide and absorb the latter in water:

$$4\,NH_3 + 5\,O_2 \rightarrow 4\,NO + 6\,H_2O \quad (13)$$

$$2\,NO + O_2 \rightarrow 2\,NO_2 \quad (14)$$

$$3\,NO_2 + H_2O \rightarrow 2\,HNO_3 + NO \quad (15)$$

the final molecule of the monoxide being recycled. The reaction needs to be conducted at high temperature (1170–1220 K) and with very short contact times (10^{-3} to 10^{-4} s), and the form of catalyst that is universally employed comprises a pad of fine gauzes (1050 apertures cm^{-2}) woven from very fine platinum–rhodium alloy wire (0.06 mm diameter). The rhodium lends mechanical stability to the wires, but recrystallization and grain growth ultimately lead to disintegration; the catalyst has then to be replaced. The process operates at superatmospheric pressure, the selectivity to nitrogen monoxide being 95–97%. In all essential respects it has remained unchanged since it was introduced in 1909, and attempts to find alternative base metal catalysts have not yet succeeded.

This reaction provides one of the simplest laboratory demonstrations of catalysis. A pre-heated platinum wire suspended over an open beaker containing a little concentrated ammonia solution will continue to glow, and clouds of the nitrogen dioxide will be seen rising from it.

Unfortunately the process of absorbing nitrogen dioxide into water is less than perfectly efficient, and small amounts of a mixture of the monoxide and dioxide (usually referred to as NO_x or NOX) escape into the atmosphere. Both are serious atmospheric pollutants, and have been implicated in the phenomena of photochemical smog and acid rain. Processes have therefore been developed to eliminate them; decomposition back to nitrogen and oxygen has proved difficult, although research in this area continues, and reduction by means of an added fuel such as methane by reactions such as

$$CH_4 + 4\,NO \rightarrow CO_2 + 2\,H_2O + 2\,N_2 \tag{16}$$

were originally hard to accomplish because of the presence of excess oxygen, with which the fuel preferentially reacted. However, catalysts have recently been developed which effect **selective NOX reduction**, *i.e.* reduction of NOX in the presence of oxygen; these are based on V_2O_5 supported on TiO_2. This breakthrough also opens the way for removal of NOX from other gaseous effluents, especially flue-gas from power stations. The catalytic chemistry of the reduction of NOX will be described more fully in Chapter 7, where its removal from vehicle exhaust will be treated in depth.

5.5 Catalysis in the Petrochemical Industry: Selective Oxidation of Hydrocarbons

The two decades following the end of the Second World War witnessed the replacement of organic products derived from coal by those obtained from crude oil, the cracking of which provided abundant supplies of the lower alkenes and aromatics: these in turn could be either polymerized to make polythene, polypropylene, polystyrene, *etc.*, or selectively oxidized to products of greater value as intermediates in the manufacture of desirable materials. The great variety of reactions which a single alkene can undergo is illustrated in Scheme 6 by reference to ethene: only its polymerization or copolymerization cannot be regarded as catalytic in the true sense.

It may be helpful first to try to classify selective oxidations in terms of the reactions taking place (see Table 4). The simplest is **oxidative dehydrogenation** in which the elements of hydrogen are removed from the reactant, with the consequent formation of water, the high enthalpy of formation of which ensures more complete reaction at lower temperature that would be possible with simple dehydrogenation. The process can be illustrated by the conversion of methanol to formaldehyde (methanal):

$$2\,CH_3OH + O_2 \rightarrow 2\,HCHO + 2\,H_2O \tag{17}$$

In a second category come reactions where oxygen is incorporated into the hydrocarbon without any fission of C–C bonds; then we have more destructive reactions in which C–C bonds are broken; and finally there is deep oxidation to carbon dioxide and water (Table 4). This last reaction is usually to be avoided,

Table 4 *Classification of catalysed reactions of hydrocarbons + oxygen*

Type I:	*Oxidative dehydrogenation*		
A	Dimerization	*e.g.*	propene → hexadiene or benzene
B	Dehydrogenation	*e.g.*	butane or butene → butadiene
C	Dehydrocyclization:	*e.g.*	hexane → cyclohexane or benzene
Type II:	*Incorporation of oxygen without C–C bond breaking*		
A	Oxygen insertion:	*e.g.*	ethene → oxirane
B	Alkene to aldehyde:	*e.g.*	ethene → acetaldelyde (ethanal) propene → acrolein (propenal)
C	Alkene to ketone:	*e.g.*	propene → acetone (2-propanone)
D	Alkene to alcohol or acid:	*e.g.*	propene → allyl alcohol (propenol) propene → acrylic acid (propenoic acid)
E	Diene or aromatic to anhydride:	*e.g.*	butadiene → maleic anhydride *o*-xylene → phthalic anhydride
Type III:	*Incorporation of oxygen with C–C bond fission*		
A	Alkene to aldehyde *etc.*:	*e.g.*	propene → ethanal + ethanoic acid
B	Aromatic to anhydride:	*e.g.*	benzene → maleic anhydride naphthalene → phthalic anhydride
Type IV:	*Deep oxidation*		
	Hydrocarbon to carbon oxides	*e.g.*	$CH_4 \rightarrow CO_2 + H_2O$

except in those situations where traces of organic vapour need to be converted into harmless products in order to avoid environmental harm.

So many and such varied catalysts have been used to try to accomplish the reactions set forth in Scheme 6 that it is not easy to discern what chemical principles are at work; but we must try to identify some of them. The noble metals have at most a limited usefulness in selective oxidations performed in the gas phase, although mild liquid-phase oxidations (*e.g.* of carbohydrates) can be carried out with platinum catalysts. The oxidations of ethene to oxirane (see Table 3, page 74, and Scheme 6) and of methanol to methanal catalysed by unsupported silver are well-established, although oxide catalysts seem to be taking over the latter process. In the one major process that appears to use supported metal, the active species is almost certainly the derived cation: this is the oxidation of ethene in the presence of acetic (ethanoic) acid to form vinyl acetate (ethenyl ethanoate):

$$C_2H_4 + CH_3COOH + O_2 \rightarrow CH_2{=}CH.OCOCH_3 + H_2O \qquad (18)$$

Catalysts for this reaction are based on palladium, but may contain gold and alkali metal salts as promoters. Reactions of this type are sometimes collectively described as **Wacker chemistry**, after the company (Wacker Chemie) that first developed palladium-catalysed oxidations, *e.g.* that of ethene to acetaldehyde (ethanal) (see Chapter 5, page 114).

Other selective oxidations are catalysed by **binary** or **compound oxides** wherein two or sometimes more different cations are combined stoichiometrically. Each cation plays a distinct and separate role in the reaction mechanism (for an

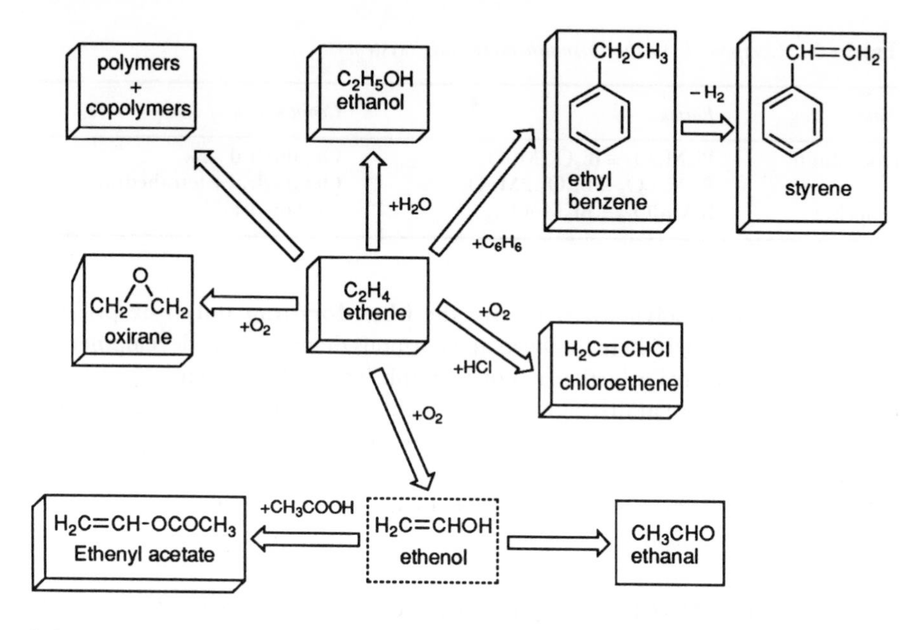

Scheme 6 *Some industrially important reactions of ethene*

example, see below), and the source of oxygen is usually lattice oxide ion rather than chemisorbed oxygen, which generally reacts non-selectively. Many of the effective binary oxides contain either vanadium, molybdenum, or antimony, and sometimes it is possible to specify in which type of coordination an ion should be for it to be effective. One of the first binary oxides to find application was the bismuth–molybdenum system, in which three different compounds can be formed (Table 5): the koechlinite phase is the most active and selective, showing that the octahedral oxo-molybdenum ion $[MoO_6]^{3-}$ is the essential ingredient. The component ions must come from elements in different groups of the Periodic Classification, typical examples being: molybdenum + iron or bismuth; vanadium + phosphorous or titanium; antimony + tin, iron, or uranium. Reactions in which C–C bonds are split usually have vanadium as one of the components.

In the space available, selective oxidation must be illustrated with a single example, that being the process which first put catalysis by binary oxides on the map. A long story has to be shortened and simplified, but its essence is as follows. In the late 1950s, work at the Sohio laboratories in the United States showed that 'bismuth molybdate' could selectively oxidize propene to acrolein (propenal):

$$CH_3.CH{=}CH_2 + O_2 \rightarrow H_2C{=}CH.CHO + H_2O \qquad (19)$$

This of itself has only limited utility, but it was soon realized that by incorporating ammonia in the reactant stream a one step synthesis of acrylonitrile (ethynyl cyanide, cyano-ethene) would be possible:

$$CH_3.CH{=}CH_2 + \tfrac{3}{2}O_2 + NH_3 \rightarrow H_2C{=}CH.CN + 3\,H_2O \qquad (20)$$

Table 5 *Compounds in the 'bismuth molybdate' system*

Name	*Formula*	*Coordination of Mo*
Koechlinite	$Bi_2MoO_6 \equiv Bi_2O_3.MoO_3$	Octahedral
—	$Bi_2Mo_2O_9 \equiv Bi_2O_3.2MoO_3$	Octahedral + tetrahedral
Scheelite	$BiMo_3O_{12} \equiv Bi(MoO_4)_3$	Tetrahedral

This compound is polymerized to give the **fibre-forming polymer polyacrylonitrile**, for which there has been an enormous demand. The development of this process represents a further major triumph for catalytic science.

These reactions are at first sight surprising, in that it is the less reactive methyl group at which oxidation takes place, rather than the normally more reactive double bond. The reason for this is as follows. The first C–H bond in the methyl group is fairly easily broken, because of the resonance energy gained in forming the allyl (propenyl) radical:

$$CH_3\text{–}CH{=}CH_2 \rightarrow H^{\cdot} + {}^{\cdot}CH_2\text{–}CH{=}CH_2 \leftrightarrow CH_2{=}CH\text{–}CH_2{}^{\cdot} \quad (21)$$

Addition of oxygen and removal of another hydrogen atom then gives the product, but it thus appears that, since the allyl radical is symmetrical, there is an equal chance of oxidizing either end; this has been confirmed by isotopic tracer experiments in which a labelled methyl carbon atom in propene appears with only 50% probablility in the methyl group of the product. A possible way in which the two components of the catalyst may collaborate is illustrated in Scheme 7.

This very short survey does less than justice to the extensive and excellent work which has been performed in this field. Many of the processes that are operated commercially do so however with only moderate selectivities, *i.e.* much of the reactant is lost as carbon oxides; much scope still remains for devising new catalysts, and advances in materials science should lead to better catalysts being found.

5.6 Catalytic Hydrogenation in the Manufacture of Fine Chemicals and in Fat Hardening

The term **'fine chemicals'** is applied to organic substances for which there is a demand of small to medium size, and which find application as pharmaceutical or medicinal products, as dyestuffs, cosmetics, foodstuffs, and food additives or as ingredients in photographic and other reprographic processes. They are usually thermosensitive, of moderate molar mass, and need to be processed in solution. The commonest type of catalytic operation applied to molecules of this type is **catalytic hydrogenation**, and in this section we quickly review the scope of the process and illustrate it with one or two examples of practical importance.

Catalytic hydrogenation is the process of adding hydrogen atoms to any unsaturated functional group capable of accepting them: the most usual are C=C, C≡C, C=C–C=C, C=O, aromatic C–C bonds, and $-NO_2$. Molecular

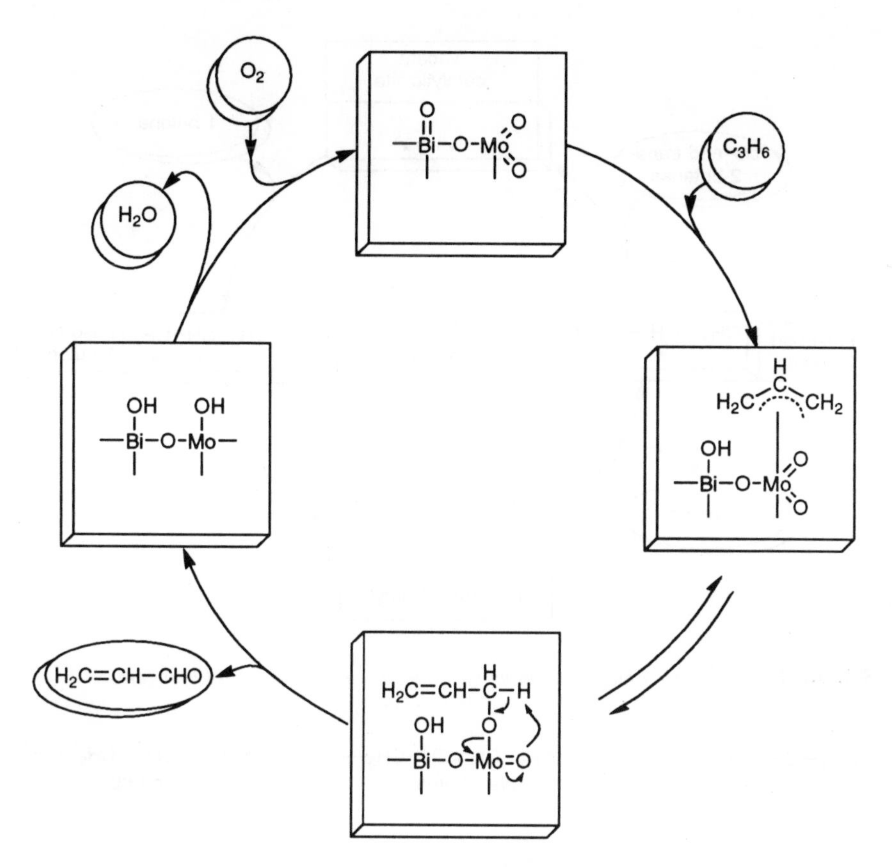

Scheme 7 *Mechanism of oxidation of propene to acrolein using a 'bismuth molybdate' catalyst*

hydrogen is the usual source of the atoms, but others including cyclohexene, formic acid, and hydrazine are occasionally employed. The most active catalysts are the noble metals of Groups 8, 9, and 10 (Table 3, page 74), which are usually used in supported form. There are few instances of where there is a simple uncomplicated reaction to perform: usually there are ancillary criteria to be met, for example avoidance of other possible reactions and achieving the desired product in high selectivity. The following examples will illustrate the sorts of complication encountered.

Let us consider first the reduction of 1-butene. The competing reaction here is **double bond migration** to *cis*- and *trans*-2-butene, due to sequential addition of a hydrogen atom and removal of another (see Scheme 8). Because the double bond is most comfortable in the central position, the equilibrium mixture will consist predominantly of 2-butenes, and if the product can desorb easily, as is the case with palladium catalysts, it will appear as a major product. By a similar sequence of reactions, *cis*-2-butene can isomerize to the stabler *trans*-2-butene.

The phenomenon of **selectivity** is also shown in the reduction of alkynes and alkadienes (see Scheme 9). In reactions of this type it is often the intermediate

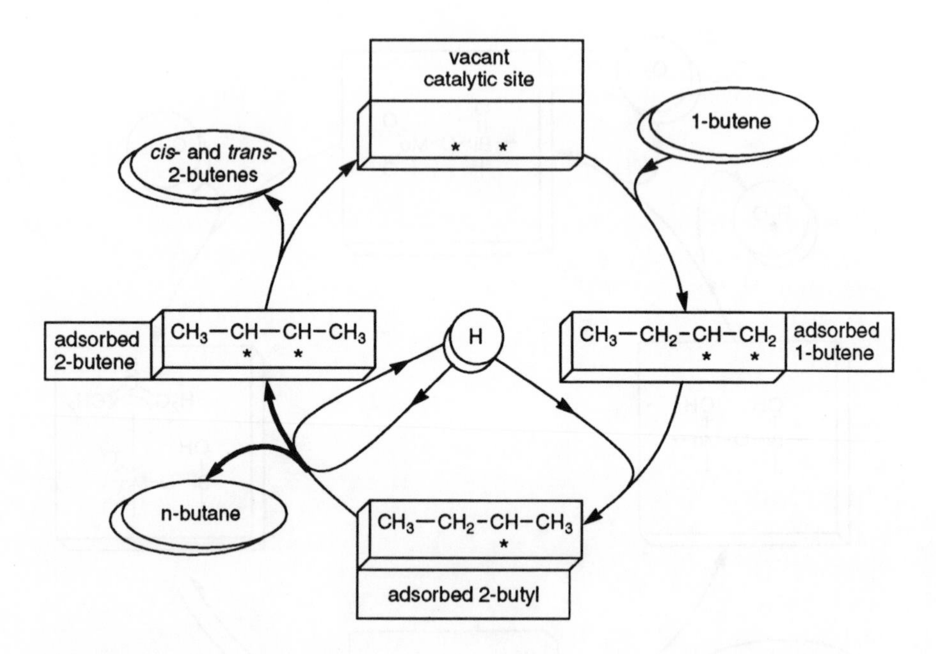

Scheme 8 *Double bond migration during the hydrogenation of unsaturated hydrocarbons*

$CH_3{-}C{\equiv}C{-}CH_3$ (2-butyne) ⟶ $CH_3{-}CH{=}CH{-}CH_3$ (*cis*-2-butene) ⟶ $CH_3{-}CH_2{-}CH_2{-}CH_3$ (n-butane)

$CH_2{=}CH{-}CH{=}CH_2$ (1,3-butadiene) ⟶ $\left\{\begin{matrix} CH_2{=}CH{-}CH_2{-}CH_3 \\ CH_3{-}CH{=}CH{-}CH_3 \end{matrix}\right\}$ (1- and 2-butenes) ⟶ $CH_3{-}CH_2{-}CH_2{-}CH_3$ (n-butane)

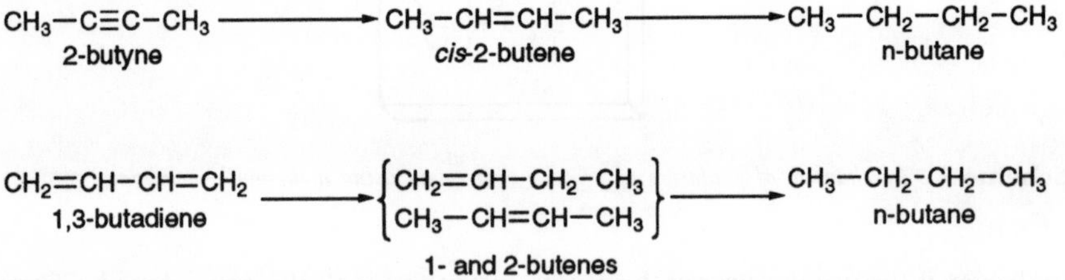

Scheme 9 *Selectivity in the hydrogenation of multiply-unsaturated hydrocarbons*

alkene which is needed, and supported palladium catalysts (if necessary, promoted or partially poisoned) can transform alkynes into *cis*-alkenes in high yields. An example is the **Lindlar catalyst** ($Pd/CaCO_3$, poisoned with Pb^{2+} and quinoline) which is used in the production of vitamin A.

The aromatic ring is much less easily reduced, and reduction to the corresponding cyclohexane is usually achieved without any intermediate being recognizable; however ruthenium catalysts are capable of forming cyclohexene from benzene with considerable selectivity.

An interesting example of selective hydrogenation is presented by the conjugated C=C–C=O system, as in crotonaldehyde (2-butenal) and cinnamaldehyde (2-phenyl-2-propenal). Here the challenge is to reduce selectively the C=O bond, converting the aldehyde into the primary unsaturated alcohol, without touching the normally more reactive C=C bond. Good results have been obtained with bimetallic (Pt–Sn, Pt–Ge) catalysts and also with platinum on graphite or on a zeolite.

Hydrogenolysis is often a competing reaction, a classic problem being the reduction of a chloronitrobenzene to the corresponding chloroaniline. The C–Cl bond is easily broken by hydrogenolysis, but success has attended the use of Pt/TiO_2 reduced at high temperature and Raney nickel poisoned with formamidine or dicyanodiamide.

Perhaps the greatest challenge in the field of catalytic hydrogenation is to effect **asymmetric reduction** of a molecule with the production of a chiral centre (see also Chapter 6). One reaction that has received much attention is the hydrogenation of an α-keto-ester such as methyl pyruvate specifically to *R*(+)-methyl lactate rather than the *S*(−) form:

$$CH_3—CO—CO—OCH_3 \longrightarrow CH_3—\overset{H\ \ OH}{C}—CO—OCH_3 \qquad (22)$$

methyl pyruvate   *R* (+)-methyl lactate

This calls for the hydrogen atoms to be added selectively to one side of the reactant as it lies flat on the surface of the catalyst, but the molecule must always lie with the same side downward if only one enantiomer is to be formed. Potentially useful optical yields have resulted when the reaction is conducted on a platinum catalyst on which is also adsorbed the alkaloid cinchonidine; it is believed that the intervening spaces can only accept methyl pyruvate molecules which adsorb in the manner which will result in *R*(+)-methyl lactate. A strange feature of this system is that the reaction in the presence of the modifier is much faster than in its absence.

We have remarked previously (Section 5.1) on the need to operate three-phase systems in a way such that the rate is not limited by diffusion of dissolved hydrogen through the liquid to the catalyst surface. Full use is made of the potential of the catalyst to effect the reaction if vigorous agitation is used to ensure this does not happen.

Nature provides an abundance of animal and vegetable oils having considerable nutritional value; they are widely used for domestic cooking and large scale food manufacture, and in animal feedstuffs. Their conversion to a butter substitute, *viz.* margarine, requires them to be subjected to **fat hardening** by hydrogenation. The oils occur naturally as glycerides, which are tri-esters of glycerol (1,2,3-trihydroxypropane) formed from long-chain fatty acids containing typically 16 or 18 carbon atoms. These long hydrocarbon chains may contain zero, one, two, or three *cis*-C=C bonds, and their partial hydrogenation, to the point where there is on average about one C=C bond per chain, affords a product having the desired physical characteristics and shelf-life of an alternative to butter. The names and structures of the principal fatty acids obtained by hydrolysis of glycerides are given in Table 6. It is necessary to lower the concentrations of the more unsaturated members to quite low levels, and fortunately this is readily achieved. What seems to happen is that by double bond migration the C=C bonds, which are initially nonconjugated, first assume the stabler conjugated configuration; the greater the number of C=C bonds, the more strongly is the molecule chemisorbed, so that linolenic acid residues are selectively hydrogenated in the initial stages. After conversion to the conjugated isolinoleic form, further hydrogenation of the dienes

Table 6 *Structures of some C_{18} carboxylic acids. Glycerides are tri-esters formed from glycerol (1,2,3-trihydroxypropane) and any combination of the carboxylic acids shown*

Trivial name	Structure
Linolenic acid	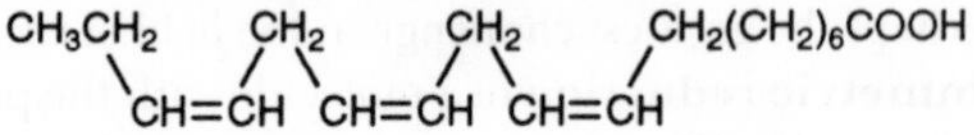 CH_3CH_2–CH=CH–CH_2–CH=CH–CH_2–CH=CH–$CH_2(CH_2)_6COOH$
Linoleic acid	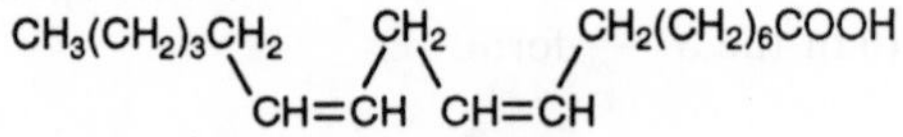 $CH_3(CH_2)_3CH_2$–CH=CH–CH_2–CH=CH–$CH_2(CH_2)_6COOH$
Oleic acid	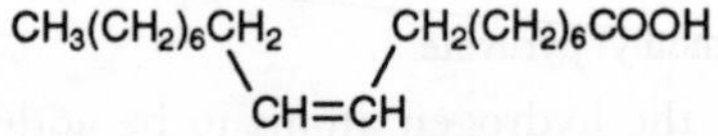 $CH_3(CH_2)_6CH_2$–CH=CH–$CH_2(CH_2)_6COOH$
Elaidic acid	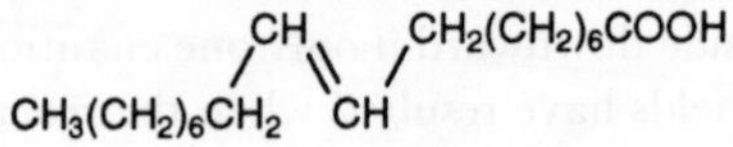 $CH_3(CH_2)_6CH_2$–CH=CH–$CH_2(CH_2)_6COOH$
Stearic acid	$CH_3(CH_2)_{16}COOH$

occurs, and when they are nearly all removed by conversion to oleic acid chains the reaction is stopped.

The process is conducted in large pressurized vessels at about 470 K in the batch mode; better control over the progress of the reaction is maintained if the supply of hydrogen to the surface is limited, so for this reason the process is deliberately operated under conditions of mass-transport limitation. The catalyst generally used is Ni/SiO_2, which luckily possesses all the necessary attributes. Small amounts of the *trans*-isomers of oleic and iso-oleic acid chains are formed, but this is acceptable because they help to raise the melting point, hence avoiding the need to continue the reaction until enough of the undesirable stearic acid is formed to obtain the required consistency. Copper and palladium catalysts have also been investigated, but have not so far found industrial use.

6 CONCLUSIONS: POTENTIAL FOR FURTHER APPLICATIONS OF HETEROGENEOUS CATALYSTS

Where do we go from here? If we first consider the needs of major industrial processes, we find that they are not necessarily for more *active* catalysts, because existing plant is usually designed to work within a fairly narrow temperature range, and vastly more active catalysts would necessitate a lower operating temperature, and hence the construction of a totally new plant. Nor is the principal requirement for materials of lower cost, because the contribution of the catalyst to total operating costs may be quite small (see Section 4.5). The two major aims in developing new catalysts for existing processes are (i) to secure a **longer lifetime**, with a slower rate of activity loss, and if possible the facility to regenerate *in situ*, and (ii) to obtain **higher yields** of the desired products and minimal formation of specific unwanted products.

There are sound economic reasons for these considerations. Loss of production time when a catalyst is being replaced, and to a lesser extent when it is being reactivated *in situ*, is expensive and adds to production costs. Misuse of valuable raw materials through their conversion to useless products is also costly, but in some processes it is the difficulty of separating a complex mixture of products which adds much to the cost of manufacture. These factors are well illustrated by the oxidation of *o*-xylene to phthalic anhydride (Table 3, page 74) where as much as 20% of the reactant is converted to carbon dioxide, and where the separation of trace amounts (~ 1%) of the intermediate phthalide and the by-product maleic anhydride from the phthalic anhydride involves unwelcome and expensive additional processing. The field of selective oxidation is one where advances in catalyst design are most urgently needed.

These considerations do not, however, apply with equal force to smaller scale batch processes of the type employed in fine chemicals manufacture. Here higher activity is more welcome, because it may allow more batches to be processed per shift, and because the catalyst contributes a greater fraction of the production cost than with large scale processes; with greater activity less is needed. Nevertheless product selectivity is still the factor of greatest importance. The development of catalysts showing an improved ability to descriminate between different reactive functions in complex organic molecules, *i.e.* **greater regiospecificity** (Section 5.6), represents a major challenge, as does the development of catalysts for **enantioselective processes** (see Chapter 6). While attention is presently focused on the use of selectively-poisoned bimetallic catalysts for the former, and of alkaloid-modified catalysts for the latter, the catalytic scientist will more and more seek to emulate the great specificity shown by enzymes, and will try to incorporate some of their essential features into the next generation of products. Indeed supported enzymes are already finding use, but it may be from the field of **supramolecular chemistry** that the next major advances in catalyst technology will come. Perhaps the fullerenes will have a role to play: they have already been tried as supports for noble metals.

The institution of a new chemical process of any size is a comparatively rare event, and only a small fraction of the exploratory and speculative work on radically new catalysts for new processes ever comes to fruition. One might hazard a guess that the area with greatest potential for application is that of **zeolites** and **pillared clays**; these materials catalyse a wealth of reactions, some of considerble complexity. It is not only the aluminosilicate zeolites that hold further promise; zeolitic frameworks containing substituted hetero-atoms such as vanadium and titanium exemplify important new concepts in catalyst design. The combination of a molecular sieving facility with that of reactive atoms or ions strategically located within the framework is one of potentially great power.

Of course anyone who could foretell the future would not spend his time writing chapters like this, and your guess is as good as mine. But whatever further directions catalysis takes, we may be sure of one thing: catalysis is here to stay.

7 REFERENCES

7.1 General Texts

The following titles are those of fairly general texts and monographs: some are clearly designed as undergraduate texts and others (marked *) are more advanced. Specialized research monographs are not cited, as these would constitute an impossibly long list.

G. C. Bond, 'Heterogeneous Catalysis: Principles and Applications', Oxford University Press, Oxford, 2nd edn, 1987.

B. C. Gates, 'Catalytic Chemistry', Wiley, New York, 1992.

I. M. Campbell, 'Catalysis at Surfaces', Chapman and Hall, London, 1988.

C. N. Satterfield, 'Heterogeneous Catalysis in Practice', McGraw-Hill, New York, 1980.

J. Oudar, 'Physics and Chemistry of Surfaces', Blackie, Glasgow, 1975.

J. W. Niemantsverdriet, 'Spectroscopy in Catalysis: An Introduction', VCH, Cambridge, UK, 1993.

*M. Boudart and G. Djega-Mariadassou, 'Kinetics of Heterogeneous Catalytic Reactions', Princeton University Press, Princeton, 1984.

*J. M. Thomas and W. J. Thomas, 'Principles of Heterogeneous Catalysis', Academic Press, London, 2nd edn, 1986.

*'Catalysis and Chemical Processes', ed. R. Pearce and W. R. Patterson, Blackie, Glasgow, 1981.

7.2 Specific References

There follows a short list of more specific references relating to the material presented above: they are mainly books or review articles. References to review articles appearing in the Specialist Periodical Report 'Catalysis' (published by The Royal Society of Chemistry, Cambridge, UK) are given as SPR 'Catalysis', with year, volume, and page numbers. Individual chapters or contributions to the books listed above are cited by the book's author or editors, plus page or chapter number.

Sections 2.1, 2.3, 3.1 and 3.2: Bond, ch. 1 and 2 (*vide supra*).

Section 2.2: Bond, ch. 2.

Sections 2.4 and 4.3: H. H. Kung, 'Transition Metal Oxides', Elsevier, Amsterdam, 1988.

Section 2.5: (a) Satterfield, ch. 7 (*vide supra*); (b) M. S. Spencer and T. V. Whittam, in SPR 'Catalysis', ed. C. Kemball and D. A. Dowden, 1980, vol. 3, p. 189; (c) S. Malinowski and M. Maczewski, in SPR 'Catalysis', ed. G. C. Bond and G. Webb, 1989, vol. 8, p. 107.

Section 2.6: (a) P. C. H. Mitchell, in SPR 'Catalysis', ed. C. Kemball, 1977, vol. 1, p. 204; (b) *ibid.*, ed. C. Kemball and D. A. Dowden, 1981, vol. 4, p. 175.

Section 4.2: (a) G. J. K. Acres, A. J. Bird, J. W. Jenkins, and F. King, in SPR 'Catalysis', ed. C. Kemball and D. A. Dowden, 1981, vol. 4, p. 1; (b) Satterfield, ch. 6; (c) Bond, ch. 7.

Section 4.4: (a) Satterfield, ch. 7; (b) A. Dyer, 'An Introduction to Zeolite Molecular Sieves', Wiley, Chichester, 1988.

Section 4.5: M. V. Twigg, in Pearce and Patterson, ch. 2 (*vide supra*).

Section 4.6: Satterfield, ch. 5.

Section 5.1: (a) P. E. Starkey, in Pearce and Patterson, ch. 3; (b) D. G. Bew, in Pearce and Patterson, ch. 4; (c) Satterfield, ch. 11.

Section 5.2: (a) Satterfield, ch. 11; (b) V. Ponec, in SPR 'Catalysis', ed. G. C. Bond and G. Webb, 1982, vol. 5, p. 48; (c) E. K. Poels and V. Ponec, *ibid.*, 1983, vol. 6, p. 196.

Section 5.3: Satterfield, ch. 9.

Section 5.4: I. R. Shannon, in SPR 'Catalysis', ed. C. Kemball and D. A. Dowden, 1978, vol. 2, p. 28.

Section 5.5: Satterfield, Ch. 8; C. F. Cullis and D. J. Hucknall, in SPR 'Catalysis', ed. G. C. Bond and G. Webb, 1982, vol. 5, p. 273: (b) W. R. Patterson, in Pearce and Patterson, ch. 11.

Section 5.6: (a) Bond, ch. 12; (b) J. M. Winterbottom, in Pearce and Patterson, ch. 12.

7.3 Literature Concerning Catalysis

There are a number of journals which are exclusively devoted to original work on catalysis; the more important are the following, the language used (if not English) being given in brackets:

Journal of Catalysis
Journal of Molecular Catalysis
Applied Catalysis A: General; and B: Environmental
Catalysis Letters
Kinetics and Catalysis Letters (mainly English)

Kinetika i Kataliz (Russian)
Cuihua Xuebao (Chinese, with English abstracts)
Shokubai (Japanese)

Publications mainly devoted to review articles include:

Advances in Catalysis
Catalysis Reviews – Science and Engineering
Catalysis Today
Specialist Periodical Reports: 'Catalysis'*

Complete coverage of the literature is obtained by perusing Chemical Abstracts, particularly the 'Catalysis – Applied and Physical Aspects' section. Significant original work also appears in the following journals:

Journal of the Chemical Society, Faraday Transactions
Journal of Physical Chemistry
Journal of the American Chemical Society
Surface Science (especially good)

Papers presented at many major conferences are also published in book form. The Faraday Discussions and Symposia enjoy a high reputation, and meetings concerned with catalysis were held in 1950, 1966, 1981, and 1989. The International Congress on Catalysis has been held every four years since 1956, most recently in 1992 at Budapest: all the proceedings have been published. Elsevier publishes a series of volumes under the general title 'Studies in Surface Science and Catalysis', many of which are conference proceedings, although a few are commissioned monographs; some 80 volumes in this series have now appeared.

* This is published approximately every one or two years in book form by The Royal Society of Chemistry, Cambridge, UK. The present editors are J. J. Spivey and S. K. Agarwal (volume 11, 1994).

CHAPTER 5

The Use of Inorganics in Large Scale Homogeneous Catalytic Processes

ADRIAN W. PARKINS

1 INTRODUCTION

There are many chemical reactions which although thermodynamically favourable do not occur spontaneously. The reaction of diamond with atmospheric oxygen is an example where the kinetic stability is useful, indeed valuable. The Chemical Industry is, of course, concerned with making reactions go, for example the oxidation of cyclohexane to cyclohexanone which is used as a nylon intermediate. Catalysis is the **acceleration of reactions** which are **thermodynamically favourable**. A catalyst is a substance which accelerates a chemical reaction without being consumed (see Chapter 4 for a general introduction to catalysis). The role of the catalyst designer is to devise a catalyst which is **selective** so that the reaction gives the required product. There are many types of catalysts, but in this chapter we focus our attention on soluble transition metal compounds which act **homogeneously** in solution, and which are involved in large scale manufacturing processes.

The function of a catalyst can be seen by reference to a free energy diagram which shows how the catalyst lowers the activation energy of a reaction. This is shown in Figure 1 of Chapter 4 (page 65).

Our understanding of the catalysis of organic reactions by transition metal complexes is based on **organometallic** and **coordination chemistry**. Organometallic chemistry has undergone enormous growth in the second half of the twentieth century, which has enabled the designer of a homogeneous catalyst to modify the coordination environment of the metal ion and so control the course of the reaction.

Table 1 Characteristics of homogeneous and heterogeneous catalysts

	Homogeneous	*Heterogenous*
Physical form	Soluble compound	Solid. Often metal or metal oxide on a support.
Utilization of catalyst material	All atoms available for reaction	Only surface atoms involved
Separation of Products	Can be difficult if involatile	Normally straightforward
Selectivity	Usually good	Variable
Temperature	Usually below 250 °C	Often high up to 1000 °C
Heat Transfer	No large temperature gradients	Local heating can lead to sintering

An excellent example[1] of this is the dimerization of propene catalysed by nickel phosphine complexes. This is shown in Equation 1.

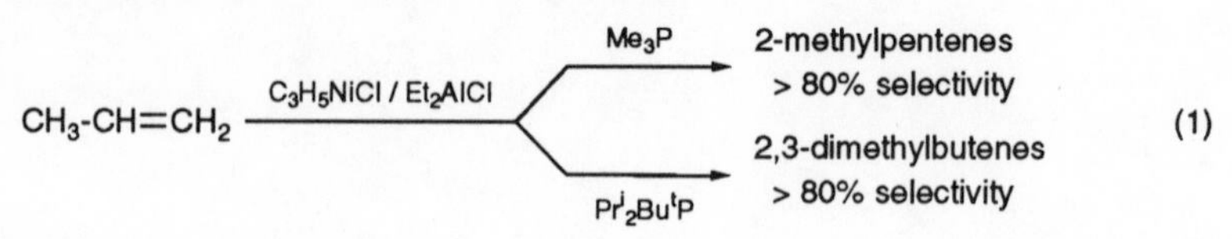

The ability to **tune the catalyst** is an advantage of a homogeneously catalysed reaction, and the reaction mechanisms are generally fairly well understood, but there are disadvantages as well and it is useful to compare the characteristics of homogeneous and heterogeneous catalysts. This is shown in Table 1.

2 CONCEPTS IN COORDINATION CHEMISTRY

The mechanisms of homogeneously catalysed reactions are usually described using the language of coordination chemistry. We need to introduce some concepts before we can look at the catalytic processes in detail.

2.1 Oxidation State and Electron Count

The oxidation state of an element is the resultant charge on the atom when the attached groups are removed in their closed shell configurations. The closed shell configuration of a non-transition element is usually the noble gas configuration. Some examples are shown in Table 2.

2.2 Oxidative Addition

Oxidative addition occurs when a transition metal complex undergoes oxidation with an increase in coordination number. The platinum group metals form a large number of square planar 16 electron complexes which provide some of the clearest examples of this type of reaction, but it is by no means restricted to them. Equation 2 shows the oxidative addition of iodomethane to a square planar Ir(I) complex giving an octahedral Ir(III) product.

Table 2 Electron counting in coordination compounds

Example 1 $CH_3Mn(CO)_5$
Oxidation state + 1
Electron count:

	number of electrons
Mn(+1)	6
5 × carbonyl groups	10
Methyl group, taken as CH_3^-	2
Total number of electrons in valence shell	18

Example 2 $[(CO)_2RhI_2]^-$
Oxidation state + 1
Electron count:

	number of electrons
Rh(+1)	8
2 × carbonyl groups	4
$2 \times I^-$	4
Total number of electrons in valence shell	16

Example 3 $(Ph_3P)_2Fe(CO)_3$
Oxidation state 0
Electron count:

	number of electrons
Fe(0)	8
3 × carbonyl groups	6
2 × triphenylphosphine	4
Total number of electrons in valence shell	18

$$trans\text{-}[Ir(CO)Cl(PMe_2Ph)_2] + CH_3I \longrightarrow [Ir(CH_3)I(CO)Cl(PMe_2Ph)_2] \quad (2)$$

oxidation state (+1) — number of electrons in valence shell = 16

oxidation state (+3) — number of electrons in valence shell = 18

2.3 Reductive Elimination

This is the reverse of oxidative addition, and so involves a reduction with a decrease in the coordination number. The final step in a catalytic cycle by which the products are obtained is often a reductive elimination. Equation 3 shows the elimination of acetyl iodide from a Rh(III) complex. This reaction occurs as a step in the rhodium catalysed carbonylation of methanol (see Section 3.1).

$$\text{[Rh(I)}_3\text{(CO)}_2\text{(COCH}_3\text{)]}^- \longrightarrow CH_3COI + \text{[RhI}_2\text{(CO)}_2\text{]}^- \quad (3)$$

oxidation state (+3)
number of electrons in valence shell = 18

oxidation state (+1)
number of electrons in valence shell = 16

2.4 Migratory Insertion

The reaction of $CH_3Mn(CO)_5$ with carbon monoxide leads to the acetyl compound $CH_3COMn(CO)_5$ (Equation 4).

$$CH_3Mn(CO)_5 + CO \rightarrow CH_3COMn(CO)_5 \quad (4)$$

A thorough stereochemical study of this reaction using isotopic labels showed that it occurs by the methyl group migrating onto a carbonyl already coordinated to the manganese. The incoming carbon monoxide becomes a metal carbonyl group, and does not form part of the acetyl group. The change in bond connectivity of the methyl–manganese bond is the insertion of carbon monoxide to give an acetyl group. For this reason, the reaction is known as **migratory insertion reaction**. In this case both the reactant and the product are stable six coordinate compounds, but in some catalytic cycles the migratory insertion may involve a change in the coordination number, and then the distinction between migration and insertion becomes less clear.

2.5 *β*-elimination

Metal alkyls which have a hydrogen atom in the *β*-position quite often undergo elimination of an alkene with the production of a metal hydride (Equation 5).

$$Bu_3PCuCH_2CD_2CH_2CH_3 \rightarrow Bu_3PCuD + CH_2{=}CDCH_2CH_3 \quad (5)$$

The mechanism is thought to involve an expansion of the coordination sphere as shown in Scheme 1.

The reaction is not restricted to alkyls, and there are several well established preparations of hydrides which involve the formation of an alkoxide intermediate. An example of this is the preparation of the catalyst, $HRh(CO)(PPh_3)_3$, which is used in the rhodium catalysed hydroformylation of alkenes (see Section 3.2, page 112). This compound can be made by reducing an ethanolic solution of rhodium

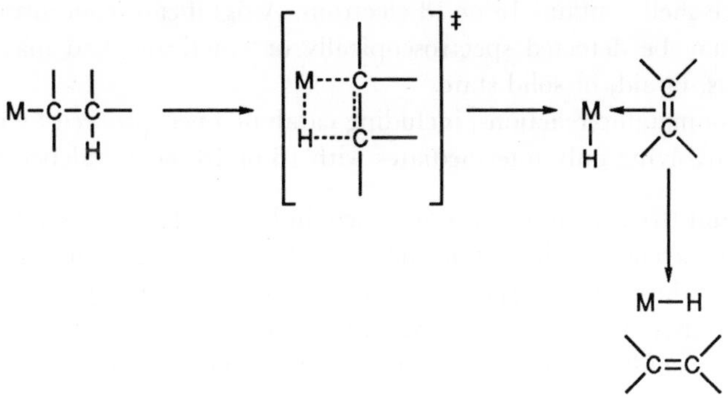

Scheme 1

trichloride in the presence of triphenylphosphine and sodium ethoxide. The hydride is formed by a β-elimination of the ethoxide as shown in Equation 6.

$$M\text{–}O\text{–}CH_2\text{–}CH_3 \rightarrow M\text{–}H + O{=}CHCH_3 \qquad (6)$$

2.6 Cone Angle

Steric effects are important in homogeneous catalysis and a measure of the steric requirement of a phosphorus ligand is the cone angle. This is measured using Corey Pauling Kolthun (CPK) space filling models assuming a metal phosphorus distance of 2.28 Å as shown in Figure 1.

3 CATALYTIC CYCLES AND THE 16 AND 18 ELECTRON RULE

Most of the commonly accepted mechanisms of reactions catalysed by transition metal complexes conform to the **16 and 18 electron rule**. The rule states that

1) 'Diamagnetic organometallic complexes of the transition metals may exist in a significant concentration at moderate temperatures only if the metals'

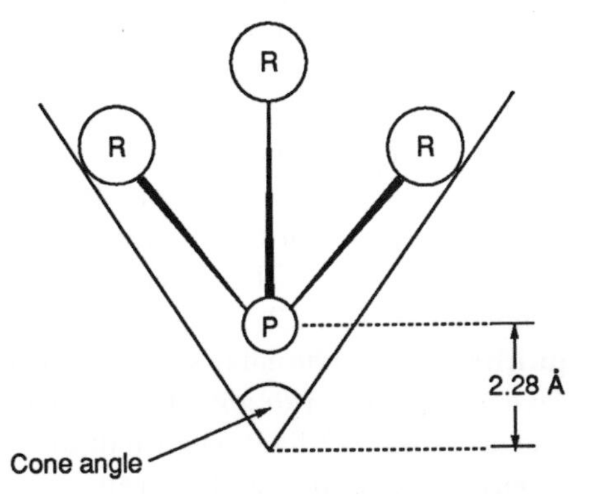

Figure 1 *Cone angle of a tertiary phosphine*

valence shell contains 16 or 18 electrons. A significant concentration is one that may be detected spectroscopically or kinetically and may be in the gaseous, liquid, or solid state.'

2) 'Organometallic reactions, including catalytic ones, proceed by elementary steps involving only intermediates with 16 or 18 metal valence electrons.'

The 16 and 18 electron rule was proposed[2] in 1972 and is very useful, but there is a growing awareness of the importance of odd electron species in organometallic processes. As the central purpose of catalysis is to provide reactive intermediates, some attention should be given to **paramagnetic species**. An illustration of how the electronic configuration can effect reactivity can be seen from the fact that the 17 electron compound $V(CO)_6$ undergoes substitution 10^{10} times more quickly than the isostructural 18 electron compound $Cr(CO)_6$.[3]

The transformation of the reactants into the products in a reaction involving a homogeneous catalyst is best understood from a **catalytic cycle**. This is a sequence of oxidative additions, migratory insertions, and reductive eliminations which form a closed loop by which the catalytic species is regenerated. The carbonylation of methanol using a rhodium iodide catalyst is a good example, because in this case the stereochemistry of the intermediates is known with reasonable certainty.

3.1 The Carbonylation of Methanol. The Monsanto Acetic Acid Process[4]

The carbonylation of methanol using a rhodium iodide catalyst provides a commercial route to acetic acid under relatively mild conditions, (30 atm/180 °C), and with high product selectivity. The overall reaction is shown in Equation 7.

$$CH_3OH + CO \rightarrow CH_3COOH \tag{7}$$

The process is generally known as the Monsanto acetic acid process and it was developed by that company, although the main operators now are British Petroleum. A similar process, which was developed by BASF, carries out the same reaction using a cobalt iodide catalyst but requires much more severe conditions (680 atm/250 °C). The catalytic cycle for the rhodium catalysed process is shown in Figure 2.

Looking at the individual steps in the cycle, A→B involves the oxidative addition of CH_3I to the square planar 16 electron complex A to give the octahedral 18 electron complex B. B is only present to the extent of about 1% of A under steady state conditions, and is unstable with respect to migratory insertion to give C, or reductive elimination to regenerate A. Strictly speaking most of the steps in a catalytic cycle are **equilibria**, but it is usual to show the reaction occurring in the thermodynamically favoured direction to give the product. C is coordinatively unsaturated: that is it has a vacant coordination site indicated by a small square. C is Rh(+3) but has only 16 electrons and so can become an 18 electron coordinatively saturated compound by coordination of a CO molecule to give D.

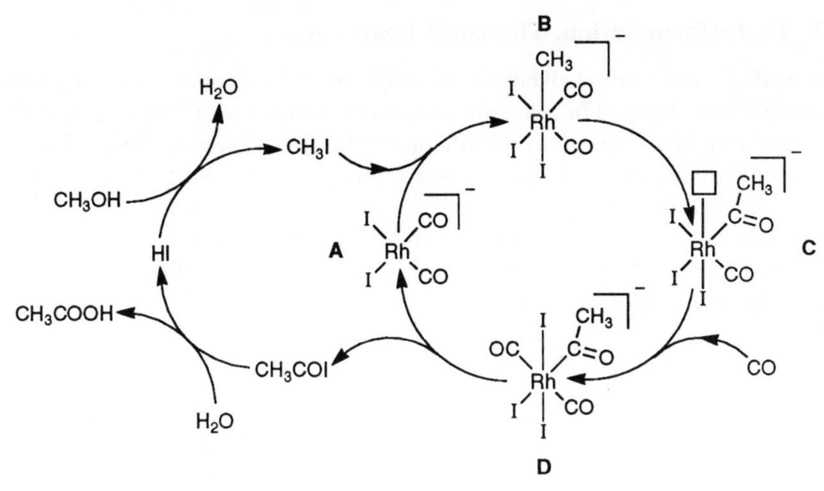

Figure 2 *Catalytic cycle for the Monsanto acetic acid process*

D undergoes **reductive elimination** to regenerate A, and gives acetyl iodide which enters the subsidiary hydrolysis/hydrogen iodide catalytic cycle.

The catalytic process used to carbonylate methanol can be modified to produce acetic anhydride from methyl acetate (Equation 8).

$$CH_3COOCH_3 + CO \rightarrow CH_3COOCOCH_3 \qquad (8)$$

The main difference in the catalytic cycle is the use of LiI as a cocatalyst which reacts with methyl acetate to give iodomethane and lithium acetate (Equation 9).

$$CH_3COOCH_3 + LiI \rightarrow CH_3I + CH_3COOLi \qquad (9)$$

The overall cycle is shown in Figure 3.

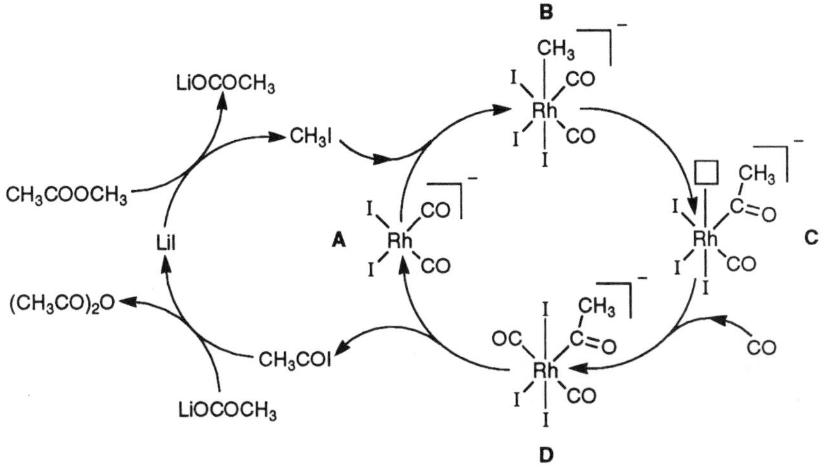

Figure 3 *Catalytic cycle for the carbonylation of methyl acetate to give acetic anhydride*

3.2 Hydroformylation. The OXO Reaction

The hydroformylation of alkenes is the addition of the elements of formaldehyde across a double bond. The reactants are carbon monoxide and hydrogen and the reaction can be catalysed by cobalt or rhodium compounds. The mixture of carbon monoxide and hydrogen is widely used in the Chemical Industry and is known as Synthesis Gas or 'Syngas'. Syngas is usually made by the steam reforming of hydrocarbons. The hydroformylation of alkenes can give two isomeric products, depending on the regiospecificity of the addition of M–H to the double bond, as shown in Equation 10.

$$R\text{-}CH{=}CH_2 + CO + H_2 \longrightarrow \begin{cases} R\text{-}CH(CH_3)\text{-}CHO \dashrightarrow \textit{iso}\text{-alcohol} & \text{Markovnikov addition} \\ R\text{-}CH_2\text{-}CH_2\text{-}CHO \dashrightarrow \textit{normal}\text{-alcohol} & \text{Anti-Markovnikov addition} \end{cases} \quad (10)$$

The reaction is a common way of introducing oxygen functions into alkenes derived from petroleum. The aldehydes are generally reduced to alcohols either directly or after aldol condensation. This allows the conversion of propene in 2-ethylhexanol which is used as the alcohol component of a plasticizer for PVC. The generally accepted mechanism for the cobalt catalysed system is shown in Figure 4. Typical operating conditions for cobalt are 200–300 atm at 130–170 °C.

Extensive studies of the stoichiometric reaction of $HCo(CO)_4$ with alkenes have been carried out in addition to the catalytic work. The stoichiometric studies show that the process involves paramagnetic intermediates, and these might be involved in the catalytic reaction.[5,6]

As mentioned above, one of the features of the hydroformylation reaction is that the addition can occur in two ways giving either the normal or iso-aldehyde. The structure of the final product depends on whether the addition of the metal hydride is Markovnikov or anti-Markovnikov. The normal aldehyde is the more desirable product for most commercial applications. The 'tuning' of homogeneous catalysts by modifying the coordination environment of the metal ion in the catalyst was described earlier. In the Shell hydroformylation process a tertiary phosphine is added to improve the normal/iso ratio, and the reaction is carried out at 50–100 atm and *ca.* 175 °C. Under these conditions some of the alkene is hydrogenated to the alkane.

An alternative approach to obtaining high normal/iso ratios involves the use of rhodium catalysts. In the Davy McKee – Johnson Matthey – Union Carbide process high normal/iso ratios are obtained under mild conditions (*ca.* 20 atm, 80–120 °C) by using triphenyl phosphine as both solvent and ligand in the initial complex which is $HRh(CO)(PPh_3)_3$. A simplified catalytic cycle is shown in Figure 5.

Recent studies[7] have shown that under hydroformylation conditions $HRh(CO)_2(PPh_3)_2$ may also be involved as an intermediate. This dicarbonyl

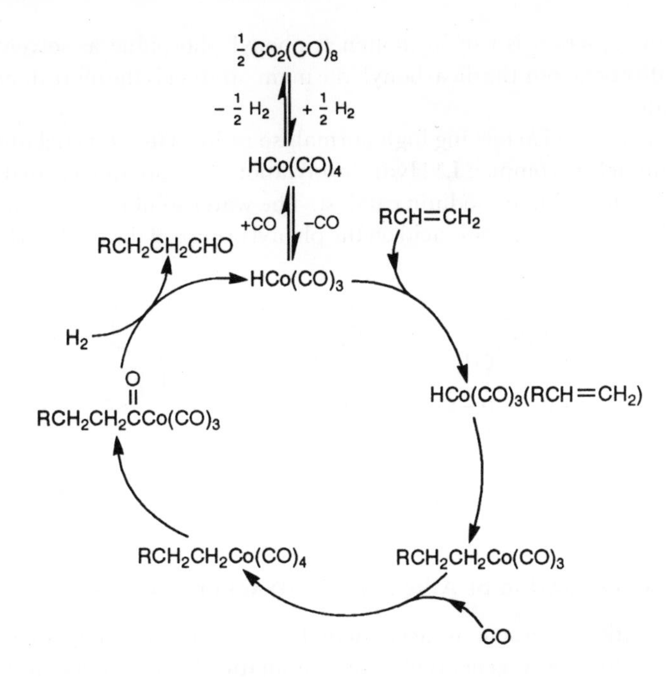

Figure 4 *Catalytic cycle for the cobalt catalysed hydroformylation of alkenes*

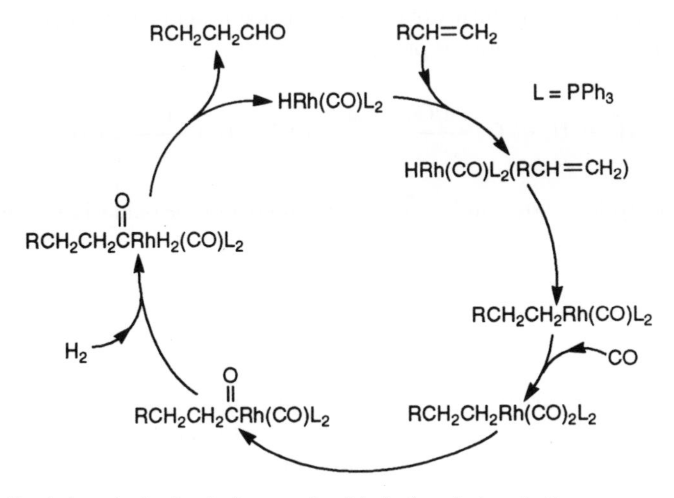

Figure 5 *Catalytic cycle for the rhodium catalysed hydroformylation of alkenes*

dissociates to give the 16 electron complex $HRh(CO)_2(PPh_3)$ which adds to the alkene with poor regioselectivity, leading to both Markovnikov and anti-Markovnikov products. The monocarbonyl, $HRh(CO)(PPh_3)_3$, dissociates to $HRh(CO)(PPh_3)_2$ which gives mainly anti-Markovnikov addition, as shown in the catalytic cycle, and so leads to the normal aldehyde. This helps explain why, when the reaction is run in molten triphenyl phosphine as solvent, conditions which do not favour the dicarbonyl, the main product is the more desirable normal aldehyde.

Another way of achieving high normal/iso ratios is to use a chelating phosphine in the dinuclear complex **I**.[8] Hydroformylation has also been carried out in water using a water-soluble rhodium catalyst. The water solubility is achieved by using the sodium salt of a sulfonic acid on the phenyl groups of the triphenylphosphine **II**.

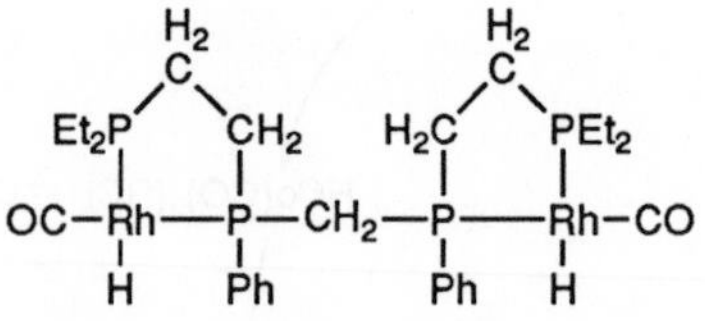

I 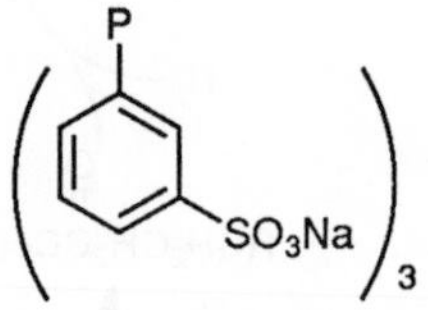 **II**

3.3 The Oxidation of Alkenes. The Wacker Process

The oxidation of ethene to acetaldehyde can be carried out using a palladium catalyst which is regenerated, not automatically as occurs in the processes described so far, but by linking the palladium/ethene cycle to a **copper based redox system**. The acetaldehyde produced may be oxidized further to acetic acid (Equation 11), but the use of this route to acetic acid is in decline following the introduction of the Monsanto process.

$$CH_2{=}CH_2 + \tfrac{1}{2}\,O_2 \xrightarrow{\text{Pd/Cu catalyst}} CH_3CHO \xrightarrow{[O]} CH_3COOH \qquad (11)$$

A catalytic cycle is shown in Figure 6. The kinetics of the oxidation conform to the equation

$$\frac{-d[C_2H_4]}{dt} = \frac{k[PdCl_4^{2-}][C_2H_4]}{[H_3O^+][Cl^-]^2}$$

and the inverse square dependence of the chloride ion concentration means that there are two consecutive equilibria (Equations 12 and 13) which determine the concentration of the complex $[C_2H_4PdCl_2OH_2]$:

$$[PdCl_4]^{2-} + C_2H_4 \rightleftharpoons [C_2H_4PdCl_3]^- + Cl^- \qquad (12)$$

$$[C_2H_4PdCl_3]^- + H_2O \rightleftharpoons [C_2H_4PdCl_2OH_2] + Cl^- \qquad (13)$$

Opinion is divided as to how the next step occurs. One possibility is that the complex $[C_2H_4PdCl_2OH_2]$ loses a proton and then rearranges within the coordination sphere to give a hydroxyethyl complex, the formation of which is the rate determining step (Equation 14).

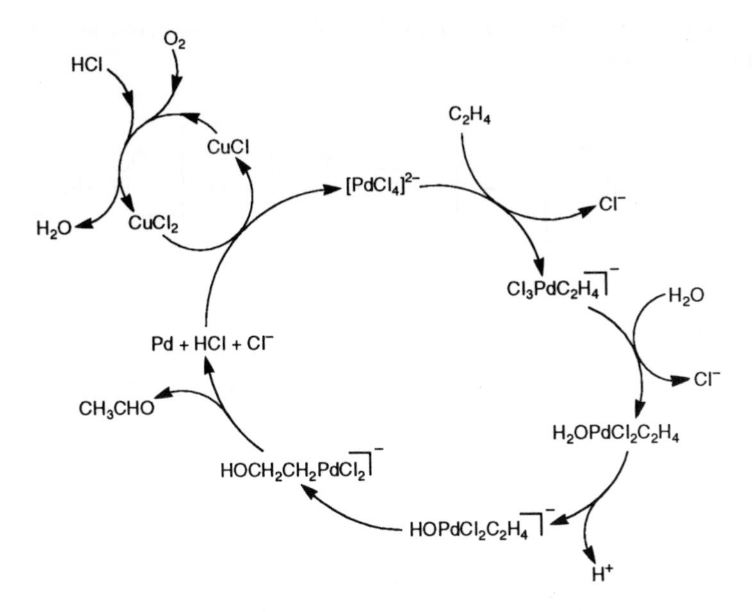

Figure 6 *Catalytic cycle for the Wacker process for the oxidation of ethene*

$$C_2H_4PdCl_2OH_2 \xrightarrow{-H^+} [C_2H_4PdCl_2OH]^- \rightarrow [HOCH_2CH_2PdCl_2]^- \quad (14)$$

Alternatively, nucleophilic attack on the coordinated ethene by a water molecule from outside the coordination sphere produces a four-coordinate hydroxyethyl compound, which loses chloride in the rate determining step (Equation 15).

$$C_2H_4PdCl_2OH_2 \xrightarrow{+OH^-} [HOCH_2CH_2PdCl_2OH_2]^- \xrightarrow{-Cl^-} HOCH_2CH_2PdClOH_2 \quad (15)$$

In both cases the hydroxyethyl compound rearranges to acetaldehyde. When the reaction is carried out in D_2O no deuterium finishes up in the acetaldehyde, and when C_2D_4 is used in H_2O the product is CD_3CDO. The mechanism given in Scheme 2 shows how the hydrogen atoms in the acetaldehyde may be obtained from the original ethene.

Evidence in favour of external attack was obtained by determining the stereochemistry of the chloroethanol obtained from *trans*-1,2-dideuteroethene at very high chloride concentration. The chloroethanol was converted into ethylene oxide which was found to be *cis* by microwave spectroscopy. The cleavage of the Pd–C bond by $CuCl_2$ and the conversion of the chloroethanol into ethylene oxide are known to occur with inversion of configuration, which implies a *threo* configuration for the chloroethanol and external attack on the ethene complex, as shown in Scheme 3.[9] However, recent work has established that the studies at high chloride ion concentration are not a valid indication of the stereochemistry under **Wacker** conditions.[10]

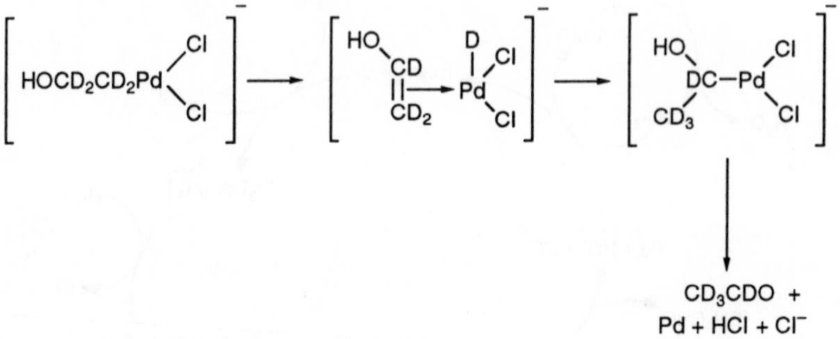

$$\left[HOCD_2CD_2PdCl_2 \right]^- \longrightarrow \left[HOCD{=}CD_2 \rightarrow PdDCl_2 \right]^- \longrightarrow \left[HO(CD_3)DC{-}PdCl_2 \right]^-$$

$$\downarrow$$

$$CD_3CDO + Pd + HCl + Cl^-$$

Scheme 2

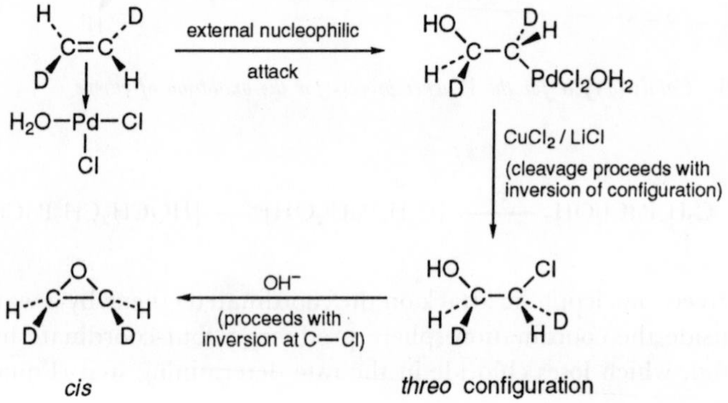

external nucleophilic attack

$PdCl_2OH_2$

$CuCl_2$ / LiCl (cleavage proceeds with inversion of configuration)

OH^- (proceeds with inversion at C—Cl)

cis

threo configuration

Scheme 3

One aspect of the Wacker mechanism which has not received much attention is that ethene has a high *trans* effect and so replacement of chloride in $[C_2H_4PdCl_3]^-$ to give $[C_2H_4PdCl_2OH_2]$ would be expected to give the *trans* complex. Most treatments which show the geometry of the complex show it as *cis*, but a *trans* configuration leading to the hydroxyethyl compound via a tetrahedral intermediate is another possibility.

The regiospecificity of the addition of nucleophiles to substituted alkenes is another interesting topic, which is related to the Wacker oxidation of alkenes. The experimental observation in this case is that the product of Wacker oxidation of a terminal alkene is the methyl ketone rather than the aldehyde.[11] Calculations by Hoffmann suggest that alkenes are activated towards nucleophilic attack when slippage occurs. The slippage causes the LUMO of the alkene to become localized on C2 rather than the terminal carbon atom. The bond between the incoming nucleophile and the alkene is thus made preferentially on C2, as shown in Scheme 4.[12]

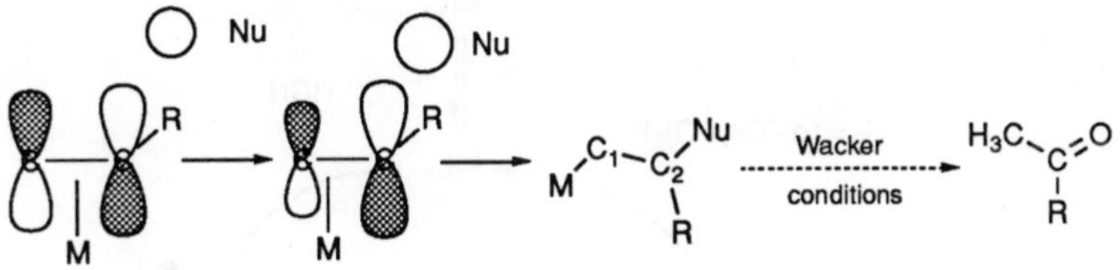

Scheme 4

If acetic acid is used as the solvent for the Wacker reaction with ethene, the product is vinyl acetate (Equation 16). This can be operated as a gas phase process with a heterogeneous palladium catalyst.

$$CH_2{=}CH_2 \xrightarrow[\text{acetic acid}]{O_2,\ Pd/Cu} CH_2{=}CHOCOCH_3 \qquad (16)$$

3.4 Hydrocyanation of Butadiene[13]

Adiponitrile $NC(CH_2)_4CN$ is an important nylon intermediate as it can be hydrogenated to give hexamethylenediamine, which gives Nylon 6,6 on reaction with adipic acid.

Dupont have developed a nickel catalysed process which brings about the double anti-Markovnikov addition of HCN to butadiene.

$$CH_2{=}CH{-}CH{=}CH_2 + 2\ HCN \rightarrow NC(CH_2)_4CN \qquad (17)$$

The reaction is carried out in three steps

1) addition of the first molecule of HCN
2) isomerization of the double bond
3) addition of the second molecule of HCN

As we have seen, the use of phosphine ligands can bring about very useful changes in the regioselectivity of catalysts and in this case the formation a π-allyl intermediate also influences the regiospecificity. A catalytic cycle for the addition of the first molecule of HCN is shown in Figure 7. The ligand in the catalyst in this case is an aryl phosphite, $(ArO)_3P$, rather than a phosphine, and this is another example where the catalyst was fine tuned by changing the substituents on phosphorus.

o-Tolyl phosphite has a large cone angle (141°) and so the fourth ligand in NiL_4 is held very weakly, giving relatively large concentrations of the reactive 16 electron complex NiL_3. The main product of the first addition is 3-pentenenitrile, which is isomerized to 4-pentenenitrile using a cationic nickel hydride. Fortunately 3-pentenenitrile isomerizes to 4-pentenenitrile, more quickly than to the thermodynamically more favourable 2-pentenenitrile (Equation 18).

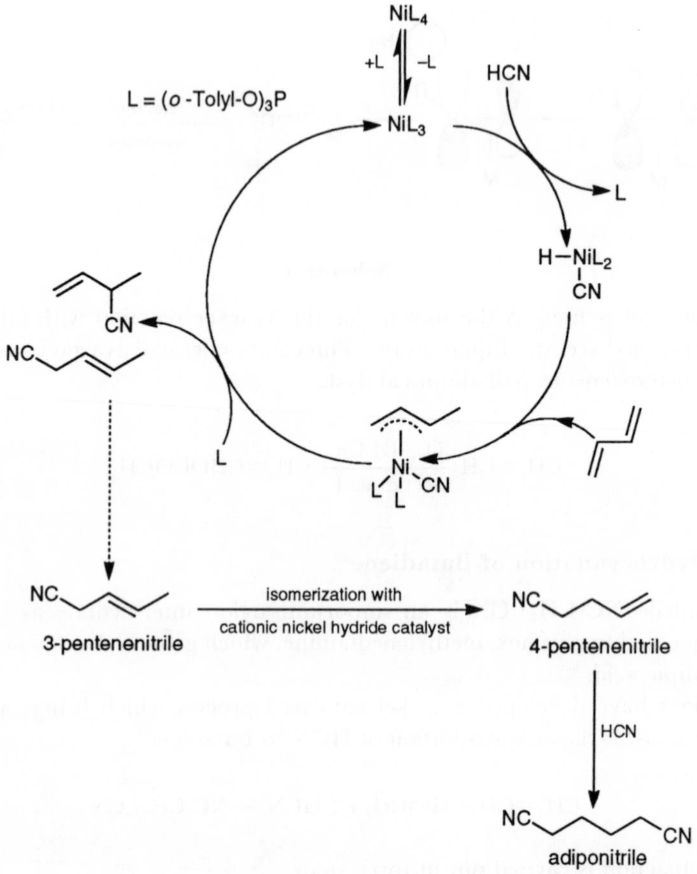

Figure 7 *Catalytic cycle for the hydrocyanation of butadiene*

$$CH_3.CH{=}CH.CH_2CN \xrightarrow{\text{cationic nickel hydride catalyst}} CH_2{=}CH.CH_2CH_2CN \quad (18)$$

The addition of the second molecule of HCN is again catalysed by a nickel aryl phosphite complex. The second addition proceeds much more quickly if a Lewis acid is present in addition to the nickel complex. Triphenyl borane was found to give high selectivity to adiponitrile in the second addition (Equation 19).

$$CH_2{=}CH.CH_2CH_2CN + HCN \xrightarrow{NiL_4/Ph_3B} NC(CH_2)_4CN \quad (19)$$

3.5 Oxidation of Saturated Hydrocarbons and Alkyl Benzenes

There are two very important manufacturing processes involving the oxidation of saturated hydrocarbyl groups. These are the oxidation of cyclohexane to a mixture of cyclohexanone and cyclohexanol (Scheme 5), and the oxidation of *p*-xylene to terephthalic acid (Scheme 6).

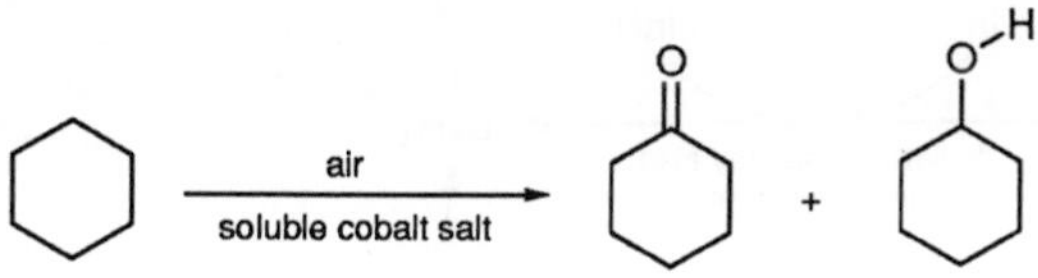

air
soluble cobalt salt

Scheme 5

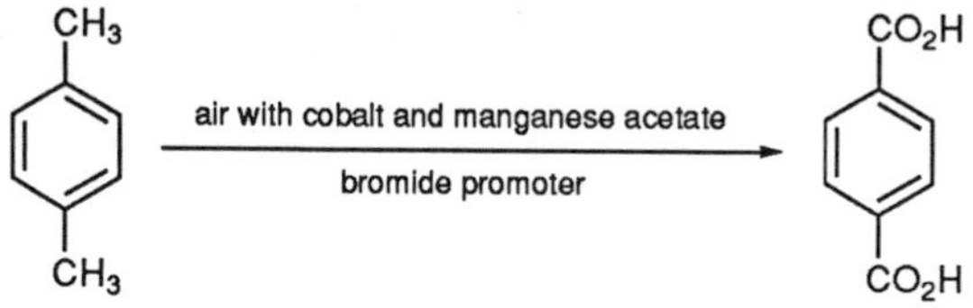

CH_3 CH_3
air with cobalt and manganese acetate
bromide promoter
CO_2H CO_2H

Scheme 6

Acetic acid is also manufactured by hydrocarbon oxidation, using either butane or naphtha as feedstock. The reaction is catalysed by transition metal ions, but uncatalysed operation is also known.

Cyclohexanone is an intermediate for caprolactam, which is polymerized to Nylon 6. Terephthalic acid is esterified with ethylene glycol to give polyester for use in fibre production. It will be obvious that if long chain polymers are to result from the terephthalic acid, then there must be complete oxidation of all the methyl groups, because if only one of the methyl groups is oxidized giving *p*-toluic acid, this could only act as an end group for the polymer chain. The use of bromide is essential in giving high purity di-acid, and in the Amoco process a mixture of cobalt and manganese acetates is used together with bromide.

A catalytic cycle for the oxidation of cyclohexane to a mixture of cyclohexanone and cyclohexanol, which is typically carried out at 10 atm/150°, is shown in Figure 8. The main role of the catalyst is to accelerate the decomposition of hydroperoxides.[14] Although this oxidation has been used for many years, there is renewed interest in the functionalization of saturated hydrocarbons using iron based systems.[15]

3.6 Other Reactions

There are other reactions involving homogeneous catalysis which are used on a large scale in the chemical industry. An interesting process is the **Shell Higher Olefin process** (SHOP) which involves catalytic oligomerization of ethene, followed by isomerization and alkene metathesis. Very little information is available in the open literature about the metathesis step, although it is probably heterogeneously catalysed.

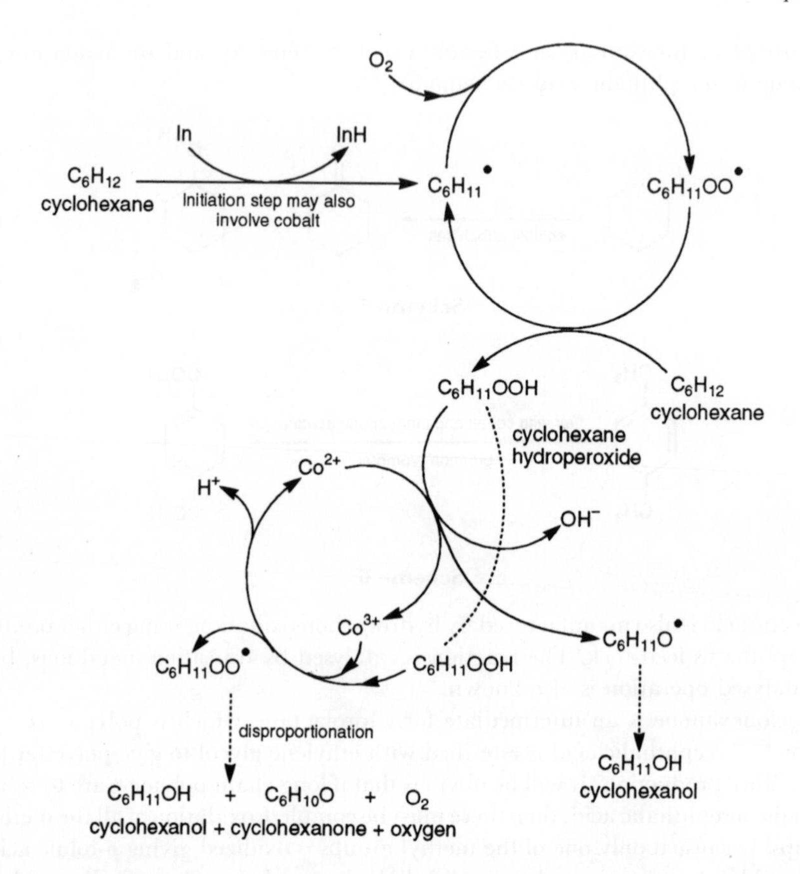

Figure 8 *Catalytic cycle for the oxidation of cyclohexane*

4 CONCLUSIONS AND POTENTIAL FOR FUTURE DEVELOPMENT OF USEFUL APPLICATIONS FOR HOMOGENEOUS REACTIONS

This survey has described large scale industrial processes and several of these have been in operation for many years and use a petroleum based feedstock. If petroleum becomes scarce or expensive, other processes can be expected to be developed. **Synthesis gas** can be obtained from coke and steam, and so processes which use synthesis gas may become even more important. Synthesis gas is readily converted into methanol using a heterogeneous catalyst (see Chapter 4, Section 5.2), and so can be converted to acetic acid using the Monsanto process. Tennessee Eastman already operate a plant producing acetic anhydride from coal.

The possibility of producing other industrial chemicals containing two carbon atoms from synthesis gas is an active research area. A process for the production of ethylene glycol from syngas has been developed by a consortium of four Japanese companies (Equation 20).

$$2\,CO + 3\,H_2 \longrightarrow HOCH_2CH_2OH \qquad (20)$$

The catalytic system is based on either rhodium or ruthenium, with a complex cocktail of additives in a mixed solvent system. Once again it was found that phosphines with a high cone angle were beneficial in giving good selectivity.[16] Earlier work at Union Carbide had suggested that rhodium carbonyl clusters were involved in ethylene glycol formation, but this now seems doubtful.

In the future the greatest impact of homogeneous catalysis will probably be in **fine chemicals** and **pharmaceuticals**. Three recent surveys of newly introduced catalytic processes give many interesting examples which demonstrate the extremely wide range of reactions for which catalysts are available[17–19] (see also Chapter 6).

5 REFERENCES

5.1 Specific References Cited in the Text

1. W. Keim, *Angew. Chem., Int. Ed. Engl.*, 1990, **29**, 235.
2. C. A. Tolman, *Chem. Soc. Rev.*, 1972, **1**, 337.
3. Qi-Z. Shi, T. G. Richmond, W. C. Trogler, and F. Basolo, *J. Am. Chem. Soc.*, 1982, **104**, 4032.
4. A. Haynes, B. E. Mann, G. E. Morris, and P. M. Maitlis, *J. Am. Chem. Soc.*, 1993, **115**, 4093.
5. M. Orchin, *Catal. Rev.*, 1984, **26**, 59.
6. T. L. Brown, *Ann. New York Acad. Sci.*, 1980, **333**, 80.
7. J. M. Brown and A. G. Kent, *J. Chem. Soc., Perkin Trans. 2*, 1987, 1597.
8. M. E. Broussard, B. Juma, S. G. Train, W.-J. Peng, S. A. Laneman, and G. G. Stanley, *Science*, 1993, **260**, 1784.
9. J. E. Backvall, B. Akermark, and S. O. Ljunggren, *J. Am. Chem. Soc.*, 1979, **101**, 2411.
10. J. W. Francis and P. M. Henry, *Organometallics*, 1992, **11**, 2832.
11. J. Tsuji, *Synthesis*, 1984, 369.
12. A. D. Cameron, V. H. Smith and M. C. Baird, *J. Chem. Soc., Dalton Trans.*, 1988, 1037.
13. C. A. Tolman, *J. Chem. Educ.*, 1986, **63**, 199.
14. L. Saussine, E. Brazi, A. Robine, H. Mimoun, J. Fischer, and R. Weiss, *J. Am. Chem. Soc.*, 1985, **107**, 3534.
15. D. H. R. Barton and D. Doller, *Acc. Chem. Res.*, 1992, **25**, 504.
16. S. Nakamura, *Chemtech*, 1990, **20**, 556.
17. M. Misono and N. Nojiri, *Appl. Catal.*, 1990, **64**, 1.
18. J. N. Armor, *Appl. Catal.*, 1991, **78**, 141.
19. N. Nojiri and M. Misono, *Appl. Catal.*, 1993, **93A**, 103.

5.2 General

1. G. W. Parshall and S. D. Ittel, 'Homogeneous Catalysis', Wiley, New York, 2nd edn, 1992.
2. B. C. Gates, 'Catalytic Chemistry', Wiley, New York, 1992.

3. 'Homogeneous Catalysis with Metal Phosphine Complexes', ed. L. H. Pignolet, Plenum Press, New York, 1983.
4. C. Masters, 'Homogeneous Transition Metal Catalysis', Chapman and Hall, London, 1981.

5.3 Core Journals

In addition to the catalysis journals indicated in the literature section of Chapter 4, articles on homogeneous catalysis occur in a wide spread of journals including:

Journal of the Chemical Society, Dalton and Perkin Transactions
Journal of the American Chemical Society
Angewandte Chemie (including the International Edition in English)
Organometallics
Synthesis
Accounts of Chemical Research
Journal of Molecular Catalysis
Journal of Organometallic Chemistry

CHAPTER 6

Catalysts for Stereospecific Synthesis

JOHN M. BROWN

1 INTRODUCTION

The synthesis of **enantiomerically pure compounds** is of increasing importance to the fine chemical and pharmaceutical industries, driven both by the demands of regulatory agencies and the desire of the end-user for **high purity** and **quality**. In part, this industrial need is being met by **asymmetric synthesis**, and has provided an impetus for high levels of academic activity worldwide.[1] From the standpoints of scale-up, controllability, and disposal of side-products, catalytic reactions are highly desirable, and have been utilized in specific industrial applications since the mid-1970s. As things stand at present, several examples of the direct utilization of metal complexes in catalytic asymmetric synthesis are recorded, and others appear to be imminent.

This Chapter provides a discussion of key examples in the context of actual or potential industrial application and attempts a rationalization of the underlying chemistry, where there is sufficient basis. The emphasis will be catalytic reduction and oxidation reactions, and it will be seen that the protocols for the two are rather distinct; much important **reductive chemistry** involves the **later transition metals** (especially rhodium and ruthenium) whereas **oxidation chemistry** tends to be associated more with the **early transition metals**, with important examples in both titanium and manganese chemistry. There is a significant exception in each class, which will be delineated.

2 CATALYTIC REDUCTIONS

2.1 Asymmetric Hydrogenation

Catalytic hydrogenation is one of the oldest successful asymmetric reactions, following the basis laid by the early discoveries of Wilkinson and coworkers for an effective protocol for homogeneous hydrogenation.[2] The mechanism of this archetypal rhodium-complex catalysed process was established over several years,

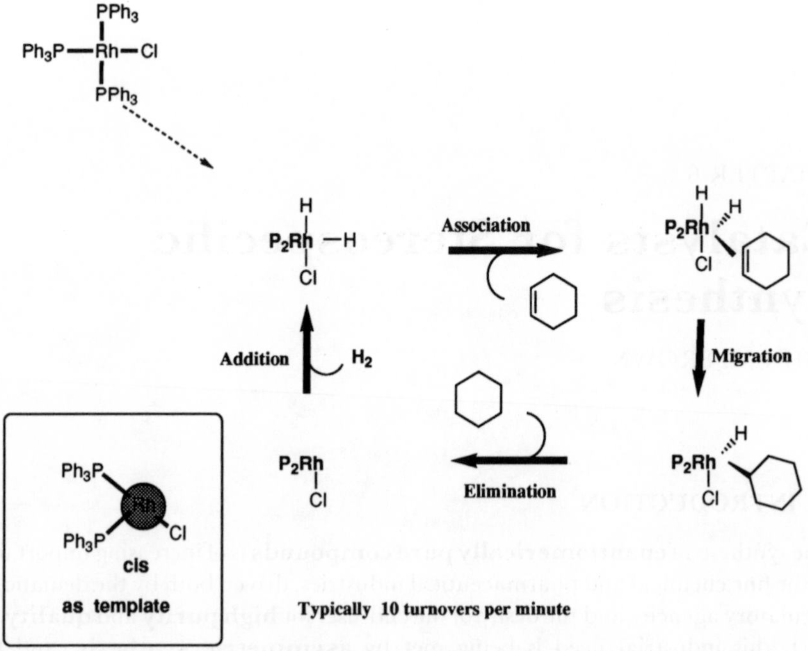

Scheme 1 *Mechanism of action of Wilkinson's catalyst*

largely through the work of Halpern, and a mechanistic consensus exists for the catalytic cycle, as shown in Scheme 1. In this, a *cis*-biphosphine **rhodium** fragment acts as a **template** which controls the coordination and reaction of the alkene and dihydrogen, through a series of intermediates which are coordinatively unsaturated, being either 2 or 4 electrons short of the next highest inert gas structure (see Chapter 5, Section 2.1).

The high turnover and quantitative yields challenged a number of workers to make chiral phosphines and diphosphines, and apply them in a potentially enantioselective variant of homogeneous hydrogenation with **prochiral alkenes.** [Prochiral alkenes generate a new stereogenic centre on hydrogenation, the minimum condition being that one terminus bears two different non-hydrogen substituents, and a mirror plane orthogonal to the central atomic plane is lacking]. Suffice it to say that the challenge was successfully met by 1972, that there is a multiplicity of ligands which can give significant **enantiomer excesses** (e.e.) in asymmetric hydrogenation, and that the chemistry now forms the basis of at least three large scale processes in the fine chemicals industry. These and related examples of potential commercial importance will be dealt with in turn, with the oldest treated first.

2.1.1 The Monsanto Process for the Production of L-DOPA. The Monsanto process for the production of **L-DOPA** (a primary drug in the treatment of **Parkinson's disease**) has been in operation since the mid-1970s, and embodies the key features

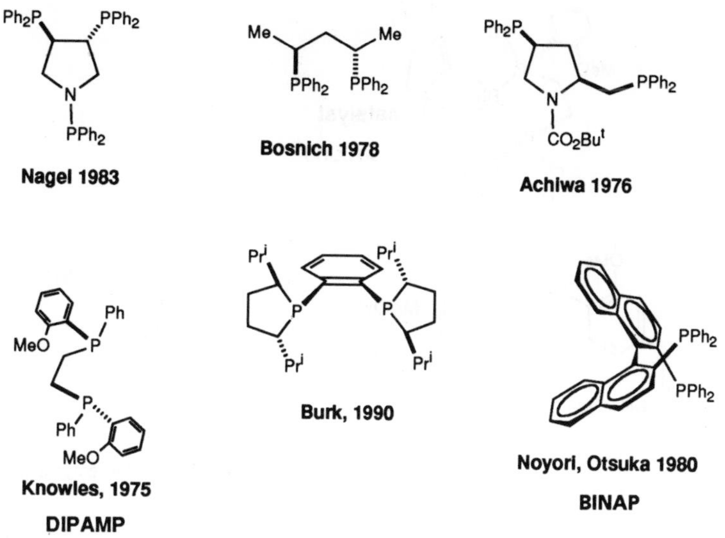

Scheme 2 *Some asymmetric ligands introduced between 1970 and 1990*

of rhodium asymmetric catalysis. Firstly, the ligand is a **chelating diphosphine**. Because of the stimulus brought about by this and other successful applications of asymmetric catalysis, well over two hundred examples of this type of ligand have been synthesized;[3] a variety of the more successful ones are collected in Scheme 2. For the Monsanto chemistry, a P-chiral diphosphine DIPAMP is employed. Although highly effective, its synthesis has proved rather difficult to emulate until quite recently, and other suitable ligands possess charility in the interphosphorus backbone rather than at the phosphorus atom itself.

The synthesis of L-DOPA by asymmetric hydrogenation is straightforward, but perhaps fortuitously effective. The reactant for this catalytic process is a dehydroamino acid derivative (Scheme 3) and both it and the reduced product bear protecting groups in the amide and catechol moieties. The hydrogenation product can be converted into the enatiomerically pure drug by recrystallization of the reduced product followed by removal of these protecting groups.[4] The hydrogenation reaction is fast and quantitative under very mild conditions, and permits *tens of thousands of turnovers* before the catalyst needs to be replenished – important factors when commercial viability is to be considered.

The novelty, intrinsic importance, and not least the **very high enantioselectivity** observed, made this an important area for mechanistic study. The questions raised have largely been resolved by the work of Halpern and Brown and their respective coworkers through a combination of X-ray analyses, NMR studies *in situ*, and kinetics; the following pathway is believed to occur.[5]

Thus the resting state of the catalyst is a coordinated enamide in which the alkene and amide carbonyl group are both bound to the rhodium as a *cis*-chelate, which is dynamic – that is to say, the reactant associates and dissociates readily on the timescale of catalysis. Because the alkene is prochiral and the ligand

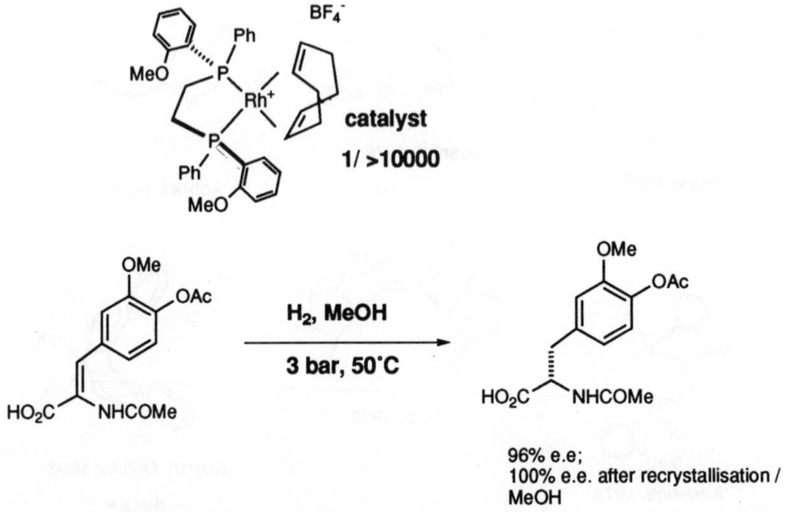

Scheme 3 *The key to the Monsanto* L-*DOPA process*

enantiomerically pure, there are two possible complexes depending on whether the olefin binds to rhodium through its *pro-R* or *pro-S* face. These interconvert through both intra- and intermolecular pathways, and at ambient pressure this process is much faster than the rate of hydrogen addition. Hydrogen addition first gives a dihydride which cannot be observed under any conditions since it rearranges rapidly to an alkylrhodium hydride which can be characterized in solution below −50 °C. The most surprising feature to emerge from mechanistic studies is that the minor diastereomer apparently carries the flux of catalysis, implying that the transition state for the addition step has a structure far removed from the square planarity of the ground state (Scheme 4).

2.1.2 Preparation of Amino Acids and Other Species Related to Natural Products. Similar catalytic chemistry is being utilized for the chemical synthesis of L-phenylalanine, an intermediate in the preparation of the dipeptide artificial sweetener **Aspartame** [L-Ala-L-Glu]. When the scientific and commercial importance of amino acids is considered, it is surprising that there are not more examples of this particular kind of catalytic chemistry in industrial usage. One drawback may well be that the enantiomer excess is 92–99% with the best catalysts and the most commonly accessible dehydroamino acids. But the configuration of the dehydroamino acid is important, since in most cases the *Z*-isomer is both more reactive and reduced with higher enantioselectivity than the *E*-isomer. Most syntheses of aliphatic dehydroamino acids are non-stereospecific and isolation of the pure diastereomers can require careful purification, hence limiting their use as reactants in asymmetric hydrogenation.

For this reason ligands based on the bis-phospholane, developed by Burk at Dupont[6] and shown in Scheme 5, are likely to be widely adopted. Among catalysts for asymmetric hydrogenation of dehydroamino acids, they provide the most

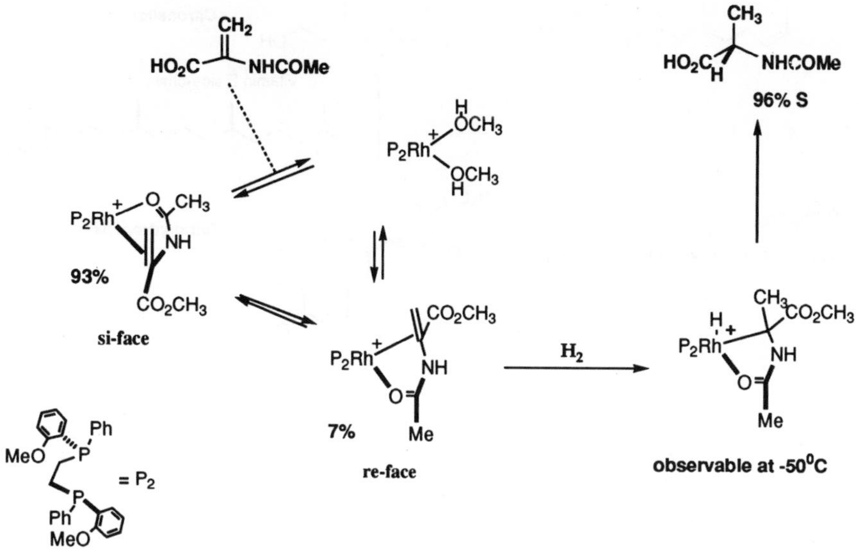

Scheme 4 *Mechanism of Rh-catalysed homogeneous hydrogenation*

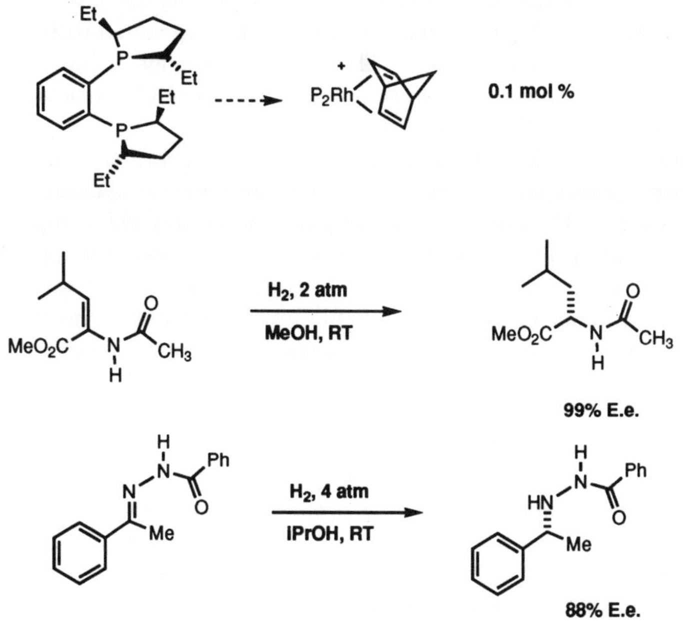

Scheme 5 *Asymmetric hydrogenation by Rh phospholane complexes*

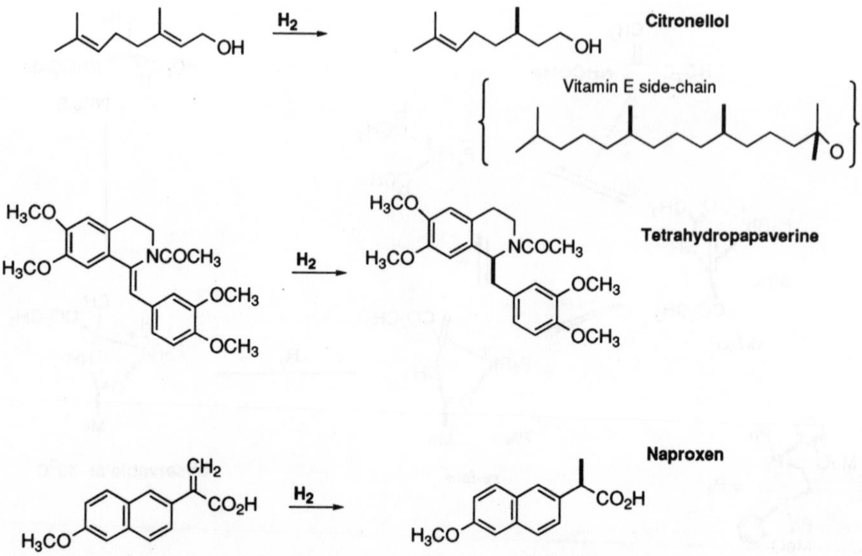

Citronellol

Vitamin E side-chain

Tetrahydropapaverine

Naproxen

Scheme 6 *Hydrogenations with Ru[(*R*)-BINAP]; all with >98% e.e.*

impressive results to date in terms of the intrinsic reactivity and enantioselectivity. In addition, the *E*- and *Z*-diastereomers are both hydrogenated with comparable e.e., to the same hand of product. This implies, although it has not yet been formally tested, that an *E*/*Z* mixture would be an acceptable reactant, and still give a product of 98–99% enantiomeric purity.

The latter result aside, rhodium catalysts have been rather eclipsed by their **ruthenium** counterparts in recent studies of asymmetric hydrogenation. The central advances were utilization of the atropisomerically chiral ligand BINAP (Scheme 2), and the discovery of 1:1 ruthenium complexes of this ligand suitable as catalyst precursors [*e.g.* $P_2Ru(OAc)_2$] – this latter advance was non-trivial since the coordination chemistry of ruthenium with chelating diphosphines was at that time seriously underdeveloped.[7] An impressive array of prochiral alkenes is hydrogenated in high enantiomer excess by this catalyst system. The first example (shown in Scheme 6), demonstrates the chemo- and enantiospecific synthesis of 2,3-dihydrogeraniol (citronellol) by reduction of the trisubstituted double bond of geraniol in proximity to the hydroxyl group; the double bond isomer nerol gives the opposite enantiomer of the same product. Attempting the corresponding rhodium chemistry leads to poor e.e.s. and correspondingly low reactivity. Related Ru(BINAP) asymmetric hydrogenation can be used in a synthesis of the side chain of tocopherol (inset), and hence affords the possibility of a commercial synthesis of **Vitamin E** in enantiomerically pure form.

The second example in Scheme 6 offers an enantioselective route to **isoquinoline alkaloids**, *e.g.* tetrahydropapaverine, in which an exocyclic enamide precursor is hydrogenated. Just as in the dehydroamino acid case, the amide oxygen acts as a clasp which helps bind the substrate to the metal. The final example in Scheme 6 is

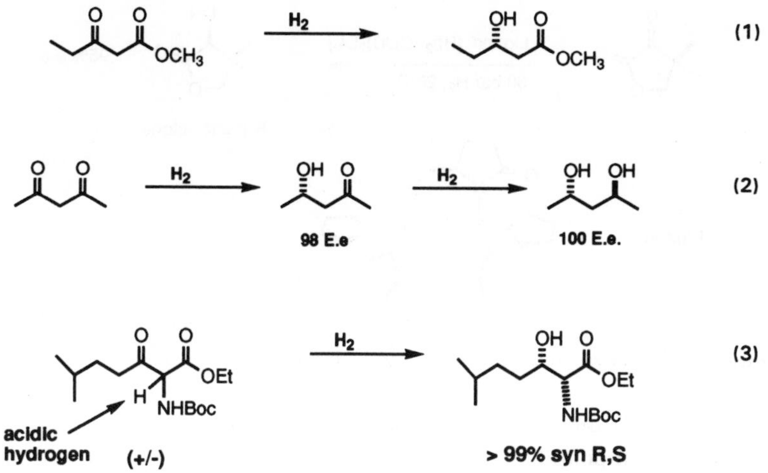

H_2 (1)

H_2 98 E.e H_2 100 E.e. (2)

H_2 (3)

acidic hydrogen (+/-) > 99% syn R,S

Scheme 7 *Hydrogenations of ketones with [(S)-BINAP]RuBr$_2$ / MeOH; generally the reactions are run at moderate pressure*

concerned with the hydrogenation of α, β-unsaturated acids and specifically with the commercial synthesis of the important **anti-inflammatory** drug **Naproxen**.[8] The reaction of the arylacrylic acid with hydrogen under BINAP catalysis is rapid, but requires moderately elevated pressures and lowered temperatures to give the highest e.e.s.

2.1.3 Reduction of Ketones. An even more important attribute of the **ruthenium catalysts** is their ability to **hydrogenate ketones** with **high enantioselectivity**. As in the corresponding alkene reductions, it is essential to present a substrate which carries a second group in proximity to the reacting centre and capable of coordinating to ruthenium throughout the catalytic cycle. In practice, the most substantial class of reactants of this type has been β-keto esters, both the resulting hydroxyesters and their transformation products providing a range of pharmacologically important targets.[9]

The first example in Scheme 7 demonstrates the general principle involved. Reaction is cleanly **chemoselective**, leading to complete reduction of the ketone carbonyl without effect on the ester carbonyl group. The ester plays an essential role, however, for in its absence reduction is slow, and enantioselectivity insignificant. Many important natural products or pharmaceutical intermediates possess the β-hydroxyester moiety, and the asymmetric synthesis of methyl 3-hydroxypentanoate (Reaction 1, Scheme 7) illustrates a straightforward analogue.

A more subtle development, **dynamic kinetic resolution**, is best illustrated here by the preparation of a key intermediate in the synthesis of a commercially important β-lactam antibiotic. In this case (Reaction 3, Scheme 7) the threonine-derived β-ketoester carries an amido group at the α-position, rendering it a **stereogenic centre**, but one which is readily racemized in the reactant (but

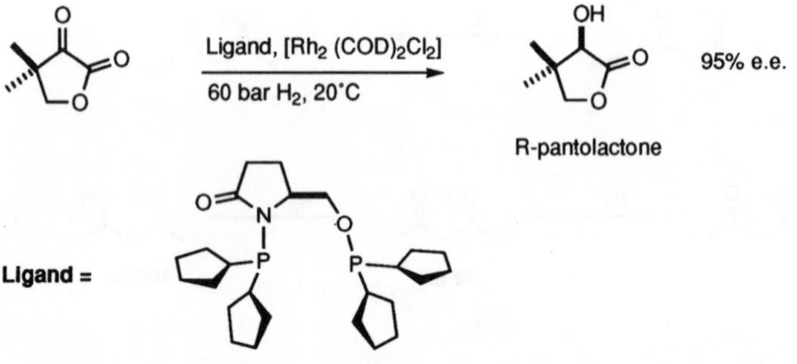

Ligand, $[Rh_2(COD)_2Cl_2]$
60 bar H_2, 20°C
95% e.e.
R-pantolactone
Ligand =

Scheme 8 *Asymmetric synthesis of pantolactone (Vitamin B_5)*

not the product!) under the conditions of asymmetric hydrogenation. Being a β-dicarbonyl compound, the reactant undergoes easy keto–enol tautomerism through ionization of the acidic hydrogen, and the intermediate anion and enol are achiral. The outcome of the reduction is that a single diastereomer of product is formed when the reaction is carried out in dichloromethane. For this to happen requires firstly that the rate of racemization is much faster than the rate of hydrogenation (so that the reaction *solvent* is critically important), and secondly that the reduction is *stereospecific*. Hydrogen adds to one face of the carbonyl group of the fast-reacting enantiomer, hence giving a single secondary alcohol which is both enantiomerically and diastereomerically pure, and the slow-reacting enantiomer does not participate to a significant extent.

The enantioselective reduction of ketones can also be effected by **rhodium** catalyst systems, provided that the controlling ligand is sufficiently basic. Thus alkyl phosphines are far more effective than the more π-acidic aryl phosphines in this regard. Although there are fewer systematic studies, one at least has led to an important procedure, with the potential for adoption on an industrial scale. The reaction in question is illustrated in Scheme 8, and involves the synthesis of pantothenic acid lactone (**Vitamin B$_5$**) by hydrogenation of the readily available prochiral precursor α-ketolactone. By systematic study, Mortreux, Petit, and coworkers were able to establish conditions for the reaction which gave the reduced product in 98% e.e., in a hydrogenation reaction which is rapid at ambient pressure and temperature. The ligand was designed after considerable 'fine-tuning' and a critical feature seems to be the presence of a pyrrolidone in the proline moiety, since the corresponding pyrrolidine is both less reactive and less enantioselective.[10]

All of the reactions described up to this point have involved hydrogenation with Rh or Ru catalysts. In classical synthetic chemistry, the reduction of carbonyl groups by **borohydride** or **aluminohydride** reagents is far more commonplace. Catalytic analogues of such reagents have been slow to emerge, but the one genuinely successful example is already extremely important with a wide portfolio of industrial interest likely to lead to effective processes. It had been recognized since the late 1970s that chiral aminoalcohols such as ephedrine were capable of modifying hydride reducing agents such that the product of ketone reduction was

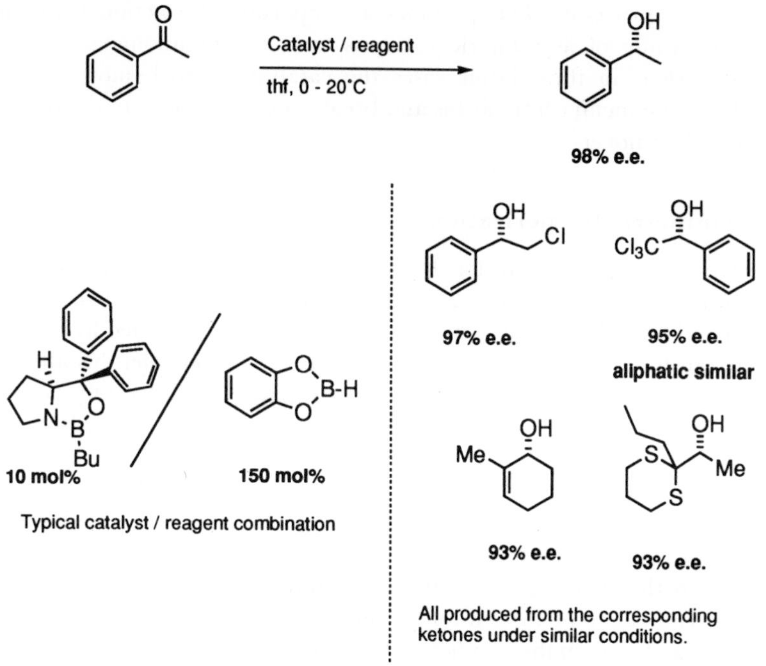

Scheme 9 *Examples of oxazaborolidine-catalysed reductions*

formed in moderate to good enantiomer excess. These reactions were stoichiometric, and a catalytic analogue was discovered in the course of developing related work by Itsuno and coworkers.[11]

The key to their discovery, which has been developed into a superbly efficient asymmetric synthetic method by Corey's group, was that **oxazaborolidines** catalysed the reduction of polar double bonds by borane (BH_3, normally as its Et_2O or THF complex). Although the initial experiments were carried out with oxime ethers, the main developments have involved the asymmetric reduction of aromatic ketones, and a sample of recent results is presented in Scheme 9. Many different oxazaborolidines have been employed in this type of reaction, not least by process development groups in industry. By such trial and error, it transpires that the most effective is also one of the first tried out by Corey, and is based on a simple proline derivative.[12] Since natural proline is of *S*-configuration, and the *R*-enantiomer is not readily available, this tends to limit the applicability of the methodology to the synthesis of products with the configurations defined in Scheme 9.

The mechanism of reaction is partly understood, insofar as there is evidence for coordination of BH_3 (or catecholborane as shown, which is a useful alternative reagent) to the nitrogen of the oxazaborolidine. This coordinated borane is the one which transfers hydride to the ketone, which is in turn complexed to the boron of the oxazaborolidine. The catalyst is thus a **template** for the association of reactant and reagent which plays no further part in the bond-making and bond-breaking processes. This provides an important distinction between the Lewis acid family of asymmetric catalysts, and the organometallic catalysts described earlier; in these latter cases the catalyst is much more intimately involved and the metal centre forms and breaks covalent bonds to the reagent in the course of turnover.

2.2 Asymmetric Isomerization

The reaction cycle revealed for homogeneous hydrogenation in Scheme 1 demonstrates that a necessary stage is the formation and breakdown of a metal alkyl, derived from the alkene complex and *en route* to the saturated hydrocarbon. If the H-transfer from metal to coordinated alkene is reversible, this affords the possibility that the alkene double bond can **migrate** if a differently positioned C–H is returned to the metal. In practice, species related to hydrogenation catalysts are known also to be effective for the isomerization of double bonds in the absence of hydrogen.

The possibility of isomerization of allylic alcohols to enols (and hence to aldehydes or ketones) had long been realized; likewise, allyl ethers can be isomerized to the thermodynamically more stable enol ethers, and this is an important synthetic route to such compounds. The isomerization of allylamines to enamines is similar, with the significant difference that it forms the basis for one of the most important reactions in industrial homogeneous catalysis. The driving force behind this chemistry is the need for a synthetic route to **enantiomerically pure menthol**, one of the most ubiquitous fragrances and food additives (**spearmint**). Much of the world's supply comes from a Chinese plant source *Mentha arvenis*, but it provides a perspective on the Takasago process to be described below that it currently rivals the natural source in output, providing *over* 10^3 *tons per annum.*[13]

Scheme 10 outlines the main features of the chemistry. The precursor tertiary allylic amine is formed by reacting the terpene diene myrcene with $HNEt_2$ / $LiNEt_2$ to give the starting materials depicted here; the stereospecificity of this nucleophilic attack on a diene is interesting. Then follows the key step – the allylamine is treated with one of several cationic (BINAP)Rh catalysts, with the 2:1 species $[(BINAP)_2Rh^+ X^-]$ preferred (in the commercial process the *p*-tolyl analogue of BINAP is used and provides for better catalyst stability and solubility, improving the degree of recycling). It is likely that one of the two diphosphine ligands dissociates under the reaction conditions, and the steric bulk of BINAP assists in this since the corresponding bis-DIPHOS complex is not catalytically active for the isomerization.

The reaction proceeds rapidly under mild conditions at slightly elevated temperatures, and most significantly, with *almost complete enantio- and chemoselectivity*. This implies that the metal-catalysed hydride shift occurs with specific removal of one hydrogen of the aminomethylene unit and replacement of that same hydrogen on the same prochiral face of the allyl, thereby defining the configuration of the new stereogenic centre. The Rh complex survives for $>10^4$ *turnovers* and the commercial attractiveness is enhanced by the possibility of recycling the catalyst.

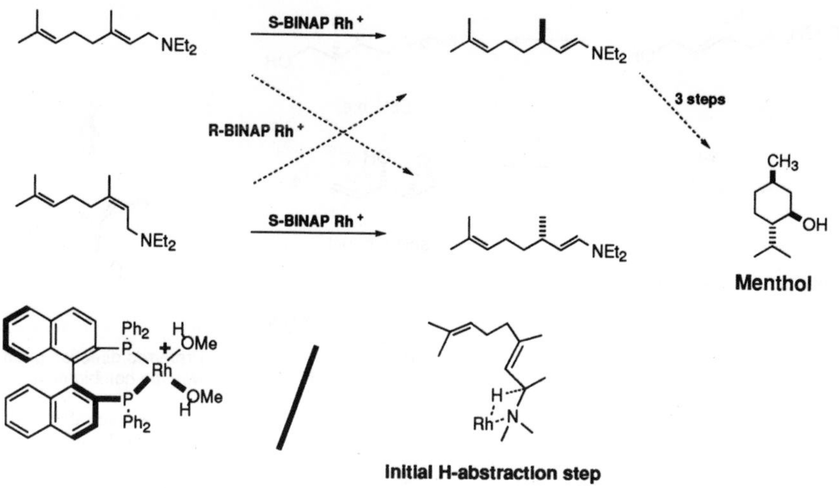

Scheme 10 *The asymmetric isomerization of enamines as a route to perfumery intermediates, including menthol*

From the resulting enamine, the route to menthol is straightforward. Acidic hydrolysis of the enamine gives citronellol in high enantiomeric purity, and further Lewis acid catalysed ene cyclization of the aldehyde is fortuitously stereospecific, giving a single diastereomer with the desired configuration. Reducing the remaining alkene double bond (with a conventional heterogeneous nickel catalyst) completes the synthesis. As well as leading to menthol, which is the largest outlet of this isomerization process by far, there are several significant offshoots leading to terpenes useful both as fragrances and elsewhere.

Ten years after its discovery this reaction remains unique in its class. There are no other examples of asymmetric isomerizations which have any commercial or even any synthetic significance.

3 CATALYTIC OXIDATIONS

There are three catalytic assymmetric oxidation reactions which have both commercial and academic significance – of these, two were discovered or developed by Barry Sharpless and the third by his student Eric Jacobsen. This represents a focal point for the development of organic synthesis in the last decade, since before then there were no catalytic asymmetric oxidations outside enzymology, nor even stoichoimetric processes with significant enantiomer excess.

3.1 Sharpless–Katsuki Asymmetric Epoxidation

The breakthrough first came through the work of Katsuki in Sharpless' group. They had been investigating different metal complexes with the hope of finding one which promoted **epoxidation of allylic alcohols**; earlier results from Mimoun and also from subsequent workers indicated that the hydroxyl group

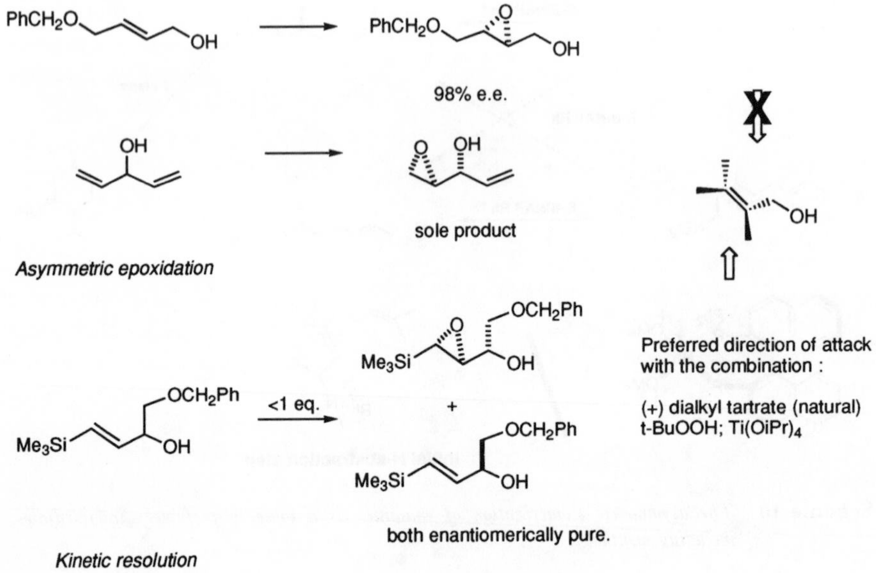

Asymmetric epoxidation

98% e.e.

sole product

Preferred direction of attack with the combination :

(+) dialkyl tartrate (natural) t-BuOOH; $Ti(OiPr)_4$

<1 eq.

both enantiomerically pure.

Kinetic resolution

Scheme 11 *Examples of Sharpless–Katsuki epoxidation of allylic alcohols*

could coordinate reversibly to the metal centre promoting oxygen transfer and eliciting stereochemical control. Titanium alkoxides, $Ti(OR)_4$, were known to promote this epoxidation reaction with *tert*-butyl hydroperoxide as the oxidant, and the Sharpless–Katsuki modification simply consisted of the addition of L-diethyl tartrate to the mixture. Under these conditions, a mixed titanium tartrate complex, probably dimeric, was formed which provided a template for oxygen transfer between the hydroperoxide and alkene.

From the standpoint of asymmetric synthesis, the first results obtained with this reactant mixture under stoichiometric conditions were far better than those from any previous report, and there have in fact only been marginal improvements to the basic reaction conditions over the intervening time.[14] Although this original reaction was stoichoimetric, the titanium complex is regenerated at the end of the reaction and so it can be used in deficiency. To achieve catalysis in practice, it transpires that traces of water must be removed, and the most practical way of doing this is to add powdered molecular sieves to the reaction mixture. Under these conditions the reactant and reagent can be used in at least 20-fold excess over the catalyst on a laboratory scale.

Basic principles of the reaction are considered first, and delineated in Scheme 11. For a given hand of the tartrate catalyst, the prochiral alkene face undergoing preferential reaction, and hence the configuration of the product, can be predicted with some degree of certainty. Most classes of alkene react readily, the exception being those where the double bond bears a substituent *cis*-related to the hydroxyalkyl group. Under the standard conditions of reaction described, the product is produced in 90–97% e.e. Since the catalyst is a powerful Lewis acid, it is capable of catalysing rearrangement of the product in many cases and care needs to be taken to avoid this. On a laboratory scale, reactions tend to be carried out at 0 °C to −20 °C; the lowered temperatures improving the chemo- and enantioselectivity at the expense of the rate. Although general for allylic alcohols as described here, the asymmetric epoxidation reaction is limited to this class of reactants; even homoallylic alcohols fail to give high e.e.s. of epoxidation product, and non-functional alkenes react far less readily. The importance of the method can be gauged from the fact that since its inception in 1980, there have been several hundred applications in the primary literature.

Further analysis of the epoxidation reaction provokes the question – what happens if the reactant itself carries a stereogenic centre? For the simplest case, consider the racemic alcohol displayed in the lower half of Scheme 11. Both hands will tend to be epoxidized on the same face, leading to diastereomerically distinct products. This means that the reactant enantiomers will have to be aligned roughly as drawn, in order to bring the hydroxyl group into contact with the titanium centre. The two hands will thus experience different internal non-bonded interactions in the resulting frozen conformation, and be expected to react at different rates. In practice the difference is quite large for a whole battery of alkenes, leading to the prospect of a **kinetic resolution** during the course of reaction.[15] It proves possible in these cases to consume one hand of the reactant, forming the epoxide stereoisomer expected on the basis of simple analogies, leaving the other one unchanged. At 55–60% reaction, the starting material is recovered in close to complete enantiomeric purity. Since the reactant and product have distinct polarities they are relatively easy to separate, *e.g.* by chromatography, making this a practical method for the resolution of enantiomerically pure allylic alcohols.

There appear to be two fully-fledged commercial applications of the asymmetric epoxidation reaction, although it is being evaluated in other cases as well. The definite ones are shown in Scheme 12, and demonstrate a contrast in scale and in the type of application.

Firstly, the epoxidation of the parent allyl alcohol produces **glycidol**, which is an important intermediate with many outlets in asymmetric synthesis. Since the alkene is monosubstituted, it is rather unreactive in electrophilic substitution, and comparative rate studies do indeed demonstrate that it is much less reactive in asymmetric epoxidation than more heavily substituted alkenes. In addition the product glycidol is quite sensitive to acidic conditions and is prone to ring-opening. To make matters worse both the reactant and product are water-soluble, making conventional work-up difficult. Suffice it to say that all these problems have been surmounted and both the enantiomers of glycidol are commercially available from asymmetric epoxidation, on a multi-ton scale. In achieving this end it is likely that the molecular sieve modification of the reaction under catalytic conditions proved to be crucial. In this particular case, but not others, it is necessary to employ 3 Å molecular sieves; the smallest zeolite pore size is necessary otherwise the reactant allylic alcohol is competitively adsorbed.

A second possible commercial outlet for asymmetic epoxidation is in the production of **Disparlure**, the sex attractant pheromone of the gipsy moth and one used extensively in **insect control** (Scheme 12). It is typical in that the

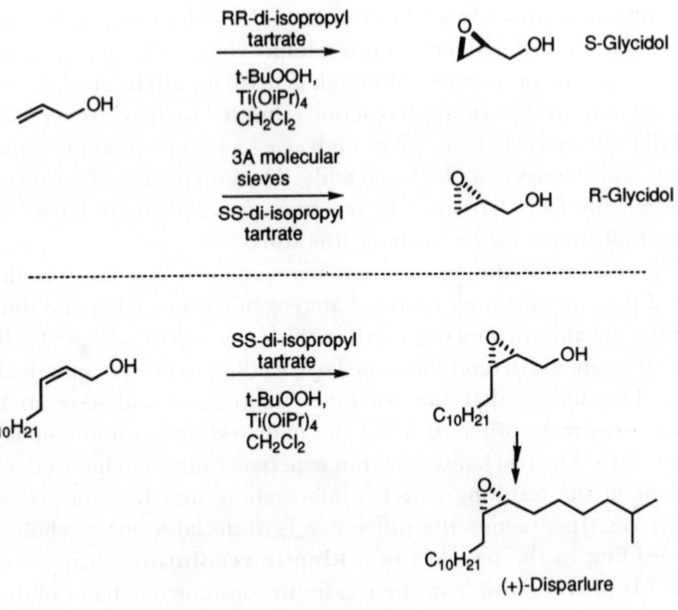

RR-di-isopropyl tartrate

t-BuOOH, $Ti(OiPr)_4$ CH_2Cl_2

3A molecular sieves

SS-di-isopropyl tartrate

OH S-Glycidol

OH R-Glycidol

SS-di-isopropyl tartrate

t-BuOOH, $Ti(OiPr)_4$ CH_2Cl_2

$C_{10}H_{21}$ OH

(+)-Disparlure

Scheme 12 *Commercial outlets for the Sharpless epoxidation reaction*

enantiomeric purity of the pheromone is critical to its application, since only the dextrorotatory hand is normally reactive and the racemic mixture may elicit a different biological response. Hence the enantiomerically pure material is employed, synthesized by a route in which asymmetric epoxidation plays a crucial part since the configuration is defined in that step. The commercial procedure is carried out on the tens of kilogram rather than the ton scale, because the dosage of this pheromone is extremely low in the field, and the demand in terms of weight therefore restricted.

3.2 Jacobsen Asymmetric Epoxidation

Although the asymmetric epoxidation described above is an extremely useful reaction, it is limited to allylic alcohols. Since all **alkenes** which lack a mirror plane orthogonal to the main atomic plane are prochiral and therefore give rise to **chiral epoxides**, the scope of a reaction which lacks this limitation is potentially enormous. It was a longstanding goal for many workers, intensified since the results of Katsuki and Sharpless were first published.

Most early efforts centred on the activation of oxidizing species by porphyrin-like catalysts, based on analogies with haem and Cytochrome P450, which activate molecular oxygen in biology. In order to provide protection of the catalyst from oxidation and at the same time provide a basis for enantioselection, Fe and Mn complexes of very bulky chiral porphyrins were prepared. In practice, their chemistry was limited to anhydrous, aprotic conditions and in these circumstances iodosobenzenes were often the most appropriate oxidants.[16] Early experience

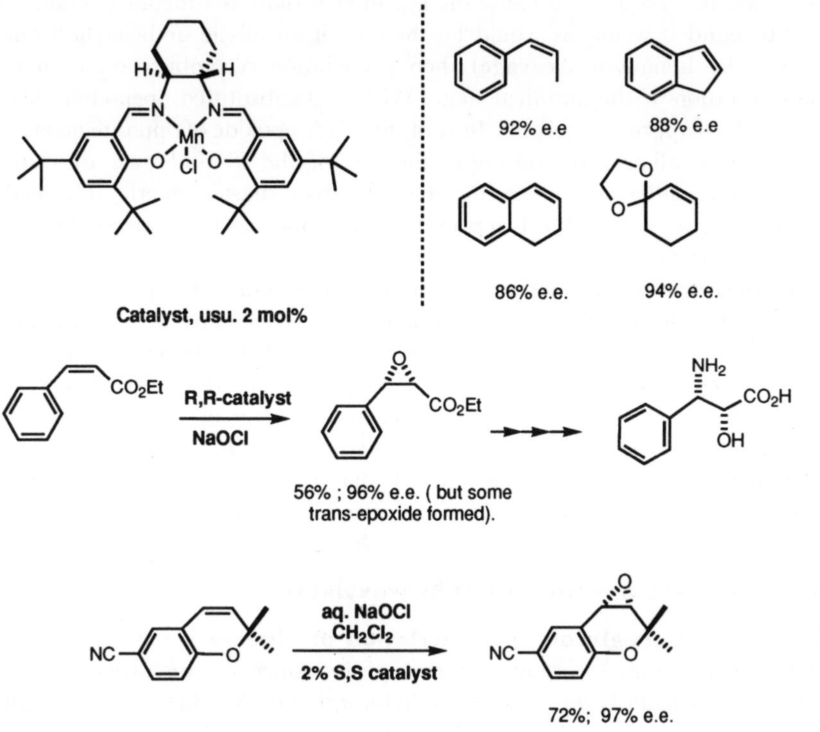

Catalyst, usu. 2 mol%

92% e.e 88% e.e

86% e.e. 94% e.e.

R,R-catalyst NaOCl

56% ; 96% e.e. (but some trans-epoxide formed).

aq. NaOCl CH_2Cl_2 2% S,S catalyst

72%; 97% e.e.

Scheme 13 *Epoxidations catalysed by manganese complexes*

indicated that the methodology was successful in terms of chemoselectivity and catalyst turnover, but unsatisfactory in terms of enantioselectivity. Since much fine-tuning has gone into the synthesis of the complexes employed in these examples it is unlikely that the approach will lead to dramatic improvement. It has been valuable, however in refining ideas about the reaction pathway in asymmetric epoxidation, and oxygen transfer chemistry in general.

A major breakthrough came with the developments of Jacobsen and coworkers, who were among the first to apply the well-known **Mn^{III}(salen) complexes** to asymmetric epoxidation catalysis.[17] An immediate advantage over the porphyrin catalysts ensued: the catalyst species were readily prepared from easily accessible starting materials. More significantly, the enantioselectivity was much higher, particularly with *cis*-disubstituted alkenes, and the catalysts appeared to be far more robust under oxidative reaction conditions; aqueous commercial bleach could be utilized as the reagent without detriment to the yield or selectivity of epoxidation.

Some examples are shown in Scheme 13, highlighting both the scope and current limitations of asymmetric epoxidation methodology. Firstly, the enantioselectivity is highest for *Z*-disubstituted alkenes, affording the only practical outlet for the reaction, whilst *E*-disubstituted alkenes give poor results (<20% e.e.) in this regard. There is a potential problem in that the reaction is not stereospecific, probably because the key intermediate is a metallooxetane. If the C–Mn bond is weak (as would be the case if an allylic or benzylic radical is formed by homolytic cleavage) then dissociation–recombination occurs with bond rotation at the diradical stage. With aryl-substituted open-chain alkenes, this leads to approximately 10–40% of the crude epoxide of opposite geometry to the starting alkene. Interestingly, the e.e. of the *cis* and *trans* disubstituted epoxides is different, indicating that the two diastereomeric intermediates relating to the two prochiral faces of the alkene undergo loss of stereochemistry to different extents.

In the two specific applications shown in Scheme 13, the first suffers the disadvantage from such isomerization and a separation step is necessary to provide the desired stereoisomer of the β-aminoalcohol. This compound has received prominence as the side-chain of the important tetracyclic **anti-cancer drug Taxol**. In the second example, both the chemo- and stereoselectivity are high, and the product is one step away from an important **hypertensive drug Chromakalim**. These examples indicate that the catalytic epoxidation chemistry described will find ready application in a variety of industrial asymmetric syntheses.

3.3 Sharpless Asymmetric Dihydroxylation

Formation of ***cis*-glycols by osmylation of alkenes** is one of the classical synthetic reactions of organic chemistry, and since the discovery of catalytic variants, particularly the use of *N*-methylmorpholine *N*-oxide as a co-oxidant, the reaction has been applied widely.[18] With increasing interest in asymmetric synthesis, many groups provided coreagents for OsO_4 additions which gave moderate to high e.e.s. in the reaction but suffered from the disadvantage that they needed to be present in 1:1 stoichiometry with the alkene. The asymmetric dihydroxylation originally discovered by Sharpless and coworkers followed this pattern, but the real breakthrough came when it was discovered that the reaction became catalytic in the presence of a **co-oxidant**.[19]

Scheme 14 indicates the essential features of the initially discovered reaction. The role of the dihydroquinidine or dihydroquinine catalyst is to ligate to OsO_4. This has two consequences for the osmylation reaction. Firstly, it provides an asymmetric environment in which the addition to one prochiral face of the alkene is favoured. Second, the addition of ligand-complexed OsO_4 to the alkene is considerably faster than the addition of the uncomplexed reagent. This latter observation of **ligand-accelerated catalysis** was an unexpected bonus, although there are long-standing precedents, and it is critical to the success of the catalytic reaction. The initial reaction, although highly interesting, is insufficiently economical in ligand or osmium to be viable on a large scale, and the e.e.s were only moderate in most cases, the best being 90% in the case of *E*-stilbene. Suffice it to say that in the seven years since the original discovery was published, the catalytic reaction has been improved quite dramatically by systematic advances in all aspects. These are described in detail below:

Firstly, it was found that there were two species in the reaction system capable of osmylation. The original catalyst is formally osmium(VIII), and after the alkene

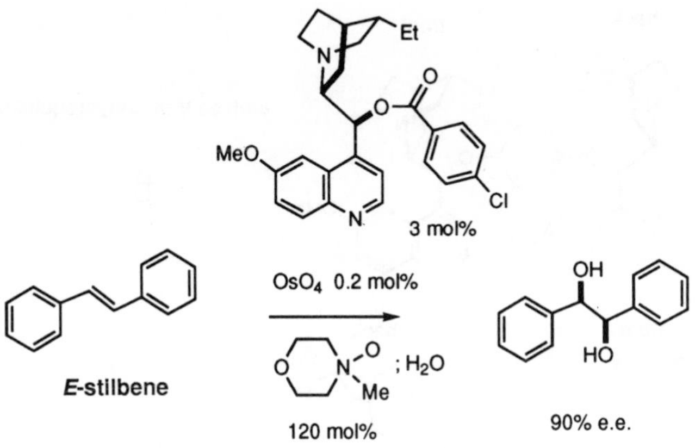

Scheme 14 *An example of the original Sharpless dihydroxylation procedure; the ligand is derived from dihydroquinidine*

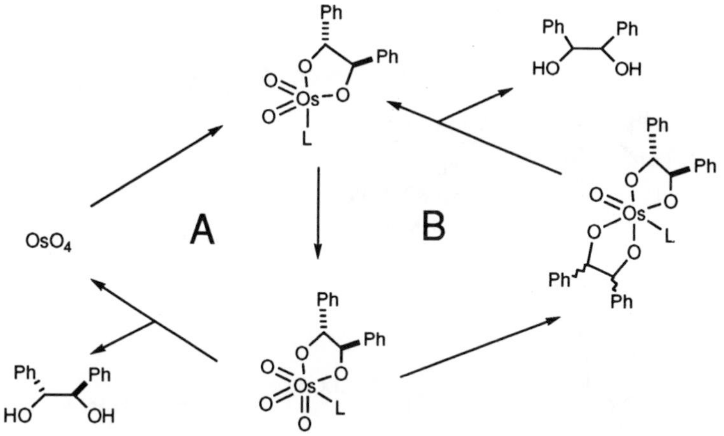

Scheme 15 *Lowering of enantioselectivity in asymmetric dihydroxylation by trapping a second alkene faster than hydrolysis in cycle B; cycle A represents the normal pathway in which high selectivity is preserved*

addition the resulting complex is formally osmium(VI). If this latter intermediate is oxidized prior to hydrolysis, it is capable of reacting with a further molecule of the alkene in a step which has intrinsically low enantioselectivity, since the catalysing ligand is displaced during the process. As seen in Scheme 15, the best way of avoiding this unwanted side-reaction is to enhance the rate of hydrolysis of osmate ester relative to alkene addition, in practice carried out by slow addition of the alkene to the reaction mixture. An even better antidote is to change the oxidant to basic potassium hexacyanoferrate(III), which permits near-complete hydrolysis before the reoxidation step; this procedure can be carried out electrochemically so that the **Fe(III) oxidant** can be **continuously regenerated**.

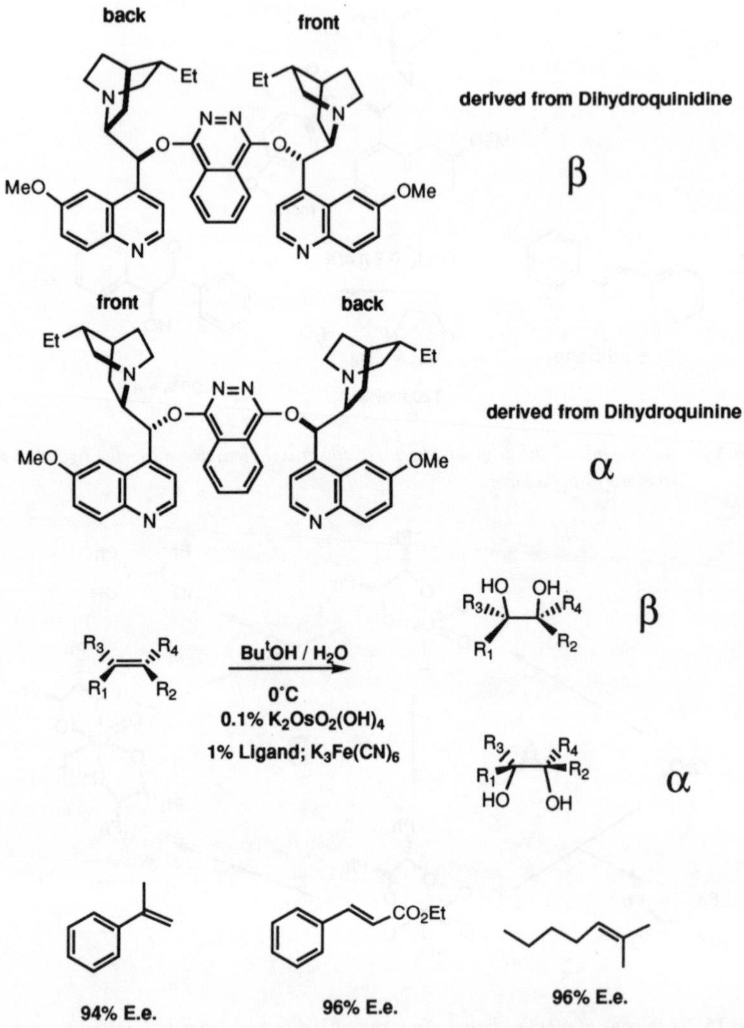

Scheme 16 *The best catalysts to date for asymmetric catalytic dihydroxylation; the two series give opposite enantiomers*

A second development also improves the balance between the desired hydrolysis of the initial product and the unwanted further oxidation, and enhances the turnover rate in cases where hydrolysis proved to be rate-limiting. This involves addition of one equivalent of methanesulfonamide for each mole of alkene, which enhances the rate of osmate ester hydrolysis of non-terminal alkene adducts, possibly because the conjugate anion of methanesulfonamide is a more powerful nucleophile than is hydroxide ion.

The third and perhaps most crucial development was to modify the catalyst structure by systematic variation of structure whilst maintaining the original dihydroquinine / dihydroquinidine entity. Of about 250 derivatives tested by the

Sharpless group, one ligand structure has emerged as the clear preference for asymmetric dihydroxylation, in both the dihydroquinine and the 'pseudoenantiomeric' dihydroquinidine series. These are the **bis-phthalazines** shown in Scheme 16, together with examples of the 'state of the art' in asymmetric dihydroxylation and the selectivities achieved.

As a potential industrial process, it holds substantial promise. There is complete tolerance of air and water, and indeed aqueous conditions are necessary to complete the cycle by hydrolytic release of the osmium. The turnover rates are high, such that the ligand is typically present at 1 mol%, and $H_2K_2OsO_4.2H_2O$ at 0.2 mol%. Given the extreme ease of usage, the procedure is likely to find its way into large-scale fine chemical synthesis. A caveat needs to be raised concerning the reputed toxicity of osmium tetroxide, and clearly any commercial process where a product of asymmetric dihydroxylation can be introduced into human metabolism needs to perfect the methodology for complete osmium removal.

4 SUMMARY AND CONCLUSIONS

These examples underscore the contemporary vigour with which reactions involving asymmetric homogeneous catalysis are being pursued. All the areas described are either exemplified by industrial application, or else such application is under active evaluation by process chemists. This is despite the considerable hurdles that a catalytic procedure must overcome before commercial viability is assured. If an expensive metal like rhodium or osmium is involved, then the catalytic procedure must be very efficient, with many thousands of turnovers required, and possibly also efficient procedures for catalyst recovery. The **enantioselectivity** is probably required to be **>95%** (although this will depend on the end usage). The catalytic procedure must be preferable on a large scale to resolution or alternative synthetic procedures, including biotransformations. The number of successful applications attests to the ability of organometallic and coordination catalysts to meet these criteria.

5 REFERENCES

1. 'Catalytic Asymmetric Synthesis', ed. I. Ojima, Verlag Chemie, Weinheim, 1993, especially Ch. 1, 2, and 4; G.A. Collins, G.R. Sheldrake, and J. Crosby, 'Industrial Asymmetric Synthesis', John Wiley, London, 1992.
2. F. R. Jardine, *Progress in Inorganic Chemistry*, 1985, **38**, 1; A. J. Birch and D. H. Williamson, *Organic Reactions*, 1976, **24**, 1.
3. H.B. Kagan and M. Sasaki, 'The Chemistry of Organophosphorus Compounds', ed. F. R. Hartley, Wiley, New York, 1990, vol. 1, ch. 3; H.B. Kagan, in 'Asymmetric Synthesis', ed. J. D. Morrison, Academic Press, New York, 1985, vol. 5, ch. 1.
4. W. S. Knowles, *Acc. Chem. Res.*, 1983, **16**, 106, and references therein.
5. C. R. Landis and J. Halpern, *J. Am. Chem. Soc.*, 1987, **109**, 1746; J. M. Brown, P. A. Chaloner, and G. A. Morris, *J. Chem. Soc., Perkin Trans. 2*, 1987, 1583; and earlier papers from these research groups.

6. M. J. Burk, J. E. Feaster, W. A. Nugent, and R. L. Harlow, *J. Am. Chem. Soc.*, 1993, **115**, 10125.
7. The first preparation of these complexes is described by R. Noyori, M. Ohta, Y. Hsiao, M. Kitamura, T. Ohta, and H. Takaya, *J. Am. Chem. Soc.*, 1986, **108**, 7117.
8. A. S. C. Chan, S. A. Laneman, and R. E. Miller, *ACS Symposium Series*, 1993, **517**, 27.
9. The subject has been extensively reviewed: R. Noyori, *Chem. Soc. Rev.*, 1989, **18**, 187; R. Noyori and M. Kitamura, *Modern Synthetic Methods*, 1989, **6**, 131; R. Noyori, *Science*, 1990, **248**, 1194; R. Noyori and H. Takaya, *Acc. Chem. Res.*, 1990, **23**, 345.
10. A Roucoux, F. Agbossou, A. Mortreux, and F. Petit, *Tetrahedron: Asymmetry*, 1993, **4**, 2279.
11. S. Itsuno, Y. Sakurai, K. Ito, A. Hirao, and S. Nakahama, *Bull. Chem. Soc. Jpn*, 1987, **60**, 39512; S. Wallbaum and J. Martens, *Tetrahedron: Asymmetry*, 1992, **3**, 1475.
12. E. J. Corey, R. K. Bakshi, S. Shibata, C. P. Chen, and V. K. Singh, *J. Am. Chem. Soc.*, 1987, **109**, 7925; E. J. Corey and C. J. Helal, *Tetrahedron Lett.*, 1993, **34**, 5227; and intervening papers.
13. S. Otsuka and K. Tani, *Synthesis*, 1991, 665; S.-I. Inoue, H. Takaya, K. Tani, S. Otsuka, T. Saito, and R. Noyori, *J. Am. Chem. Soc.*, 1990, **112**, 4897.
14. T. Katsuki and K. B. Sharpless, *J. Am. Chem. Soc.*, 1980, **102**, 5974; B. E. Rossiter, in 'Asymmetric Synthesis', ed. J. D. Morrison, Academic Press, New York, 1985, vol. 5, ch. 7; S. S. Woodard, M. G. Finn, and K. B. Sharpless, *J. Am. Chem. Soc.*, 1991, **113**, 106; M. G. Finn and K. B. Sharpless, *J. Am. Chem. Soc.*, 1991, **113**, 113.
15. V. S. Martin, S. S. Woodard, T. Katsuki, Y. Yamada, M. Ikeda, and K. B. Sharpless, *J. Am. Chem. Soc.*, 1981, **103**, 6237; Y. Gao, R. M. Hanson, J. M. Klunder, S. Y. Ko, H. Masamune, and K. B. Sharpless, *J. Am. Chem. Soc.*, 1987, **109**, 5765; *cf.* H. B. Kagan and J. C. Fiaud, *Top. Stereochem.*, 1988, **18**, 249.
16. J. T. Groves and R. S. Myers, *J. Am. Chem. Soc.*, 1983, **105**, 5791; K. Konishi, K-i. Oda, K. Nishida, T. Aida, and S. Inoue, *J. Am. Chem. Soc.*, 1992, **114**, 1313, and intervening references cited therein.
17. W. Zhang, J. L. Loebach, S. R. Wilson, and E. N. Jacobsen, *J. Am. Chem. Soc.*, 1990, **112**, 2801; E. N. Jacobsen, W. Zhang, A. R. Maci, J. R. Ecker, and L. Deng, *J. Am. Chem. Soc.*, 1991, **113**, 7063; L. Deng and E. N. Jacobsen, *J. Org. Chem.*, 1992, **57**, 4320; *cf.* R. Irie, K. Noda, Y. Ito, N. Matsumoto, and T. Katsuki, *Tetrahedron: Asymmetry*, 1991, **2**, 481, and earlier references.
18. V. Van Rheenen, R. C. Kelly, and D. Y. Cha, *Tetrahedron Lett.*, 1976, **17**, 1973.
19. For key papers see: E. N. Jacobsen, I. Marko, W. S. Mungall, E. Schroeder, and K. B. Sharpless, *J. Am. Chem. Soc.*, 1988, **110**, 1968; K. B. Sharpless, W. Amberg, Y. L. Bennani, G. A. Crispino, J. Hartung, K.-S. Jeong, H.-L. Kwong, K. Morikawa, Z.-M. Wang, D. Xu, and X.-L. Zhang, *J. Org. Chem.*, 1992, **57**, 2768.

6 CORE JOURNALS

The journals indicated in Chapters 4 and 5 sometimes contain articles on the present topic, but the following are especially relevant when seeking papers on stereospecific synthesis using homogeneous solution chemistry:

Tetrahedron: Asymmetry
Tetrahedron Letters
Topics in Stereochemistry
Journal of Organic Chemistry
Journal of the American Chemical Society
Journal of the Chemical Society, Perkin Transactions

CHAPTER 7

Automobile Catalysts

MICHAEL BOWKER AND RICHARD W. JOYNER

1 INTRODUCTION

1.1 Air Pollution

The types of pollution of our atmosphere vary depending on the altitude, and the most significant pollutants in the troposphere (up to ~20 km altitude) can be different from those in the stratosphere (up to ~50 km), where ozone chemistry dominates our considerations. This chapter is concerned only with **tropospheric pollution**, especially in the biosphere (the near-surface region of the earth where life prevails), although this in turn has an impact on the upper atmosphere.

Air pollution is not a new problem, indeed the ancient atmosphere of earth was extremely poisonous to life. In more recent times many people have died from natural air pollution incidents, especially in the wake of volcanic eruptions. For instance, a cloud of toxic gas/ash is thought to have killed most of the inhabitants of Pompeii after the Vesuvius eruption in AD 79. Man made pollution incidents affecting large numbers of people only began to occur in the recent past, perhaps the most notable early cases being the so-called 'pea-souper' smogs of Victorian London which were responsible for the deaths of many thousands of people. Major irritants in these smogs were **smoke** and **sulfur dioxide** (and sulfurous or sulfuric acid resulting from its dissolution in mist droplets and soot particles). The SO_2 was produced from the widescale burning of **high sulfur coal** in domestic hearths. In 1956 these problems resulted in the introduction of a Clean Air Act of Parliament (UK), which permitted the burning of only 'smokeless' fuels in specific zones of high pollution risk.

The next area of note where pollution became a severe public nuisance was in Los Angeles in the 1950s, where the source was the large and ever-increasing number of cars powered by **internal combustion engines**. Los Angeles was unique at that time with respect to its density of cars and its geographical situation, being surrounded by hills over a significant circumference of the city and being subject to continuous, high levels of sunshine. These effects led to the formation of atmospheric inversion layers (often with an inverted temperature profile, that is, with colder air at lower altitude close to ground level which serves to prevent

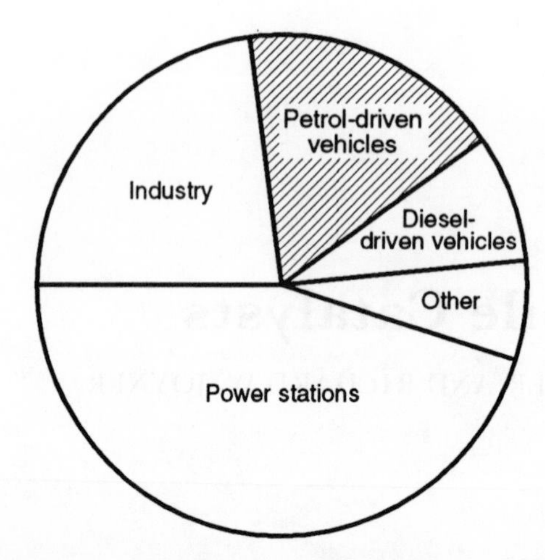

Figure 1 *Pie diagram of the sources of NO_x emissions, showing most emanating from power stations, but with more than a quarter being contributed by mobile sources*[1]

normal air circulation), which kept and concentrated the pollution close to ground level.

There are many toxic gases which are present in such smogs and we will deal with only a few in what follows. **NO_x** is a mixture of oxidized nitrogen molecules, the major one released by the engine being NO. This is produced by direct reaction of nitrogen and oxygen in the higher temperature region of the combustion chamber. **Sulfur oxides** are produced by oxidation of organic sulfur compounds, present typically at up to 300 ppm in petrol in the USA and Europe. **Hydrocarbons** result from unburned and partially combusted molecules emitted from the engine. The 'smog' is a soup of these molecules and others produced by homogeneous and heterogeneous reactions in the atmosphere, many being the result of **photochemical reactions**. As an example, peroxyacetyl nitrate (PAN) is a chemical which is toxic at the sub-ppm level, and causes severe bronchial problems especially for those with respiratory difficulties already. It is formed by photochemical production of the peroxyacetyl radical which attacks nitrogen dioxide.

NO_x and SO_x are also responsible for the phenomenon of '**acid rain**', and for the extensive forest destruction and lake acidification which is seen globally. However, the major sources of these pollutants are static sources such as industrial plant and power stations (Figure 1).[1] Acid rain is mainly due to the hydrated forms of these molecules, nitrous/nitric and sulfurous/sulfuric acids. Technology is currently in place, and in further development, to remove NO_x and SO_x from such stationary sources. NO_x can be removed catalytically by reaction with ammonia to produce nitrogen and water (the 'deNO_x' process). SO_x can be removed by passing the effluent through reactive solids, usually alkaline earth oxides, which convert to sulfates and are subsequently used by the building industry for brickmaking.

As a result of the LA smogs, the Californian authorities took the courageous and inventive step of **forcing improvements by legislation**, requiring significant

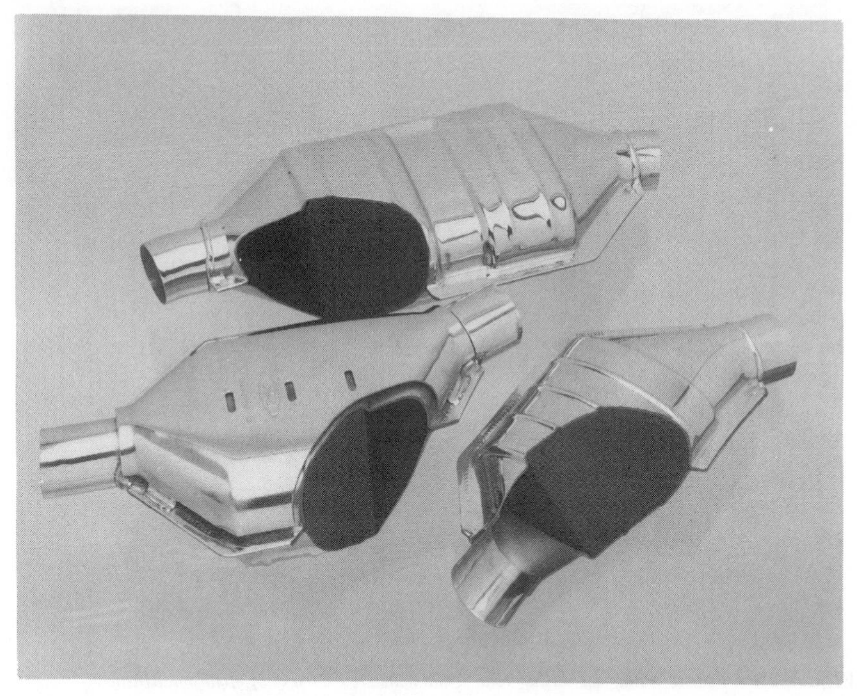

Figure 2a *The catalytic convertor. The cutaway shows the ceramic monoliths inside. Courtesy of Johnson Matthey*

reductions in the three major gaseous emissions – hydrocarbons, carbon monoxide, and NO_x – and the need for new technologies to meet the legislation (see Section 4 and Table 3 for legal limits on current and future emissions in the US). The successful new technology which emerged was to convert these compounds using precious metal catalysts,[2] an example of the device used, the **catalytic converter**, being shown in Figure 2a. It consists of a cylinder attached in the exhaust line of the vehicle, in which is contained a ceramic monolith, preformed as a honeycomb which offers minimal obstruction of the gas flow. A so-called 'washcoat' is deposited onto the monolith base and this consists of a highly porous alumina, which in turn is doped with a variety of active components, the main ones being precious metals, usually **platinum** and **rhodium**. The sequence of development of such catalysts is described in the following section in more detail.

1.2 Development of the Three-way Catalyst

Legislation left car makers free to decide how to reduce emissions, but it rapidly became clear that the best approach was to introduce a catalyst between the engine manifold where the exhaust gases leave the cylinders, and the silencer box (US, muffler). The catalytic reactions need a minimum temperature to start, or **'light-off'**, and in this location the heat of the exhaust gases is used to achieve this.

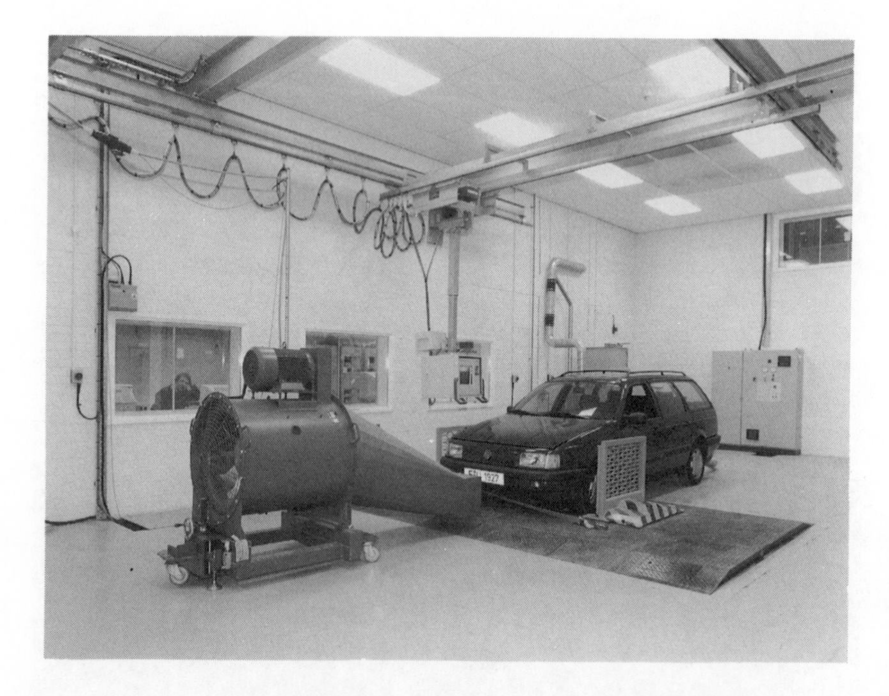

Figure 2b *Vehicle testing on rolling road dynamometer. Courtesy of Johnson Matthey*

It was also obvious from the start that two of the pollutants, carbon monoxide and unburnt hydrocarbons, could be removed by oxidation:

$$CO + \tfrac{1}{2} O_2 \rightarrow CO_2 \quad (1)$$

$$C_nH_{2n+2} + m\,O_2 \rightarrow n\,CO_2 + (n+1)\,H_2O \quad (2)$$

These reactions are relatively easy to catalyse, and can be carried out by a number of **noble metal** or **oxide** catalysts. The first automotive catalysts to be introduced were therefore called 'two-way' catalysts, as they performed these two oxidation reactions. They consisted typically of monolithic supports, or pellets perhaps 2 mm in diameter, containing a high surface area of alumina used as a support for small particles of the noble metals platinum and/or palladium. Ruthenium also has good catalytic properties for these reactions and for NO reduction (see below). Its use was discounted, however, because ruthenium(IV) oxide is volatile at the higher temperatures reached, and is very toxic.

At this time, *ca.* 1970–75, a wide range of catalytic materials was examined including a number of oxides. Oxides are often efficient catalysts, but with comparatively high light-off temperatures. The light-off temperature and the precise location of the catalyst determines the time taken after a cold start for the catalyst to become effective, and this is of considerable significance because of the way in which automobile catalysts are tested. Figure 2b illustrates the testing of a vehicle on a rolling road dynamometer. The catalyst is judged by the total amount of pollutants emitted over a specified test cycle. In practice the catalyst normally operates with >90% efficiency once it has heated up, producing most emissions during the warming up period.

Since only oxidation was required at this stage, the car engine and control system required little modification. All that was needed was to ensure that the **fuel to air ratio** did not fall below the stoichiometric value required for complete combustion – the so called equivalence point, often referred to as $\lambda = 1$. It soon became clear, however, that nitrogen oxide pollution needed to be treated, because of its role in promoting photochemical smog. Catalytic removal of these nitrogen oxides (referred to as NO_x) represented a much more difficult challenge. Oxidation to NO_2/N_2O_4 occurs slowly in the gas phase, and is clearly undesirable. The most effective means of NO_x treatment was found to be to promote the reaction with CO, which is always present in excess over NO_x:

$$2\,CO + 2\,NO \rightarrow 2\,CO_2 + N_2 \qquad (3)$$

This reaction is also catalysed by the noble metals platinum, palladium, and rhodium, but **rhodium** is by far the most effective. As a result it became and remains an essential component of automotive catalysts, even though it is normally considerably more expensive even than platinum. [Precious metal prices vary from day to day, and rhodium prices have recently ranged between $30–300 per gram. Platinum is typically $13 per gram.] The automotive catalyst now facilitates three reactions and so is called a **'three-way' catalyst**.

Figure 3 indicates how the emission levels of the three major pollutants from petrol engines change as a function of air/fuel ratio, and shows that in order to achieve good conversion of all three, *very precise control* of the air/fuel ratio is required. The need to remove NO_x as well as CO and unburnt hydrocarbons makes it necessary to operate the engine at the stoichiometric composition for complete combustion ($\lambda = 1$). If the mixture is too rich in fuel, ($\lambda < 1$), the oxidation reactions cannot occur; if the mixture is too lean ($\lambda > 1$), the catalyst becomes poisoned for NO_x treatment. The typical conversion profile of the pollutants is shown in Figure 4. This need for precise control of the air/fuel ratio necessitated changes to the design of the car and the introduction of an **engine management system**. The composition of the fuel/air mixture introduced to the cylinders is now controlled electronically, using a feedback loop which uses a sensor to monitor the oxygen content of the exhaust, and adjusts the mixture to keep it as close to stoichiometric as possible. Typically the air/fuel ratio is held at $\lambda = 1.00$, with a cyclic deviation of around 0.01.

Pelleted catalysts are simple to manufacture and install. They are prone however to problems of blocking and bypassing. Pellets may settle in use, and allow gas to flow over or around the catalyst bed, reducing its effectiveness. Very occasionally under severe conditions the pellets may fuse together, with highly undesirable consequences.

An alternative to pelleted catalysts is to use a **ceramic monolith** as the catalyst support, and this approach developed by the Dow–Corning Co. is now very generally used. A schematic view of the monolith catalyst is shown in Figure 5. The monolith structure is formed by extrusion of **cordierite**, a mineral with the composition $Al_3Mg_2(Si_5Al)O_{18}$ and a structure similar to beryl, and subsequent

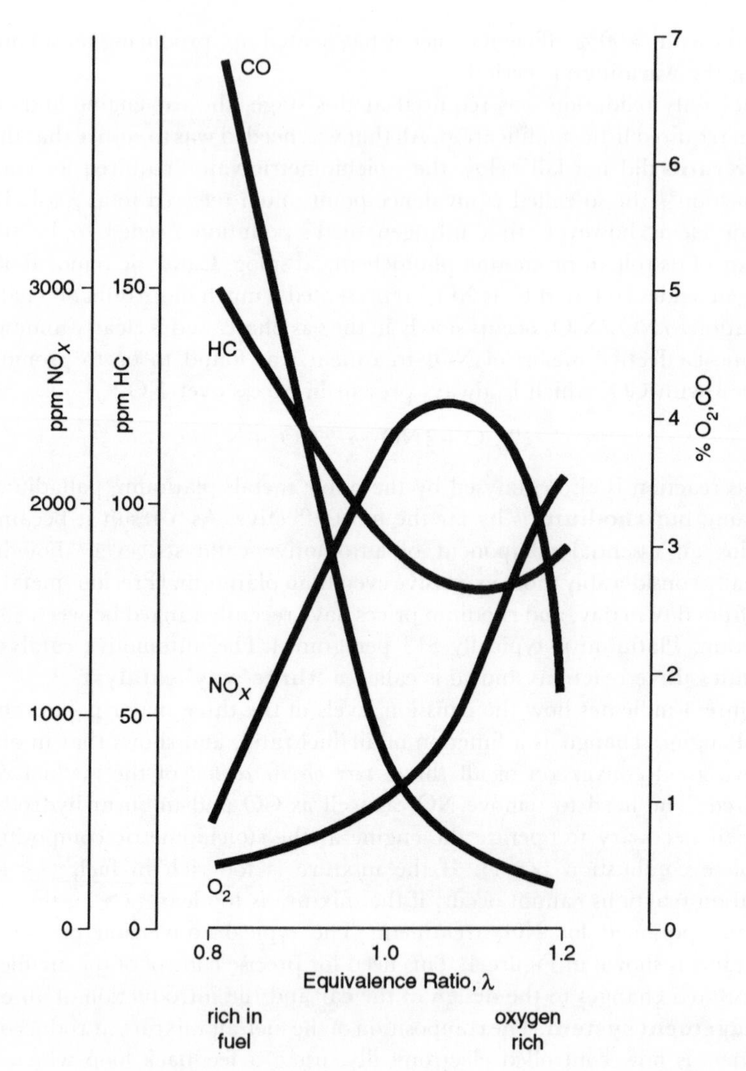

Figure 3 *Diagram showing the composition of pollutants in exhaust gases (before catalytic conversion) as a function of equivalence ratio, courtesy of Johnson Matthey*

firing of the extrudate at 1600–1700 K, using technology originally developed by ICI. The monolith in a typical family car is about 15 cm long by 10 cm in diameter. The exhaust gases traverse it longitudinally. The alumina catalyst support is located around the walls of each individual channel, and introduced by a 'dip and drain' technique followed by a calcination, in an operation which is commercially very sensitive. The techniques of precious metal addition and the exact amounts added are also commercial secrets, but typical car catalysts contain 1–3 g of precious metals, which are typically introduced as salts, from solution, followed by high temperature calcination to give the active metal constituents.

The automotive catalyst operates in an environment which, compared to most

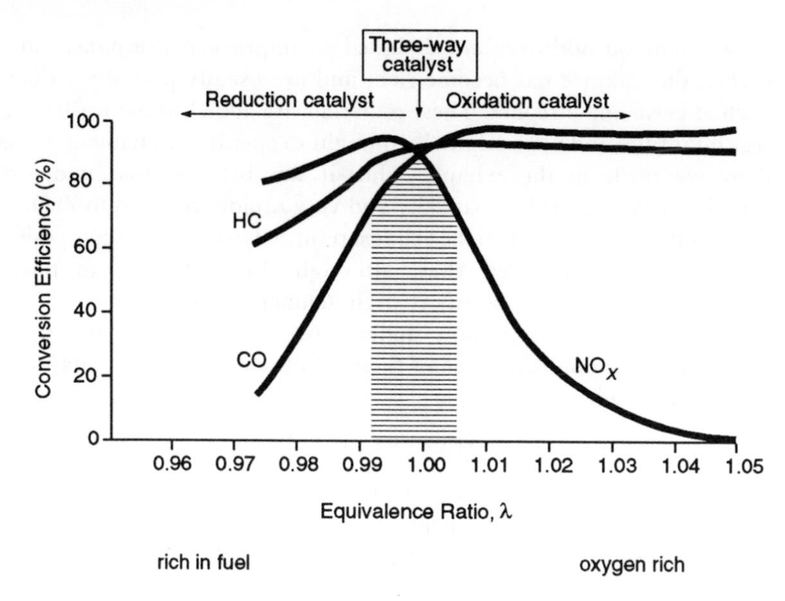

Figure 4 *Diagram showing the conversion of pollutants through a car catalyst as a function of equivalence ratio, courtesy of Johnson Matthey*

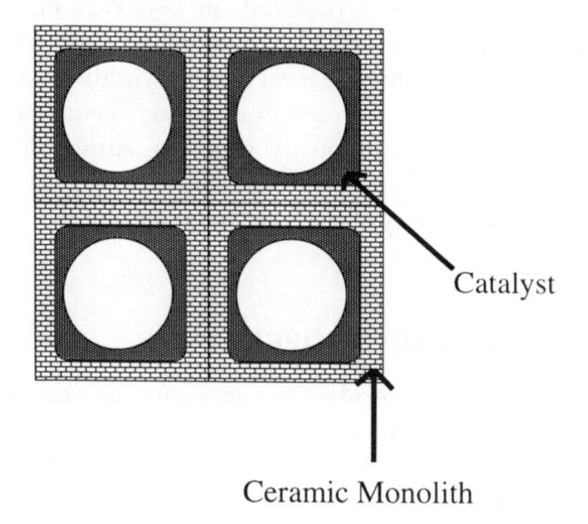

Figure 5 *Schematic section through part of a ceramic monolith catalyst; the gas flow is perpendicular to the plane of the paper*

industrial catalysts, is constantly changing. Sudden, **rapid acceleration** causes the mixture to be **rich in fuel** for a short time, until the engine management system responds by adjusting the air/fuel ratio. By contrast, during **deceleration** the mixture will be **oxygen rich**, or **lean** for a period. The legislative specification is now so tight that the catalyst must respond to these changes without more than momentary loss of effectiveness, and this has led over the last ten years to the development of additives which promote the catalyst's performance during these excursions away from the stoichiometric feed ratio.

The most common additives are designed to improve performance during periods when the mixture has become rich, and are usually partially reducible oxides such as ceria and zirconia. These materials are added to the wash coat in quite large quantities and were originally thought to operate by giving up oxygen when there was little in the exhaust. Thus it was believed that under rich conditions, for example, ZrO_2 is partially and very rapidly reduced to $ZrO_{(2-x)}$, becoming rapidly reoxidized when the exhaust returned to its normal composition. Recent studies have suggested that, although this reduction to form a non-stoichiometric oxide can occur at high temperatures, it does not occur sufficiently rapidly to be important under most conditions.[3] A more likely explanation for the undoubted value of these additives is that they catalyse the **water gas shift reaction**:

$$CO + H_2O \rightarrow CO_2 + H_2 \tag{4}$$

which reduces the CO concentration in the exit gases by reaction with water, which is of course a major component of the exhaust stream. The hydrogen content of the exhaust is usually negligible, so it is not proscribed by legislation.

The modern automobile catalyst is thus a very sophisticated piece of advanced technology, containing up to 7 components. Under USA law, catalysts must now be **effective for 100 000 miles**, which is greater than the lifetime of the average European car. Catalysts are now required on new cars in the USA, Japan, Australia, and the European Union (EU), comprising a very high fraction of all new cars sold. Catalyst improvement is a very active worldwide area of research; in particular much effort has been devoted to replacing the precious metals and in particular the most expensive component, rhodium. Although there are claims, particularly from the Ford Motor Co., for effective catalysts containing only platinum and palladium, the day of the base metal automotive catalyst still seems to be a long way off.

2 PRINCIPLES OF CATALYST OPERATION

The three-way catalyst must be able to carry out the reactions indicated in Equations 1–3. The major active components are Pt and Rh: the former is an excellent oxidation catalyst, while the latter has the ability to activate NO efficiently with minimized nitrous oxide production. The comparative ability of Pt and Rh for NO dissociation (a prerequisite for NO conversion) is shown in Figure 6,[4] showing the yield of nitrogen after NO adsorption on a metal crystal surface and then measuring the nitrogen evolution upon heating, using mass spectrometry. As already indicated, the real catalyst is a very complex piece of technology, and the wide variety of requirements for the catalyst is now discussed.

2.1 Sulfur Tolerance

In Europe and the USA (but not in Japan) petrol contains significant quantities of sulfur, up to ~300 ppm by weight, equivalent to 20 ppm of SO_2 in the exhaust. Sulfur is a major poison of all types of catalyst, diminishing activity by blocking

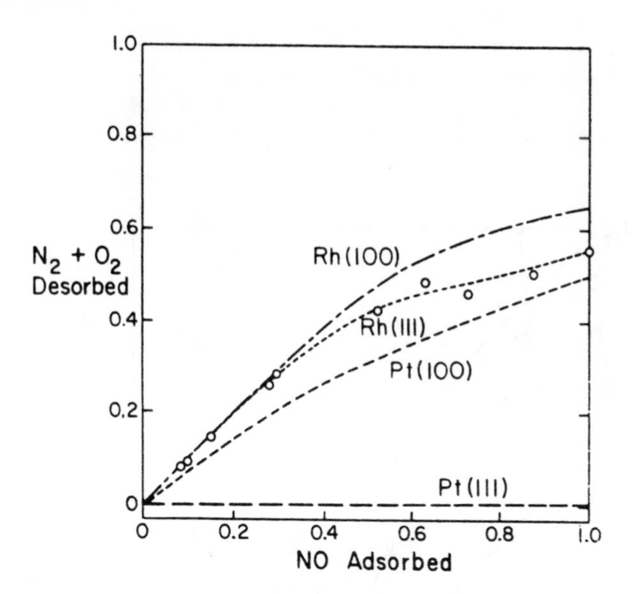

Figure 6 *A comparison of the abilities of pure metal crystal surfaces to dissociate NO in terms of measured nitrogen yield evolving from the metals.*[4] *These yields are relative to the total monolayer capacity of the surface for NO. Rh gives better yields than Pt*

active sites on the metal component. Thus it is important to be able to avoid this poisoning and the oxidic components (especially ceria) achieve this by 'mopping up' the sulfur in the form of sulfate or sulfite and re-releasing it as SO_2. In this way it is thought that these components act as sinks for these otherwise deactivating components of the exhaust gases.

2.2 Oxygen Storage

The operating system results in a feedback response cycle of about 1 second, during which the air/fuel ratio cycles close to the stoichiometric value, but through a range of oxygen concentrations. One role of the oxygen storage component is to maintain surface oxygen concentration during fuel rich operation for combustive oxidation and to 'mop' it up during the lean cycle, although other chemical properties may also be important as described above (Section 1.2). Ceria has this capability at its surface due to its dual valence state Ce^{3+}/Ce^{4+} and is used in exhaust catalysts at high concentration.

2.3 'Light-off'

Car catalysts do not work at low temperatures due to kinetic considerations and if the temperature of the catalyst is increased, increase in conversion from zero to high levels occurs over a short temperature range (Figure 7); the inflexion point of this curve is often called the **'light-off' temperature**. It is important that this temperature is as low as possible for efficient conversion quickly after the cold start

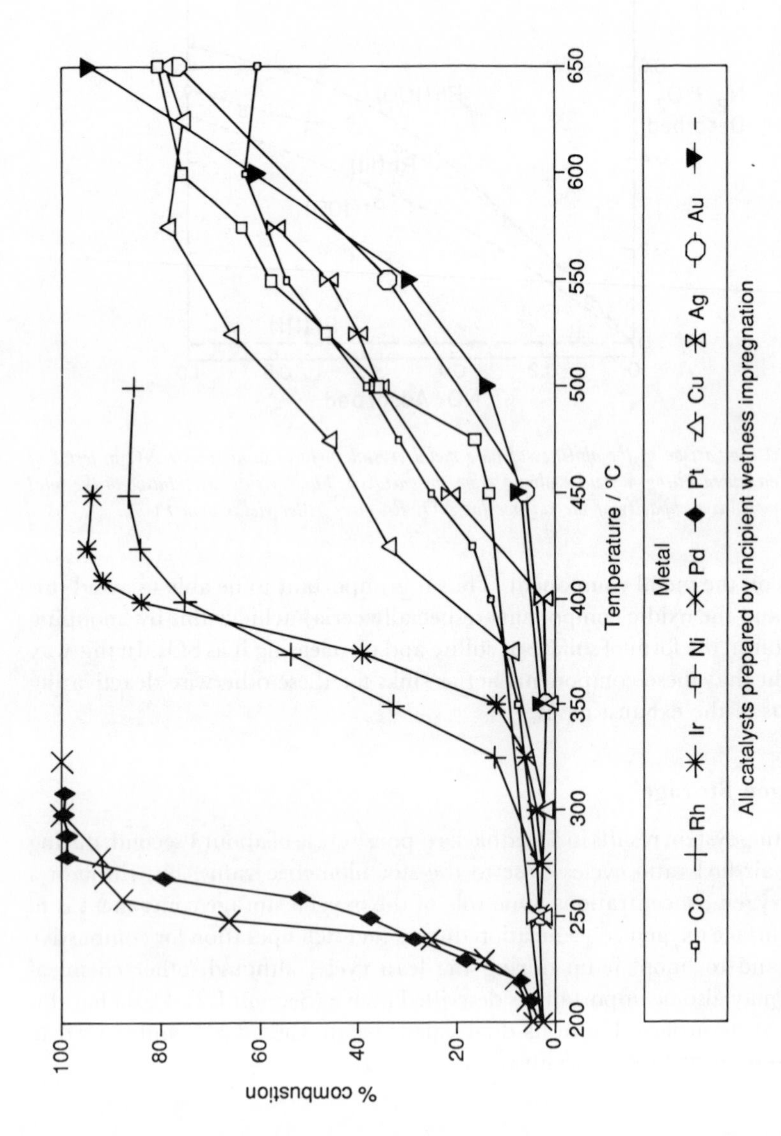

Figure 7 *An example of 'light-off' curves for the combustion of a hydrocarbon, in this case propene,*[5] *on supported metal catalysts. This shows the excellent oxidation abilities of Pt and Pd, these showing the lowest temperature for 'light-off'*

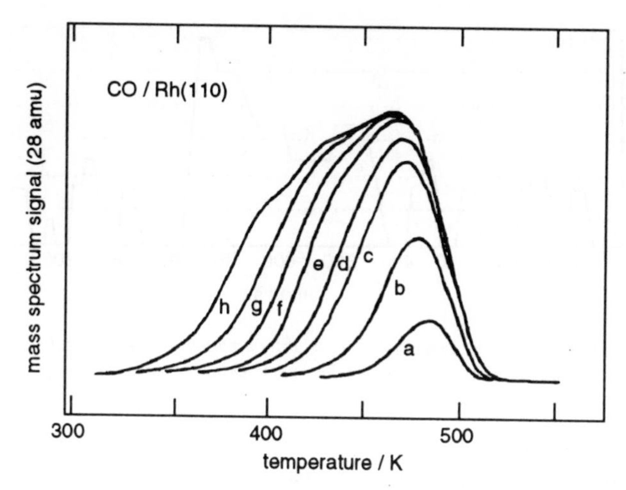

Figure 8 *Showing the desorption of molecules from a single crystal Rh surface*[6] *with desorption beginning at* ~*350 K. The curves a to h are for increasing amounts of CO adsorbed on the surface prior to heating and desorption. Curve h is saturation coverage (one monolayer)*

of the engine. This is especially so since the testing cycles described below require the catalyst to achieve total averaged conversion from a cold start and include cycles of low temperature operation. Precious metals are highly active materials which encourage low temperature light-off, but below this temperature the metals tend to be covered by carbon monoxide and light-off only occurs when this begins to desorb from the surface and liberate surface sites for reaction (see Figure 8 and Section 3.1). Additives such as ceria may also lower the light-off temperature.

2.4 Thermal Stability

The catalysts can cycle from cold start to temperatures as high as 900 °C, and this is a very demanding situation not usually experienced by catalysts used in chemical processing (see Chapter 4). Thus both **mechanical and chemical stability** are required. Mechanical strength and thermal shock resistance are provided by the monolith support, a high stability ceramic based on cordierite, as described above, with other additives, such as Ba for enhanced performance. Clearly the chemically active components of the catalyst must also be thermally stable and this eliminates many possible elemental components, including low melting point metals and oxides which may be reducible to more volatile species during the engine cycle. Pt and Rh are relatively high melting point metals, Cu and Ag are certainly not. Ce also plays a role here in helping to prevent metal sintering which would result in activity loss.

2.5 Chemical Nobility

Another feature of Pt in particular is its 'nobility', that is, it is very difficult to oxidize under these conditions. A material which oxidized would be relatively

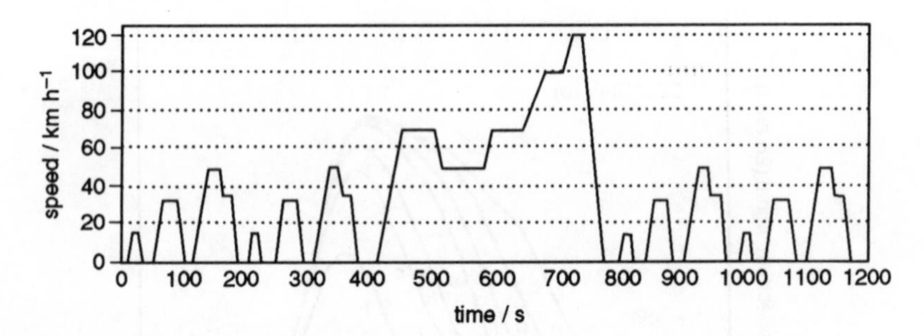

Figure 9 *Speed–time testing procedures for the determination of overall catalyst emissions, to test compliance of emission-regulated cars with legislated emission levels*[7]

inactive and one which cycled between oxidized and reduced states would be likely to **sinter**, reducing the surface area available, so resulting in deactivation. Rh is a little less resistant to oxidation which is partly why NO_x conversion decreases under lean burn conditions, due to overoxidation of the surface layers.

2.6 Lifetime

The catalyst must have considerable longevity and maintain high conversion over at least 50 000 miles operation in Europe (about 6 years average use) or 100 000 miles in the USA. Thus all the materials used are also designed for great **time stability** and many of the features described above aid in this feature of the catalyst engineering.

Thus there is a very difficult set of conditions to be satisfied, but by scientific cleverness, skill, and by the continual evolution/development of catalyst systems, designers have been able to keep up with constantly tightening legislative requirements. The test procedures for the catalytic convertor *in situ* in the car are quite rigorous and slightly different methodologies are used in the United States and Europe (an example of a test cycle is shown in Figure 9). These are designed to simulate city operating cycles during normal driving with many stop–start sequences, although they are also configured to allow simple, repetitive, computer-controlled testing procedures. The efficiency of the catalyst is then determined from the overall emissions during the extreme variations of driving conditions during this test.

3 SURFACE CHEMISTRY

In this section we consider the surface chemistry by which the three main catalytic reactions occur: carbon monoxide and hydrocarbon oxidation, and nitric oxide reduction by CO to yield dinitrogen. The chemistry of these reactions is understood in considerable detail, often as a result of studies using idealized models of the catalysts. The chemistry of the promoter oxides, zirconia and ceria, is much less well understood and will not be discussed.

Table 1 *Reactivity with rhodium surfaces*

	Oxygen	*CO*	*CO_2*
Nature of adsorption	**dissociative** forming oxygen **adatoms**	molecular	molecular
Initial sticking probability	high, >0.7	high, >0.7	zero
Saturation coverage	>1 monolayer	~0.8 monolayers	zero
Heat of adsorption (kJ mol^{-1})	>200	110–130	small
Surface mobility	small	high	very high

3.1 CO Oxidation

In many ways the **chemistry of carbon monoxide oxidation** is the simplest, so it will be described first, but the other reactions occur in a sufficiently similar way that this reaction also provides a very good starting point. The CO oxidation reaction occurs by a very similar mechanism on all three metals, so we will take rhodium as our example.

Let us first consider the reactivity of a rhodium surface with each of the three individual species involved in the reaction: carbon monoxide, oxygen, and carbon dioxide. We need to know the extent to which each reacts with the rhodium surface and how this may be influenced by the presence of other *adsorbed* species. We also need to know the nature of the surface reaction and the strength of bonding to the surface. This information is summarized in Table 1, for each of the three species. From this it can be seen that there is a high probability that a CO molecule will become adsorbed on striking a rhodium surface; that it will retain its molecular identity as a CO molecule, and that it will be held to the surface with a bond strength of *ca.* 120 kJ mol^{-1}. This is enough to hold the molecule on the surface indefinitely at room temperature. When the temperature is raised, however, the CO molecule will desorb (leave the surface) when the temperature exceeds *ca.* 460 K (Figure 8). This temperature is below that where the automotive catalyst normally operates, and that is an important feature of the catalyst. Even while it is on the surface the **CO molecule is mobile**, diffusing quite easily from one adsorption site to another.

Table 1 indicates that the behaviour of each of the three reactant species on the rhodium surface is quite different. Thus oxygen, like CO, reacts readily with the surface, with a high sticking probability. However, unlike CO, the **oxygen molecule dissociates** into two **adatoms**, immediately on hitting the surface. The oxygen–oxygen bond of the molecule is permanently lost, and the adatoms now react independently. Both adatoms are **very strongly bound to the surface**, and the heat of adsorption is so high that desorption does not occur until *ca.* 900 K, *above* the temperature at which the auto-catalyst often operates. There is another interesting feature of the rhodium/oxygen interaction, namely that adsorption of oxygen changes the *structure* of the rhodium surface. The change influences only the top two or three layers of surface atoms, but is readily

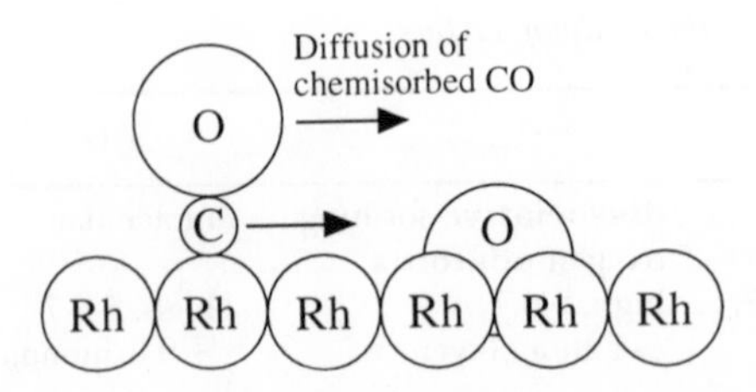

Figure 10 *Schematic view of the reaction between an oxygen adatom and a diffusing CO molecule*

detectable on model catalysts using methods such as low energy electron diffraction (LEED) or scanning tunnelling microscopy (STM).

Carbon dioxide interacts only **weakly** with the rhodium catalyst surface. The heat of adsorption is very small, so that once a CO_2 molecule is formed it desorbs from the surface immediately. This is clearly very desirable, as is the low sticking probability, which means that carbon dioxide in the gas phase will not adsorb and block the catalyst surface.

Knowing how the individual components interact with the catalyst surface, we can now consider the mechanism of catalytic reaction.[8] Since CO_2 interacts only very weakly with rhodium, we must look in detail only at oxygen and carbon monoxide. Careful kinetic and mechanistic studies have shown that the **reaction** occurs between an **adsorbed carbon monoxide molecule** and an **adsorbed oxygen atom**. The oxygen adatom is held in a position where it will be coordinated to several surface rhodium atoms. The carbon monoxide molecule is chemisorbed with the carbon atom bound to surface rhodium, but is able to diffuse rapidly across the surface.

The surface reaction, shown schematically in Figure 10, involves what would be described in homogeneous catalysis as a migratory insertion of oxygen into a Rh–C bond: the CO molecule 'climbs onto' the adsorbed oxygen atom, and metal–CO and metal–oxygen bonds are broken at the same time as the carbon–oxygen bond of the carbon dioxide molecule is formed. **The CO_2 molecule leaves the surface immediately it is formed**, carrying away with it some of the 283 kJ mol^{-1} exotherm of reaction, the remaining heat being absorbed by the catalyst. Spectroscopic analysis of the desorbed CO_2 show that it carries excess energy in vibrational, rotational, and translational modes before further collisions in the gas phase or with the surface.

The importance of this surface reaction between two adsorbed species was first recognized independently by Irving Langmuir, in the General Electric Co. Laboratories in the USA, and by Cyril Hinshelwood in Oxford University, England, and so has become known as a Langmuir–Hinshelwood reaction. The kinetics of a Langmuir–Hinshelwood reaction should be of the form:

$$\text{Rate} = A.\theta_a.\theta_b.\exp[-E(\theta_a\theta_b)/RT] \quad (5)$$

where θ_a and θ_b are respectively the coverages of a and b, (expressed as fractions of an adsorbed monolayer), and $E(\theta_a\theta_b)$ is the activation energy for reaction, with

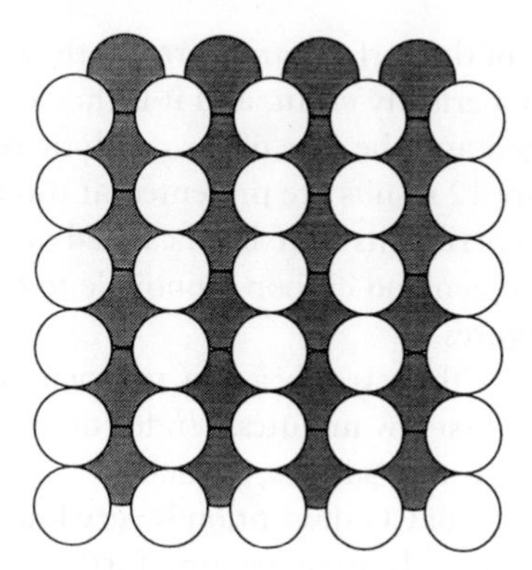

Figure 11 *Schematic view of the Rh(110) surface. Open circles represent the topmost layer of atoms, while the filled circles are the next layer below the surface. The diameter of a rhodium atom is 269 pm (2.69 Å)*

the term in brackets indicating that this will normally vary with the coverages of both a and b. A is a pre-exponential factor, which is assumed to be constant. For the CO oxidation reaction we may thus write:

$$-\mathrm{d}\theta_{\mathrm{O}}/\mathrm{d}t = 2F_{\mathrm{O_2}}.S_{\mathrm{O_2}}(\theta) + R_{\mathrm{CO_2}} \tag{6}$$

$$-\mathrm{d}\theta_{\mathrm{CO}}/\mathrm{d}t = F_{\mathrm{CO}}.S_{\mathrm{CO}}(\theta) + R_{\mathrm{CO_2}} - A_1.\theta_{\mathrm{CO}}.\exp(-E_{\mathrm{des}}/RT) \tag{7}$$

$$R_{\mathrm{CO_2}} = \mathrm{A}_2.\theta_{\mathrm{CO}}.\theta_{\mathrm{O}}.\exp[-E(\theta_{\mathrm{CO}}\theta_{\mathrm{O}}/RT] \tag{8}$$

where F_{O_2} and F_{CO} are respectively the fluxes of oxygen and CO impinging at the catalyst surface, S_{O_2} and S_{CO} are the coverage-dependent sticking probabilities of oxygen and CO, and R is the rate of production of CO_2. E_{des} is the activation energy for the desorption of CO, and A_1 *etc.* are pre-exponential factors.

The equation describing the **surface coverage of oxygen** and its variation with time contains only two terms: oxygen is adsorbed from the gas phase onto the catalyst surface and is removed by reaction to form CO_2. The first term shows that adsorption is proportional both to the sticking probability $S(\theta)$, which is a function of coverage, and the number of oxygen molecule impacts at the catalyst surface, which is related directly to the oxygen pressure. The heat of *adsorption* of oxygen is high, (see Table 1) and so *desorption* of molecular oxygen does not occur in the temperature range of importance for automotive catalysis. By contrast, the equation for CO contains three terms, the last showing that the rate of CO desorption from the catalyst surface may be rapid compared to the rate of reaction to form CO_2. E_{des} is the activation energy for CO desorption, which is approximately equal to the heat of adsorption. The equation for CO_2 formation is simply the Langmuir–Hinshelwood equation discussed earlier.

We now consider the consequences of these kinetics for the operation of the automotive catalyst for CO oxidation and we use some results obtained on the **model catalyst**, a single crystal of rhodium with the (110) surface structure shown in Figure 11. In most situations, the rate of this reaction is not very dependent on the details of the surface structure. In these experiments the single crystal catalyst is initially perfectly clean, and it is then exposed to an equimolar mixture of CO and oxygen and the rate of the catalytic reaction monitored as a function of time. In Figure 12 results are presented at three temperatures, 340 K, 400 K, and 450 K. The experiments have been carried out in a way which allows the surface coverage of oxygen and carbon monoxide to be monitored, and these are also shown in the Figures.

At the lowest temperature there is only a very transient reaction, and no further CO_2 is produced after the first few minutes. Under these conditions there is very little adsorbed oxygen on the surface, which is covered by CO. At this temperature, 340 K, the rate of CO desorption is very low, and carbon monoxide molecules block the catalyst surface, preventing further reaction. Increasing the temperature by only 60 K brings about a dramatic change: reaction now becomes very rapid and continues indefinitely. The single crystal is now behaving as a catalyst, and is effective enough to convert 50% of all CO molecules impinging on it into CO_2. The active area of rhodium in this experiment is about 7 mm^2 and the impingment rate of CO and oxygen is equivalent to a pressure of *ca.* 5×10^{-8} mbar.

Note that at 400 K the surface coverages have now altered considerably compared to those at 340 K. The coverage of CO has dropped to <0.2 monolayer (ML), while that of oxygen is now *ca.* 0.3 ML. The kinetic analysis indicates that the main factor which opened up the surface to allow catalytic reaction is that CO *desorption* is now occurring at a finite rate. At the lower temperature CO completely blocks the surface, preventing the dissociative adsorption of oxygen.

Under preferred operating conditions **carbon monoxide desorption** is often found to be **rate limiting**, as shown by Fisher *et al.*[9] in parallel studies of the Rh(100) single crystal plane and of Rh/Al_2O_3 supported catalysts. This results in kinetic expressions which are of order -1 in carbon monoxide, indicating that carbon monoxide is poisoning the surface, as was the case in our example at 340 K.

The comparison of the single crystal reactivity with that of the supported catalyst is interesting. Figure 13 shows that the turnover number, defined as the number of CO molecules reacting per second *per surface rhodium site*, is the same on both catalysts. It is noteworthy that the agreement extends over *ca.* 60 K, the full range where rates of reaction can be measured on the supported catalyst. This is taken to indicate that the **reaction is structure insensitive**, in other words it proceeds at the same rate on rhodium surfaces with different structures.

The CO oxidation reaction proceeds in a similar manner on all three of the noble metals in the catalyst, following the same general kinetics. Carbon monoxide interacts in a broadly similar way with all three metals, but rhodium is the most oxophilic and platinum the least of the three noble metals used. **Platinum** is thus the most effective catalyst for the CO oxidation reaction, although detailed comparative rates under relevant conditions are not available.

For palladium and rhodium catalysts, the rate of reaction slows in excess oxygen. On rhodium the effect is dramatic once the oxygen coverage reaches a full chemisorbed monolayer, and the mechanism of reaction changes. Studies using the recently developed scanning tunnelling microscope (STM) show that the reaction now must nucleate at steps on the catalyst surface, and only then is it able

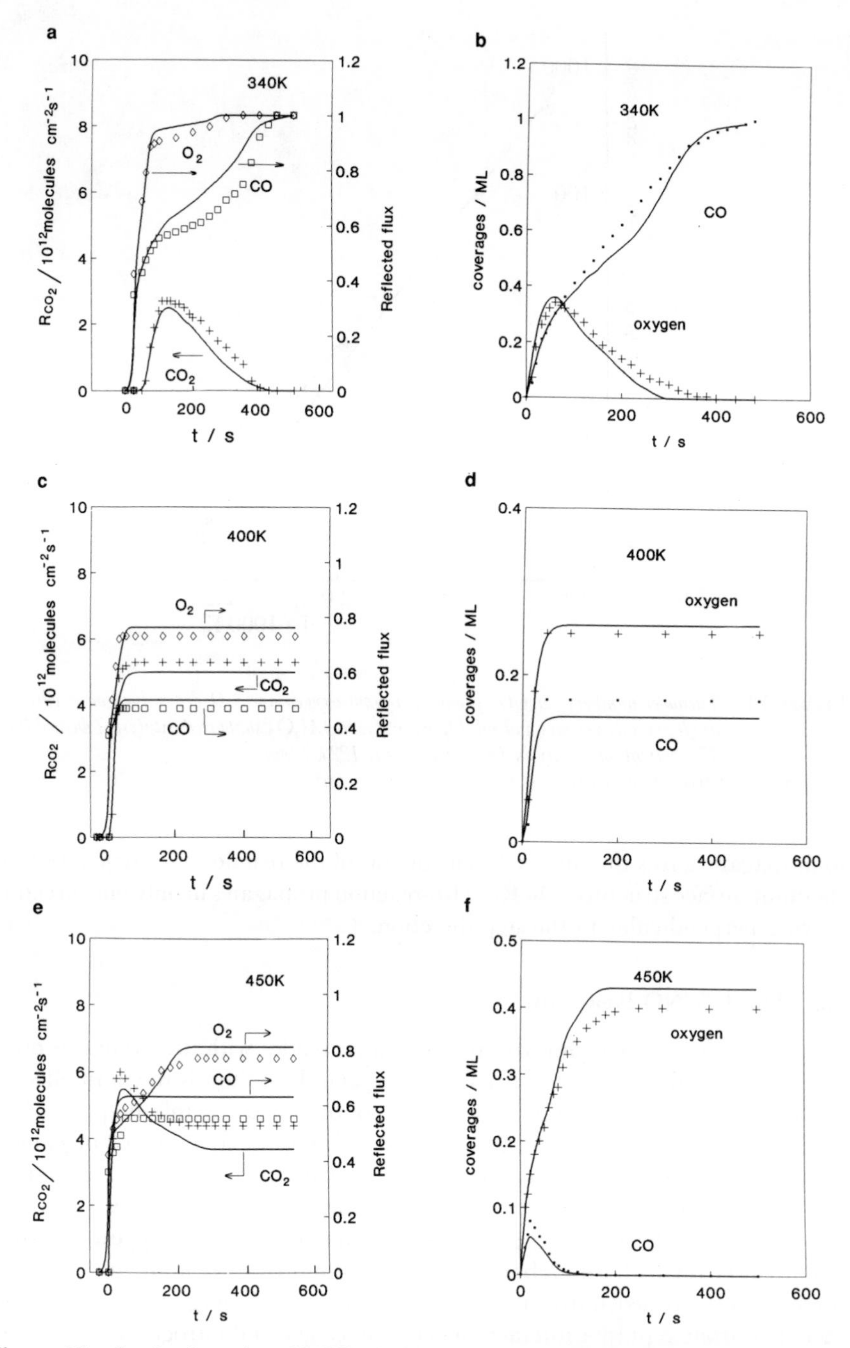

Figure 12 *Results of exposing a Rh(110) single crystal to a molecular beam containing equal amounts of CO and oxygen at three different temperatures. Figures a, c, and e show the rate of production of CO_2. Figures b, d, and f show how the coverages of CO and oxygen on the rhodium surface change as the reaction proceeds. In all cases the lines represent the experimental results, while the symbols result from the kinetic model described in the text.*

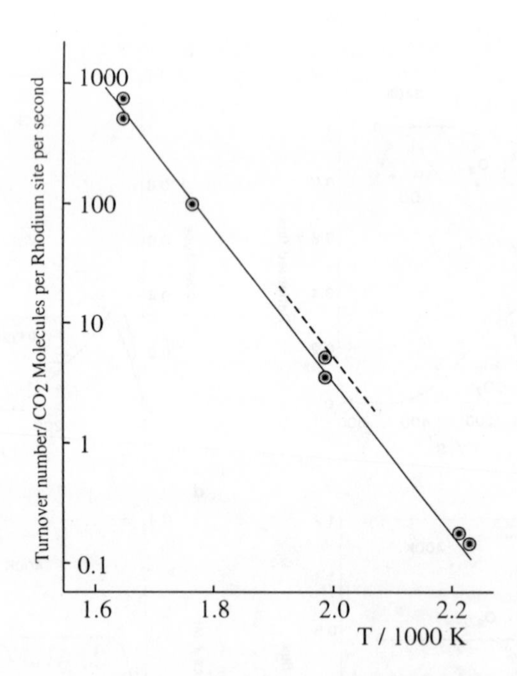

Figure 13 *Turnover numbers for CO_2 production from oxygen and CO determined on a Rh(100) single crystal, (circles and solid line) and a Rh/Al_2O_3 supported catalyst, (dashed line). The activation energy in both cases is ca. 122 kJ mol^{-1}*
(Reproduced from Oh, Fisher, *et al.*[9] with permission)

to propagate across the surface.[10] The details of the reaction now depend on the rhodium surface structure. On Rh(110) reaction propagates in only one direction, moving perpendicular to the step direction.

3.2 The CO/NO Reaction

The interaction of **nitrogen monoxide** (nitric oxide) with the catalyst is much more complex than that of either CO or oxygen. Rhodium is the most effective metal for the CO/NO reaction and owes its continuing use to this, despite its high cost. The interaction with the catalyst surface is complex, as indicated in Table 2. At low coverages, dissociative adsorption into nitrogen and oxygen adatoms predominates, but at higher coverages molecular adsorption takes over, with the molecule bound through nitrogen, in a similar way to CO. The **oxygen adatoms** may then **react** with **adsorbed CO**, as they do in CO oxidation, although the presence of adsorbed nitrogen modifies the activation energy for this reaction; there is a strong repulsive interaction between oxygen and nitrogen adatoms. The nitrogen adatoms may be removed from the surface in two ways:

$$N_{(ads)} + N_{(ads)} \rightarrow N_{2(g)} \quad (9a)$$

$$NO_{(ads)} + N_{(ads)} \rightarrow N_{2(g)} + O_{(ads)} \quad (9b)$$

Table 2 *Reactivity of nitrogen-containing species at rhodium surfaces*

	NO	*Nitrogen*
Nature of adsorption	**dissociative**, forming N+O adatoms, and **molecular**	adatoms
Initial sticking probability	high, 0.67	zero
Saturation coverage	~1 monolayer, initially dissociative, then molecular	not known
Desorption temperature	400 K (as NO)	550–700 K (as N_2)

The first route predominates at higher temperatures, ($T >$ *ca.* 450 K), while the second is more important in the lower temperature regime. Like CO_2, the nitrogen molecule does not adsorb on the catalyst surface, so desorbs as soon as it is formed. The temperature of nitrogen desorption depends strongly on which other species are present, particularly oxygen, and occurs over a broad band of temperature from 550–700 K.

3.3 Hydrocarbon Oxidation

Hydrocarbon oxidation has not been studied in as much detail as the other two main reactions in automotive catalysis. The outline of the mechanism may, however be deduced from what is known about metal/hydrocarbon interactions. All hydrocarbons adsorb on the three metals present in the catalyst, but the unburnt **aromatic** components of the fuel will **adsorb associatively** at ambient temperature. On warming in the absence of oxygen, benzene would decompose with desorption of hydrogen, leaving carbon on the catalyst surface. This decomposition will still occur in the presence of oxygen, but the hydrogen is then converted to water which will rapidly desorb, and the carbon is oxidized to CO_2. Saturated **aliphatic** hydrocarbons are **not adsorbed** at ambient temperature. Higher temperatures are required to remove the first hydrogen atom compared with unsaturated molecules, and this is the rate determing step:

$$C_nH_{(2n+2)} \rightarrow H_{(ads)} + C_nH_{(2n+1)(ads)} \quad (10)$$

Once this reaction has occurred, the hydrocarbon will fall apart on the surface as the temperature is increased, ending with all the hydrogen being stripped from the molecule, as was the case for benzene. Hydrogen is again readily oxidized to water, and the molecular fragments are converted to CO_2 and water. Of the three noble metals present in the catalyst, palladium is considered to be the best hydrocarbon oxidation catalyst, while rhodium is thought to be the least effective.

Under stoichiometric conditions, hydrocarbons must compete with CO for the available oxygen. In these conditions another mechanism of hydrocarbon removal may come into play, *i.e.* **steam reforming**:

$$C_nH_{(2n+2)} + n\ H_2O \rightarrow n\ CO + (2n+1)\ H_2 \quad (11)$$

Rhodium is an excellent steam reforming catalyst.

3.4 Sulfur Chemistry

Under stoichiometric conditions, ($\lambda = 1.00$), noble metals in the catalyst **oxidize SO_2 to SO_3**, much of which forms sulfate groups (SO_4^{2-}) on the alumina support. If the mixture is rich, ($\lambda < 1.00$) sulfur dioxide may be **reduced to H_2S**, resulting in the exhaust transiently having the characteristic smell of bad eggs. Sulfur in the exhaust generally acts as a catalyst poison, by adsorbing on the noble metal surface and blocking sites for reaction. Rhodium appears to be most strongly influenced by sulfur poisoning and is the least active of the noble metals of interest in SO_2 oxidation. **CO oxidation is suppressed** at low temperatures, while the response in hydrocarbon oxidation is complex. Sulfur has some unexpected benefits, for example it can reduce the susceptibility of the catalysts to lead poisoning (gasoline still contains trace levels of lead).

3.5 Catalyst Architecture

The automotive catalyst assembly is required to operate with very high efficiency at very high gas flows. As a result much care must be taken in the design and construction of the catalyst to ensure that these requirements are met. Of particular importance are the **pore size distribution** in the catalyst, the **metal particle size**, and the **distribution of the active metals within the pores**. The distribution of the reducible oxide promoters, such as ceria or zirconia, must also be optimized. Catalyst manufacturers concentrate strongly on these aspects of catalyst design, and precise details of optimum designs are closely guarded commercial secrets.

Two main factors influence the design: the requirement for effective mass transport so that the reactant gases can reach the catalysts and the products can be efficiently removed; and chemical considerations, to ensure that catalysis is effectively carried out with the minimum need for expensive resources, particularly precious metals. The mass transport needs usually require a network of both macropores ($d > 500$ Å) which carry most of the gas load, and mesopores (20 Å $< d < 500$ Å) where most of the precious metal component is located and catalysis occurs. Questions of catalyst effectiveness determine where the precious metals are located with respect to each other, and with respect to promoters such as zirconia or ceria.

4 FUTURE DEVELOPMENTS

Control of emissions from mobile sources represents a **very big business**, and one which is set for **continued growth**. However, this growth will also be very challenging because **legislation** is driving the pollution limits continually downward (see Table 3) with the ultimate aim of so-called zero emission vehicles (ZEVs) – normally electric cars powered by storage batteries. California law requires that 2% of all cars sold by 1998 be ZEVs. A more complete review of these challenges to catalyst designers is given by Cooper.[11]

Work on the improvement of current systems is continuing apace, with

Table 3 *US and California emission standards*

	Pollutant, g mile^{-1}		
Year and standard	*Hydrocarbons*	*Carbon monoxide*	*Nitrogen oxides*
1990 US	0.41	3.4	1.0
1994 US	0.25	3.4	0.40
1993 California	0.25	3.4	0.40
1994 (TLEV)	0.125	3.4	0.40
1997 (LEV)	0.075	3.4	0.20
1997–2003 (ULEV)	0.040	1.7	0.20

palladium-only catalysts showing utility as enhanced durability catalysts which can operate at higher average temperatures, close to the engine manifold. This ensures fast catalyst warm up upon starting the engine, enabling quicker light-off and therefore overall better conversion efficiency. Future systems may have current three-way Pt/Rh catalysts, combined with such Pd based convertors.

Another area of current research is the development of catalysts to operate with **lean-burn engines**, which work at high values of λ, while giving good fuel consumption and lower total emissions. The difficulty here is in achieving NO_x conversion under such oxidizing conditions. Some catalysts have been developed which can do this, in particular Cu-doped zeolites (which are crystalline microporous aluminosilicates), but they have short lives in the real exhaust situation due to hydrolytic destruction of the zeolite crystallinity at high temperatures.

Vehicles are being developed which use alternative, cleaner fuels, such as methanol or methane, in order to reduce total unconverted emissions. These too present pollution problems, for instance aldehyde emission, but many of these can be overcome with simple oxidation catalysts.

It is now recognized that, contrary to initial claims, **diesel engines** contribute significantly to urban pollution. They emit a toxic cocktail, perhaps the most significant component of which is the particulates – soots with a variety of chemicals present on their surface including heavy metals, sulfates and partially combusted hydrocarbons. Control of these pollutants represents a continuing problem for the industry, and one which has not yet been satisfactorily solved.

The major challenge of the future may be **CO_2 emissions**, especially if the **greenhouse effect** is recognized to be taking on disastrous proportions. Then legislation may mitigate against CO_2-producing artificial processes. In the near future so-called **'zero emission' vehicles** will be present on our roads. These will take the form of battery or fuel-cell powered units, but neither are truly zero emission since the former energy depends on power station supply (generally from oil, gas, or coal), while the latter case depends on the use of methanol as a source of hydrogen. These all produce CO_2 from combustion or reforming. Solutions to this problem may not involve catalysis. For instance, electric power can be provided by non-polluting sources (hydroelectric or 'alternative' energy), although probably not universally at current vehicle demand. However, the quest for the ZEV may yet involve catalytic technology. The ultimate approach is to use **hydrogen as the power source**, the combustion process converting it back to water again. Hydrogen may be made from water, by a pollution free route, using sunlight and a precious metal/photoactive supported catalyst to split water into its elements (see Chapter 18). However these photocatalysts are not yet efficient enough to represent a cheap and fast source of hydrogen. Development of these alternatives are continuing apace, with the aim of providing a clean source of power for our transport in the future.

Acknowledgements

The authors are grateful to a number of people from Johnson Matthey plc for help in the preparation of this article and in particular Drs Andy Walker, Jack Frost, and Alan Diwell.

5 REFERENCES

5.1 Specific References

1. P. Brimblecombe, 'Air Composition and Chemistry', Cambridge University Press, Cambridge, UK, 1986, p. 97.
2. See, for instance, K. C. Taylor, in 'Catalysis – Science and Technology', ed. J. R. Anderson and M. Boudart, Springer-Verlag, Berlin, 1984, vol. 5, p. 119.
3. A. F. Diwell, R. R. Rajaram, H. A. Shaw, and T. J. Truex, in 'Catalysis and Automotive Pollution Control II', ed. A. Crucq, Elsevier, Amsterdam, 1991, p. 139.
4. T. Root, L. D. Schmidt, and G. B. Fisher, *Surf. Sci.*, 1983, **134**, 30.
5. P. Millington and R. Burch, to be published.
6. M. Bowker, Q. Guo, and R. W. Joyner, *Surf. Sci.*, 1991, **253**, 33.
7. See, for instance, C. Cucchi and M. Hubin, in 'Catalysis and Automotive Pollution Control II', ed. A. Crucq, Elsevier, Amsterdam, 1991, p. 44.
8. M. Bowker, Q. Guo, and R. W. Joyner, *Surf. Sci.*, 1993, **280**, 50.
9. S. H. Oh, G. B. Fisher, J. E. Carpenter, and D. W. Goodman, *J. Catal.*, 1986, **100**, 360.
10. F. Leibsle, P. Murray, S. Francis, G. Thornton, and M. Bowker, *Nature (London)*, 1993, **363**, 706.
11 B. J. Cooper, *Platinum Met. Rev.*, 1994, **38**, 2.

5.2 Reviews in the Area

K. C. Taylor, *Catal. Rev. – Sci. Eng.*, 1993, **35**, 457.

Also by K. C. Taylor: *Chemtech*, 1990, 551; reference 2 above; and 'Automobile Catalytic Converters', Springer-Verlag, Berlin, 1984.

'Catalysis and Automotive Pollution Control' and 'Catalysis and Automotive

Pollution Control II', Studies in Surface Science and Catalysis volumes 30 and 71, Elsevier, Amsterdam, 1987 and 1991. These are the proceedings of the first two international conferences in this field and contain both general and detailed papers.

'Catalysts for the Control of Automotive Pollutants', ed. J. McEvoy, American Chemical Society, Washington, DC, 1975.

B. Harrison, M. Wyatt, and K. G. Gough, in 'Catalysis', ed. G. C. Bond and G. Webb, Specialist Periodical Reports, The Royal Society of Chemistry, London, 1982, vol. 5, p. 127.

5.3 Relevant Journals

Journal of Catalysis
Applied Catalysis (especially part B – Environmental)
Surface Science
Applications of Surface Science
Platinum Metals Review
Catalysis Letters
Catalysis Reviews – Science and Engineering
Advances in Catalysis
Catalysis Today

CHAPTER 8

Fast Ion Conductors

ROBERT C. T. SLADE

1 INTRODUCTION

In most solids ions can migrate only via inherent thermodynamic point defects (vacant sites, the inclusion/generation of which raises the entropy of solids at temperatures greater than 0 K), such as those left by ions which have moved to otherwise unoccupied interstitial sites (Frenkel defects), or generated by omission of stoichiometrically equivalent numbers of anions and cations from the structure (Schottky defects). Such materials have temperature-dependent defect (and charge carrier) concentrations, high activation barriers to ion migration, and consequently have low ionic conductivities.

In materials known as **fast ion conductors** or **superionic conductors**, at least one ion type (cation or anion) is present with a high charge carrier concentration and usually with small activation barriers. Ionic conductivities are consequently much higher for fast ion conductors (the terminology is somewhat confusing; the ions may not be exceptionally mobile, and there are no 'superions' present). Where ion migration is the sole electrical conduction mechanism, such materials are termed **solid electrolytes**. Where conduction occurs both ionically and electronically (either within a band or by electron-hopping), those materials are termed **mixed conductors**. The potential for application of fast ion conductors has led to intense activity internationally. The subject area is widely known as **solid state ionics** (also the title of a dedicated international journal and of regular international conferences) and is promoted by the International Society for Solid State Ionics (the address of which is given at the end of this chapter).

Both solid electrolytes and mixed conductors have their own applications (some of which are outlined in greater detail later in this chapter), and when combined they offer the possibility of all-solid-state electrochemical devices. Industrial applications for solid electrolytes include **battery systems**, **fuel cells** (*e.g.* producing electricity by electrochemical reaction of hydrogen and oxygen), and **sensors** (for analytical determination of gases such as hydrogen and oxygen).

Use of cell reactions involving alkali metals gives batteries with high cell emfs; sodium β-alumina is an Na^+ conductor targeted as the ceramic electrolyte membrane for the Na/S battery. H_2/O_2 fuel cells employ either H^+ or O^{2-}

conductors; in autonomous fuel cells at near ambient temperatures (such as might be used in transportation or other applications requiring transportability) the H^+-conducting fluoropolymer NAFION® is used, whilst in utility applications (*e.g.* power stations) cells at 900–1000°C with O^{2-} conducting yttria-doped zirconia ceramic are employed. Various electrolytes are applied in potentiometric sensors (*e.g.* zirconium hydrogen phosphate in hydrogen sensors). Mixed conduction occurs on insertion of cations into bulk materials functioning as cathodes, with the charge balance being maintained by electrons passing through the external electrical circuit. Discharge of a cell containing an Li anode and a TiS_2 cathode (and an Li^+-conducting non-aqueous electrolyte) forms a compound Li_xTiS_2 (x increasing with extent of reaction), which has a wide composition range and is an Li^+ and electronic conductor.

A cell reaction inserting M^+ (M=H, alkali metal) into a thin WO_3 film (colourless) to produce M_xWO_3 (blue) provides a basis for **electrochromic displays** and **'smart windows'** – writing/colouration and erasing/bleaching are accomplished by passage of a current to perform the insertion and reverse reactions respectively. An Anglo–Danish project led to the development of an **all-solid-state battery system** with a Li alloy as anode, a Li^+-conducting polymer electrolyte, and V_6O_{13} as cathode (prior to discharge/Li^+-insertion).

This chapter will first outline some basic principles in solid state ionics. Examples of fast ion conductors classified in terms of the mobile ion are given. An overview of the techniques used in the characterization of fast ion conductors at both the bulk and atomic levels are presented. A section is devoted to the application of solid state ionics in electrochemical devices ('solid state ionic devices'). Finally, an overview of the current state of the art; and future trends, directions, and prospects is given.

2 SOME BASIC PRINCIPLES OF SOLID STATE IONICS

2.1 Characteristics of Fast Ion Conductors

Classification of a material as a fast ion conductor, or otherwise, is commonly made on a phenomenological basis, by consideration of the **ionic conductivity σ_i** of that material at a temperature of interest for applications. The choice of a minimum acceptable σ_i value is arbitrary and depends on the application; the minimum conductivity required for use in a potentiometric device (*e.g.* a sensor) can be orders of magnitude lower than in a high current device (in which ohmic losses due to resistive heating must be minimized). In his text on solid state chemistry, West[1] classifies fast ion conductors as those materials for which $\sigma_i \geq 10^{-3}$ S cm^{-1} (S = Siemens ≡ ohm^{-1}), the cut-off being the conductivity characteristic of NaCl (in which ion migration is via inherent thermodymamic defects) at just below its melting temperature (801 °C). A conductivity as low as 10^{-6} S cm^{-1} may, however, be acceptable in sensor applications. Van Gool[2] suggested that for a fast ion conductor to have applications potential, the activation energy for ionic conduction should be much less than the energy for point defect formation ($\sim$100 kJ mol^{-1} in a close-packed ionic solid).

Materials that can function as fast ion conductors include crystalline solids, amorphous materials, glasses, polymers, and ceramics. The ionic charge carrier can be univalent or divalent, and the fast ion conductor can be either a solid electrolyte or a mixed conductor. When selecting materials for device applications, the intended operating temperature and the thermal stability of the fast ion conductor are very significant considerations. Some materials functioning as fast ion conductors at, or near, room temperature decompose at elevated temperatures, while others which are essentially insulating (have very low σ_i) at ambient temperature become fast ion conductors at high temperatures.

A set of basic requirements for a material to be a potential fast ion conductor has been set out[3] as follows:

(i) a large number of potential charge carriers/mobile ions
(ii) an excess of acceptable sites for those ions
(iii) a small energy difference between ordered and disordered distributions of those ions in the structure
(iv) a low activation barrier for the motion of charge carriers between sites
(v) a rigid framework structure through which the ions can migrate
(vi) a highly polarizable framework
(vii) thermal and chemical stability in the intended device environment

Known solid electrolytes conform to these conditions to varying degrees.

The situation in mixed conductors is more complicated than that in solid electrolytes. The electronic conductivity can be much higher than the ionic conductivity, making the latter contribution difficult to establish. Further, in device applications it is the chemical diffusion coefficient, D_c, (pertaining to ion migration under an electrochemical potential gradient) that is of primary interest, and the transport of ions to the cathode particles can reflect processes both in the intergranular region (the detailed microstructure and packing of particles within the electrode) and at the electrolyte–cathode interface.

2.2 Ionic Conductivity at the Bulk and Atomic Levels in Solid Electrolytes

It can be far from trivial to establish unequivocal links between the ionic conductivity (a bulk phenomenon) of a solid electrolyte and atomic level processes (such as intersite jumps of mobile ions). The fast ion conductor is rarely studied as a single crystal. The empirical ionic conductivity can contain contributions from conduction via **particle surfaces** and via **intergranular regions**. The techniques of electrochemical ac impedance spectroscopy may assist in assigning the conductivity as being intragranular (within the bulk of the solid) or via particle surfaces / grain boundaries. Apparent inconsistencies between conductivity results and predictions based on results revealing intraparticle atomic-level detail can often be traced to failure to consider other conduction pathways. A further complication is that ionic conductivities measured by ac techniques involve no passage of electrical charge overall, while those values measured by dc techniques (more directly related to intended device applications) involve passing charge.

Ionic conductivities measured by ac and dc techniques may appear to differ; this inconsistency can arise if phenomena at electrolyte–electrode interfaces and cell polarization in dc studies are not considered.

The case of **ionic conduction** by ion migration **in a single crystal** solid electrolyte can be considered as an illustration of the possible link between the atomic-level jumping of charge carriers and the bulk-level ionic conductivity. The ionic conductivity σ_i is related to the ionic self-diffusion coefficient D_i (that pertaining to ion migration in a material with uniform charge carrier concentration) by the **Nernst–Einstein equation**:

$$\sigma_i = \left[\frac{Nq^2}{kT}\right] D_i H_R \tag{1}$$

where N is the concentration (ions per unit volume) of charge carriers, q is the charge per carrier, k is the Boltzmann constant, T is the temperature in Kelvin, and H_R is the Haven ratio (a correlation factor). H_R lies in the range 0–1 and depends on the detail of the conduction mechanism; for a solid electrolyte with a large excess of available sites for the mobile ion, $H_R \approx 1$. For an isotropic random walk of ions through the crystal, D_i is related to the jump frequency, ν, by the equation

$$D_i = \frac{\lambda^2 \ \nu}{6} \tag{2}$$

where λ is the distance between neighbouring sites. For an activated jump

$$\nu = \nu_o \exp\left(\frac{-E}{RT}\right) \tag{3}$$

where E is the jump activation barrier and R is the gas constant. Combining equations (1)–(3):

$$\sigma_i T = \left[\frac{Nq^2\lambda^2 H_R \nu_o}{6k}\right] \exp\left(\frac{-E}{RT}\right) \tag{4}$$

Ionic conductivities of solid electrolytes often display Arrhenius-type temperature-dependencies:

$$\sigma_i = A \exp\left(\frac{-E}{RT}\right) \tag{5}$$

i.e. there is a linear variation of $\ln(\sigma_i)$ with $1/T$. An activated-hop conduction mechanism would, on the basis of the preceeding paragraph (equation (4) in particular), lead to

$$\sigma_i T = A\ exp\left(\frac{-E}{RT}\right) \tag{6}$$

i.e. a linear variation of $\ln(\sigma_i T)$ with $1/T$. Experimental data can often be fitted satisfactorily to both equations (5) and (6) over the restricted experimental temperature ranges accessed, with slight differences in empirical E values. There is no convention concerning illustration of temperature dependencies; some authors plot $\ln(\sigma_i)$ versus $1/T$ {following (5)}, while others plot $\ln(\sigma_i T)$ versus $1/T$ {following (6)}.

The discussion so far has assumed that the charge carrier density does not vary with temperature in solid electrolytes; as will be seen in some of the examples below, this is not always the case. In the presence of a temperature-dependent carrier concentration, the empirical E will include a contribution from that temperature dependence, and hence will not be directly related to the activation barrier for charge migration.

3 EXAMPLES OF FAST ION CONDUCTORS

The first record of conductivity measurements on fast ion conducting solids (not recognized as such at the time) are to be found in the diaries of Michael Faraday, in which he reports high conductivities for hot Ag_2S (1833) and for hot PbF_2 (1834).[4] In 1914 Tubandt and Lorentz provided the first direct evidence that ions in a solid can carry an electrical current; the changes in mass of silver electrodes separated by AgI and between which a known charge was passed were in accord with theoretical predictions.[5]

In this section examples of fast ion conductors are presented, with an emphasis on the structural aspects of those materials. The examples are classified according to the conducting ion.

3.1 Silver Ion Conductors

Silver iodide is a poor conductor at room temperature, the structure then containing tetrahedrally coordinated Ag^+ ions in a close-packed array of I^- ions. At $T \geq 146\,°C$ the array of I^- ions is body centred cubic, with Ag^+ ions distributed over an excess of interlinked 3- and 4-coordinated sites (see Figure 1) and highly mobile. The high temperature form (α-AgI) is a fast ion conducting solid electrolyte ($\sigma_i \sim 1\ S\,cm^{-1}$, comparable with liquid electrolytes).

It has proved possible to stabilize a disordered arrangement of Ag^+ ions to lower temperatures than for α-AgI by substitution of some of the Ag^+ by a variety of other monovalent cations. For instance, the compound $RbAg_4I_5$ has σ_i (25 °C) = 0.25 S cm^{-1}. In this compound Rb^+ and I^- ions define a rigid structure containing an excess of tetrahedral sites through which Ag^+ ions can move.

While **silver halides** are **solid electrolytes**, **silver chalcogenides** and related derivates are **mixed conductors**. Thus the phenomena reported by Faraday for high temperature α-Ag_2S pertained to a mixed conductor.

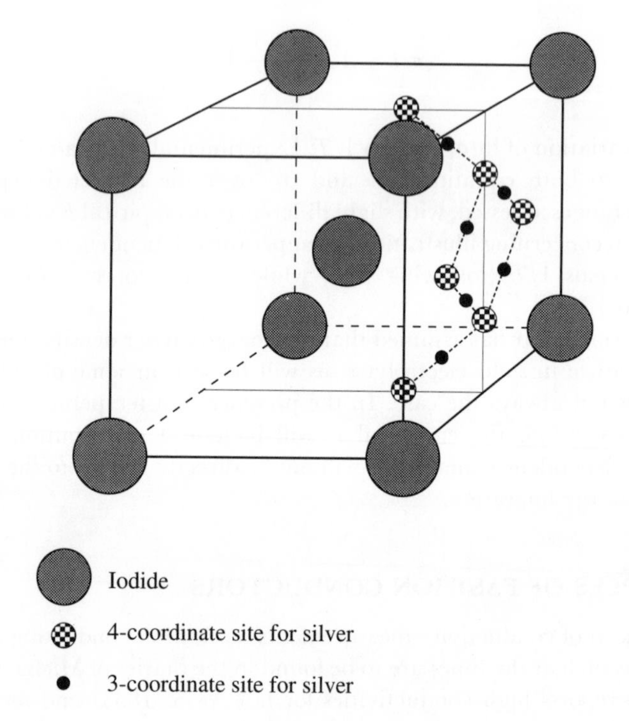

Figure 1 *The structure of α-AgI. Ag^+ ions migrate between an excess of 3- and 4-coordinated sites defined by the body centred cubic array of immobile iodide ions. All faces of the cube are equivalent*

3.2 Alkali Ion Conductors

Alkali ion conductors can be used as **solid electrolytes** in **battery systems** based on the reactions of alkali metals. A major impetus in the area of alkali ion conductors came with studies in the 1960s, by the Ford Motor Company and others, of β-aluminas[6] (sodium β-alumina being a potential ceramic electrolyte for the sodium–sulfur battery, see later).

β-aluminas are compounds of the general formula $M_2O.nAl_2O_3$ with $5 \leq n \leq 11$, the name for these compounds arising from the original belief that a new polymorph of alumina had been found. The structures (illustrated in Figure 2) of these compounds contain spinel-like oxide blocks (close packed oxygens with aluminiums located in octahedral and tetrahedral interstices), which are held apart by oxygen pillars. The interblock regions are the location for the mobile charge carriers, M^+, and for their migration in 2 dimensions; they are therefore described as the conduction planes. There are 2 different structures for such compounds, the β and β'' types (see Figure 2), which differ in the stacking of the spinel-like blocks. For the β-type the 'ideal' structure corresponds to $n = 11$, and lower values of n correspond to insertion of extra M_2O into the conduction planes. **Sodium β-aluminas** have the **highest conductivities** (Na^+ being the mobile

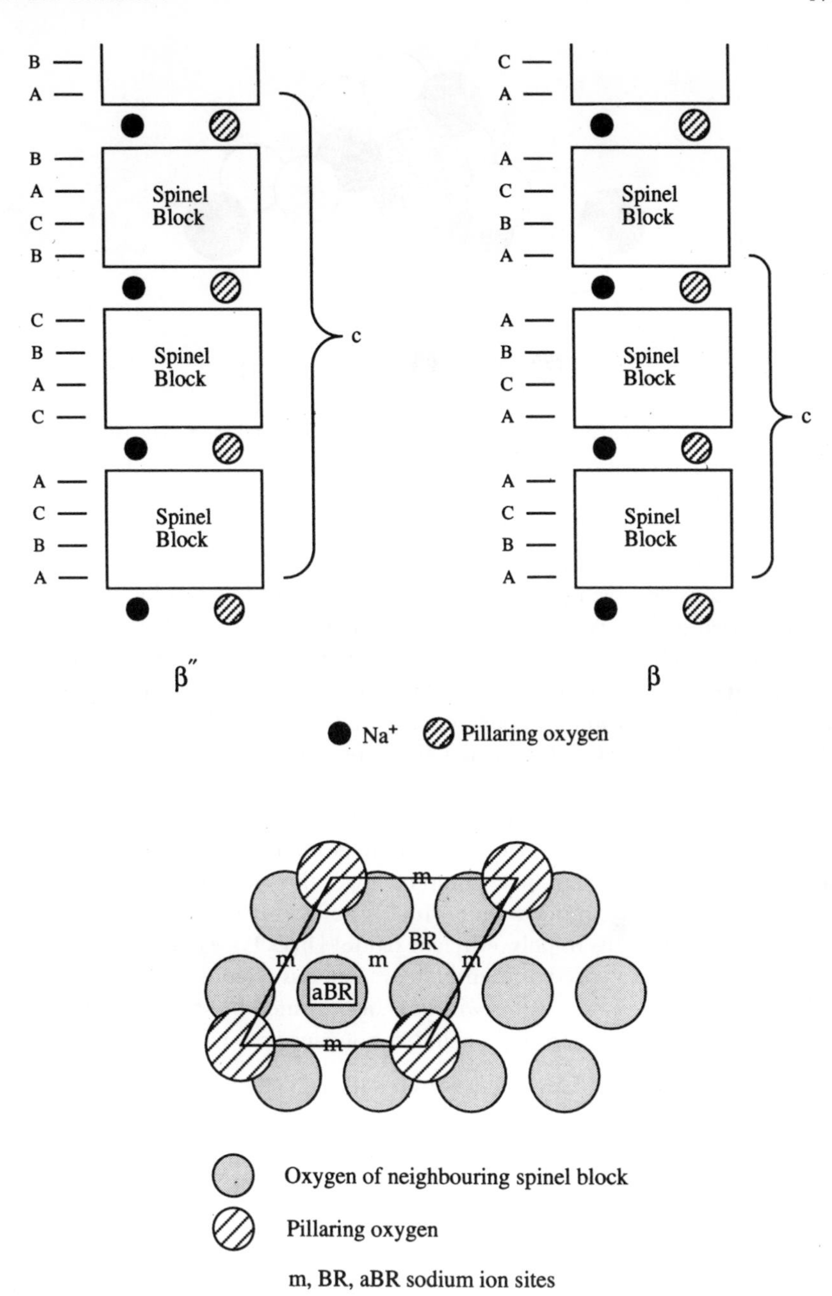

Figure 2 *The structures of sodium 'β-aluminas'.*
TOP: The β and β″ forms differ in the stacking of the spinel blocks, and consequently have different repeat distances (c) in the stacking direction. A, B, C define the stacking of hexagonally-packed oxygen layers.
BOTTOM: The interblock region (the conduction plane) in the β form contains an excess of sodium ion sites of three types; Beevers–Ross (BR, lowest in energy), anti-Beevers–Ross (aBR), and mid-oxygen (m)

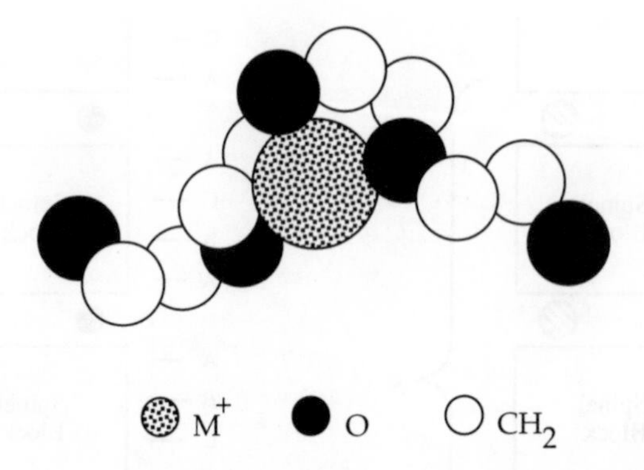

Figure 3 *Coordination of a metal ion M^+ by a poly(ethylene oxide) chain. M^+ migration involves segmental motion of the polymer backbone and transfer of M^+ to a new coordination site*

ion). Ambient temperature conductivities are not suitable for battery applications, but these materials have potential at *ca.* 300 °C (see later).

There is considerable interest in materials with high Li^+ conductivities, which could be used in cells with lithium anodes, giving EMFs higher than analogous sodium-based cells. The silicates and germanates of lithium are **moderate Li^+ conductors** at 300–400 °C, with Li^+ moving between face-sharing polyhedra defined by isolated tetrahedral silicate/germanate ions. $Li_{14}ZnGe_4O_{16}$ (LISICON, LIthium SuperIonic CONductor) is a highly promising Li^+ conductor [σ_i 300 °C) ≈ 1 S cm^{-1}].

Investigation of **solid polymer electrolytes** is exciting much contemporary interest;[7] the processability of organopolymers offers the prospect of thin film electrolytes (for which resistances could be lower than with crystalline materials) for battery or display applications. Much interest in this area has centred on solutions of lithium salts in poly(ethylene oxide) [P(EO)], *e.g.* $P(EO)_8LiCF_3SO_3$. Li^+ conduction occurs predominantly in amorphous regions of the polymer film, with Li^+ coordinated to oxygens within helices formed by the host polymer (see Figure 3). Criteria for choosing a suitable polymer host have been stated[7] as

(1) Formation of coordinate bonds between the polymer host and the guest cation
(2) A low barrier to the bond rotation necessary for segmental motion of the polymer chain
(3) A distance between coordinating atoms that will allow more than one polymer-to-cation bond to form.

Mixed conductors (conducting both alkali metal ions and electrons, see earlier) can be formed by chemical or electrochemical insertion (intercalation) of alkali metals into reducible hosts, such as TiS_2 (which has the layered CdI_2-type structure). Thus Li_xTiS_2 ($0 \leq x \leq 1$) can be prepared by reaction of the parent chalcogenide with n-butyllithium, or by discharging a cell containing a lithium anode, a TiS_2 ($x = 0$) cathode (x increases reversibly as discharge progresses), and an Li^+-conducting electrolyte (which can also be a solid). The inserted Li is present as Li^+ in the interlayer region, with the electrons to maintain charge

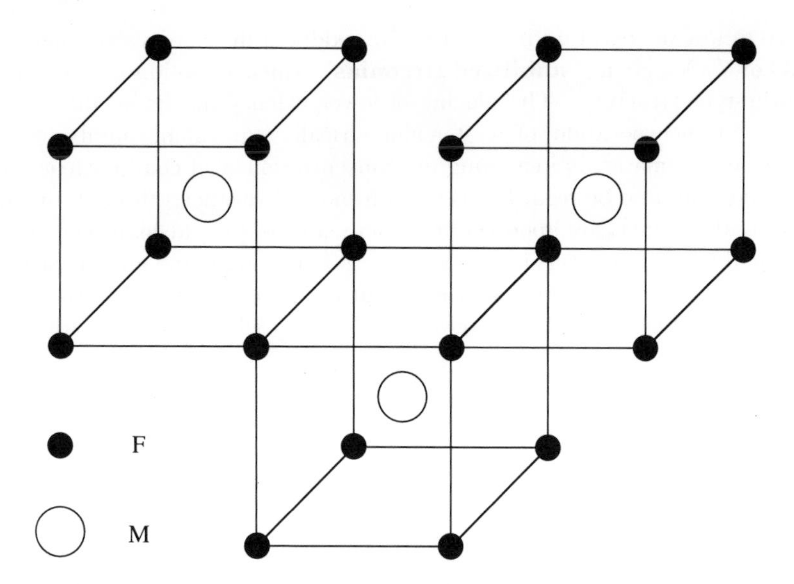

Figure 4 *The fluorite (MF_2) structure, showing the presence of alternate empty cubes defined by a primitive cubic array of fluorines*

balance being injected into the conduction bands of the TiS_2 host. Mixed conductors can also result from deintercalation of alkali metal ions from ternary oxides, *e.g.* Li_xCoO_2 formed by electrochemical deintercalation of Li^+ from $LiCoO_2$. The parent $LiCoO_2$ has a layered structure with Co and Li located in alternate layers of octahedral sites; reversible Li deintercalation occurs in planes separated by CoO_2 layers.

3.3 Anion Conductors

Fast anionic conduction typically occurs at much higher temperatures than is the case for metal ion conduction. This is because the conduction is via a **defect mechanism** and because **activation energies** for ion migration are generally higher for anions than for cations (typically the smaller of the two ionic species present). The best known anionic conductors (such as lead(II) fluoride and stabilized zirconia, see below) are based on the fluorite (CaF_2) structure type. In that structure the anion array is primitive cubic, and alternate cubes are filled by cations (as shown in Figure 4); the remaining cubes are empty in the ideal structure (which is only at thermodynamic equilibrium at 0 K). Anion migration can occur following displacement of an anion into an empty cube (onto an interstitial site, creating a Frenkel defect), leaving vacant a site into which another anion can migrate. The defect concentration is temperature dependent (and hence so is the charge carrier concentration $\mathcal{N}$). PbF_2(s) displays a 'superionic' conductivity at $T \geq 500\,°C$ ($\sigma_i \sim 5$ S cm^{-1}).

The structure of zirconia (ZrO_2) varies with temperature. The high temperature polymorph has a cubic fluorite structure. The cubic structure can be stabilized down to low temperature by doping with oxides of di- or tri-valent metals (*e.g.* CaO or Y_2O_3), giving **'stabilized zirconias'** (which are solid solutions with the fluorite-type structure). The placing of lower valency metals on the Zr^{IV} sites results in incorporation of a stoichiometrically equivalent number of O^{2-} vacancies, dramatically increasing the concentration N of charge carriers (O^{2-} migration can also be regarded as the transport of vacancies through the solid). Stabilized zirconias are good ceramic conductors of O^{2-} at high temperatures, despite the charge carriers carrying two negative charges and energy barriers to migration being high. For instance, conductivities of calcia-stabilized zirconias (*e.g.* 0.15 CaO : 0.85 ZrO_2) are σ_i (1000 °C) $\approx 5 \times 10^{-2}$ S cm^{-1} with $E \approx 130$ kJ mol^{-1}.

3.4 Proton Conductors

Fast conduction of electrical charge through solids by proton (H^+) migration has been much less widely publicized than other topics within the solid state ionics field, and has often been omitted from general discussions of ionic conduction phenomena. The underlying science is, nonetheless, approaching maturity, and application of proton-conducting materials in devices such as **fuel cells, sensors, and electrochromic displays** is being very seriously investigated.[8] Protonic conduction is often present in systems in which it may not, at first glance, be anticipated: **smectite clays** conduct protons due to the acidity of hydrated interlayer cations, the operation of the glass membrane **(pH) electrode** depends on **protonic conduction in soda glass**, and protonic conduction is even found in some oxide ceramics (see below). Protonic conduction, and devices employing it, can be considered in 3 temperature ranges: (a) near ambient T, (b) medium T (150–350 °C), and (c) high T (>600 °C). Until *ca.* 1980 materials for use at medium and high temperatures were considered unlikely.

Protonic conductors at near-ambient temperatures are hydrated (with H^+ migration by protons hopping between H_2O and H_3O^+ or OH^-) and include materials as diverse as H-form inorganic ion exchangers, heteropolyacids, and NAFION®. In inorganic systems conduction via the surfaces of particles often dominates conduction through a sample. NAFION® is a fluorocarbon polymer with sulfonic acid side-chains, and has found application in solid polymer fuel cells (SPFCs). The chemical formula of NAFION® and the nature of the conducting region are illustrated in Figure 5.

Studies in the medium temperature range have established protonic conduction in H_3O^+-exchanged and phosphoric-acid-bonded forms of known alkali ion conductors such as NASICON ($Na_{1-x}Zr_2Si_xP_{3-x}O_{12}$, $0 \le x \le 3$) and in anhydrous acid salts such as $CsHSO_4$.

The occurrence of moderate protonic conductivities in ceramics at high temperature was contrary to the accepted wisdom at the beginning of the 1980s, but systems such as M_2O_3-doped (M = Y, rare earth) $BaCeO_3$ ceramics with a perovskite structure do show promise. Protonic conductivity in such materials is a result of reactions with moisture and/or H_2. For instance, the oxygen vacancies introduced on inclusion of the dopant oxide can result in reaction of some of the

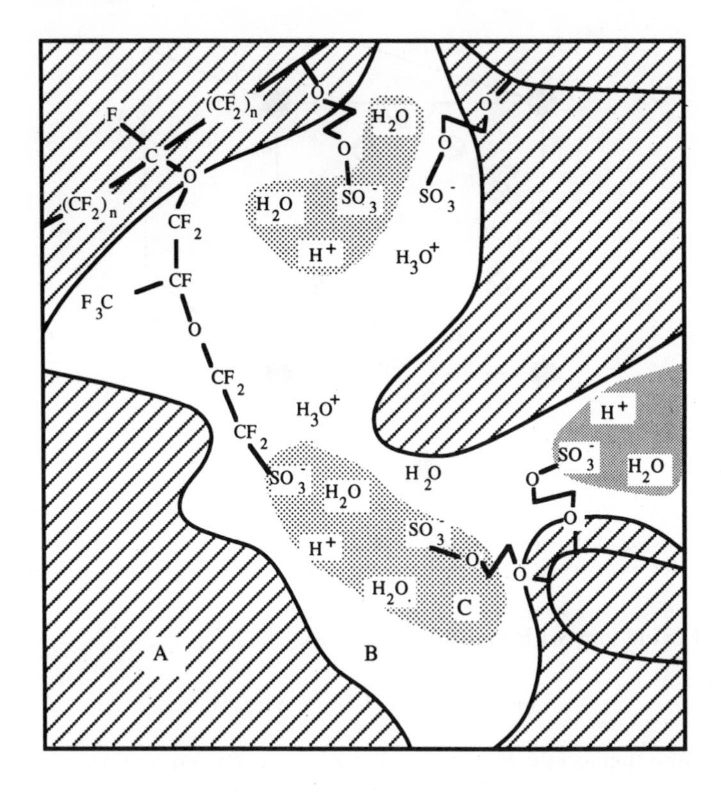

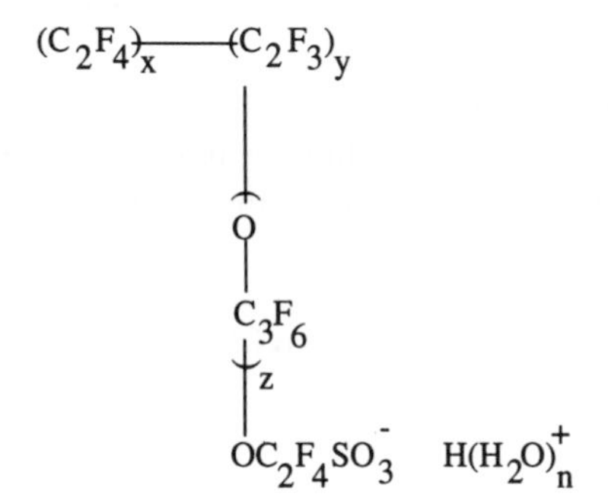

Figure 5 *The nature of the hydrogen form of NAFION®. BOTTOM: Chemical formulation of the polymer. TOP: A structural model defining 3 regions –* A *is the fluorocarbon backbone,* B *is a largely void volume containing some charge carriers,* C *is an ion-cluster region containing sulfonate groups and hydrated protons*

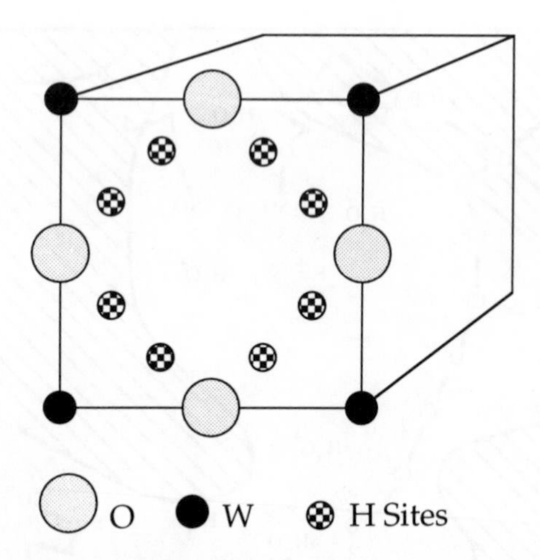

Figure 6 *Idealized structure of H_xWO_3 : x <0.5, all faces are equivalent, and no oxygen carries more than one hydrogen. There is a large excess of hydrogen sites between which migration can occur*

structural oxide ions with moisture as follows

$$\textit{vacancy} + O^{2-} + H_2O(g) \rightarrow 2\ OH^- \quad (7)$$

H^+ migration then occurs by hopping between sites corresponding to attachment to neighbouring oxide ions (*viz.* $O_a{}^{2-} + O_bH^- \rightarrow O_aH^- + O_b{}^{2-}$). The equilibrium position of Reaction (7) lies to the left; H^+ concentrations are low and temperature-dependent, but ionic mobilities are very high at high temperature. Under working conditions of technological interest, the conductivity is dominated by migration of H^+, with the contribution due to O^{2-} migration (by hopping into vacancies; *i.e.* in a manner similar to that in the stabilized zirconias, which have a different structure) being much smaller.

Mixed conduction (ionic + electronic) in which protons are the migrating ions occurs in the nonstoichiometric **hydrogen oxide bronzes** and related compounds. The bronzes are formed from parent oxides by chemical reduction (*e.g.* with zinc and hydrochloric acid) or by electrochemical insertion of H^+ (accompanied by addition of e^- into the conduction band of the oxide host to maintain charge balance). These bronzes are intensely coloured. For instance, formation of hydrogen tungsten bronze is described by

$$\underset{\textit{yellow}}{WO_3} + xH^+ + e^- \rightarrow \underset{\textit{blue}}{H_xWO_3} \quad (8)$$

The sites for H^+ within the bronze, and between which ion migration occurs, are illustrated in Figure 6 and correspond to attachment to oxygens of the ReO_3-type framework (*i.e.* as tungsten-coordinated hydroxyls, –OH). The analogous molybdenum bronzes are of a different structure (the parent oxide being layered) and H can be present both in –OH and in molybdenum-coordinated water

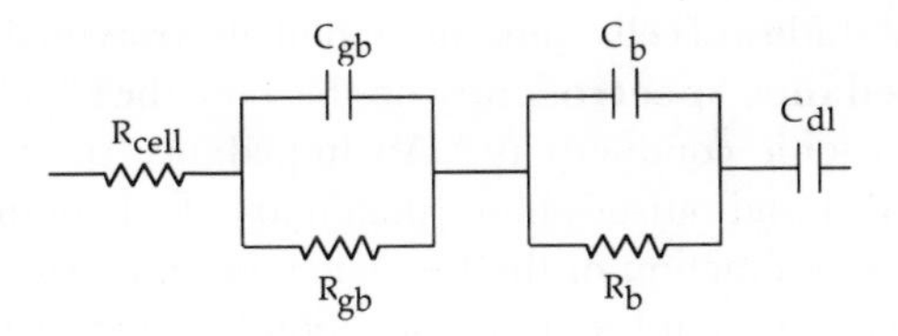

R_{cell} 250 ohm : R_{gb} 1750 ohm : R_b 3000 ohm
C_{dl} 75 μF : C_{gb} 0.05 μF : C_b 1 μF

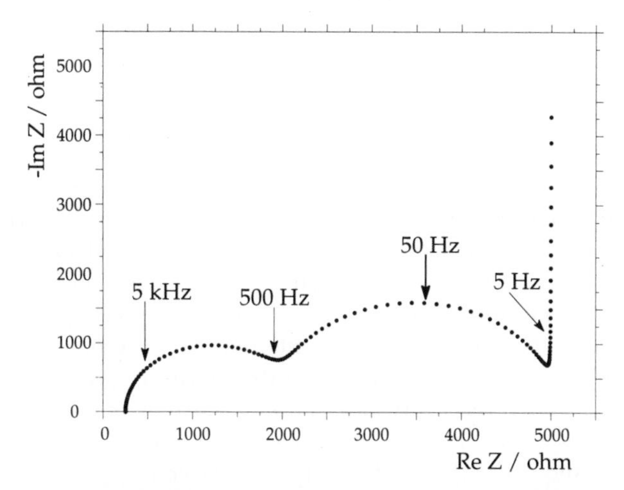

Figure 7 *The impedance spectrum for a defined equivalent circuit. The frequencies shown are those giving the points indicated. In an electrochemical cell formed by attaching electrodes to a solid electrolyte the significance of each component is as follows:* C_{dl} *is the electrode–electrolyte double layer capacitance,* R_b *is the resistance associated with the electrolyte particles,* C_b *is the geometrical capacitance of the cell,* R_{gb} *and* C_{gb} *are the resistance and capacitance associated with grain boundaries between electrolyte particles, and* R_{cell} *is a resistance associated with aspects of cell design*

molecules. Analogous mixed conduction and colourations are characteristic of the insertion chemistry of related hydrated parents, *viz*. the tungstic and molybdic acids $MO_3.nH_2O$ (M = W, Mo; n = 1, 2).

4 CHARACTERIZATION OF FAST ION CONDUCTORS

4.1 Characterization at the Bulk Level

Ionic conductivity of bulk specimens can be studied directly by both ac and dc electrochemical techniques. The ionic self-diffusion coefficient can also be determined directly by tracer diffusion measurements and, in favourable cases, by pulsed field gradient NMR.

ac impedance studies of cells containing solid electrolytes were first published in the 1960s. **Impedance spectroscopy** has become the standard ac technique for investigations of ionic conductivity.[9] An impedance spectrum is a plot of the in-phase [real, Re] and out-of-phase [imaginary, Im] components of the vector impedance ($\boldsymbol{Z}$) as a function of the frequency of an ac electrical stimulus; each point generated in the complex plane corresponds to a measurement at a different frequency. In such measurements the cell and its impedance spectrum are discussed and modelled in terms of an equivalent circuit of resistors and capacitors, each of which corresponds to a different process in the cell (the spectrum for a specified equivalent circuit is shown in Figure 7). It is not necessary for electrodes attached to the electrolyte to be electrochemically reversible with respect to the ionic charge carrier (*i.e.* able to inject and remove charge-carrying ions); reversible and 'blocking' electrodes simply correspond to different electrical components. Depending on the form of the spectrum, it may be possible to assign values to electrode capacitance, intraparticle resistance, grain-boundary resistance and capacitance, and electronic conductivity. Very detailed information on the functioning of cells is therefore, in principle, available.

dc electrochemical investigations of cells containing solid electrolytes are more directly related to intended device applications. The passage of current requires electrodes reversible to the ionic charge carrier; the use of blocking, or partially blocking, electrodes results in a decaying dc current (a capacitative component in the response to the applied dc voltage). Such a response gives information concerning the total cell, but can obscure information concerning the conductivity of the electrolyte itself. **The four-probe dc method** seeks to overcome any problems of partial reversibility of electrodes; current is passed through two (outer) electrodes, voltage drop is sensed at a second (inner) pair of electrodes, and the resistance (and hence the conductivity) of the electrolyte follows from Ohm's law and the geometry of the cell.

The Nernst–Einstein equation (1) links ionic conductivities at the bulk level with ionic self-diffusion coefficients. Measurement of the latter therefore provides further information on the bulk conductivity and, if both σ_i and D_i are known, a value for $N.H_R$ ($\approx N$). In tracer diffusion studies, one face of a piece of fast ion conducting material is coated with a thin film of a compound containing a radioactive isotope of the ionic charge carrier (*e.g.* Na^{+*}). The evolution of the activity profile (counts versus distance from the coated face) through the material is then followed as a function of time and temperature, and temperature-dependent D_i values are calculated.

NMR techniques can be used to probe ionic motions involving nuclei of a wide range of species of interest (*e.g.* ^1H, ^7Li, ^{17}O, ^{19}F, ^{23}Na, ^{65}Cu). In pulsed NMR studies of systems with low translational mobilities, the free induction decay (FID) is related to the structure of the material in which the probe nucleus is sited. In the case of fast ion conductors, the FID can have a characteristic time $T_2{}^* > 200\ \mu s$, and the FID then reflects the dephasing of spins due to inhomogeneities in the static magnetic field of the spectrometer magnet. The spins can be rephased by a standard $90°_x$–τ–$180°_y$ sequence (τ is the time between radiofrequency [RF] pulses) to give a 'spin echo' at time $t = 2\tau$. In pulsed field gradient (PFG) NMR

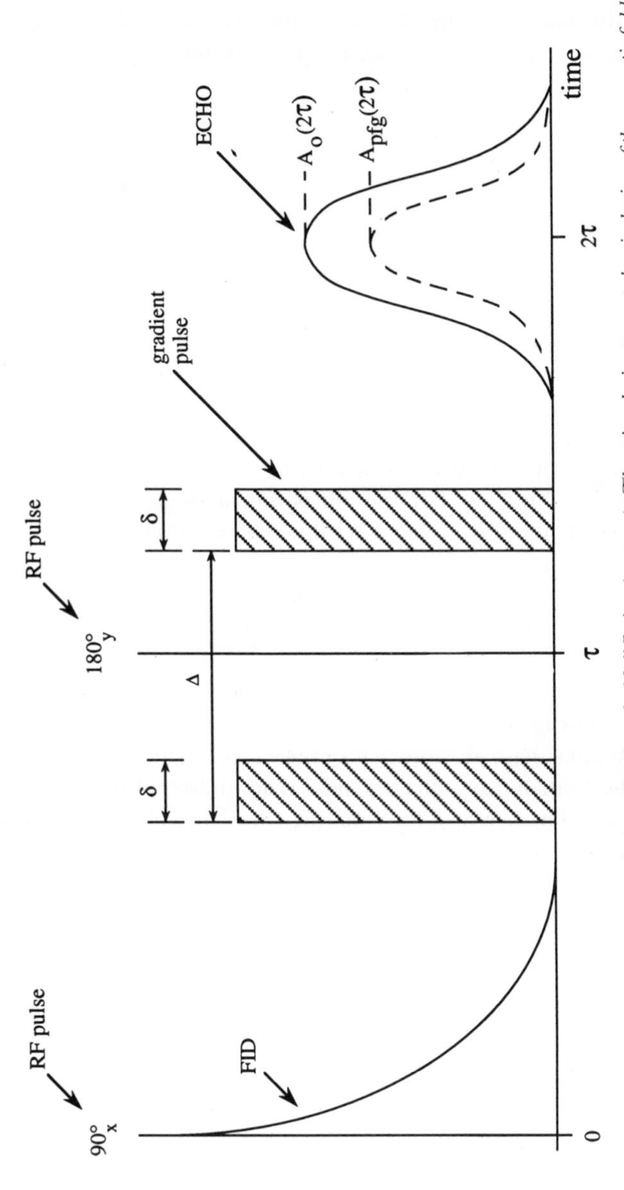

Figure 8 *The pulsed field gradient NMR technique for investigation of self-diffusion (see text). The spin echo is attenuated on inclusion of the magnetic field gradient pulses*

studies[10] the spin echo pulse sequence is modified by insertion of short magnetic field gradient pulses (with a time interval Δ between pulses each of duration δ) before and after the $180°_y$ pulse. The pulse sequence and the observed free induction decay (FID) and echo are illustrated in Figure 8. In the presence of isotropic self-diffusion involving the probe nucleus, the amplitude A_{pfg} of the echo in the presence of the gradient pulses is related to the amplitude A_0 observed in their absence by

$$\log_e\left(\frac{A_{\text{pfg}}}{A_0}\right) = -\gamma^2 g^2 \delta^2 D_{\text{pfg}}(\Delta - \delta/3) \tag{7}$$

where g is the pulsed field gradient. For instrumental reasons this technique is restricted to measurements of self-diffusion coefficients $D_{\text{pfg}} > 10^{-8}\,\text{cm}^2\,\text{s}^{-1}$. Results have to be considered with care. The self-diffusion studied is that within particles. The technique actually monitors the self-diffusion of the probe nucleus, and D_{pfg} may therefore differ from D_{i} (that for ionic charge) if, as in hydrated proton conductors, the probe nucleus can translate in neutral, as well as charged, species.

Chemical diffusion coefficients, D_{c}, pertaining to ionic charge carriers in mixed conductors in electrochemical cells can be evaluated using a variety of electrochemical techniques. For motion within a single crystal, chemical and self-diffusion coefficients are related by the Darken equation:

$$D_{\text{c}} = D_{\text{i}}\left(\frac{\text{d}\ln(a)}{\text{d}\ln(c)}\right) \tag{8}$$

where a and c are the activity and concentration of the mobile species. As discussed above, the D_{c} value measured in a cell may, however, reflect complications such as intergrain and interphase phenomena. In the widely-used current pulse relaxation method[11] a short current pulse (τ seconds) is administered to a cell to inject ions into the cathode. The decay of the resulting transient voltage (ΔE) with time (t) enables calculation of D_{c} via the relaxation law:

$$\Delta E(t) = \frac{miV_{\text{m}}\tau}{FA(\pi D_{\text{c}}t)^{1/2}} \tag{9}$$

where i is the current, m is the rate of change of voltage with composition during constant current discharge, V_{m} is the molar volume of the cathode, F is the Faraday constant, and A is the area of the electrolyte/electrode contact.

4.2 Characterization at the Atomic Level

The atomic-level detail of ionic conduction within particles of fast ion conductors has been probed with the whole range of physicochemical techniques available to the materials chemist. Among techniques that have found to be particularly useful are X-ray and neutron powder diffraction, nuclear magnetic resonance, and optical spectroscopies (IR and Raman).

Diffraction studies of crystalline fast ion conductors yield the symmetry, the detailed atomic coordinates, the effects of thermal motion, and the site occupancies (number of atoms per site; 1 corresponds to every such site in the crystal being occupied) for the average unit cell. Those data provide basic information for examination of potential models/conduction pathways for intraparticle ionic conduction. Where single crystals are available, **X-ray techniques** are often sufficient for full structure determination. X-rays are scattered by the electrons of the atoms present and use monochromatic radiation; atomic scattering powers increase with increasing atomic number and decrease rapidly at high scattering angles (the latter is a consequence of the similarity of the X-ray wavelength and the dimensions of the atomic electronic charge clouds). X-ray techniques can consequently be poor in determining the positions of light atoms (*e.g.* H) in the presence of heavy atoms; in the presence of heavy metals even oxygen can sometimes be regarded as a light atom in this sense. The determination of full structures from powder diffraction profiles (counts versus scattering angle) took a major step forward with the introduction of the profile modelling techniques of Rietveld.[12] High resolution X-ray powder diffraction studies using central synchrotron facilities have become increasingly routine.

Neutron diffraction offers some advantages, particularly in the location of light atoms. Neutrons are scattered by nuclei (which effectively act as point scatterers, leading to a constant scattering power with angle), and variations in scattering power with nuclear mass number are not systematic. Thus, light atoms make a significant contribution to the diffraction profile, and their positions can be determined accurately. In locating hydrogen atoms by neutron diffraction it is normal to use deuteriated (^{2}H) analogues; ^{1}H itself is a particularly powerful incoherent scatterer (omnidirectional scattering not related to diffraction/Bragg scattering), and this can lead to unacceptably high background scattering in diffraction profiles (that incoherent scattering is useful in itself for other purposes, as will be discussed later).

NMR techniques can also yield basic information for examination of potential models and conduction pathways. Measurements of absorption spectra can provide structural information concerning local environments of probe nuclei in solids, and measurement of NMR relaxation times as a function of temperature can provide information on dynamic processes such as ion migration. Classical wide-line absorption studies of spin-$\frac{1}{2}$ nuclei (*e.g.* ^{1}H, ^{19}F) provide this information via the dipolar spectrum (arising from internuclear interactions) and its second moment, M_2, both of which are related to crystal structure. For spin-1 nuclei (*e.g.* ^{2}H) the wide-line quadrupolar spectrum reflects the electric field gradients about individual nuclei. The advent of high resolution magic angle spinning NMR (MAS-NMR) has led to the availability of chemical shift information comparable to that available for solution-state samples. Measurement of the NMR relaxation times T_1, T_2, $T_{1\rho}$, and T_{1D} can yield information on motions with characteristic frequencies in the range $10 < \nu/\text{Hz} < 10^{11}$. A full study over a wide temperature range can be time-consuming unless automated, and data interpretation can be complex if several motions and interactions are present. Relaxation studies have, nonetheless, yielded valuable information for motions of a wide range of nuclei (*e.g.* ^{1}H, ^{2}D, ^{7}Li, ^{17}O, ^{19}F, ^{23}Na, ^{65}Cu). As with pulsed field gradient NMR (Section 4.1), relaxation studies do not, in themselves, establish that the moving particle is charged.

Dynamic information is also available from experiments utilizing **incoherent scattering of neutrons**. The ^{1}H nucleus is the most powerful incoherent scattering nuclide. This results in techniques using that scattering being of particular use in studying motions involving hydrogenic species, even in the presence of motions of heavy atoms and of complex frameworks which could dominate optical spectra. The particular sensitivity towards motions involving ^{1}H can also be considered to be a property of NMR relaxation studies, ^{1}H being the most sensitive stable nuclide in NMR experiments. In many respects NMR and incoherent neutron scattering (the latter requiring access to central facilities) are complementary in the information they provide about dynamics in proton-conducting systems.[13] In the presence of rotational or translational (migrational) motion with characteristic time $10^{-8} > \tau / s > 10^{-11}$ in hydrogen-containing systems, broadening of part or all of the elastic peak (quasielastic neutron scattering, QNS) in the incoherent scattering spectrum can be detected. Examination of that broadening as a function of elastic scattering vector magnitude Q_{el} can reveal the nature (reorientation and/or translation) and geometry of the motion, and measurements as a function of temperature may give activation parameters for the various motions detected. At larger energy differences from the elastic peak an inelastic neutron scattering (INS) spectrum is observed. This constitutes a vibrational spectrum revealing energy transfer to modes (particularly those involving hydrogen) of the material under investigation. Such spectra are prone to combination and overtone bands, and to low intensities at energy transfers $> 1800\ cm^{-1}$, but are uniquely powerful in highlighting hydrogenic modes in complex systems.

Optical vibrational spectroscopies (infrared and Raman) provide information concerning vibrational modes of the system under investigation. Subject to the modes of interest being observable and correctly assigned, they supply detailed information concerning local environments in fast ion conductors. X-ray absorption spectra, and in particular **extended X-ray absorption fine structure** (EXAFS) studies carried out at central synchrotron facilities, can provide detailed information concerning the local arrangement around absorbing atoms. EXAFS is an atom-specific probe (the atom being defined by the absorption edge investigated), and is applicable to both crystalline and non-crystalline (*e.g.* glassy) systems. This technique can yield local structural information (such as nearest neighbour atoms and bond lengths) that may not be readily available from other techniques, particularly in the case of non-crystalline solids.

5 SOME APPLICATIONS OF FAST ION CONDUCTORS (SOLID STATE IONIC DEVICES)[14]

The applications of fast ion conductors that have been explored and, in some cases, commercialized are primarily those employing the fast ion conducting material as a component in an **electrochemical cell** at ambient or elevated temperature. Cells can be constructed to employ solid, as opposed to liquid, electrolytes. Mixed conductors can be employed as electrode materials in contact with an electrolyte (solid or liquid) which can accept or donate the ion which is mobile in the mixed conductor. Such electrodes are commonly known as **insertion (or intercalation) cathodes**; the term 'solid solution electrodes' is also used, the intercalation compound ideally being a single phase across the entire composition range, and current therefore not being limited by the kinetics of phase conversion (that is, however, not always the case).

In this section examples of the actual or potential application of fast ion conductors in batteries, fuel cells, electrochemical sensors, and electrochromic displays are presented.

5.1 Batteries

The term battery is commonly used to describe an electrochemical cell in which EMF and current are consequent on a spontaneous chemical reaction of the electrodes between which the ionic current is carried by the electrolyte (strictly speaking the term 'battery' refers to an interlinked set of such cells). **Primary batteries** are based on irreversible electrochemical reactions, and are **discarded** after discharge. **Secondary batteries**, in contrast, are based on reversible reactions and can be recharged (reversing the cell reaction electrochemically), making them useful as electrochemical **stores of energy**. The development of fast ion conductors has offered the prospect of battery systems based on design concepts different from those currently widely utilized, *e.g.* solid electrolyte membranes can be employed, insertion electrodes can be considered, and all solid state batteries (with potential for miniaturization) are possible. Battery systems based on the reactions of alkali metals offer the possibility of high cell emfs (E, volts), and high energy and power densities ($J\,kg^{-1}$ and $W\,kg^{-1}$ respectively).

The **sodium–sulfur battery** has been widely investigated and is approaching commercialization; an idealized sodium–sulfur cell is shown in Figure 9. The properties of the sodium β-alumina as a solid electrolyte (see earlier) and the electrodes used lead to an operating temperature of $\approx 300\,°C$, at which temperature the sodium and sulfur electrodes are molten. The discharge of this secondary battery corresponds to the reaction

$$2\,Na + x\,S \rightarrow Na_2S_x \qquad (10)$$

where the product is soluble in the molten sulfur cathode. E for a fully charged cell is 2.08 V at 300 °C. The operating temperature might seem a major disadvantage, but the sodium–sulfur battery has real potential for energy storage applications, *e.g.* the storage and release of energy in levelling the load in electrical power generation from fossil and nuclear fuels.

The application of insertion cathodes in secondary batteries based on the reactions of lithium is being pursued. The **lithium–titanium disulfide battery** operates at room temperature ($E \approx 2.8\,V$). The functioning of the cathode (Li_xTiS_2, $0 \leq x \leq 1$) was discussed in Section 3.2. The overall cell reaction is

$$TiS_2 + x\,Li \rightarrow Li_xTiS_2 \qquad (11)$$

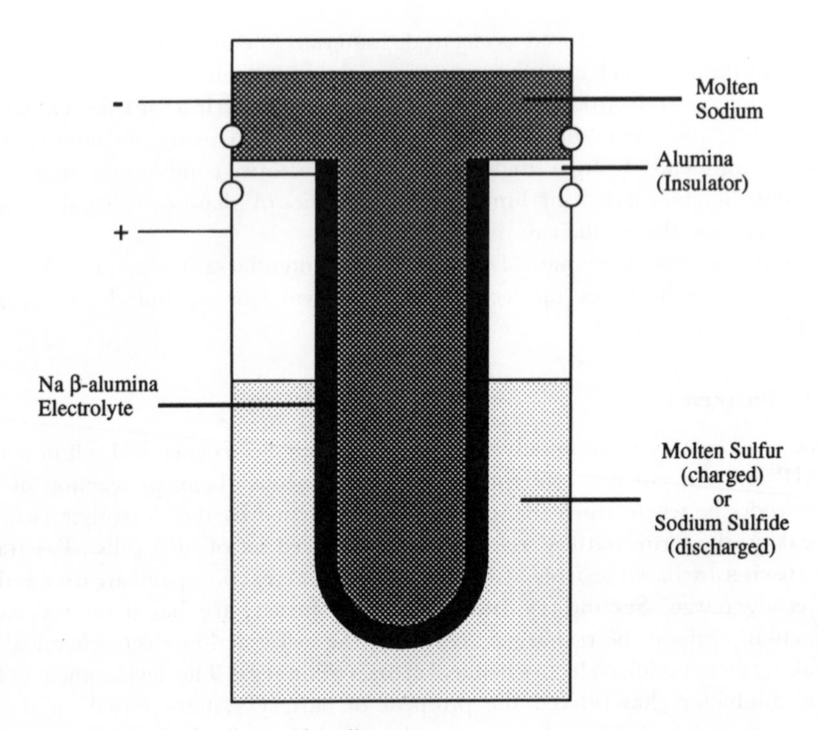

Figure 9 *An idealized sodium–sulfur cell*

The electrolyte for the cell is a solution of a lithium salt (*e.g.* $LiAsF_6$) in a non-aqueous solvent (*e.g.* propylene carbonate). An Anglo–Danish project resulted in developments in the area of all-solid-state batteries. An Li^+-conductive solid polymer electrolyte (see Section 3.2) functions in a battery with a lithium anode and a V_6O_{13} insertion cathode. The conductivity of the electrolyte is, however, too low for room temperature application and the battery is therefore maintained at *ca.* 120°C ($E \approx 3.1$ V). The achievement of an all-solid-state system for ambient temperature application would be a strategic development. A possible design for an all-thin-solid film battery is illustrated in Figure 10.

A primary battery of considerable importance in the medical sphere, and in heart pacemakers in particular, is the Li/I_2 cell. Li^+-conducting LiI electrolyte first forms chemically as an interfacial phase. LiI is itself the product of the cell reaction, and the electrolyte layer thickens during discharge of the battery. The 'I_2' cathode is in fact a charge transfer complex of iodine with poly(2-vinyl pyridine), to ensure that the electrode is sufficiently electronically conducting. Such cells operate at low current density (with consequently lower conductivity required of the electrolyte) over periods > 10 years (an obvious benefit to the patient).

5.2 Fuel Cells

In a fuel cell, electrical energy is generated by **electrochemical reaction of a fuel and an oxidant**, which are continuously fed to porous electrodes

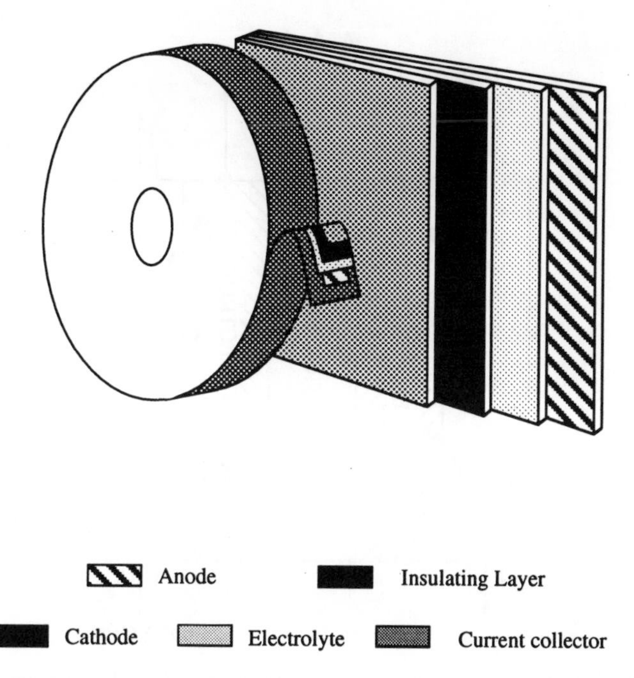

Figure 10 *Concept of a possible all-thin-solid-film battery. Layers are stacked in the sequence on the right, and formed into the roll on the left*

interconnected by an electrolyte (which can be either a solid or a liquid). Solid electrolytes conducting H^+ or O^{2-} find application in fuel cell technology. The principles of power generation in a fuel cell are illustrated in Figure 11. Working fuel cells use hydrogen as fuel and oxygen (or air) as oxidant, giving water as the reaction product (in either the oxidant or the fuel gas stream, depending on whether an H^+- or O^{2-}-conducting electrolyte is used). Other fuels have been investigated, but electrochemical problems (in particular the reaction sequence at the fuel electrode becomes complicated for fuels other than hydrogen) have largely prevented their utilization. There is, however, currently considerable interest in methanol as a fuel in transportation-related applications.

The advantages of the fuel cell for power generation are the **higher efficiency** (for energy production), relative to the combustion of fossil fuels, and the **absence** of the production of **pollutants**. The situation is, however, a little more complex than is immediately apparent; the hydrogen fuel must be produced either electrochemically (*e.g.* using electricity generated conventionally at a remote nuclear or fossil-fuel-based facility) or on site, the latter option necessitating an associated chemical facility (*e.g.* for production of hydrogen by reforming a petrochemical feedstock). Similar cells can be used for the reverse reaction (*i.e.* the electrolysis of water), and for the electrochemical pumping of hydrogen or oxygen gas across solid electrolyte membranes (*e.g.* for purification purposes).

Transportable fuel cells utilizing H^+-conducting solid polymers (**solid polymer fuel cells**, SPFCs) have found limited application. Such a fuel cell is displayed in

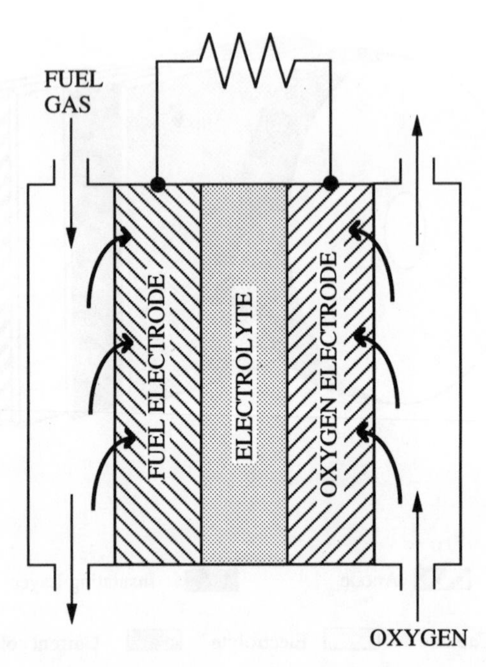

Figure 11 *Principles of operation of a fuel cell. In the case of an H^+-conducting electrolyte, ions migrate to the oxygen electrode, forming product H_2O in the oxygen gas stream. In the case of an O^{2-}-conducting electrolyte, ions migrate to the fuel electrode, forming product H_2O in the fuel gas stream*

Washington DC in the Smithsonian Museum of Air and Space Travel, fuel cells having been widely employed in the US space programme. Early electrolyte membranes were made from sulfonated polystyrene, but later devices have used much more stable fluoropolymers such as NAFION® (see Figure 5). The electrodes are commonly platinized carbon (platinum as electrocatalyst) included in an applied film with poly(tetrafluoroethylene) as binder. Other possible applications for such fuel cells include use in transportable low-emission generators for power generation 'in the field', use in cars, and use in submarines.

Solid oxide fuel cells (SOFCs) employ an oxide ceramic as electrolyte. The device temperature is high ($\approx 1000\,°C$ if stabilized zirconia is the electrolyte), but such devices have potential in utility power generation (*i.e.* in power stations). SOFCs are suitable for cogeneration of both power and heat, the latter for either industrial or neighbourhood heating. Cogeneration itself is a well-established concept and increases the efficiency of generating plant, *e.g.* cogeneration based on a conventional coal-fired power station provides domestic heating for sections of the city of Odense (Denmark). The high operating temperature is consequent on the electrolyte chosen (stabilized zirconias are O^{2-}-conductors, see earlier) and entails severe materials problems, including necessary slow heating to operating temperature (thereby preventing fracture of components, joints and seals). These problems would be obviated in devices operating at lower temperatures, and this

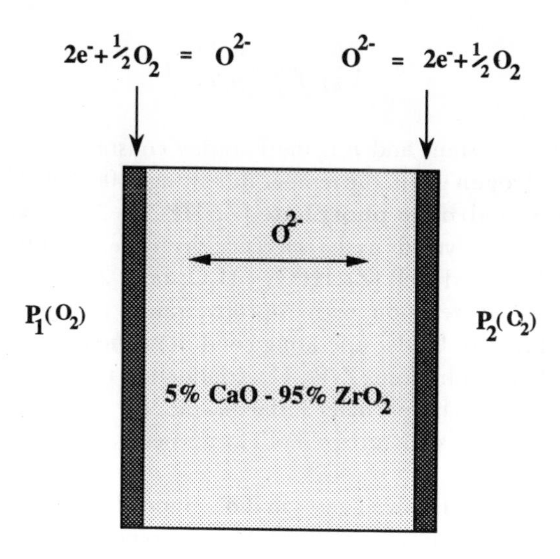

Figure 12 *Principle of operation of a potentiometric sensor for oxygen gas. Stabilized zirconia ceramic electrolyte separates porous electrodes exposed to differing partial pressures of oxygen*

has been the stimulus for examination of alternative ceramic electrolytes. Doped barium cerate (an H^+-conductor, see earlier) is one type of material that is being considered.

5.3 Electrochemical Sensors

Cells employing solid electrolyte membranes can be used to determine the partial pressures of electrochemically-active gases such as oxygen and hydrogen, or the concentrations of those gases in solution in liquids. These analytical methods are potentiometric (*i.e.* involve measurement of the emf of the cell). In such cells the potential of one electrode (the reference electrode) is fixed (by maintaining the composition of that electrode fixed), and the emf of the cell then varies with the amount of the gas of interest (the analyte gas) in the environment in which the second electrode (the indicator electrode) is placed. The circuit is completed by the electrolyte membrane which conducts the ion to which the electrodes are reversible (O^{2-} or H^+). Such sensors can be used in monitoring the positions of equilibria (*e.g.* $CO_2 \rightarrow CO + \frac{1}{2}O_2$) and the composition of process or exhaust gases.

Stabilized zirconias may be used in cells monitoring the partial pressure of **oxygen** at high temperatures ($T \geq 500\,°C$). Such a cell can be formed from an electrolyte tube which is coated inside and out to provide porous metal (*e.g.* Pt) electrodes and through which oxygen of a known partial pressure $P_1(O_2)$ flows; the principles of such a sensor are illustrated in Figure 12. The emf of the cell (which is functioning as a concentration cell) then depends on the partial pressure $P_2(O_2)$ in the environment through which the tube passes. For a cell obeying Nernst's law the emf E is given by

$$E = \left(\frac{RT}{4F}\right)\log_e\left(\frac{P_1(O_2)}{P_2(O_2)}\right) \quad (12)$$

where R is the gas constant and F is the Faraday constant.

A sensor for **hydrogen** gas in the temperature range 100–350 °C employs a thin film of **zirconium hydrogen phosphate** [$Zr(HPO_4)_2$, abbreviated to ZrP] as electrolyte, and a metal hydride as the reference electrode. The film is produced by delamination of hydrated ZrP [$Zr(HPO_4)_2.H_2O$; both ZrP and $ZrP.H_2O$ have layered structures] by reaction with n-propylamine, and film deposition on subsequent acidification. Under operating conditions the ZrP film is anhydrous. In both hydrated and anhydrous ZrPs, H^+ conduction through a bulk specimen occurs predominantly via particle surfaces, the H^+ mobility between the intraparticle layers of the structures of both forms being very much less than that on the particle surfaces.

The concentrations of dissolved gases in molten metals can be monitored using sensors containing ceramic electrolytes. The reference electrode has a fixed pressure of the analyte gas, the electrolyte conducts the ion to which the reference electrode is reversible (O^{2-} or H^+), and the ceramic probe is dipped into the melt, which itself acts as the indicator electrode. Stabilized zirconias are used in determining dissolved oxygen in molten steels, and doped barium cerate can be used in determining dissolved hydrogen in molten aluminium (too much dissolved hydrogen results in formation of bubbles on cooling, and consequent embrittlement).

5.4 Electrochromic Displays

A cell reaction inserting M^+ ($M = H$, Ag, or Li) into thin film (colourless) insertion cathodes results in a colour change in the cathode. This provides a basis for electrochromic devices – writing/coloration and erasing/bleaching being accomplished by passage of a current to perform the insertion and reverse reactions respectively. The time response of such displays (the 'switching time') has not been reduced sufficiently to be viable in rapidly changing displays, such as watches or calculators. There is, however, considerable potential for application in 'smart' windows and mirrors in houses and cars, and in large scale displays (*e.g.* in transportation termini).

Thin film WO_4 or MoO_3 cathodes have been much investigated for electrochromic applications. The insertion reaction generates a blue coloration on formation of an oxide bronze (*e.g.* formation of H_xWO_3 in equation (8), Section 3.4). The structural chemistry of the thin film differs somewhat from that of bulk compounds; the films are of low crystallinity and are partially hydrated (having some features in common with tungstic/molybdic acids). The electrolyte, from which the mobile cation is inserted into the cathode, can be liquid (*e.g.* an ethanolic solution of H_2SO_4 or a solution of a lithium salt in a non-aqueous solvent) or solid (*e.g.* a heteropolyacid or an Li^+-conducting polymer such as $P(EO)_8LiCF_3CO_2$). A possible cell design involving insertion/removal of Li^+ is illustrated in Figure 13.

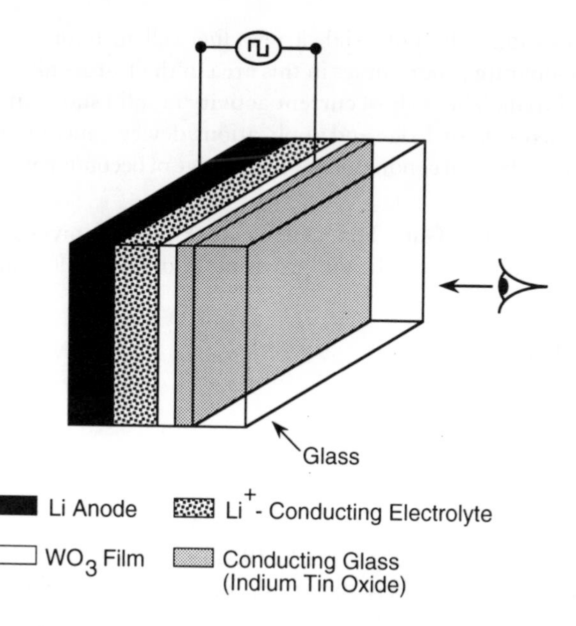

Figure 13 *Design of an electrochromic device based on electrochemical insertion of Li^+ ions into a tungsten trioxide film. The symbol in the external circuit denotes the ability to both 'write' and 'bleach', on insertion of Li^+ and removal of Li^+ respectively*

6 AN OVERVIEW OF THE STATE OF THE ART AND FUTURE PROSPECTS

Our understanding of the fundamental science underlying fast ion conductivity is approaching maturity. Many fast ion conductors are known and have been studied both fundamentally and in actual or potential applications, but it remains problematical to design and tailor new materials to have a desired conductivity for a particular ion at a particular temperature (in the case of solid electrolytes) or to have a suitable cation chemical diffusion coefficient and fully reversible electrochemistry (in the case of mixed conductor/insertion cathodes). For instance, the kinetic problems at the methanol electrode in a methanol/O_2 fuel cell would be largely obviated by operating such a cell at $T \approx 100\,^{\circ}C$, but fully satisfactory H^+-conducting membranes for such a system have yet to be developed. There continues to be intense activity worldwide in fundamental investigations of fast ion conduction; and in the design, synthesis, fabrication, and testing of new materials.

Many applications of fast ion conductors have been envisaged, and (as will have been seen in this chapter) some have been realized or are close to realization. Scientists and technologists worldwide are seeking to exploit and further develop new technologies employing fast ion conductors, in particular in **batteries, fuel cells, sensors, and displays** (as discussed in this chapter). For instance, thirty years of research and development have been expended both on the sodium–sulfur battery and on the solid oxide fuel cell (SOFC); both devices are nearing commercialization. The potential of new fuel cell technology is evident in the existence of separate programmes in this area in the European Union, the United States, and Japan. The scale of current activity in solid state ionics, together with the state of current knowledge and applications/devices, encourages optimism that the utilization of fast ion conductors is on the verge of becoming more commonplace.

Acknowledgement. Drs Gary Hix and Kevin Young (University of Exeter) are thanked for their assistance in the provision of diagrams for this chapter.

7 REFERENCES

7.1 Numbered References

1. A. R. West, 'Solid State Chemistry and Its Applications', Wiley, Chichester, 1984.
2. W. van Gool, *Ann. Rev. Mater. Sci.*, 1974, **4**, 311.
3. A. Hooper, *Contemp. Phys.*, 1978, **19**, 147.
4. M. Faraday, in 'Faraday's Diaries', ed. T. Martin, G. Bell & Sons, London, 1932.
5. G. Tubandt and E. Lorentz, *Z. Phys. Chem.*, 1914, **87**, 513.
6. J. T. Kummer, *Progr. Solid State Chem.*, 1972, **7**, 141.
7. F. M. Gray, 'Solid Polymer Electrolytes: Fundamentals and Technological Applications', VCH, Cambridge, UK, 1991.
8. 'Proton Conductors: Solids, Membranes and Gels – Materials and Devices', ed. P. Colomban, Cambridge University Press, Cambridge, UK, 1992.
9. W. I. Archer and R. D. Armstrong, in 'Electrochemistry', Specialist Periodical Reports, The Chemical Society, London, (now The Royal Society of Chemistry, Cambridge, UK), 1980, vol. 7, p. 153.
10. E. O. Stejskal and J. E. Tanner, *J. Chem. Phys.*, 1965, **42**, 288.
11. S. Basu and W. L. Worrell, in 'Fast Ion Transport in Solids', ed. P. Vashista, J. N. Mundy, and G. K. Shenoy, North–Holland, Amsterdam, 1979, p. 149.
12. 'The Rietveld Method', ed. R. A. Young, Oxford University Press, Oxford, 1993.
13. R. C. T. Slade, *Solid State Commun.*, 1985, **53**, 927.
14. 'Solid State Ionic Devices', ed. B. V. R. Chowdari and S. Radhakrishna, World Scientific, Singapore, 1988.

7.2 Relevant Texts

Solid State Chemistry

A. R. West, 'Solid State Chemistry and Its Applications', Wiley, Chichester, 1984.

L. Smart and E. Moore, 'Solid State Chemistry: An Introduction', Chapman and Hall, London, 1992.

Solid State Ionics

'Fast Ion Transport in Solids', ed. W. van Gool, North–Holland, Amsterdam, 1973.

'Proton Conductors: Solids, Membranes and Gels – Materials and Devices', ed. P. Colomban, Cambridge University Press, Cambridge, UK, 1992.

F. M. Gray, 'Solid Polymer Electrolytes: Fundamentals and Technological Applications', VCH, Cambridge, UK, 1991.

S. Kudo and K. Fueki, 'Solid State Ionics', VCH, Cambridge, UK, 1990.

'Solid State Ionic Devices', ed. B. V. R. Chowdari and S. Radhakrishna, World Scientific, Singapore, 1988.

Techniques

'Impedance Spectroscopy', ed. J. R. Macdonald, Wiley, Chichester, 1987.

'Neutron Scattering at a Pulsed Source', ed. R. J. Newport, B. D. Rainford, R. Cywinski, Adam Hilger, Bristol, 1988.

R. K. Harris, 'Nuclear Magnetic Resonance Spectroscopy', Pitman, London, 1983.

M. Bée, 'Quasielastic Neutron Scattering: Principles and Applications in Solid State Chemistry, Biology and Materials Science', Adam Hilger, Bristol, 1988.

'The Rietveld Method', ed. R. A. Young, Oxford University Press, Oxford, 1993.

'Solid State Chemistry: Techniques', ed. A. K. Cheetham and P. Day, Oxford University Press, Oxford, 1987.

7.3 International Society for Solid State Ionics

The Society can be contacted through its secretary: W. Weppner, Technical Faculty, Chair for Sensors and Solid State Ionics, Christian Albrechts University, D-24098 Kiel, Germany

7.4 Journals for Fast Ion Conductors

Chemistry of Materials
Electrochimica Acta
Materials Research Bulletin
Journal of Applied Electrochemistry
Journal of the Electrochemical Society
Journal of Materials Chemistry
Journal of Power Sources
Journal of Solid State Chemistry
Solid State Ionics*

* The central journal for the field.

CHAPTER 9

Inorganic Chemicals and Metals in the Electronics Industry

PHILIP D. GURNEY AND RICHARD J. SEYMOUR

1 INTRODUCTION: MATERIALS FOR ELECTRONICS

This chapter provides an introduction to and overview of the applications of inorganic chemicals and metals in the electronics industry. Perhaps more so than in any other area, electronic devices have become virtually synonymous with applications for one special element – **silicon**. With terms like 'silicon chip' now in general usage, it is true that many people associate electronic technology with this element alone. However, it is important to appreciate that it is actually the subtle, and often complex, interactions between silicon and many other elements which give rise to the useful effects which form the basis for electronic devices and that in some applications other semiconductor materials, such as **gallium arsenide**, are preferred to silicon.

This section provides an outline of the types of materials that can be found in modern electronic devices and their functions. The importance of the material properties are emphasized and then an overview of some of the most important manufacturing processes is given. These processes are referred to again at several places in the subsequent sections.

1.1 Concepts

The range of inorganic materials used in electronics is extremely broad. Table 1 gives an indication of this range. Many of these material types are dealt with in other parts of this book, such as the chapters on magnetic materials (page 239), superconducting materials (page 275), solar energy conversion (page 457), and ceramics, glasses, and hard metals (page 325). This chapter will concentrate on the chemistry underlying the manufacture of **semiconductors, conductors, and dielectrics** and their use in producing devices and circuitry. Before we begin, it will be useful to outline some of the more important terms used in this industry.

The fundamental component of modern digital electronics is the silicon **integrated circuit**. This makes use of interconnected transistor structures at the surface of the silicon chip which process electronic signals. The transistor structures are built up both by diffusing elements into the top surface of the silicon

Table 1 *Types of material used in the electronics industry*

Material	*Effect/Device*	*Applications (examples)*
Silicon	Diodes, transistors	Integrated circuits, power electronics
III–V semiconductors (*e.g.* GaAs)	LEDs, laser diodes MESFETs	Displays, optical fibre communications High speed integrated circuits
II–VI semiconductors (*e.g.* HgCdTe)	Photovoltaics Infrared detectors	Solar cells Avionics, night-sights
Oxides (*e.g.* $LiNbO_3$, $BaTiO_3$, $Bi_{12}GeO_{20}$, $YBa_2Cu_3O_7$)	Nonlinear optical materials, dielectrics, phosphors, high temperature superconductors	Laser frequency doubling, capacitor dielectrics, display devices, lossless interconnects
Metals (*e.g.* Al, Cu, Ag, Pd)	Conductors	Conductive interconnects on chip or at the circuit scale

and by adding further layers on top. These layers, which are typically one micron (10^{-6} m) or less in thickness are deposited by a variety of **thin film deposition** methods, which are discussed later.

Many different ways of packaging a silicon chip are used. The chip, or **die**, as it is sometimes known, is normally attached to a substrate with a conductive adhesive and connected to the external circuitry through very thin metal wires. This external circuitry may be just a **chip carrier**, which enables the chip to be inserted or mounted by soldering on a **printed circuit board**, or it may be a complete circuit in itself. In the latter case, the circuit is normally prepared by the **screen printing** of conductor, resistor and dielectric inks onto the substrate on which the die is mounted. These layers are usually several microns thick and the term **thick film processing** is used to differentiate it from the thin film processes used in semiconductor device manufacture. This circuit may also contain other mounted components such as **chip capacitors** and **chip resistors**. The entire circuit containing these components is often termed a **hybrid circuit**, as it is a hybrid of many different technologies.

Semiconductors are also used as the basis for many optical devices, since **compound semiconductors** such as **gallium arsenide** can both emit light (as **lasers** and **light-emitting diodes**) and sense it (as **photodetectors**).

In addition, this chapter provides an overview of **sensor technology**, notably where the mechanism of action relates to a chemical reaction at the sensor–environment interface. Finally, future trends in electronic device technology are discussed in relation to new developments in chemistry and materials science.

1.2 Properties

The properties of the individual materials in an electronic device and their interaction will determine the overall functioning of the device. These properties will be a combination of the intrinsic properties of the material and the properties conferred on it by a series of processes.

For semiconductors, for instance, in addition to their intrinsic properties such as **band gap** and **electron mobility**, the type and level of dopant and the degree of crystalline perfection obtained from the crystal growth process used will have a large effect on the material properties.

The use of **dopants** at the part per million level or less means that undesirable impurities in the bulk semiconductor must first be reduced to very low levels. 99.9999 to 99.999999% pure semiconductor materials (six nine's to eight nine's purity) are often required.

In the same way as purity is a crucial factor in determining semiconductor properties, many materials must be produced as single crystals, free from microscopic defects, to perform effectively as electronic materials. As a result, great care is taken in the growth of crystalline material to ensure the production of high quality material.

Controlled deposition of materials (elements, compounds, alloys, and multilayers) in **thin film** form is a vital part of most electronic production processes. These thin films may be **single crystals**, **polycrystalline**, or **amorphous** depending on the requirements of the device. A wide range of factors can affect the properties and quality of the final film; these include the substrate material used, whether the substrate is a single crystal and, if so, its crystal orientation, the deposition method, the processing conditions, and the nature of the film precursor materials.

The properties of **thick film** materials are very dependent on the processing and properties of the original inks that are deposited by screen printing. Important factors here are the **rheology** (flow properties) of the inks, the **particle size distributions** of the dispersed phases and the drying and firing schedules of the printed inks.

1.3 Processes

1.3.1 Bulk Purification Processes

Semiconductor devices make use of **dopants**, or additives at the part per million level, in order to control their properties. This means that the background level of impurities must be of a sufficiently low level so as not to interfere with this process. **Purity** levels commonly required in semiconductor materials and their precursors are therefore at the **sub part per million level**. The purification processes used will depend on the properties of the element or compound to be purified, but will typically include such techniques as fractional crystallization, distillation, sublimation, electrochemical processes, and zone refining. Frequently, the process will include a chemical reaction to form another compound which is more amenable to purification before re-converting it back to its original form. In addition, it is important not to introduce any further impurities during the processing stages, and high purity solvents and reagents are required, as well as containment vessels and apparatus that are inert to the chemicals present.

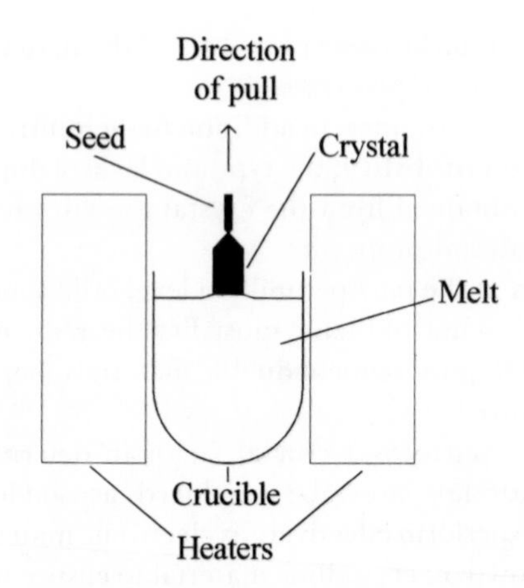

Figure 1 *Czochralski crystal growth apparatus*

1.3.2 High Purity Crystal Growth

Semiconductors used in the manufacture of transistors, integrated circuits, or optical devices such as light-emitting diodes (LEDs) and lasers are normally required in single crystal form. Single crystals of elemental semiconductors such as silicon, or compound semiconductors such as gallium arsenide, are prepared by slow cooling from the melt and may also use a seed crystal to initiate the growth process.

The most widely used technique for bulk single crystal growth here is the **Czochralski process**, shown in Figure 1. A rotating seed crystal is brought into contact with the surface of the molten semiconductor material, and slowly withdrawn. As it is withdrawn the surface of the melt crystallizes onto the seed. The speed of rotation of the seed and its withdrawal speed are used to control the dimensions of the sausage shaped **single crystal**, known as a **boule**, that is formed over several days of growth. The **semiconductor wafer** is prepared by transverse sawing of the boule using a diamond saw. The wafer surface is then both mechanically and chemically cleaned before it is ready for use in a semiconductor device fabrication line. Single crystal silicon, gallium arsenide, and indium phosphide are all commonly made by the Czochralski process.

Another important bulk crystal growth method is the **Bridgman technique**. Here the molten semiconductor material, held in a crucible, is solidified into a single crystal by the movement of a solid/liquid interface from one end to the other as the crucible is slowly cooled. Figures 2 and 3 show the two most common configurations of horizontal and vertical crystal growth. Bridgman crystal growth can either use a seed crystal or will often be designed so that nucleation takes place in a confined region at one end of the crucible, which results in growth from only one grain occurring in the bulk of the crucible. Single crystal gallium arsenide and cadmium mercury telluride are produced in this way.

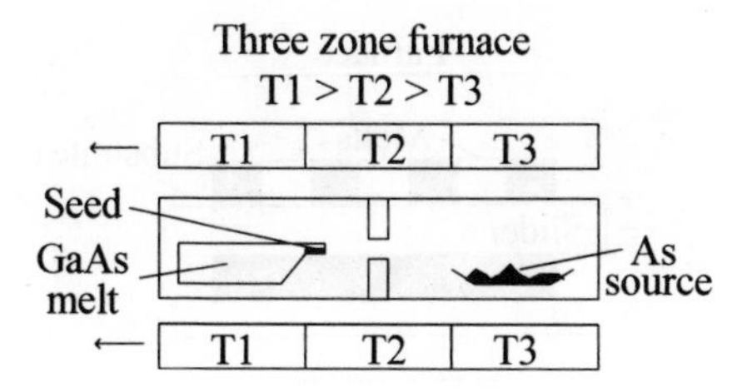

Figure 2 *Horizontal Bridgman crystal growth apparatus*

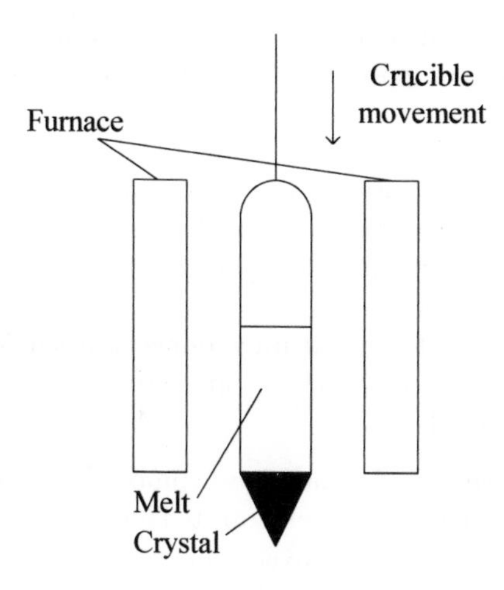

Figure 3 *Vertical Bridgman crystal growth apparatus*

1.3.3 Deposition of Thin Film Materials

An electronic device will commonly require one or more thin films of material to be deposited on the semiconductor surface during processing. If crystal perfection in the thin film is required, then single crystal films can be deposited by **epitaxial growth*** onto the substrate. This can most easily be achieved when the substrate material is a single crystal and a thin film of an identical material is to be grown on the top surface (as in the growth of a thin single crystal film of silicon on a silicon wafer). Where this is the case, the process is known as **homoepitaxy**. Where a single crystal film is grown on a different single crystal substrate, the process is known as **heteroepitaxy**. Heteroepitaxy is more likely to be successful the more closely the lattice constants of the substrate and growing film match. In the growth of $Al_{1-x}Ga_xAs$ on GaAs, for instance, the fact that the lattice constant of AlAs is very close to that of GaAs means that the ternary compound can be grown lattice matched onto a GaAs substrate for any value of x. However, another ternary, $In_{1-x}Ga_xAs$, varies widely in lattice constant as x decreases from 1 to zero. Only one composition is lattice matched to an available substrate, that is $In_{0.47}Ga_{0.53}As$,

* Epitaxial growth is the growth of single crystal layers from the vapour or liquid phase onto crystalline substrates.

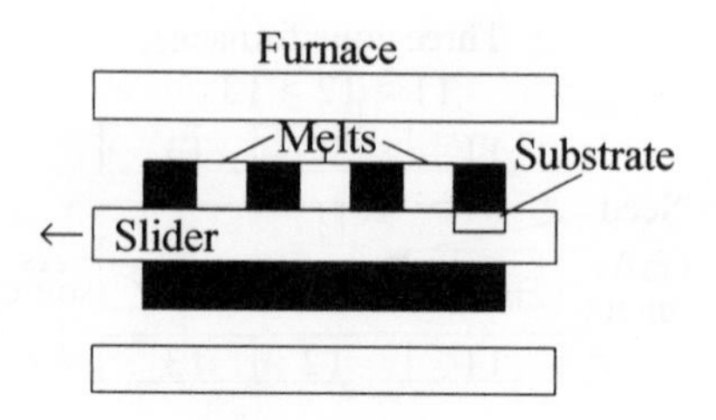

Figure 4 *Apparatus for liquid phase epitaxy (LPE)*

which has a lattice constant of 5.87 Å, the same as indium phosphide. This is a major restriction on the compositions of ternary and quaternary semiconductors that can readily be grown.

The four main types of epitaxial growth methods in use today are **liquid phase epitaxy** (LPE), **chloride vapour phase epitaxy** (chloride VPE), **metal–organic vapour phase epitaxy** (MO-VPE) and **molecular beam epitaxy** (MBE). Chloride VPE has been largely superseded by the more recently-developed MO-VPE.

In many cases, such as for metal interconnects, a single crystal film is not required and other techniques such as sputtering, evaporation, or chemical vapour deposition are used.

1.3.3.1 Liquid Phase Epitaxy. Liquid phase epitaxy (LPE) is a process in which epitaxial films are deposited from supersaturated solutions. The solvent is always a high purity metal which has a relatively low melting point and a low vapour pressure. The most common metals chosen are tin, tellurium, gallium, and indium. The semiconductor material is dissolved in the liquid metal so that it forms a saturated solution and then cooled by a few degrees so that the solution becomes supersaturated. This supersaturated solution is then brought into contact with the substrate so that the single crystal material can begin to grow. Once a suitable thickness has been grown, then the liquid is separated from the solid.

This process takes place usually by means of a moving slider which pushes solutions over the surface of a substrate, held in a recess. A typical apparatus is shown in Figure 4. Although LPE has been used for the growth of elemental semiconductors, it is most widely used for the production of compound semiconductor thin films, where one of the elements of the semiconductor can be used as the solvent. This restricts the introduction of unwanted impurities and enables high quality material to be produced.

There are two main disadvantages of LPE growth. Firstly, the very thin layers often required in modern laser and transitor structures are difficult to grow controllably. Secondly, some complex semiconductor alloys have a miscibility gap at desirable compositions, which cannot therefore be produced by this method. However, the LPE method has been widely used for the production of a range of semiconductor materials over the years, in particular **GaP LEDs** and **$Al_{1-x}Ga_xAs$ lasers**.

1.3.3.2 Vapour Phase Epitaxy. Vapour phase epitaxy (VPE) is a process in which

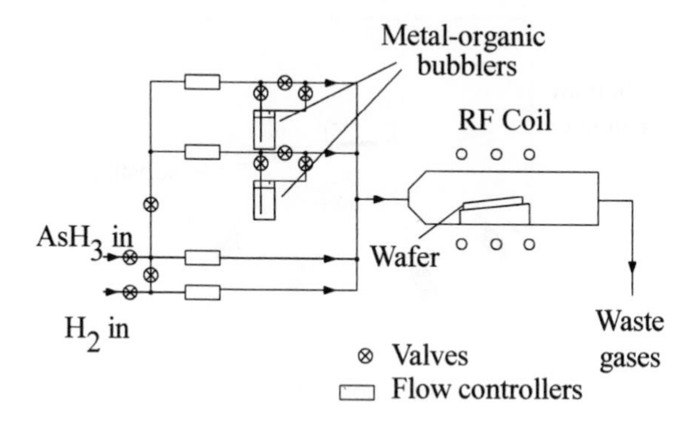

Figure 5 *Metal–organic vapour phase epitaxy (MO-VPE) reactor*

the constituents of the epitaxial film are transported to the substrate surface in the vapour phase, where they react to form the single crystal. While the constituents may be present as the elements, often they are introduced as volatile compounds. There are two main types of VPE that have commonly been used, **chloride VPE** and **metal–organic VPE** (MO-VPE). Chloride VPE was first developed during the 1960s and typically used volatile chlorides of the Group IIIA or Group VA elements* to transport the semiconductor components to the substrate. The development of more sophisticated growth techniques such as MO-VPE and MBE during the 1980s enabled more complex heterojunction structures such as quantum well lasers and high electron mobility transistors to be developed.

MO-VPE is distinguished from chloride VPE in that some or all of the constituents of the epitaxial film are transported in the vapour phase as metal–organic compounds. The term metal–organic is used as it encompasses both organometallic compounds (with a metal–carbon bond) and other compounds, such as metal alkoxides which do not have metal–carbon bonds and which are not truly organometallics. MO-VPE is most widely used for the production of single crystal thin films of **III–V and II–VI semiconductors*** such as GaAs, AlGaAs, InP, and CdHgTe, the classical reaction being between trimethylgallium and arsine at 500–700 °C, under a hydrogen ambient atmosphere, to give gallium arsenide.

$$(CH_3)_3Ga + AsH_3 \rightarrow GaAs + 3\,CH_4 \tag{1}$$

A typical MO-VPE reactor is shown in Figure 5. The reactants, or precursors, are usually liquid, held in a temperature controlled container through which a gas, normally hydrogen, can be bubbled. Precursor vapour is transported by the gas from the bubbler into the quartz reactor and over the surface of the single crystal wafer which has been heated, usually by an RF coil, to a temperature that will induce decomposition and reaction of the vapour phase constituents. MO-VPE is now used as a production process for the manufacture of a wide range of compound semiconductor devices.

* In this chapter only, old group nomenclature is used, and the widely used notation for III–V and II–VI semiconductors is retained. *cf.* footnote to page 3.

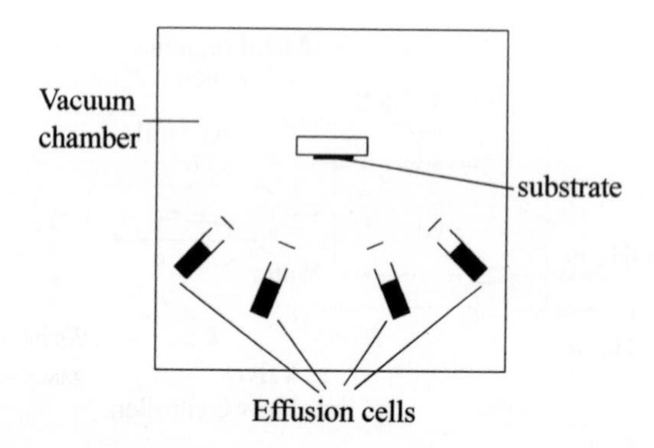

Figure 6 *Molecular beam epitaxy (MBE) reactor system*

1.3.3.3 Molecular Beam Epitaxy. Molecular beam epitaxy (MBE) is a high vacuum evaporation process in which high purity elemental sources are contained in individual heated effusion cells with independently operated shutters. MBE is an expensive technique, complicated by the use of ultra-high vacuums, but it has a clear advantage over VPE and LPE methods of epitaxial growth. The use of elemental sources means that extrinsic impurities can be limited, while the relatively low growth temperatures means that more abrupt interfaces between layers can be obtained using the lower diffusion rates.

A diagram of an MBE reactor system is shown in Figure 6. As the shutters open, a molecular beam of the material is directed at the heated substrate. The high vacuum and the lack of impurity elements present in the system means that **very high purity epitaxial thin films** can be produced. Controlling the opening and closing of the shutters enables close control of film composition and enables the production of heterostructures with very abrupt interfaces.

One problem with MBE is controlling the flux from the effusion cells in the case of high melting point elements, such as silicon, or where vapour pressures are very high, as with phosphorus and arsenic. This has led to the development of a derivative of the MBE technique, called **chemical beam epitaxy** (CBE) or **metal–organic molecular beam epitaxy** (MO-MBE) in which the properties of the flux are controlled by using compounds with controlled vapour pressures and decomposition routes.

MBE and its derivatives can produce semiconductor thin films of the highest quality, but the cost and slow throughput of the method means that it is only used as a production method when high purity films with sharp interfaces are demanded.

1.3.3.4 Sputtering. Sputter deposition is a versatile method of forming thin films where the material to be deposited is abraded from a source (the target) by high energy gas ions. These gas ions are formed from a **plasma** struck in a vacuum chamber in the presence of an inert gas, usually argon. dc plasmas can be used with metal or semiconductor targets, but insulators require the use of a plasma formed by an RF field. A wide range of thin films can be grown by sputtering, including metals, semiconductors, oxides, and even polymers.

Two important developments of sputtering have been made. Magnetron sputtering uses a strong permanent magnet behind the target material to constrain the movement of the ions to the region between the target and substrate. This gives a much faster deposition rate. Reactive sputtering replaces some of the inert gas with a more reactive gas such as oxygen or nitrogen. This allows for the direct formation of oxide and nitride thin films from elemental targets.

Sputtering is widely used for a multitude of thin film deposition process. Its only significant disadvantage is a tendency to give shadowing effects when substrates with pits or grooves are to be coated.

1.3.3.5 Evaporation. Evaporation is the simplest method of thin film deposition. It is based on the fact that, in a vacuum chamber, a relatively high vapour pressure of many metals can be obtained by heating the evaporation source. Any colder surface within the vacuum chamber will provide a suitable site for condensation of the film. This method has the advantage of simplicity and the fact that, as with MBE, very high purity films can be produced. However, the technique in this form is limited to those metals that can give a suitable vapour pressure on heating and the use of a cold substrate means that epitaxial growth is difficult to achieve. The problem of vapour pressure has been largely overcome by the development of **electron beam evaporation** which heats the evaporation source with an intense beam of electrons. High melting point elements and compounds are readily evaporated in this method which is used commercially in the deposition of conductive metal tracks.

1.3.3.6 Chemical Vapour Deposition. Chemical vapour deposition (CVD) is a widely used technique for the deposition of thin films from the vapour phase. Typically the constituent or constituents of the film to be deposited are carried as a volatile compound to the place where the film is required, at which point a reaction is induced that decomposes the gaseous precursor(s) so that film formation can occur. Classically, the method used has been **thermal decomposition**, brought about by heating the substrate on which the film is to be deposited. Many other means have been employed, however, especially the use of plasmas (**plasma-enhanced CVD**) and light beams (**photolytic CVD** and **laser CVD**). In the case of laser CVD, decomposition may be either thermal, photolytic, or a combination of the two.

CVD presursors are designed to have a sufficiently high vapour pressure so that they can be transported easily to the deposition area, while being able to readily decompose when desired. Thus the elements of the film are deposited cleanly, with the remainder of the precursor being flushed out of the system without contaminating the film. Hydrides, halides, and metal–organic compounds have all been widely used as CVD precursors for the production of **silicon, metals dielectrics, and insulators**. This process has many similarities with the MO-VPE process, and can often be distinguished only by the fact that CVD processes are not usually intended to produce single crystal films.

1.3.4 Doping

The doping of bulk semiconductors is normally carried out by one of two methods,

thermal diffusion or **ion implantation**. Thermal diffusion involves heating the wafers to a high temperature in a flowing gas stream containing the dopant. The dopant diffuses into the exposed parts of the semiconductor, the concentration profile being determined by its diffusion coefficient and the process conditions. Ion implantation relies on the bombardment of the semiconductor surface by energetic dopant ions. This method gives improved control over dopant concentrations compared to thermal diffusion methods.

Where semiconductor thin films are being deposited by methods such as silicon CVD, MO-VPE or MBE, then it is possible to incorporate the dopant in the growing film by including it as a minor additive to the flux of thin film precursor species. Thus, in silicon CVD, low levels of arsine gas can be added to the silane to introduce arsenic into the growing film and render it **n-type**. (Section 2.1.1).

1.3.5 Deposition of Thick Film Materials

Thick film materials, used in the manufacture of hybrid circuits, are made as **screen printable inks** which can be printed onto the substrate as conductor lines, resistor areas, or as dielectric overlays. The inks normally consist of a solvent/resin mixture to give the desired printing properties, a glass or oxide adhesion agent, and a functional phase in the form of a fine powder. In the case of a conductor ink the powder is a metal, in an insulating or dielectric ink it is often a mixture of oxides, and in a resistor ink it is ruthenium dioxide.

The substrates are normally made of alumina and this, together with the printed film is normally dried at 100–150 °C to remove the solvent and then fired at between 700–900 °C. If several layers are produced then several print–fire repetitions will be made. This area is described in more detail elsewhere (page 227).

2 ELEMENTAL SEMICONDUCTORS

Elemental semiconductors are found in Group IV of the Periodic Table*. Carbon, silicon, germanium, and the grey form of tin in this Group form giant molecules based on the tetrahedral 'diamond' structure, and show an increasing electron mobility as the atomic weight increases. Carbon is an insulator; **silicon, germanium, and grey tin** are semiconductors; while white tin, which has a distorted octahedral structure, is a metal. These trends are summarized in Table 2.

2.1 Silicon

Silicon is the most important of the elemental semiconductors. The intrinsic properties of its band gap and ability to form a stable oxide, combined with its abundance and ease of processing make it pre-eminent in its use for electronic semiconductor device applications.

At one time it had been thought that much silicon technology would be replaced by compound semiconductors such as gallium arsenide, which have a much higher electron mobility than silicon and can therefore switch much faster. Rapid

* See footnote on page 203.

Table 2 *Properties of elemental semiconductors*

Element	*Band gap (eV)*	*Structure*	*Type*
Carbon (diamond)	6.0	Diamond	Insulator
Silicon	1.12 (indirect)	Diamond	Semiconductor
Germanium	0.66 (indirect)	Diamond	Semiconductor
Grey Tin (stable below 13.2°C)	0.1	Diamond	Semiconductor
White Tin (stable above 13.2°C)	0	Distorted octahedral	Metal

improvements in silicon device technology have meant that much of this replacement has not happened. New developments in silicon technology, such as the production of ultra high speed silicon–germanium transistors, are likely to maintain silicon's technical and commercial lead. Silicon–germanium alloys and heterostructures are described in Section 2.2.

The fact that silicon has an indirect band gap, however, means that light-emitting devices are very difficult to achieve due to the fact that photon emission here would also require simultaneous phonon emission. Direct band gap semiconductors such as gallium arsenide and related compounds are most important here. However, recent work on the still controversial subject of 'porous silicon' does seem to give the promise of light-emitting silicon devices in the future. This subject is discussed further in Section 2.3.

2.1.1 Properties

Normally, silicon is a hard, brittle, grey, crystalline material of low density. It crystallizes into the diamond cubic structure when cooled from the melt at 1415°C. Silicon can also form an amorphous phase under certain circumstances, such as when thin films are deposited by low temperature chemical vapour deposition.

Highly pure silicon is an **intrinsic semiconductor**, that is, most moving electrons and holes result from thermal fluctuations rather than from impurities. The number of electrons and holes will be the same and will be equal to the intrinsic carrier concentration, n_i.

When **impurities** are added to silicon then the effect is to add charge carriers. Impurities from Group VA of the Periodic Table, such as phosphorus or arsenic have five electrons in their outer shell and therefore have one electron not involved in bonding into the silicon lattice. This electron is only weakly bound to the impurity and can move easily through the lattice. Such an impurity is called a **donor** because it donates electrons to the conduction band of the lattice. Impurities from Group IIIA of the Periodic Table, such as boron and aluminium, have only three electrons in their outer shell and therefore require one extra electron to satisfy their bonding requirements. A bonding electron can move from an adjacent bonding site to the impurity site giving the net effect of a positive 'hole' moving in the opposite direction. These impurities are called **acceptors**.

Silicon which has been doped with acceptor impurities is called **p-type** silicon, while silicon which has been doped with donor impurities is called **n-type** silicon. Semiconductor devices are operated by the interaction of p- and n-type silicon at p–n junctions.

2.1.2 *Manufacture*

2.1.2.1 Production and Purification. Elemental silicon is obtained by the chemical reduction of silicon dioxide, obtained from a range of silicate minerals such as quartzite. Reduction usually takes place in electric arc furnaces using coke as the reducing agent. This produces silicon with a purity of about 98%. The required purity level for semiconductor grade silicon is 99.99999%, and purification is normally undertaken by conversion of the silicon into volatile silanes which are then isolated by fractional distillation and reduced back to silicon. The most common process is the Siemens process which produces $SiHCl_3$ by reaction of the silicon with HCl. After the fractional distillation step, the halide is reduced with hydrogen at around 1000 °C to produce polycrystalline silicon (**polysilicon**) rods.

The overall process may be represented by the following equations:

$$SiO_2 + 2\,C \rightarrow Si + 2\,CO \quad (2)$$

$$Si + 3\,HCl \rightarrow SiHCl_3 + H_2 + \text{impurities} \quad (3)$$

$$SiHCl_3 + H_2 \rightarrow Si + 3\,HCl \quad (4)$$

2.1.2.2 Crystal Growth. The fundamental requirement for most silicon devices is that the silicon should be single crystalline with a very low number of defects. The Czochralski technique (see Section 1.3.2) is the most commonly used process for single crystal silicon production. Here, polysilicon is heated in a silica crucible under an inert gas atmosphere to around 1500 °C and a rotating seed crystal is used to pull the single crystal silicon boule slowly from its melt.

Dopants are usually added to the melt to control the electrical properties of the silicon. These include boron (as a p-type dopant); and arsenic, phosphorus, or antimony (as n-type dopants).

2.1.3 *Preparation of Silicon-based Semiconductor Devices*

The **integrated circuit** (IC) which is the fundamental component of so many electronic devices today, is based almost exclusively on silicon. Figure 7 shows the structure of a typical semiconductor device used in integrated circuit fabrication, the metal-oxide-semiconductor field effect transistor (MOSFET). As can be seen, this transistor structure is composed of several layers which are produced by a range of chemical processes. The basic principle of building up this complex series of layers lies in the repetition of three basic steps for each layer. These are **deposition**, **patterning**, and **etching**.

In the deposition stage one of a range of techniques is used to deposit a layer of material onto the substrate material. Most commonly available techniques such as sputtering and chemical vapour deposition are employed. Chemical vapour

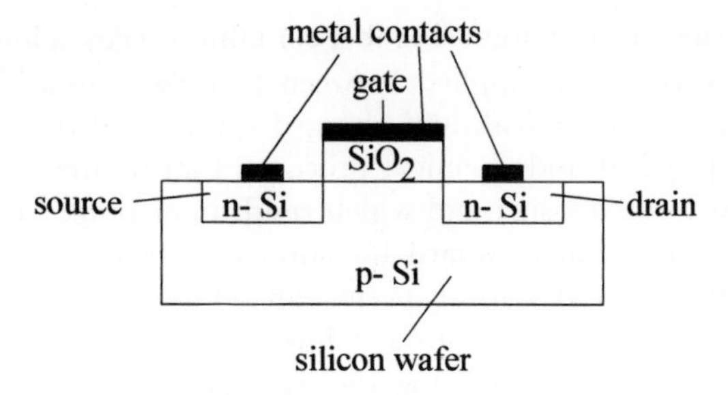

Figure 7 *Structure of the metal-oxide-semiconductor field effect transistor (MOSFET)*

Table 3 *Typical CVD reactions in semiconductor device manufacture*

Reactants	*Product*
SiH_4	Si
$SiH_4 + O_2$	SiO_2
$SiH_4 + N_2$	Si_xN_y
$WF_6 + H_2$	W

deposition (CVD) is often used to deposit further layers of silicon or silicon compounds, using silane (SiH_4) gas as the prime reactant. Metals such as tungsten which have volatile precursor compounds are also deposited by chemical vapour deposition. Some common CVD reactions are given in Table 3.

This layer then has a pattern laid on top by means of **lithography** where a radiation-sensitive polymeric film is spin coated onto the substrate surface and exposed to light through a patterned mask. Depending on whether the resist is a positive resist or a negative resist, then either the illuminated areas or the un-illuminated areas can be dissolved away by the use of a **developer**. The exposed regions of the top layer can then be removed by etching.

There are two main types of etching that are used: wet etching, and dry or plasma etching. **Wet etching** acts isotropically, that is it etches equally in all directions. It is less suited therefore to applications where vertical edges are required and is mainly used for bulk substrate removal in, for instance, the chemomechanical polishing of semiconductor wafers prior to device manufacture or the thinning of wafers for use in electro-optical devices. Acids or alkalis can be used to etch silicon, as shown by the equations below:

$$3\,Si + 4\,HNO_3 + 18\,HF \rightarrow 3\,H_2SiF_6 + 4\,NO + 8\,H_2O \qquad (5)$$

$$Si + 2\,KOH + H_2O \rightarrow K_2SiO_3 + 2\,H_2 \qquad (6)$$

Plasma etching, on the other hand, is a highly directional process and is commonly used for all etching processes in device manufacture that involve the production of patterns of small dimensions.

A plasma is formed in a vacuum chamber containing a gas at low pressure when a high frequency voltage is applied between two electrodes. The plasma will contain neutral, neutral radical, and charged species. All these species can be involved in the physical and chemical processes that occur at the surface of a material exposed to the plasma and which result in etching. Thus charged and neutral radical species can bombard the surface at high energies and remove material by sputtering, and neutral species can act as a simple chemical etchant and form a volatile product at the surface thus also removing material. However, the combination of these processes together can have a synergistic effect to give fast, directional etching.

The most common plasma etchants used are volatile oxidants such as O_2, F_2, Cl_2, and Br_2. F_2 is widely used as an isotropic etchant for silicon. Fluorine radicals attack the surface of the silicon to form SiF_2 species which may either be removed as a gaseous species or may become stabilized as an $-SiF_2-$ surface film. In this latter case, further fluorination leads to the formation of gaseous SiF_4. The full mechanism of this process is still not entirely understood, but may be represented overall by the following equations:

$$F_2 \rightarrow F^{\cdot} + F^{\cdot} \tag{7}$$

$$2\,F^{\cdot} + Si \rightarrow SiF_2\ (g) \tag{8}$$

$$2\,F^{\cdot} + Si \rightarrow SiF_2\ (c) \tag{9}$$

$$2\,F^{\cdot} + SiF_2\ (c) \rightarrow SiF_4\ (g) \tag{10}$$

The toxicity of fluorine gas means that alternatives such as CF_4 and SF_6 are sometimes preferred. Fluorine is highly selective for silicon compared to silicon dioxide, with a selectivity of 40:1, and other species may be used to selectively etch other surfaces. CF_4/H_2 combinations have a high selectivity for silicon dioxide over silicon and chlorocarbon or fluorocarbon gases are normally used to etch metals. CCl_4 is often used for aluminium, widely used as a metal interconnect on silicon, with the addition of BCl_3 to scavenge any trace water or oxygen from the system.

2.2 Si/Ge Alloys and Heterostructures

A relatively new area of work on silicon, but one that is likely to result in exciting new devices, is the current research on silicon–germanium systems. First investigated at the AEG research laboratories in Germany during the 1970s, more recent work has shown that very high electron mobilities can be obtained in these systems, opening the prospect of super-fast silicon–germanium transistors. Such transistors would have the benefit of compatibility with much of current silicon processing technology, unlike gallium arsenide high speed devices.

Both silicon and germanium have an indirect band gap and this means that they are not used as optical emission devices since a radiative transition in such a material requires that the electron transition must simultaneously produce a photon and a phonon – a relatively unlikely event. However, it has been found that when short period multilayer structures of the two elements are deposited, a quasi-direct band-gap is formed. This brings forward the prospect of **silicon–germanium optoelectronic devices**.

A wide range of Si–Ge structures has now been produced, including SiGe alloys, Si–Ge ultrathin multilayers (superlattices), and Si–SiGe heterojunction devices. SiGe structures have been made by MBE and a range of CVD methods. The main problem with growing good quality epitaxial material is the different lattice constants of the semiconductors, and this can lead to strain in the deposited films that can result in distortions and dislocations in the crystal lattice. This problem can be overcome by growing SiGe buffer layers between the silicon substrate and the Si–Ge device.

Some of the most important results to date have been in the improvement of the performance of silicon bipolar transistors by the addition of $Si_{1-x}Ge_x$ heterojunctions to form Si/SiGe heterojunction bipolar transistors. Such transistors have been shown to operate at frequencies up to 94 GHz.

2.3 Porous Silicon

In Section 2.1.1 we discussed how the indirect band gap of silicon made it so inefficient as a source of photons that light-emitting devices were extremely difficult to conceive. In 1990, however, the Defence Research Agency in the UK showed for the first time that porous silicon could be used to emit light.[1]

Porous silicon is made by partial electrochemical dissolution of the top surface of a silicon wafer using hydrofluoric acid based electrolytes. Large numbers of pores with widths less than 20 nm are etched into the surface to form a skeletal framework of silicon wires with a high volume fraction of voids. Such structures exhibit **photoluminesence** in the visible region, the wavelength of which varies according to the diameter of the silicon wires. The smaller the diameter of the wires, the shorter the wavelength. Such a result is in fact in accordance with work on semicon ductor nanostructures (structures with dimensions on an atomic scale) which predicts that as these structures are reduced to very low dimensions the band gap will increase. If the electronic charge carriers are completely confined in one or more dimensions, then this effect is even more marked. A silicon wire of two nanometres diameter or less shows carrier confinement in two dimensions and is expected to have a band gap of between 2 and 3 electron volts, *i.e.* in the visible region.

Electroluminescence in porous silicon has also been demonstrated and the first light-emitting diodes have been produced. These devices infiltrate the porous region with liquid electrolytes containing persulfate $S_2O_8^{2-}$ ions by dipping the silicon wafer, with porous silicon regions, into a $H_2SO_4/Na_2S_2O_8$ solution. Electrical contact is made between the back of the wafer and a platinum counter electrode. Electroluminescence throughout the visible region has been demonstrated, with the emitting colour being controlled by the applied voltage.

Much work remains to be done in this area, in particular elucidating the full mechanism of operation of both porous silicon and porous silicon devices. There are concerns also about the lifetime of these devices. The prospect, however, is for a range of silicon-based optoelectronic devices, compatible with current silicon technology.

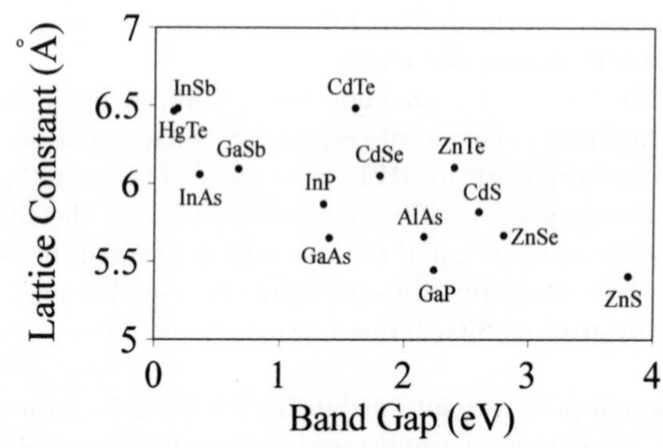

Figure 8 *Band gap vs lattice constant for III–V and II–VI semiconductor materials*

3 COMPOUND SEMICONDUCTORS

3.1 Introduction to Compound Semiconductors

The simplest and technologically most important class of compound semiconductors are formed from the equiatomic mixing of a Group IIIA element (B, Al, Ga, and In) and a Group VA element (N, P, As, and Sb).* These **III–V semiconductors** crystallize in the so-called 'zinc blende' (sphalerite) structure. This structure is equivalent to the diamond structure of silicon and germanium, with the Group IIIA and Group VA elements occupying adjacent sites in the crystal lattice. The most well known members of this Group are gallium arsenide (GaAs) and indium phosphide (InP).

A key property of III–V semiconductors is their ability to emit light in the visible and near infrared, the wavelength corresponding to the direct band gap of the material. This has made this class of material the predominant choice for the production of optoelectronic devices such as **light-emitting diodes** (LEDs), lasers, and photodetectors. The **band gap** of a binary III–V semiconductor can be **tuned** by adding another binary to give either a ternary semiconductor (as in GaAs and AlAs) or a quaternary semiconductor (as in GaAs and InP) with intermediate properties. The range of band gaps and lattice constants available from the common III–V and II–VI semiconductors is shown in Figure 8. For the III–V ternaries, such as $Al_{1-x}Ga_xAs$, the band gap range attainable is represented by a line joining the GaAs and the AlAs positions in the figure. This is also equivalent to the wavelength range possible from lasers or LEDs produced from this system, in this case, from approximately 600 to 900 nm. The fact that the GaAs–AlAs line is horizontal means that all compositions will be closely lattice matched to the GaAs substrate and should be able to be grown epitaxially. The availability of suitable substrates for epitaxial growth is one of the main limitations imposed on epitaxial compound semiconductor growth.

In addition to their useful optical properties, III–V semiconductors have much higher electron mobilities and peak electron velocities than silicon and this has led to their use in a number of high speed semiconducting devices.

* See footnote on page 203.

The second important class of compound semiconductors are those prepared from the equiatomic mixing of a Group IIB element (Zn, Cd, and Hg) with a Group VIA element (O, S, Se, and Te). **II–VI semiconductors** also tend to crystallize in the sphalerite form, although an alternative wurtzite form can be produced. As with the III–V semiconductors, ternary or quaternary II–VI semiconductors can also be produced from the intermixing of the simple binaries to give materials with properties intermediate between the parent compounds. The range of band gaps and lattice constants available is shown, alongside the III–V semiconductors in Figure 8.

The II–VI semiconductors can be subdivided into two groups, the wide bandgap II–VIs such as ZnS, ZnSe, and CdS, and the narrow gap II–VIs such as HgTe and CdTe. The wide bandgap II–VIs are used as phosphors and are showing great promise for blue and green LEDs and lasers. The narrow gap II–VIs, exemplified by the ternary compound $Hg_{1-x}Cd_xTe$, are extensively used as long wavelength infrared detectors.

A third class of semiconducting materials can be formed as a variation on III–V or II–VI semiconductors. In the first variation, alternate Group IIIA elements in a III–V crystal lattice are replaced with a Group IIB and a Group IVA element. The result is a II–IV–V_2 semiconductor such as $ZnSiP_2$. In the second variation, alternate Group IIB elements in a II–VI crystal lattice are replaced with a Group IB and a Group IIIA element. The result is a I–III–VI_2 semiconductor such as $CuInSe_2$ or $CuGaSe_2$. II–IV–V_2 and I–III–VI_2 semiconductors are referred to as the **chalcopyrite semiconductors**.

The main properties of members of the III–V, II–VI, and chalcopyrite semiconductors are given in Tables 4 and 5.

3.2 III–V Semiconductors

3.2.1 Purification of the Elements

Compound semiconductors are most commonly made by reaction of the elements at elevated temperatures. The purification steps normally precede compound formation and involve purification of the individual elements.

Gallium is normally obtained as a by-product of aluminium or zinc production and initial purification is performed by electrolytic refining from alkaline solution to give 99.99% gallium. Further purification can be achieved by converting the gallium to gallium trichloride which is then zone refined, before being reduced back to the metal again. Purification of **indium** can also be achieved through electrolysis after which a further step of vacuum distillation is often used.

Aluminium produced by normal electrolytic processes can be as high as 99.9% pure, but significantly higher purities are required for semiconductor starting materials. Electrolytic cells can produce aluminium with a purity of 99.995%, but zone refining this grade of metal can give aluminium with a purity better than 99.999%.

Arsenic is normally obtained as an impure lead–arsenic–antimony alloy as a by-product of lead refining. The high purity metal is obtained by vacuum distillation of the alloy, followed by zone refining.

Table 4 *Properties of III–V compound semiconductors*

Compound	*Energy gap at 300 K (eV)*	*Band gap type*	*Lattice constant (Å)*	*Structure*	*Melting point (°C)*
GaN	3.39	Direct	$a=3.189$ $c=5.186$	Wurtzite	~1500
GaP	2.24	Indirect	5.447	Zinc blende	1467
GaAs	1.40	Direct	5.654	Zinc blende	1238
GaSb	0.67	Direct	6.095	Zinc blende	707
InN	1.9	Direct	$a=3.533$ $c=5.693$	Wurtzite	927
InP	1.35	Direct	5.869	Zinc blende	1062
InSb	0.18	Direct	6.479	Zinc blende	525
AlN	6.2	Direct	$a=3.111$ $c=4.978$	Wurtzite	~2200
AlAs	2.16	Indirect	5.661	Zinc blende	1700

3.2.2 Crystal Growth

Single crystals of compound semiconductors can be **grown in bulk or as thin films** by a wide variety of methods. Some of the more popular ones for each semiconductor are summarized in Table 6.

3.2.2.1 Bulk Crystal Growth. Large scale bulk single crystal growth of III–V semiconductors is mainly restricted to **gallium arsenide** and **indium phosphide**. Gallium arsenide is prepared by both the horizontal Bridgman (HB) and Czochralski methods (described in Section 1.3.2). The horizontal Bridgman growth of gallium arsenide is achieved by the reaction of gallium in a silica boat containing a seed crystal, with arsenic vapour obtained from solid arsenic sealed together inside a quartz ampoule. This method can produce single crystal GaAs boules of up to several kilogrammes weight. Wafers cut from these crystals are of very high quality, with less than 10^2 dislocations per square centimetre. There are, however, two main disadvantages to HB GaAs. Firstly, semi-insulating material can only be obtained through chromium doping, which tends to out-diffuse during any subsequent heat treatments. Secondly, they are not circular in shape and this can make them difficult to integrate with semiconductor processing techniques derived from silicon technology.

The Czochralski method of GaAs crystal growth is made difficult by the high partial pressure of the molten components at the crystal growth temperature (above 1200°C) which can result in loss of stoichiometry. This can be overcome, however, by the addition of a molten capping layer (usually boron trioxide) to the surface of the GaAs, which prevents loss. This variation is called the **liquid encapsulated Czochralski** (LEC) **method** and has enabled the production of 80 mm diameter circular cross-section **high quality GaAs boules of up to several kilogrammes weight**. High purity gallium and arsenic (or polycrystalline GaAs) is added to a pyrolytic boron nitride (PBN) crucible. An excess of arsenic is usually added to compensate for any likely to be lost during the process.

Table 5 *Properties of II–VI and chalcopyrite compound semiconductors*

Compound	*Energy gap at* 300 K (eV)	*Band gap type*	*Lattice constant* (Å)	*Structure*	*Melting point* (°C)
ZnO	3.4	Direct	4.630 a=3.250 c=5.207	Zinc blende Wurtzite	1977
ZnS	3.8	Direct	5.409 a=3.820 c=6.260	Zinc blende Wurtzite	1827
ZnSe	2.8	Direct	5.669 a=4.003 c=6.540	Zinc blende Wurtzite	1417
ZnTe	2.4	Direct	6.104 a=4.270 c=6.990	Zinc blende Wurtzite	1295
CdS	2.6	Direct	5.820 a=4.137 c=6.716	Zinc blende Wurtzite	1477
CdSe	1.8	Direct	6.050 a=4.298 c=7.015	Zinc blende Wurtzite	1239
CdTe	1.6	Direct	6.482	Zinc blende	1092
HgS	2.1		5.852	Zinc blende	1750
HgSe	semimetal	—	6.084	Zinc blende	797
HgTe	semimetal	—	6.462	Zinc blende	670
$CuInSe_2$	1.04	Direct	a=5.782 c=11.564	Chalcopyrite	1327

Table 6 *Bulk and epitaxial thin film crystal growth methods for compound semiconductors*

Compound Semiconductor	*Method*					
	Czochralski	*Bridgman*	*LPE*	*Halide VPE*	*MO-VPE*	*MBE*
GaAs	✓	✓	✓	✓	✓	✓
GaN					✓	✓
GaP	✓		✓	✓	✓	✓
InP	✓		✓	✓	✓	✓*
InSb		✓			✓	✓
AlGaAs			✓		✓	✓
InGaAsP			✓	✓	✓	✓*
CdS		✓			✓	✓
ZnSe			✓		✓	✓
CdTe	✓	✓	✓		✓	✓
HgCdTe			✓		✓	✓
$CuInSe_2$					✓	

* gas source MBE using phosphine (see page 219)

A disc of boron trioxide is then placed over the surface of the reactants. As the crucible is heated, the boron trioxide melts, flows over the reactants, and encapsulates them. The whole system is heated above the melting point of gallium arsenide at which point the rotating single crystal seed is lowered through the boron trioxide into the molten gallium arsenide. The seed is slowly withdrawn over a period of several days as the crystal grows.

The dislocation density of LEC wafers tends to be higher than HB wafers and they are mainly used for simple optoelectronic devices and as substrates for further epitaxial growth. The LEC method has also been successfully applied to the production of indium phosphide and gallium phosphide single crystals.

3.2.2.2 Thin Film Deposition. **Epitaxial methods** are normally used for thin film deposition, as the perfection of a single crystal film is required to obtain the desired device properties. Liquid phase epitaxy (**LPE**) (outlined in Section 1.3.3.1) is a process in which epitaxial films are deposited from supersaturated solutions. It is a process most suited to the growth of III–V semiconductors as the Group III elements gallium or indium can be used as the solvent, restricting the introduction of impurities.

LPE GaAs is made from gallium solution which has been saturated with arsenic at a growth temperature of between 700–800 °C. Doping of the epitaxial layer is achieved by adding Group IIA and IIB elements (*e.g.* Mg, Zn) for p-type layers and Group VIA elements (*e.g.* S, Se, Te) for n-type layers. Group IVA elements are amphoteric and can occupy either the gallium or the arsenic lattice sites depending on the growth temperature. Ternary III–V layers for LEDs and lasers are also grown, but although the quality of the layers is very high, the high growth rates and possibility of meltback means that it is difficult to grow structures with very sharp and multiple interfaces. This, combined with the relatively low throughput compared to MO-VPE, means that it is becoming less widely used.

Metal–organic vapour phase epitaxy (**MO-VPE**, Section 1.3.3.2) has widespread commercial use for the production of compound semiconductors. Normally, each element of the semiconductor is delivered to the growth area as a separate metal–organic or hydride compound in the vapour state, where it forms the desired semiconductor composition on the surface of the heated wafer.

The main **precursor materials** for III–V growth are given in Table 7. Although a large number of precursors have been investigated, normally the simple **trimethyls or triethyls of the Group III elements** and the **hydrides of the Group V elements** are used. The main exception to this is the use of trimethyl- and triethylantimony instead of the unstable hydride, stibine.

All the III–V semiconductors can be formed by apparently simple reactions which are performed under a hydrogen atmosphere, such as those represented by the following equations:

$$(CH_3)_3Ga + AsH_3 \rightarrow GaAs + 3\,CH_4 \quad (11)$$

$$(CH_3)_3Al + AsH_3 \rightarrow AlAs + 3\,CH_4 \quad (12)$$

$$(CH_3)_3In + PH_3 \rightarrow InP + 3\,CH_4 \quad (13)$$

Table 7 *MO-VPE precursors for III–V semiconductors*

Deposited product	*Precursor*	*Formula*	*Boiling point* (°C)
Aluminium	Trimethylaluminium	$(CH_3)_3Al$	126
	Triethylaluminium	$(CH_3CH_2)_3Al$	194
Gallium	Trimethylgallium	$(CH_3)_3Ga$	56
	Triethylgallium	$(CH_3CH_2)_3Ga$	143
Indium	Trimethylindium	$(CH_3)_3In$	88*
			136
	Triethylindium	$(CH_3CH_2)_3In$	184
Nitrogen	Ammonia	NH_3	
Arsenic	Arsine	AsH_3	−63
	Tertiarybutylarsine	$(CH_3)_3CAsH_2$	65
Phosphorus	Phosphine	PH_3	−88
	Tertiarybutylphosphine	$(CH_3)_3CPH_2$	54
Antimony	Trimethylantimony	$(CH_3)_3Sb$	81
	Triethylantimony	$(CH_3CH_2)_3Sb$	160

* melting point

$$(CH_3)_3Ga + (CH_3)_3Sb \rightarrow GaSb + 6\,CH_4 \qquad (14)$$

In fact, the chemical reactions that occur both in the gas phase and on the surface of the wafer are characterized by a high degree of complexity and this can lead to difficulties in controlling the reaction pathway and in the incorporation of impurities in the growing film. One example of this was experienced in the preparation of indium phosphide by MO-VPE where a low temperature premature gas phase reaction between the trimethylindium and phosphine precursors gave an involatile polymer which deposited on the walls of the reaction vessel. By the time the trimethylindium reached the surface of the wafer where it was to undergo thermal decomposition with phosphine to form indium phosphide, its concentration in the gas phase had become severely depleted.

One solution to this problem is to block the low temperature reaction by preforming a Lewis acid–base adduct from trimethylindium and triethylphosphine, which only breaks down at high temperatures to regenerate the trimethylindium when it is required to react with the phosphine. The trimethylindium–triethylphospine adduct is formed by simple addition, as shown in Equation 15, and this compound, which still retains a relatively high vapour pressure, is then transported into the MO-VPE reactor. Thermal breakdown of the adduct at the wafer surface allows the reaction of trimethylindium with phosphine to occur, which at high temperatures results in complete elimination of the organic groups and deposition of the indium phosphide:

$$(CH_3)_3In + (CH_3CH_2)_3P \rightarrow (CH_3)_3In.P(CH_2CH_3)_3 \qquad (15)$$

$$(CH_3)_3In.P(CH_2CH_3)_3 + PH_3 \rightarrow InP + (CH_2CH_3)_3P + 3\,CH_4 \qquad (16)$$

One important benefit of MO-VPE is the ability to control film composition very simply, by control of the gas flow rates through the precursor bubblers. Thus the **ternary semiconductor $Al_{1-x}Ga_xAs$** can be made as the combination of Equations 11 and 12 above, with x determined by the relative gas flow rates through the trimethylgallium and trimethylaluminium bubblers.

The key feature that has made MO-VPE such a successful method has been the ability to use chemical processes to obtain **extremely pure metal–organic precursors**. Trimethylindium, as an example, is prepared by the alkylation of indium trichloride by any one of several methods, but the use of methyllithium in the presence of an ether as solvent is typical:

$$3\,CH_3Li + InCl_3 \rightarrow (CH_3)_3In + 3\,LiCl \tag{17}$$

The trimethylindium prepared by this method usually contains trace quantities of several elements, including silicon and zinc, as well as the remainder of the solvent, all of which need to be removed. The reaction of the indium compound with a high molecular weight Lewis base, such as 1,2-bis(diphenylphosphino)ethane (diphos), gives a Lewis acid–base adduct that can be purified easily by recrystallization. Subsequent heating of the adduct regenerates the purified adduct-free trimethylindium.

GaAs prepared from trimethylgallium and arsine by MO-VPE is grown with an excess of the Group V precursor in the gas phase as the Group V elements have an appreciable volatility at the normal growth temperatures. This V/III ratio is usually three or more. The highest purity material is obtained at the lowest possible growth temperatures, typically between 600 and 650 °C. The 77 K electron mobilities of GaAs prepared under these conditions are typically better than 100 000 $cm^2\,V^{-1}\,s^{-1}$. As the growth temperature increases, so more carbon is incorporated into the film, which acts as an electron acceptor. If GaAs is prepared under low pressure conditions with triethylgallium as the Group III source then the carbon concentration is much reduced.

Both n- and p-type GaAs can be made by the addition of a range of compounds to the gas stream. Silane is a popular n-type dopant source, while compounds such as diethylzinc are used for p-type doping.

$Al_{1-x}Ga_xAs$ has been a difficult material to grow in high purity form because of the reactivity of aluminium, which forms strong bonds with carbon and oxygen. However, care in removing all traces of oxygen from the system via *in situ* gas purification and the inclusion of oxygen getters has enabled a high photoluminescence efficiency to be obtained.

InP is commonly grown from either trimethylindium and phosphine at atmospheric pressure or triethylindium and phosphine at low pressures. As with GaAs, the Group V precursor is usually present in large excess and the growth temperature is around 600 °C. Both methods can give high quality material with electron mobilities well in excess 100 000 $cm^2\,V^{-1}\,s^{-1}$.

GaSb, InSb, and their derivatives can be produced by using organo-antimony compounds as the Group V source. Sb has a much lower partial pressure than As or P and so V/III ratios close to unity are preferred to obtain stoichiometric material.

A point about the **health and safety aspects of MO-VPE** should be made here. The metal–organics commonly used are extremely air and moisture sensitive and will spontaneously combust if exposed to air. **Extreme caution** must be exercised when handling these materials. The reactants are normally supplied in stainless steel bubblers ready to be fitted to the MO-VPE equipment. Arsine gas is one of the most poisonous chemicals known and suitable precautions, including sensor and alarm systems must be installed when it used. Phosphine gas is both highly toxic and highly inflammable.

The dangers inherent in the use of arsine and phosphine have led to much research into **less toxic replacements** that can give equally good results. The best results to date have been achieved with compounds such as tertiarybutylarsine, $(CH_3)_3CAsH_2$, and tertiarybutylphosphine, $(CH_3)_3CPH_2$. Gallium arsenide with a 77 K electron mobility of 160 000 $cm^2 V^{-1} s^{-1}$ has been obtained from triethylgallium and tertiarybutylarsine precursors.

MBE growth of the III–V semiconductors has mainly concentrated on the GaAs/AlGaAs system with high quality homo- and heterostructures being obtained. One of the major problems associated with the production of GaAs by MBE has been the generation of oval shaped defects on the surface of the film. However, this effect has been shown to be associated with the gallium source, and improvements in gallium purity and design of the gallium effusion cell have led to a reduction in defect levels from several thousand per cm^2 to 50 or less. GaAs prepared with ultra-high purity sources has given record high electron mobilities.

MBE growth of phosphorus-containing compounds has been fairly limited for several reasons, including difficulties in obtaining high purity elemental phosphorus and lack of control of the P flux arising from the presence of different phosphorus allotropes. One solution to this problem is to use phosphine gas which can be thermally decomposed before reaching the substrate. This method has been called **gas source MBE** (GSMBE). This technique has been developed further into **metal–organic MBE** (MO-MBE), where the same precursors that were developed for MO-VPE are used in a high vacuum environment. MO-MBE has the advantage that, as with MO-VPE, film concentrations can be easily controlled by controlling the flow of the metal–organic vapour into the reaction chamber.

3.2.3 III–V Devices

The main applications of compound semiconductors are shown in Table 8. III–V semiconductors are the most widely used, finding **optoelectronic applications as lasers, light-emitting diodes, and photodetectors**. The high electron mobility of compound semiconductors has meant that gallium arsenide in particular is often used in **high frequency devices** and **integrated circuits**.

Semiconductor lasers are diodes in which photons are generated at a p–n junction and the external surface of the laser chip is cleaved or polished at two opposing ends. This structure is called a Fabry–Perot cavity. As current flow increases, a threshold is reached at which stimulated emission occurs and a coherent beam of light is emitted from the junction. Control of the nature of the **p–n junction** can control the **threshold current**, and very low thresholds can be

Table 8 *Applications of compound semiconductors*

Compound	*Applications*
GaAs	Bipolar integrated circuits, diode and transistor devices, photodetectors, solar cells
GaN	Blue LEDs and lasers
GaP	LEDs
InP	Transistor devices, solar cells
InSb	Photodetectors
AlGaAs/GaAs	High speed transistors and integrated circuits, solar cells Laser diodes for 780–1000 nm wavelength range
InGaAsP	Laser diodes for 780–1000 nm wavelength range
ZnO	Varistors Electronic frequency filters
ZnS	Cathode ray tube phosphors Infrared transparent lenses and windows
ZnSe	Infrared transparent lenses and windows
CdS	Photovoltaic cells
CdSe	Photovoltaic cells
CdTe	Photovoltaic cells Electro-optic modulators
$Hg_{1-x}Cd_xTe$	Infrared detectors

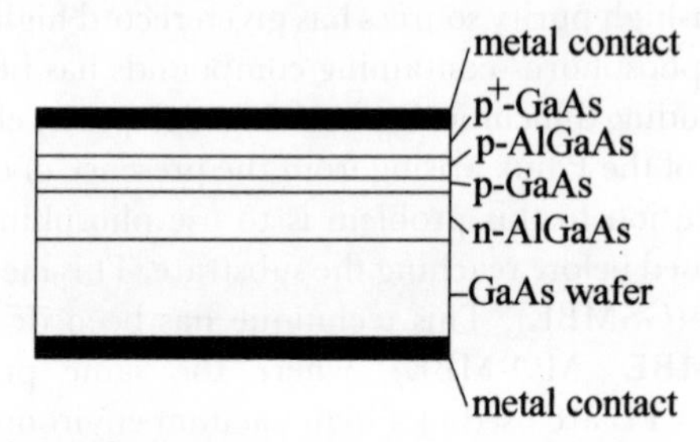

Figure 9 *Structure of an AlGaAs laser*

obtained from the double heterostructure laser, as shown in Figure 9. Here the p–n junction is between n-AlGaAs and p-GaAs, with a further layer of p-AlGaAs to confine the carriers by heterojunction potential barriers and the optical field by the refractive index change between the AlGaAs and GaAs compositions. New laser structures are continuously being developed to optimize laser performance with **graded refractive index** (GRIN) structures as well as AlGaAs/GaAs superlattices called **multiple quantum wells** (MQWs).

Reference to Table 4 shows that III–V compounds based on the nitrides, and in particular GaN and AlN have a band gap that would enable device emission in the blue region of the visible spectrum. However, the production of epitaxial films of these nitrides has been difficult due to their wurtzite structure and a lack of suitable substrates with a close match of lattice constant and thermal expansion coefficient. This is now being overcome with the development of heteroepitaxial growth techniques that enable the growth of high quality nitride films on mismatched

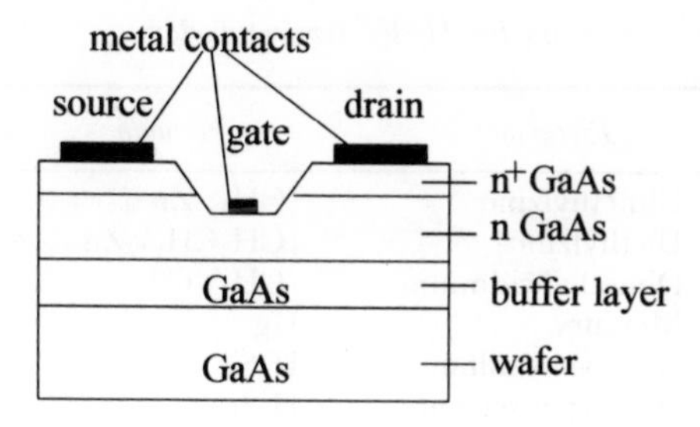

Figure 10 *Structure of the metal-gated field effect transistor (MESFET)*

substrates by using annealed GaN or AlN buffer layers. MO-VPE process using trimethylgallium and ammonia as the precursors have now been used to produce GaN p–n junction light-emitting diodes with both blue and ultraviolet emission.

Gallium arsenide high speed **integrated circuits** are manufactured from the metal-gated field effect transistor (**MESFET**), shown in Figure 10. This device is up to **five times faster** than equivalent silicon circuits, but can only be manufactured at a lower density of integration on the wafer compared to silicon.

One area of current interest is the heteroepitaxial growth of gallium arsenide onto silicon wafers. The lattice mismatch between the two materials can be mitigated by the use of strain-relieving buffer layers. Such structures could lead to the development of optical and silicon circuits on the same chip.

3.3 II–VI Semiconductors

3.3.1 Purification of the Elements

High purity **tellurium** is prepared by a combination of chemical purification followed by one or more physical methods. Several methods of chemical purification of tellurium are known. These include the recrystallization of tellurium compounds such as tellurium nitrate and telluric acid obtained by dissolving the metal in acid solution, or the distillation of tellurium tetrachloride obtained by the reaction of metal with chlorine. In all cases, the purified compound is converted to tellurium dioxide which is reduced at high temperature to the metal. Electrodeposition of tellurium from both acid and alkaline solutions also yields the high purity metal. Production of semiconductor grade tellurium can then be obtained by either vacuum distillation or zone refining of the metal.

Cadmium is normally purified by the dissolution of cadmium compounds in excess aqueous ammonia. Addition of ammonium sulfide will precipitate heavy metals which can be removed by filtration. Heating the solution removes the ammonia. **Mercury** is purified by distillation of the metal under reduced pressure.

3.3.2 Crystal Growth

Common bulk and thin film crystal growth methods for II–VI semiconductors are listed in Table 6.

Table 9 *MO-VPE precursors for II–VI semiconductors*

Deposition product	*Precursor*	*Formula*	*Boiling point* (°C)
Zinc	Dimethylzinc	$(CH_3)_2Zn$	46
	Diethylzinc	$(CH_3CH_2)_2Zn$	118
Cadmium	Dimethylcadmium	$(CH_3)_2Cd$	106
Mercury	Mercury	Hg	357
Sulfur	Hydrogen sulfide	H_2S	
Selenium	Hydrogen selenide	H_2Se	
Tellurium	Diethyltelluride	$(CH_3CH_2)_2Te$	137
	Diisopropyltelluride	$\{(CH_3)_2CH\}_2Te$	—*

* sublimes at 49 °C (14 torr)

3.3.2.1 Bulk Crystal Growth. Bulk single crystals of many of the binary II–VI semiconductors have been grown from the melt, often using the vertical Bridgman method described in Section 1.3.2. This has been widely used in the preparation of **cadmium telluride**, the wafer material for the growth of the ternary semiconductor mercury cadmium telluride, $Hg_{1-x}Cd_xTe$. Cadmium telluride, however, often suffers from a high dislocation density as well as a high degree of brittleness. To combat this, a small percentage of zinc, which bonds more strongly to the tellurium than cadmium, is often added.

3.3.2.2 Thin Film Deposition. LPE is the most popular epitaxial technique at present for the production of thin film $Hg_{1-x}Cd_xTe$, used for **photodiode** arrays. The semiconductor is grown from a tellurium-rich solution using graphite sliding boats. Film growth takes place at temperatures in excess of 400 °C and care must be taken with this method to allow for possible mercury loss. This can be done by providing a mercury overpressure through the addition of HgTe to separate wells in the equipment. In addition, the grown layers can be annealed for several hours in a mercury ambient to remove mercury vacancies from the lattice.

II–VI semiconductors can also be grown by MO-VPE. The main precursors for II–VI growth are given in Table 9. One of the most studied processes here is the production of the narrow gap semiconductor mercury cadmium telluride, $Hg_{1-x}Cd_xTe$. The high volatility of mercury metal means that it can be used as an elemental precursor in mercury cadmium telluride growth and the original work used this, dimethylcadmium, and diethyltelluride as the precursors, as shown by the reaction below:

$$x\,(CH_3)_2Cd + (1-x)\,Hg + (CH_3CH_2)_2Te + (1+x)\,H_2 \rightarrow Hg_{1-x}Cd_xTe + 2x\,CH_4 + 2\,C_2H_6 \quad (18)$$

Dimethylcadmium and diethyltelluride, however, have very different thermal stabilities, with dimethylcadmium decomposing at a much lower temperature. Simultaneous decomposition of these precursors in the MO-VPE reactor leads to inhomogeneous films. Several approaches have been used to overcome this problem such as the growth of HgTe–CdTe multilayers, which are then interdiffused to obtain a homogeneous alloy. This process can produce very high quality material, as good as that made by LPE.

Another approach has been the use of new organotellurium precursors with lower decomposition temperatures, such as dimethylditelluride, diisopropyltelluride and ditertiarybutyltelluride. The use of diisopropyltelluride in particular has been shown to give $Cd_{1-x}Hg_xTe$ of LPE quality, and the flexibility of the MO-VPE process means that this could become an important production method in the future.

Many methods for the epitaxial growth of the wide band gap II–VI semiconductors have been investigated, but the most popular at present are MO-VPE and MBE. In MO-VPE, the most widely used processes are represented by the following reactions:

$$(CH_3)_2Zn + H_2Se \rightarrow ZnSe + 2\,CH_4 \qquad (19)$$

$$(CH_3)_2Cd + H_2S \rightarrow CdS + 2\,CH_4 \qquad (20)$$

Work on these materials is helped by the fact that ZnSe has a lattice constant very similar to that of GaAs, and that epitaxial growth can be achieved if a GaAs substrate is used.

3.3.3 II–VI Devices

$Hg_{1-x}Cd_xTe$ is an important material for **infrared detection** since its spectral response characteristics can be varied from 0.8 μm to 30 μm depending on the value of x. Materials with $x = 0.7$ can be used in the 3–5 μm region, while those with $x = 0.3$–0.4 are effective in the 1.3 to 1.55 μm region, which is used for fibre optic communications. They are also used in thermal imaging equipment where very high quality images can be obtained. However, the $Hg_{1-x}Cd_xTe$ detector array needs to be cooled to liquid nitrogen temperatures ($-196\,°C$) and the cost and complexity of the technology has meant that it has remained essentially a military technology.

Very little work has been performed to date on narrow band gap light-emitting devices, although the successful growth of $Hg_{1-x}Cd_xTe$ infrared diode lasers has been reported. Grown by MBE, these lasers had emissions wavelengths between 2.9 and 5.3 μm.

Reference to Figure 8 shows that while the III–V semiconductors in current use have band gaps that centre the optical devices around the near infrared, there are several II–VI materials, such as **ZnSe** and **CdS** that have band gaps greater than 2.5 eV and therefore have the potential to operate in the **visible region**. While this has been recognized for many years, it has proved too difficult to produce the p–n junctions in these materials that are the basis of LEDs and lasers. These problems have been due to a combination of the poor crystalline quality of the thin films and the presence of impurities in the films.

Current work with the MBE growth of selenide heterostructures using ultra-pure elements as the starting materials is resulting in the first **blue–green lasers and LEDs** from this materials system. In 1993, the first continuous-wave blue–green laser operating at room temperature was reported by Sony, using a ZnCdSe/ZnSe/ZnMgSSe heterostructure grown on gallium arsenide. Commercialization of this technology is still likely to be several years away.

4 PROCESSING METHODS FOR ELECTRONIC MATERIALS

This section describes the techniques which are used for the preparation of oxide electronic materials (*e.g.* for dielectrics) and one of the main applications of metals in the electronics industry, namely solders.

4.1 Sol-gel Processing of Oxides

The sol-gel process (SGP) is becoming an increasingly important route to the advanced glass and ceramic systems required in electronic applications.[2] The SGP is a means of manipulating molecular precursors to form bulk oxide materials. Controlled hydrolysis–condensation reactions of the molecular precursors (*e.g.* oxides or alkoxides) gives successively: dimers, oligomers, polymers, and a **sol** (a colloidal suspension of solid particles < 1000 nm in size) (see Chapter 13, Section 4). In turn, the sol particles join together to form a **gel**, which is a highly viscous network of metal oxide bonds containing trapped solvent molecules. The gel may be heated to remove the solvent, followed by further heating to remove residual organic groups. Finally, increasing the temperature further may cause conversion of the amorphous residue to crystalline material. Usually the hydrolysis and condensation reactions occur at room temperature, permitting metastable phase formation and the incorporation of organics to produce hybrid or composition materials. Two types of approach are used:

i) **dispersion of colloidal particles** (often oxide) in a liquid to give a sol, which, upon manipulation of pH or concentration, undergoes gelation,
ii) **preparation of an organometallic precursor** in solution (often an alkoxide) which upon the addition of water undergoes gelation.

The organometallic route is the most versatile because the range of chemical precursors and the conditions for sol-gel synthesis provide many options; also the use of high-purity alkoxides permits the production of ceramic materials with low impurity defect levels, and homogeneous multi-component ceramics can be formed when different alkoxide precursors are used.

The gelation products may take on several different physical forms depending on the gelation and drying technique. Drying by evaporation at room temperature and pressure produces a xerogel, which through the action of capillary pressure tends to shrink and crack. A major goal of SGP research is to minimize the degree of cracking, particularly for thin film applications. Drying under supercritical conditions in an autoclave removes the liquid/vapour interface, so there is no capillary pressure and minimal shrinkage. The resulting solid is called an aerogel and may have volume fractions of solid as low as 1%. Table 10 indicates the

Table 10 *Electronic materials prepared by sol-gel processing*

Type of material	*Applications*	*Example*
High-T_c superconductors	Thin film/bulk devices	$YBa_2Cu_3O_7$ (YBCO)
Dielectrics	Capacitors, sensors, phase shifters, dynamic RAMs	$BaTiO_3$
Piezoelectrics	Acoustic transducers, waveguides, nonlinear optics, micro-actuators	$Pb(Zr,Ti)O_3$, $LiNbO_3$
Pyroelectrics	Pyrodetectors	$PbTiO_3$
Ferroelectrics	Non-volatile memories	$Pb(Zr,Ti)O_3$

diverse range of electronic materials prepared by SGP (See also Table 2, Chapter 13).

The nature of the alkoxide precursor or added ligand and the hydrolysis conditions are the most important factors determining the physical properties of the thin film or bulk material produced by SGP. The precursor may be chosen on the basis of solubility in the chosen medium (usually alcohol), and its availability or its ease of purification. Otherwise the precursor may be designed to stabilize against too rapid hydrolysis so that it forms a more homogeneous sol with other, less rapidly hydrolysed components. Stabilizing agents include methoxyethoxides, diols, and carboxylates, although carboxylates may be too tightly bound to the metal to be removed in the hydrolysis step.

It is becoming increasingly important for inorganic chemists to incorporate into one molecule all of the metal components, in the desired stoichiometric ratio. In general, a higher water ratio leads to a greater degree of hydrolysis, which removes more organic groups and leads to lower 'burn-out' temperatures. The amount of water also influences the solution species: a high proportion of solvent will produce linear polymers desirable for fibre spinning, while a lower proportion of solvent will give branched polymers and a higher solution viscosity suitable for thin film deposition. Complete hydrolysis produces a high surface area and porosity in the final ceramic. The pH of the hydrolysing medium also has a major effect, basic media generally giving dense ceramics.

4.2 Joining Methods

4.2.1 Solders

Soldering is the most widely used technique for assembling semiconductor components, and for connecting components to each other to form the circuit required. Selecting the right solder material for the intended application is thus very important, and consideration of a number of intended properties of the finished bond – mechanical, electrical, resistance to extreme temperatures, and corrosion resistance – must be taken into account. The principal solder types, along with their melting temperatures, are listed in Table 11.

Table 11 *Solders for electronic applications*

Solder	*Approximate melting temperature* (°C)
Tin–lead	180–320
Tin–zinc	190–370
Tin–silver	220–240
Tin–antimony	240
Cadmium–zinc	270–390
Lead–silver	300–360
Zinc–aluminium	380
Cadmium–silver	390
Gold–silicon	370
Gold–tin	280

4.2.2 Silver-bearing Solders

Tin–lead is an extremely good solder and is still widely used. The problems sometimes experienced with this solder are overcome by the adoption of more specialized solders. For example, the problem of silver scavenging when soldering onto thin silver layers (silver plates and pastes) resulting from silver migration into the solder, causes the silver layer to decrease or, in extreme cases, even disappear, and gives rise to a reduction in the mechanical strength of the bond and in quality of the electrical properties. Silver scavenging increases with greater tin content in the solder and with higher soldering temperatures. Thus solders with both a low tin content and low soldering temperature are desirable, but it is difficult to obtain both these conditions simultaneously. The alternative is to minimize scavenging by **alloying tin–lead solders with silver**, usually in the range 1.5 to 5%.

4.2.3 Gold-based Solders

Gold solders have excellent corrosion resistance, thermal and electrical conductivity, fluidity and wettability, and they readily form eutectic bonds with silicon. This makes them an ideal choice for high-reliability electronic applications. Typical uses are for sealing packages and in die bonding of silicon chips to substrates. **Die bonding applications** demand the following performance: sufficient mechanical strength of the bond, good ohmic contact, high heat dissipation, and no chip breakage during or after bonding.

4.2.4 Gold Wire Bonding

Packaging methods can be broadly classified into two types: the **resin sealing type** which uses gold wire, and the **hermetic sealing type** which uses aluminium wire. Gold is the most suitable metal because it is extremely stable, does not oxidize, forms stable spheres when melted, absorbs little gas, is ductile, and can be readily formed into very thin wires (20–50 μm diameter). It is the third most conductive metal (after silver and copper). The gold ball generated at the thermocompression point is plastically deformable, having a hardness optimal for bonding. Hence a highly reliable bond to vapour-deposited aluminium electrodes can be achieved on the silicon chip.

At the bonding point, a gold–aluminium alloy layer is formed. However, if the gold–aluminium bond is heated for extended periods up to 300 °C, a purple-coloured intermetallic compound ($AuAl_2$) is formed, which causes severe degradation and it is therefore desirable to keep the temperature as low as possible and complete the bonding process as quickly as possible.

It is important to control trace impurity elements in the manufacture of gold wire to give it consistent mechanical properties. Very high purity gold itself is not suitable because it is too soft. However, the addition of trace levels of elements such as silicon or nickel is used to increase the softening temperature.

In terms of mechanical and physical properties, gold bonding wires should have high dimensional precision and homogeneity, smooth surface and metallic lustre, and be scratchfree, as well as having the necessary tensile strength and elongation. **Strict quality control** must therefore be observed from raw material to final product. This is of particular importance because gold wire bonding is an assembly process close to the final processing of the semiconductor device, and this operation will therefore have a significant effect on the cost of the product.

4.2.5 Conductive Adhesives

Silver-based conductive adhesives are being increasingly used in the electronics industry for a range of applications including **silicon die bonding**. This latter case is less expensive than eutectic die bonding using gold–silicon preforms. Some adhesives contain resins such as epoxy or polyimide, and the use of polyimide is increasing because of its heat resistant properties. However, silver polyimide pastes have not yet been used for hermetic ceramic packages due to out-gassing and moisture discharge during sealing at 450 °C. Other adhesives contain glass as the bonding agent, and silver–glass pastes are used widely as it is the lowest cost and best mass production method for hermetic packaging. The die can be attached at about 400 °C with a greater stress absorbing capacity than is possible with gold–silicon eutectic bonding. Also there is none of the out-gassing and moisture discharging disadvantages of resin-type silver adhesives.

5 THICK FILM AND PACKAGING MATERIALS

5.1 Thick Film Materials – Resistors

Thick film technology utilizes **screen printing** techniques to deposit patterned resistive, conductive, or insulating films to form miniature electronic components or circuits. The **printing inks** (or **pastes**) used consist of viscous dispersions of finely divided inorganic powders in organic liquids, which exhibit pseudoplastic rheological properties under the various shear rates employed. During the

Table 12 *Compositions of commercial thick film inks*

Functional phase		
Conductor	*Resistor*	*Dielectric*
Gold	$Bi_2Ru_2O_7$	Glass
Platinum	RuO_2	Glass/ceramic
Silver		Glass/alumina
Palladium		
Pt–Pd–Ag		
Copper		
Nickel		
Aluminium		
Binder		
Fritted	Borosilicate or aluminosilicate glass	
Reactively-bonded	CuO, CdO	
Mixed bonded	Oxide/glass	
Vehicle		
Volatile phase	Terpineol	
Non-volatile phase	Ethyl cellulose	

printing process, the inks are deposited through a finely woven stainless steel mesh onto which a pattern has been delineated. A squeegee applied across the surface of the screen is used to ensure an even print. The supporting substrate onto which the print is applied is usually a ceramic material such as alumina, which is chosen for its inertness and ability to withstand the multiple firing stages which follow paste deposition.

After deposition, the print is allowed to settle, is then dried at 100 to 150 °C, and is then fired at between 700 and 900 °C. Although extremely stable devices are produced by this process, the chemistry involved is complex and poorly understood. Consistent results can only be achieved by careful control of the physical and chemical properties of the starting materials, as well as accurate process control.

Thick film technology is widely used throughout the industry, covering applications ranging from simple chip resistors costing a few pence to advanced multilayer hybrid circuits costing hundreds of pounds. Table 12 shows the constituents commonly used in commercial thick film formulations.

Nearly all **thick film resistors** are based on a conductive phase of **ruthenium** in oxide form. Such materials represent an excellent compromise of high temperature oxidation resistance over base metal alternatives such as molybdenum and titanium compounds, and significant cost savings over competing conductive phases based on compounds of iridium and rhodium. The ruthenium systems also offer superior temperature coefficient of resistance (TCR) and enhanced long term stability, particularly over base metal alternatives.

The microstructure of a fired thick film resistor is such that the ruthenium oxide is concentrated at the boundaries of the glass particles, thereby forming a **conductive chain inside an insulating matrix**. The resistance depends not only on the volume fraction of RuO_2 present but also on the particle size and morphology of the conducting and insulating phases, as well as the chemical nature of the glass itself. During firing, the organic liquid burns off and the film sinters to a dense thick film. Accurate control of firing temperature, time, and atmosphere is necessary to ensure the resistance value obtained is close to that required. A final post-processing adjustment is normally required as typically resistance values of $\pm 20\%$ from specification are obtained.

In some applications, use is made of minor additives such as niobium pentoxide. This is added to the resistor formulation to counteract any oxygen deficiency which may occur during the firing cycle. Such additives can also have a beneficial effect on the TCR.

5.2 Thick Film Conductors

Resistor paste constituents tend to vary only marginally. However, in the case of conductors much greater variations are found in the composition of the functional phase, depending on the **technical requirements** and **cost sensitivity** of the application. Apart from the obvious function of providing low resistivity interconnects from point to point in a circuit, thick film conductors are used for several more specialized purposes including terminations for screen printed resistors, pads for lead frame and device attachments, and pads for wire bonding connections to silicon chips. Like resistors, fritted conductor inks consist of a functional phase of conducting particles in a glass matrix. Fluxing agents such as bismuth oxide are sometimes included to aid sintering and adhesion. Alternatively, higher firing temperatures are employed together with low levels of added metal oxides and this leads to the formation of an alumina spinel at the metal–substrate interface.

The microstructure of fired thick film conductors varies markedly from resistors. The film thickness is much less and the loading of the metallic conducting phase is much higher. Both precious metal and base metal formulations are used. Precious metals and alloys are widely used due to their excellent stability and the fact that precious metal thick film pastes can be fired under an oxidizing atmosphere, which enables efficient removal of all the organic species. Binary combinations of **palladium–silver** are widely used. The presence of palladium tends to inhibit the leaching of silver out of the conductor through the action of solder, and reduces silver migration in hot or humid conditions. The ratio of silver to palladium can vary from as much as 12:1 or just under 2:1 for higher reliability applications. However, **gold** is chosen as the functional phase for circuits demanding the **highest performance and reliability**.

Base metal conductor pastes containing copper or nickel have been developed with organic formulations that allow for a clean burn-off in the absence of oxygen. These pastes must be fired in neutral (nitrogen) or reducing (nitrogen/hydrogen) atmospheres as the properties of the fired film are seriously degraded if the furnace atmosphere becomes contaminated with oxygen. Base metal pastes are less expensive than precious metal pastes and are used for less demanding applications where cost is the major factor.

5.3 Multilayer Ceramic Capacitors

Multilayer ceramic capacitors (MLCs) pose different requirements for the conductive material used for the electrodes. MLCs pack a large electrode surface area into a **small volume package** by **stacking alternate layers of electrode and dielectric**. Barium titanate, doped with various oxides to modify its sintering characteristics, is the dielectric material most often used in MLCs. The barium titanate is comminuted to a fine powder and cast into a flexible tape with the aid of a plasticizer and acrylic binder. The electrodes are screen printed onto this tape using a paste containing fine metal particles dispersed in an organic vehicle. After drying, the layers of tape are laminated together, pressed, cut and fired in air at around 1300 °C.

Unlike other thick film conductor inks, MLC electrode materials contain only the metal and organic vehicle, the metal of choice being restricted to **palladium** or a **palladium-rich palladium–silver alloy**. These metal powders must be produced to a very high level of purity and controlled morphology, with spherical unagglomerated particles of low surface area and size between 0.5 and 2 μm generally providing the best results. This is because the powders must be screen printed to the minimum possible thickness and must shrink on firing to form a coherent electrode structure at roughly the same rate as the dielectric which surrounds them. Considerable effort has been expended in the search for high permittivity dielectrics which will sinter at lower temperatures and thus allow the use of low melting point alloys containing higher levels of silver. Some glass/ceramics are able to fulfil this role, but the bulk of commercial applications still rely on the barium titanate/palladium system.

5.4 Packaging Materials

A substantial amount of effort is being devoted to the development of materials and systems for **microelectronic packaging**. While silicon chip technology itself has advanced at a tremendous rate over the past 10–15 years, it is well to keep in mind that a high-density IC chip itself is useless unless it can be suitably packaged and interconnected to the "outside world". During the 1980s, a typical microelectronic packaging scheme comprised the following stages. The IC chip was first interconnected to a packaging device known as a **dual in-line package** (DIP) by means of microscopic gold wires or solder bumps. Such devices were then plugged into a printed circuit board, thus enabling the chips to 'talk' to each other. Finally, individual circuit boards were plugged into the master panel, which connected them to each other and to other components inside the black box. This hierarchy of interconnection is similar whether the end product is a TV set or a personal computer.

The DIP, though a widely used method for interconnecting ICs to printed circuit boards, was only able to provide a limited number of input/output leads. A newer and more versatile IC packaging device is the **ceramic chip carrier**. This occupies about the same surface area as a DIP but uses all four sides for interconnecting to IC chips, thus providing many more input/output leads. Also as the chip carrier may be surface bonded directly to the printed circuit, the heat dissipation problem from the sensitive IC chip is alleviated.

Today, however, the major thrust in the development of new and improved interconnections is based on the concept of the **multi-chip module** (MCM). It has become apparent that the performance of today's high-speed electronic systems is dominated by the **packaging** and **interconnection** systems external to the ICs. This has resulted in increased emphasis being placed on the development of interconnection and packaging schemes that minimize inter-device spacing, electrical signal paths, parasitic capacitance, and cross talk. The MCM eliminates the single chip package level, by mounting and interconnecting the chips onto a higher-density, fine pitch substrate that provides all of the chip-to-chip interconnections. Since the chips are only one tenth of the area of the packages, they can be placed closer together providing for both higher density assemblies and shorter/faster interconnects. Key technical issues include the development of conductor materials which can be screen printed and etched down to line sizes of 15–20 μm, and dielectric materials in which holes (or vias) can be made down to 50 μm.

A number of alternative materials systems have been developed for the manufacture of MCMs. Examples of these are advanced PCB technology (MCM-L), low temperature tape co-fired structures (MCM-C), and polymer dielectrics with thin film conductors (MCM-D). However, a new MCM technology recently developed can achieve circuit densities approaching those of MCM-D whilst retaining much of the convenience, reliability, and **low cost of thick film technology**. This advanced thick film system is based on a combination of etched gold conductors and a novel thick film dielectric.[3] The latter is formulated using inorganic powders produced from chemical precursors using sol-gel processing techniques (see Section 4.1). These are dispersed in a photosensitive vehicle to produce a screen printable paste. The combination of screen printing and photoprocessing of the dielectric is able to give via resolutions smaller than 50 μm. This compares with the smallest dielectric via sizes that can normally be achieved using screen printing of around 250 μm. Fired films have low permittivity ($\sim$4), combined with very low dielectric loss and high hermeticity.

As the density of the active silicon is increased, heat generation and dissipation problems are becoming more difficult to resolve. Not only are the individual chips increasing from 10 to 40 watts, but the assemblies to 25 watts or more per square inch. In some assemblies there are up to twenty-five 40-watt chips on a 4 inch square interconnect module. The design of suitable heat sink materials and structures is therefore becoming of greater importance.

6 CHEMICAL SENSORS

Over the past 15 years there has been much interest in the development of chemical and **gas sensors** fully compatible with silicon chip technology. The basic principle behind this research is the idea that if the response of a transistor can be made proportional to the concentration of gas in the environment of interest, then the modified signal can be processed readily (either directly, or via

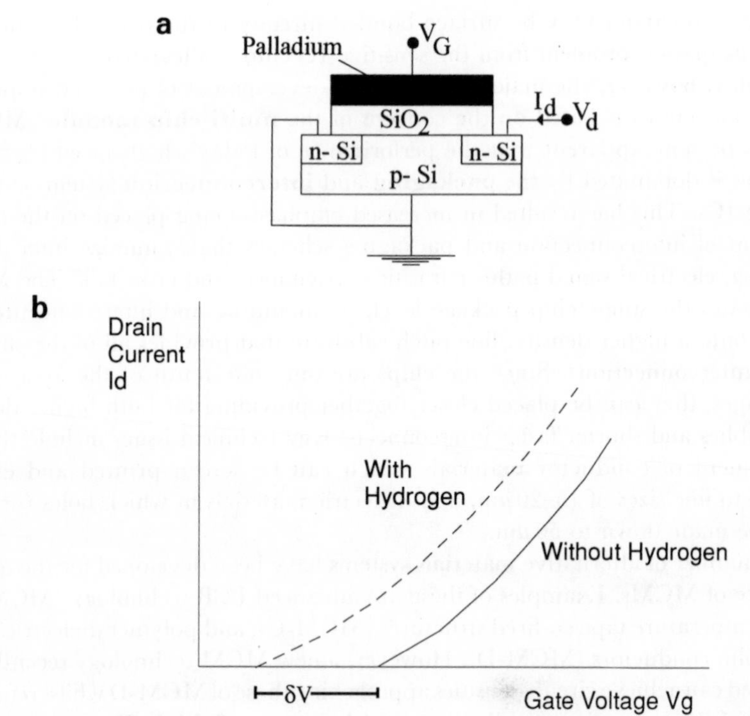

Figure 11 *(a) Schematic of a MOSFET gas sensor. (b) Response curves in normal and hydrogen-containing atmospheres*

techniques such as pattern recognition) to provide a visual display of the concentration of gas components in the mixture.

Much of the work has centred around the hydrogen sensitive **metal-oxide-semiconductor field effect transistor (MOSFET)**, originally reported by Lundström *et al.*,[4] which is based on the combination of a catalytic material (such as palladium) as the metal layer. Figure 11 shows a schematic of such a device together with a typical response curve. Hydrogen molecules from the surrounding environment are dissociated and adsorbed at the palladium gate region; the atoms diffuse through the thin metal film to the metal–insulator interface. Here the atoms form a dipole layer which produces a voltage drop, and this in turn produces proportional variations in the drain current I_d of the device. The operational characteristics of the palladium MOSFET device, in both normal and hydrogen-containing atmospheres, are shown in the accompanying response curve (Figure 11b). The change in response can be calibrated to indicate the concentration of hydrogen present in the atmosphere. It has been found that detection limits for hydrogen in air are of the order of 0.5 ppm and applications include leak detectors, fire alarm systems, and a variety of other industrial, medical, and scientific uses.

Much work has been undertaken to determine the factors which influence the sensitivity and selectivity of MOSFETs with catalytic metal gates. Three factors are of primary importance: (i) operating temperature, (ii) nature and surface of the catalytic metal, and (iii) microstructure of the catalytic metal. By controlling these variables it is now possible to fabricate devices which are selectively sensitive to gases other than hydrogen, such as ethanol, ethylene, and ammonia. In the latter case, high sensitivity and selectivity has been achieved by a number of different approaches: a sputtered palladium–iridium alloy gate; platinum-, iridium-, and lanthanum-modified palladium MOSFET devices; and an evaporated platinum gate MOSFET.

The cross-sensitivity between different gases, at one time regarded as a disadvantage, can be overcome through the use of advanced signal processing methods such as pattern recognition. In one example of this reported recently,[5] a gas sensor array consisting of three pairs of palladium gate MOSFETs and platinum gate MOSFETs is exposed to a multiple component gas mixture consisting of hydrogen, ammonia, ethylene, and ethanol. Each pair of sensors is operated at a different temperature. The signals from the six sensors are analysed with both linear and nonlinear partial least squares (PLS) models. The calculations of the PLS models are based on signals obtained from calibration experiments. By using pattern recognition, it is possible to accurately predict hydrogen concentrations in the presence of the other interfering gases.

The sensitivity of catalytic metals such as palladium and platinum has also been utilized to modify the response of **surface acoustic wave** (SAW) devices. In this case the resonant frequency of the device can be changed in accordance with the amount of gas in the atmosphere. Proposals have also been put forward for fibre optic devices in which the fibre optic tip is coated with a gas-sensitive layer. Gases are sensed through refractive index changes which occur in the sensitive layer on absorption.

Several technical problems remain to be solved, particularly those related to long-term stability. Changes can take place at the metal surface and the metal–insulator interface, especially as a result of contamination and poisoning of the catalytic film. Different responses can occur in films prepared under different conditions. Drift problems have also been pointed out in the literature.

Not all gas sensors by any means are based on MOSFET technology. However, catalytic metals such as platinum or palladium are widely used in these other technologies, such as the **tin oxide semiconductor gas sensor**. Tin oxide is used because it demonstrates a wide variability in conductance and because it responds to both oxidizing and reducing gases. Such a general response leads to severe interference effects so in almost all practical cases it is used in a modified form. This modification usually involves a metallic additive, such as palladium, which produces an enhanced response or increased selectivity.

Many other sensors are under development, for application across the whole range of measurands (temperature, pressure, position, displacement, magnetic and electric field, *etc.*). A large number involve the use of specialized inorganic materials in the active sensor element.

7 INORGANIC NONLINEAR OPTICAL MATERIALS

Nonlinear optical (NLO) materials are a class of material whose **optical properties change under the influence of light or an electric field**. Such materials are polarized under the influence of an electromagnetic field. The polarizability (P) of the material can be defined in terms of a power series of the applied field (E) and the field dependent susceptibility of the material (χ).

$$P = \varepsilon_0.[\chi^{(1)}.E + \chi^{(2)}.E^2 + \chi^{(3)}.E^3 + ...] \quad (21)$$

The $\chi^{(1)}$ term corresponds to linear optical properties such as the index of refraction, absorption, and birefringence. The $\chi^{(2)}$ term corresponds to second-order effects such as second harmonic generation, optical rectification, parametric mixing, and the Pockels effect. These effects occur in non-centrosymmetric crystals. The $\chi^{(3)}$ term corresponds to third-order effects such as third harmonic generation, and the Brillouin and the Kerr effects. These second- and third-order susceptibilities, once determined for a material, may be used to describe the important nonlinear optical properties given above.

One of the key effects of second-order nonlinearity is that of second harmonic generation. If two intense light beams of frequency f_1 and f_2 pass through an NLO crystal with a large $\chi^{(2)}$ term, then new frequencies are generated at $f_1 + f_2$. If $f_1 = f_2$, as in the case of a single high intensity light beam, then frequency doubling of the light occurs.

The main types of nonlinear optical materials include compound semiconductors, **ferroelectric oxides**, and polymers.[6] This discussion will concentrate on ferroelectric oxides.

Ferroelectric oxides are the most widely used materials for bulk nonlinear optical applications. Their prime applications are in the frequency doubling of laser light and as waveguides. The most well known material in this class is **lithium niobate** ($LiNbO_3$), which has a large second-order susceptibility, $\chi^{(2)}$. However, its relatively low laser damage threshold is a limitation for use in high power laser systems. The more recently-discovered **potassium titanyl phosphate** ($KTiOPO_4$), also known as KTP, has a much higher laser damage threshold and is increasingly being used as a replacement for lithium niobate.

Lithium niobate is produced as a single crystal material by the Czochralski method (see Section 1.3.2), normally from lithium carbonate and niobium pentoxide precursors. $LiNbO_3$ wafers are widely available. KTP crystals are normally grown from a flux of potassium phosphate. Both of these crystals have potential for the future in the frequency doubling of laser diodes to produce blue laser light for improved optical storage and laser printers.

Compound semiconductors from the III–V group also exhibit nonlinear optical effects and these have much promise for optical computing elements in the future, as they can be completely integrated with other devices such as GaAs integrated circuits, and lasers.

8 CONCLUSIONS AND DIRECTIONS FOR THE FUTURE

In this chapter we have shown how developments in electronics have been underpinned by a knowledge of and intensive research into the chemistry of key materials. The developments in silicon technology would not have been possible without development of the purification processes that enable control of its electrical properties. Equally, the current state of the art of silicon chip design would have been impossible without the work on the plasma chemistry of etchants.

In the field of III–V semiconductor materials, the major process today for the manufacture of AlGaAs **lasers** for use in compact disc players is MO-VPE, a process whose existence is dependent upon the original chemical research into organometallic preparation and purification.

The leading edge developments of the inorganic chemistry of electronic materials show very clear trends for the future. Developments in **silicon chip technology** are centred round increasing the integration level of the chip. This is dependent on the reduction of feature size to submicron levels which will in turn require new developments in the chemistry of etchants to achieve this.

Research into compound **semiconductor materials** will depend on improvements in purity of the source materials and on developments in the techniques themselves. Advanced techniques such as MO-VPE and MO-MBE will rely on improvements in the understanding of the chemical reactions taking place and on the scope for the development of new chemical precursors that can undergo a thermal, photo-induced, or other decomposition process to generate a thin film of the required composition and structure. As these technologies develop, we shall see the preparation, characterization, and exploitation of new semiconductor materials. The current work on wide band gap semiconductors such as gallium nitride and zinc selenide is the first step in this direction.

In the field of **hybrid circuit manufacture**, the understanding and control of high temperature chemical reactions which occur on firing will remain a key issue. New technologies based on novel chemistry will be used to address issues such as circuit miniaturization through the reduction in conductor track width. One approach being investigated is the use of gold organometallic inks instead of metal powders. Others are looking at using laser CVD writing of gold and platinum organometallics from the vapour phase.

In other emerging areas such as **sensors**, it is the chemical reactions occurring at the surfaces and interfaces of semiconductor devices which form the basis of their mode of action. Here, the chemistry of catalysis is likely to be widely exploited in the development of new technologies.

In all these fields, the physics of the materials may be of crucial importance, but only with a full understanding of their *chemistry* will future developments be made.

9 REFERENCES

9.1 Specific References

1. L. T. Canham, *Appl. Phys. Lett.*, 1992, **61**, 108.
2. G. R. Lee and J. A. Crayston, *Adv. Mater.*, 1993, **5**, 434.
3. G. Shorthouse, A. Berzins, T. Funnell, and J. Smyth, *Gold Bull.*, 1993, **26**, 127.
4. I. Lundström, *Sens. Actuators*, 1981, **1**, 403.
5. H. Sundgren, I. Lundström, and F. Winquist, *Sens. Actuators*, 1990, **B2**, 115.

6. R. Dorn, D. Baums, P. Kersten, and R. Regener, *Adv. Mater.*, 1992, **4**, 464.

9.2 Further Reading

F. Meijer, 'Advanced Materials in the Electronics Industry', *Adv. Mater.*, 1991, **3**, 332.

C. Weyrich, 'Materials R&D in the Electrical and Electronics Industry', *Adv. Mater.*, 1993, **5**, 416.

N. M. Davey and R. J. Seymour, 'The Platinum Metals in Electronics', *Platinum Met. Rev.*, 1985, **29**, 2.

G. B. Stringfellow, 'Organometallic Vapour-Phase Epitaxy: Theory and Practice', Academic Press, 1989.

P. J. Holmes and R. G. Loasby, 'Handbook of Thick Film Technology', Electrochemical Publications, 1976.

'Fine Chemicals for the Electronics Industry II: Chemical Applications for the 1990s', ed. D. J. Ando and M. G. Pellatt, The Royal Society of Chemistry, Cambridge, UK, 1991.

S. M. Sze, 'Physics of Semiconductor Devices', John Wiley & Sons, 1981.

9.3 Core Journals

Journal of Materials Chemistry
Chemistry of Materials
Journal of Crystal Growth
Advanced Materials
Sensors and Actuators, Parts A & B
Journal of Applied Physics
Applied Physics Letters
IEEE Transactions on Components, Hybrids, and Manufacturing Technology
Materials Science Reports
Advanced Materials for Optics and Electronics

ABBREVIATIONS

CBE	chemical beam epitaxy
CVD	chemical vapour deposition
DIP	dual in-line package
GRIN	graded refractive index
GSMBE	gas source MBE
HB	horizontal Bridgman
HEMT	high electron mobility transistor
IC	integrated circuit
LEC	liquid encapsulated Czochralski

LED	light-emitting diode
LPE	liquid phase epitaxy
MBE	molecular beam epitaxy
MCM	multi-chip module
MESFET	metal-semiconductor field effect transistor
MLC	multilayer ceramic capacitor
MO-MBE	metal–organic molecular beam epitaxy
MOSFET	metal-oxide-semiconductor field effect transistor
MO-VPE	metal–organic vapour phase epitaxy
MQWs	multiple quantum wells
SAW	surface acoustic wave
SGP	sol-gel process(ing)
TCR	temperature coefficient of resistance
VPE	vapour phase epitaxy

CHAPTER 10

Magnetic Materials

ANDREW HARRISON

1 INTRODUCTION

The first material to be used as a **magnet** was probably a form of magnetite (Fe_3O_4) called lodestone or loadstone.[1] The name of this mineral is a corruption of 'leading stone', a reference to its use in the first compass, and the word 'magnet' may be a corruption of 'Magnesia', the region where much of it was mined. The next generation of magnetic materials was produced either by stroking iron or steel with a piece of lodestone, or cooling the metals while they were aligned in the Earth's magnetic field. Throughout the nineteenth century, the work of Ørsted, Ampère and Davy, Faraday, Maxwell, and many others established the relationship between **electricity** and **magnetism** and led to the production of **magnetically polarized materials** using a solenoid. Once more, the material to be magnetized was selected from a small number of naturally-occurring materials or steels.

It is only in this century that scientists have understood the relationship between the structural, electronic, and magnetic properties of solids sufficiently well to design new magnetic materials. Technologists and society in general have matched this progress with a clamour for better magnets, tailored for specific tasks. The diversity of applications of such materials is illustrated by the wide range of magnets found in a car: all electrical motors and generators; the ignition system, speedometer, and electrical relays; as well as any loudspeakers or cassette player all rely in different ways on magnetically active solids.

In this chapter we start by establishing what a magnet is, and how its characteristic properties relate to its chemical composition, structure, and bonding; then we describe the important roles that chemistry and crystallography play in the design of magnetic solids.

2 DIAMAGNETISM AND PARAMAGNETISM IN ISOLATED ATOMS AND MOLECULES

Magnetic materials are composed of magnetic atoms or molecules. There are two fundamental types of magnetism in such elementary bodies: **diamagnetism** and

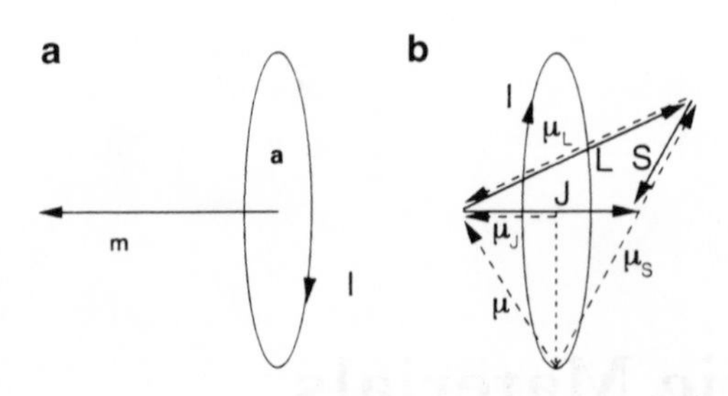

Figure 1 *(a) The magnetic moment* **m** *arising from a current* I *in a loop of wire of radius* a. *(b) The magnetic moment* **μ** *produced by electronic angular momentum* **J** *as a result of coupling orbital and spin contributions* $\boldsymbol{\mu}_L$ *and* $\boldsymbol{\mu}_S$ *respectively*

paramagnetism. These effects may be distinguished by their response to an applied magnetic field: a diamagnet is repelled, while a paramagnet is attracted. The origin of these phenomena may be appreciated by considering the relation between magnetic flux and current in a loop of an electrically conducting wire. According to **Lenz's Law**, a change in flux through the loop induces a circulation of charge which produces an opposing flux. All atoms and molecules show this effect, and its magnitude is proportional to the mean squared radius of rotation of charge in the particle: diamagnetism is most significant for heavy elements or large, conjugated molecules.[2]

Any ion or molecule containing unpaired electrons, and therefore a net **electronic angular momentum**, may show a much larger, paramagnetic response to an applied magnetic field. The origin of this effect may be illuminated with some accuracy by drawing an analogy with the magnetic moment that is produced when a current I flows round a loop of wire under the influence of an external potential [Figure 1(a)].

The magnetic effect of the current is called the **magnetic field,** $\boldsymbol{H}$, and its strength at distance r from the loop of radius a is proportional to Ia^2/r^3. According to the SI, $\boldsymbol{H}$ has units of A m^{-1}. When $\boldsymbol{H}$ is generated in a medium by such a current, the response of the medium is called the **magnetic induction** or **magnetic flux,** $\boldsymbol{B}$. When the medium is a vacuum, $\boldsymbol{B}$ has the symbol $\boldsymbol{B}_0$, and is given by

$$\boldsymbol{B}_0 = \mu_0 \boldsymbol{H} \tag{1}$$

where μ_0 is the **magnetic permeability** of a vacuum and has the value $4\pi \times 10^{-7}$ J m^{-1} A^{-2} or $4\pi \times 10^{-7}$ N A^{-2} and $\boldsymbol{B}_0$ has units of Tesla (T) or J m^{-2} A^{-1}. When $\boldsymbol{H}$ is applied to a material, a magnetic polarization called the **magnetization** may be induced, and Equation 1 is modified to

$$\boldsymbol{B} = \mu_0(\boldsymbol{H} + \boldsymbol{M}) \tag{2}$$

where $\boldsymbol{M}$ is the magnetization per unit volume: it is positive for a paramagnet, and negative for a diamagnet. The moment of the magnetic dipole generated by the current is the vector $\boldsymbol{m}$ with magnitude $I.a$. There is much confusion associated with the units for magnetic properties: we shall use the current standard of rationalized SI units throughout this article, but those readers who become confused when they read some of the references should consult the review by Quickenden and Marshall.[3]

At an atomic level, the size of the magnetic moment μ_J produced by angular momentum $\boldsymbol{J}$, which is in turn produced by the **Russell–Saunders coupling** of orbital and spin contributions, respectively $\boldsymbol{L}$ and $\boldsymbol{S}$, is given by

$$\mu_J = |\boldsymbol{\mu}_J| = g\beta\sqrt{J(J+1)} \tag{3}$$

where g is the Landé g value, defined as

$$g = \frac{3J(J+1)+S(S+1)-L(L+1)}{2J(J+1)} \tag{4}$$

and β is the Bohr magneton, defined as

$$\beta = \frac{|e|h}{4\pi mc} \cong 9.274\times10^{-24}\ \mathrm{J\,T^{-1}} = 9{,}274\times10^{-24}\ \mathrm{A\,m^2} \tag{5}$$

c is the speed of light, e and m are respectively the charge and mass of an electron and h is Planck's constant. g reflects the fact that although $\boldsymbol{L}$ and $\boldsymbol{S}$ couple to produce $\boldsymbol{J}$, as do μ_L and μ_S to produce μ; $\boldsymbol{J}$ and μ are not parallel to one another because the value of g for a spin moment is approximately 2 while that for an orbital moment is 1. g then gives the projection of μ on $\boldsymbol{J}$, and this is what we take to be μ. The relation between the different vectors is given in Figure 1(b).

The calculation of μ is most straightforward for atoms in which Russell–Saunders coupling applies and where there are no additional perturbations such as a ligand field to disrupt the distribution of energy levels. The best example of such a case is provided by complexes of lanthanides in which the valence orbitals, 4f, are to a good first approximation unperturbed by the coordinating ligands.

2.1 Paramagnetism in Lanthanide Ions

The electronic ground state of a gaseous tripositive lanthanide ion may be calculated using **Hund's three rules.**[4] We shall demonstrate how this is done for Pr^{3+} which has the outer electron configuration $4f^2$. Hund's **first** rule tells us that the ground state has the maximum spin multiplicity ($S=1$); Hund's **second** rule tells us that of the triplet spin states, the one with the largest value of L is lowest in energy ($L=5$); finally, Hund's **third** rule tells us that the lowest-energy J state for Pr^{3+}, which has a less-than-half-filled 4f sub-shell, takes the value $L-S=4$. Thus, the ground term is $^3H_4$ and g may be calculated from a simple application of Equation 4.

When Pr^{3+} is placed in magnetic flux $\boldsymbol{B}_0$ the degeneracy of the ground J state is lifted: it splits into $(2J+1)m_J$ states separated in energy by $g\beta B_0$. If we assume that these levels are the only ones that are thermally accessibly (*i.e.* kT is very much smaller than the difference between the ground J state, and the next excited J state), then the change in energy of the ion may be calculated from the Boltzmann-weighted average of the moments contributed by every m_J level. The contribution to μ_J from the m_J level is $g\beta m_J$ and its energy is $g\beta m_J B_0$, so the Boltzmann sum becomes

$$\left\langle \mu_J \right\rangle = \frac{\sum_{m_J=-J}^{+J} g\beta m_J B_0 \exp(-g\beta m_J B_0/kT)}{\sum_{m_J=-J}^{+J} \exp(-g\beta m_J B_0/kT)} \quad (6)$$

It may be shown that when $kT \ll g\beta m_J B$, Equation 6 may be approximated by

$$\left\langle \mu_J \right\rangle = \frac{g^2\beta^2 J(J+1)B_0}{3kT} \quad (7)$$

Commonly, we measure the magnetization $\boldsymbol{M}$ induced per unit volume of sample when $\boldsymbol{H}$ is applied. The ratio of the response $\boldsymbol{M}$ to the stimulus $\boldsymbol{H}$ is called the **magnetic susceptibility**, χ, and when Equation 7 is valid it may be expressed per mole of sample as the **Curie Law**:

$$\chi_m = \frac{C}{T} \quad (8)$$

where C is called the Curie constant and is equal to $\frac{g^2\beta^2 J(J+1)\mu_0 N_A}{3k}$, where N_A is Avogadro's number. At high temperatures, the different m_J levels may be populated to similar degrees, giving almost equal projections of moment with and against $\boldsymbol{B}_0$: the nett polarization is consequently very small and so is χ. At low temperatures, only the lowest m_J levels may be filled and the nett polarization is large. This effect is demonstrated in Figure 2 which shows the **Zeeman splitting** of the 9 m_J levels for Pr^{3+}, and the contribution these levels make to $\boldsymbol{M}$ at low and high temperatures.

Our derivation has involved the following implicit and explicit assumptions:

(i) Only the ground J state is thermally accessible: this allows us to calculate the Boltzmann distribution for just the m_J levels of this J state.
(ii) J is a good quantum number.
(iii) The ions behave independently of one another, *i.e.* there is no preference for a particular orientation of the moments relative to one another.

The last assumption is only valid if the magnetic ions are well isolated from one another in the sense that their wavefunctions do not mix either directly through space or indirectly through covalent bonds with intermediate atoms. In Section 3 we shall consider what the consequences are for the magnetic properties when this is not so. The first two assumptions are clearly dubious for transition metal ions

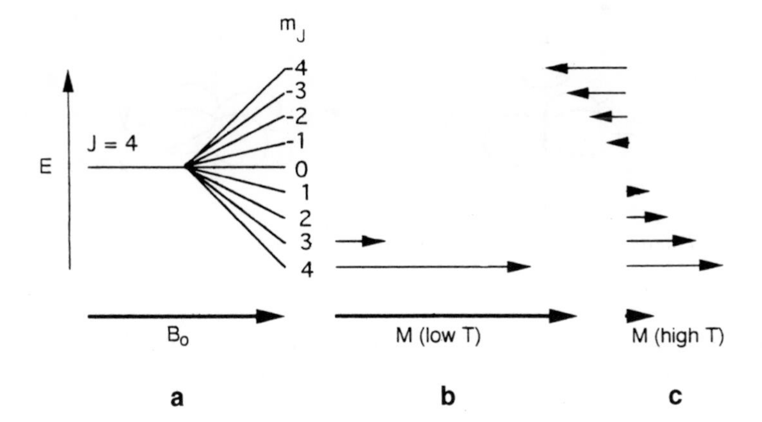

Figure 2 *(a) The Zeeman splitting of the 9 m_J levels of the J = 4 state in a magnetic flux* **B**. *(b) and (c) illustrate the contributions of the moments from the individual m_J levels at low and high temperatures, and the overall magnetization* **M**. *The lengths of the arrows would be larger for (b) than depicted here, and have been reduced to fit the available space*

with an incomplete d-subshell. In the case of first-row transition metal ions, the ligand field greatly perturbs L and prevents us from treating its value as if it applied to a free ion in the gas phase. Second and third-row transitional metal ions are further complicated by the strength of the spin–orbit coupling which invalidates Russell–Saunders coupling, and compels us to couple $\boldsymbol{l}$ and $\boldsymbol{s}$ individually to produce $\boldsymbol{j}$, rather than first forming $\boldsymbol{L}$ and $\boldsymbol{S}$. We will focus on first-row elements because they are simpler to treat and also because their compounds are more common and find a wider range of applications.

2.2 Paramagnetism in First-row Transition Metal Ions

The ligand field may act on the $m_L = 0, \pm 1, \pm 2$ levels of the orbital part of the wavefunction that describes the 3d orbitals and mixes them to produce composite wavefunctions. In the case of an octahedral or tetrahedral ligand field, we find the states that result may be described in terms of the free-ion m_L states $|0\rangle$, $|\pm 1\rangle$, and $|\pm 2\rangle$ as the following three-fold degenerate set with t_{2g} symmetry and the two-fold degenerate set with e_g symmetry:

$$d_{yz} = \frac{i}{\sqrt{2}}(|-1\rangle + |+1\rangle) \qquad d_{x^2-y^2} = \frac{1}{\sqrt{2}}(|+2\rangle + |-2\rangle)$$

$$d_{xz} = \frac{1}{\sqrt{2}}(|-1\rangle - |+1\rangle) \qquad d_{z^2} = |0\rangle$$

$$d_{xy} = \frac{1}{i\sqrt{2}}(|+2\rangle - |-2\rangle)$$

$$t_{2g} \qquad\qquad e_g$$

The precise value of L that we should take in combination with S to calculate J

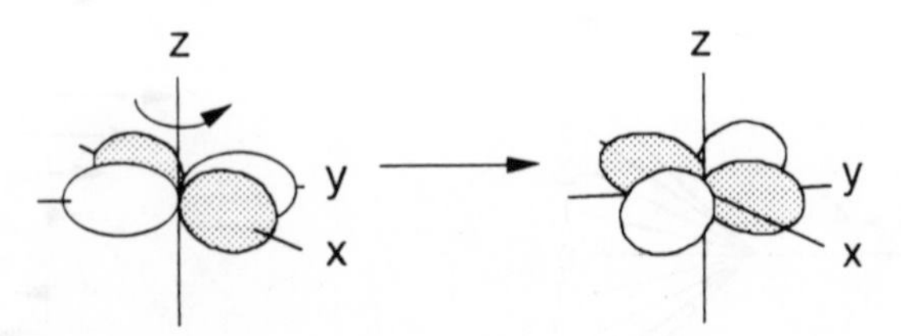

Figure 3 *Transformation of $d_{x^2-y^2}$ into d_{xy} by rotation about the z-axis*

depends on the ligand field. This, combined with the secondary influence of spin–orbit coupling, discourages rapid, back-of-an-envelope calculations of the magnetic moment. Fortunately, there is one great simplification that provides a serviceable way around this obstacle for many 3d ions.

2.2.1 Spin-only Moments. Let us consider first-row transition metal ions in an octahedral environment. We may use a pictorial representation of the orbital angular momentum that appears somewhat woolly, but is in fact representative of the quantum mechanical principles. The orbitals $d_{x^2-y^2}$ and d_{xy} are drawn in Figure 3. The transformation that takes the $d_{x^2-y^2}$ orbital into the d_{xy} orbital looks like a rotation about the quantization axis z; likewise the conversion of the d_{xz} into the d_{yz} orbital. Both transformations are associated with orbital angular momentum about this axis. The d_{z^2} orbital corresponds to the $m_l=0$ orbital and may not be taken into any other orbital by such an operation: it has no orbital angular momentum about this axis.

The orbitals will only contribute orbital angular momentum if they are occupied by electrons. Let us see how their populations change as the t_{2g} and e_g orbitals are progressively filled across the first transition series. A single electron in the t_{2g} set may occupy a d_{xz} orbital and be rotated into the d_{yz} orbital. Two electrons will, by Hund's first rule, sit most stably with parallel spins and the vacancy may be in any three of the members of the set. Thus, it is still possible to arrange the electrons so that an electron in the d_{yz} orbital may be rotated into the d_{xz} orbital with no cost in energy and an orbital contribution to the moment is to be expected. However, three electrons in the t_{2g} set have parallel spins and therefore, according to the **Pauli Exclusion Principle**, occupy each of the member orbitals singly. The transfer of an electron from the d_{xz} to the d_{yz} orbital may only occur if the spins have opposite polarizations. This costs considerable energy and reduces the significance of the contribution of the term to the moment. We say that the **orbital contribution** to the moment is **quenched**.

If we repeat this scrutiny of the occupancy of orbitals and their changing character for high-spin d^4 and d^5 configurations we find quenching in these cases too. High-spin d^4 possesses a half-filled t_{2g} set, which we have observed not to contribute, and an electron that may reside in either the $d_{x^2-y^2}$ or the d_{z^2} orbital. There is no rotational transformation that interconverts these two orbitals. Thus d^4 ions have no orbital contribution to the moment; similar considerations reveal d^5 to be similar in this respect, though of course a more straightforward consideration of the orbital angular momentum for a half-filled 3d subshell would have brought us to the same conclusion.

The rule of thumb is that terms with A and E symmetry have no orbital contribution to the moment while states of T symmetry may have an orbital contribution whose magnitude must be calculated for that particular case. The reader who is familiar with the group-theoretical basis of selection rules may wish to note that since the operator $\boldsymbol{L}$ ($= \boldsymbol{L_x} + \boldsymbol{L_y} + \boldsymbol{L_z}$) for orbital angular momentum transforms as T_{1g} in the point group O_h, then the matrix element $\langle \Psi | \boldsymbol{L} | \Psi \rangle$ is only non-zero when Ψ has the symmetry T_{1g} or T_{2g}.

When orbital quenching occurs, J may be approximated by S and g takes the value for pure electron spin. The formula for the moment now adopts the so-called **spin-only** form:

$$\mu_{so} = g\beta\sqrt{S(S+1)} \approx \beta\sqrt{n(n+1)} \qquad (9)$$

where n is the number of unpaired electrons on the ion. Conversely, terms with T symmetry may have an orbital contribution and require explicit evaluation of L and the manner of its coupling with S to produce a moment. This is beyond the scope of this chapter, and further details may be found in the books by Figgis, by Drago, and by Mabbs and Machin, to which we refer at the end of this chapter. You may wish to run through the remainder of the transition series and show that spin-only moments may also be expected for high-spin d^8, d^9, and of course d^{10}, as well as for low-spin d^6.

2.2.2 Deviations from Spin-only Moments. In Table 1 we provide calculated and experimental values of the moment for ions with partially-filled 3d subshells. You will notice that there are in fact significant deviations from the predicted spin-only moments even for ions with E and A ground terms. This is because we have assumed that the ground state is unadulterated with contributions from other electronic states. Consider the case of the d^7 ion Co^{2+} in a tetrahedral environment. We expect the ground state to have a high-spin configuration because the ligand-field splitting Δ will be relatively small. The ground term is $^4A_2$ and we might therefore expect a spin-only moment appropriate for 3 unpaired electron spins, that is $\sqrt{3}\beta$. The promotion of an electron from a t_{2g} to an e_g orbital produces two orbital and spin triplet states, $^3T_1$ and $^3T_2$

In the strong-field approximation these lie at an energy Δ above the ground state and may be mixed into it through spin-orbit coupling, λ. Thus, there may be an orbital contribution to the ground state which is proportional to λ/Δ and we modify the spin-only formula (Equation 9) as follows:

$$\mu = \mu_{so}\ (1 - \alpha\lambda/\Delta) \qquad (10)$$

where α depends on the composition of the ground and excited states in terms of the free-ion m_L levels: it is 4 for an A_2 term and 2 for an E term. λ is negative when the d-subshell is greater-than-half-filled, and positive otherwise. The magnetic moments in a series of $[CoX_4]^{2-}$ complexes in which changes in the halide X

Table 1 *Electronic configurations, calculated and experimental moments for 3d transition metal ions in an octahedral ligand field and, where appropriate, in high- and low-spin states. μ_J is calculated using Equations 3 and 4, and the free-ion values of L and S and μ_{so} with the aid of Equation 9*

Number of 3d electrons	*Ion*	*Free-ion ground term*	*Ground configuration in O_h symmetry (high, low spin)*	*Ligand field ground term (high, low spin)*	$\mu_J = g_J\{J(J+1)\}^{1/2}\beta$	$\mu_{so} = \{2(S(S+1)\}^{1/2}\beta$	μ_{exp} (300 K)
1	Ti^{3+}, V^{4+}	$^2D_{3/2}$	t_{2g}^1	$^2T_{2g}$	1.55	1.73	1.7–1.8
2	V^{3+}	$^3F_2$	t_{2g}^2	$^3T_{1g}$	1.63	2.83	2.6–2.8
3	V^{2+}, Cr^{3+}	$^4F_{3/2}$	t_{2g}^3	$^4A_{2g}$	0.77	3.87	~3.8
4	Cr^{2+}, Mn^{3+}	$^5D_0$	$t_{2g}^3e_g^1$, t_{2g}^4	^5E_g, $^3T_{1g}$	0.00	4.90	~4.9
5	Mn^{2+}, Fe^{3+}	$^6S_{5/2}$	$t_{2g}^3e_g^2$, t_{2g}^5	$^6A_{1g}$, $^5T_{2g}$	5.92	5.92	~5.9
6	Fe^{2+}	$^5D_4$	$t_{2g}^4e_g^2$, t_{2g}^6	$^5T_{2g}$, $^1A_{1g}$	6.70	4.90	5.1–5.5
7	Co^{2+}	$^4F_{9/2}$	$t_{2g}^5e_g^2$, t_{2g}^6 e_g^1	$^4T_{1g}$, ^2E_g	6.63	3.87	4.1–4.2
8	Ni^{2+}	$^3F_4$	$t_{2g}^6e_g^2$	$^3A_{2g}$	5.59	2.83	2.8–4.0
9	Cu^{2+}	$^2D_{5/2}$	$t_{2g}^3e_g^3$	^2E_g	3.55	1.73	1.7–2.2

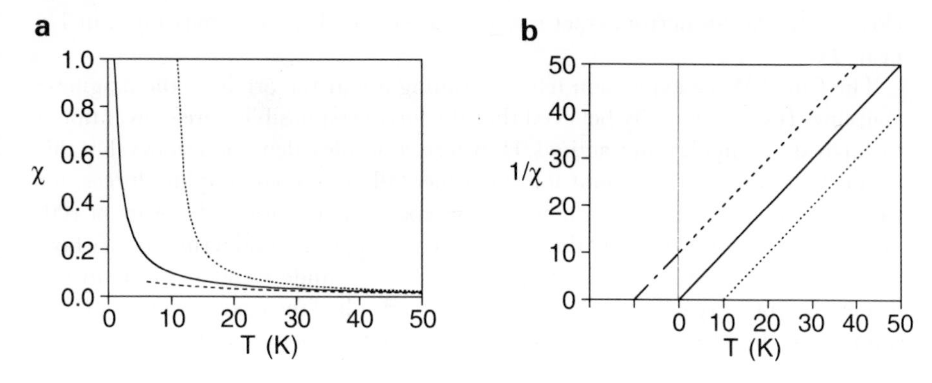

Figure 4 *Graphs of (a) χ against* T *and (b) $1/\chi$ against* T *for a paramagnet (solid line), a ferromagnet (dotted line) and an antiferromagnet (dashed line), all of which have the same magnitude of atomic moment. In 4(b) the values of $1/\chi$ below* T_N *have been extrapolated from those at higher temperatures, and these are depicted by a dot–dash line*

produces changes in Δ provides one of many cases where Equation 10 successfully accounts for the observed values of μ.

3 CO-OPERATIVE MAGNETISM

Magnetic materials in which the moments appear to behave in isolation to one another are called paramagnets: when the constituent ions also conform to assumptions (i) and (ii) in Section 2.1, a characteristic linear dependence of χ on $1/T$ according to the Curie Law (Equation 8) is observed, and this is illustrated in Figures 4(a) and 4(b). Most technologically important magnetic materials are not paramagnets, but possess forces between the moments that favour a particular **orientation of the moments**. A material in which a parallel alignment of moments is favoured is called a **ferromagnet**: such a material shows a greater magnetic polarization for a given B than would be the case with no co-operative effect, and χ is higher [Figure 4(a)].

Alternatively, when the forces favour antiparallel alignment, M and hence χ are suppressed for a given B and we call the material an **antiferromagnet**. The change in the graph of $1/\chi$ against T is shown in Figure 4(b) for these two cases. If such materials are cooled sufficiently, the moments may freeze to an ordered array. This occurs at the **Curie** and **Néel temperatures**, T_C and T_N, for ferro- and antiferromagnets respectively. Above these critical temperatures, the Curie Law must be replaced by the **Curie–Weiss expression**:

$$\chi = \frac{C}{T - \theta} \tag{11}$$

where the sign of the Curie–Weiss constant θ depends on whether the material is a ferromagnet or an antiferromagnet, and its magnitude provides a measure of the strength of the magnetic coupling. It is positive for a ferromagnet and takes a value close to T_C. An antiferromagnet has a negative θ and shows a maximum in $1/\chi$ near T_N.

The Curie–Weiss expression tells us nothing about the origin of the magnetic coupling. It was originally believed that the forces responsible were conventional through-space dipolar interactions. However, a simple calculation shows that this would be too weak to account for experimentally observed coupling forces: for strong ferro- and antiferromagnets the discrepancy is a factor of the order of 100. We now consider the origin of the magnetic coupling in general terms and see how it applies to insulators and metals. If you do not understand basic quantum mechanics you may wish to skip to the last paragraph of the next section where we summarize the effect of the electrostatic interaction between electrons on the relative polarization of their spin.

3.1 Magnetic Exchange[5,6]

The coupling of moments on two atoms arises from the electrostatic repulsion between the electrons concerned and the Pauli Exclusion Principle. If the treatment of the coupling is to be in terms of the angular momenta on the atoms, it must be a relatively small perturbation so it is appropriate to discuss the problem using a valence-bond rather than a molecular orbital model. We start with the simplest case – the coupling between electrons 1 and 2 in the 1s orbitals of hydrogen atoms A and B with wavefunctions $\psi_A(1)$ and $\psi_B(2)$ respectively for isolated atoms. One possible arrangement of the electrons in the collection of particles is described by $\psi_A(1)\psi_B(2)$. The arrangement described by $\psi_B(1)\psi_A(2)$ is equally likely so the true wavefunction must contain equal portions of both and we write it as:

$$\psi = c_1\psi_A(1)\psi_B(2) + c_2\psi_A(2)\psi_B(1) \qquad (12)$$

The symmetry of the problem tells us that $c_1^2 = c_2^2$. The form of the Pauli Exclusion Principle that was used in Section 2.2.1 is actually a special case which states that no two electrons in an atom may have an identical set of quantum numbers. This is a consequence of a more general form of the Principle which states that a collection of electrons, as **fermions**, has a **wavefunction** that is *antisymmetric* with respect to interchange of any pair. Therefore, if we swap electrons 1 and 2 in the expression for ψ, its sign must change. This implies that $c_1 = -c_2$. However, we have omitted one important factor – electron spin. The total wavefunction is a product of the spatial part above, and a spin part whose symmetry depends on the relative spins of 1 and 2. There are two possibilities: if we denote 'up' and 'down' spins by α and β respectively, then a triplet state is obtained with combinations such as $\alpha(1)\alpha(2)$ and $\beta(1)\beta(2)$ and a singlet state may be represented by $\alpha(1)\beta(2)$ or $\beta(1)\alpha(2)$. The **triplet spin** wavefunction *is symmetric* with respect to interchange of 1 and 2 so it must be combined with an *antisymmetric spatial* combination of wavefunctions for the Pauli Exclusion Principle to be satisfied; alternatively, the **singlet spin** wavefunction is *antisymmetric* with respect to interchange of electrons and requires a *symmetric* combination of *spatial* wavefunctions. Thus, the wavefunction ψ_+ which is a spin singlet has a spatial part with $c_1 = c_2$, and the wavefunction ψ_- which is a spin triplet has a spatial part with $c_1 = -c_2$. The energies of ψ_+ and ψ_- may be obtained from the Schrödinger Equation with the Hamiltonian H expressed as

$$H = H_a + H_b + H_{ab} \tag{13}$$

where H_a and H_b pertain to the interactions between each electron and an isolated hydrogen atom and H_{ab} accounts for the additional interactions introduced by bringing the two atoms together, *i.e.*

$$H_{ab} = -\frac{1}{r_{a2}} - \frac{1}{r_{b1}} + \frac{1}{r_{12}} + \frac{1}{R_{ab}} \tag{14}$$

The first two terms represent the energy of attraction between each electron and the second nucleus, and the last two terms the repulsion between the electrons and the nuclei respectively. Two solutions are found are found for the Schrödinger Equation in this case[7,8] – the spin singlet and the spin triplet states whose respective energies E_+ and E_- are given by

$$E_\pm = 2E_h + \frac{Q \pm \mathcal{J}}{1 \pm S^2} \tag{15}$$

E_h is the energy of an isolated atom–nucleus pair, S is the overlap integral and Q is the **Coulomb integral** which may be expressed as

$$Q = \frac{e^2}{R} + \iint \frac{e^2}{r_{12}} \psi_a^2(1)\psi_b^2(2)\,d\tau_1 d\tau_2 - \int \frac{e^2}{r_{a2}} \psi_b^2(2)\,d\tau_2 - \int \frac{e^2}{r_{b1}} \psi_a^2(1)\,d\tau_1 \tag{16}$$

The terms in this expression describe the additional classical electrostatic attractions and repulsions that are introduced when the atoms are allowed to interact: the terms taken from left to right describe the nucleus–nucleus repulsion, the repulsion between electron charge clouds and the attraction between each electron and the further nucleus. Binding will only occur if the last two terms are larger than the first two.

The term $\mathcal{J}$ in Equation 15 has no equivalent in classical electrostatics and is called the **exchange energy**. It accounts for the additional energy that arises when we introduce the requirement that the wavefunction is antisymmetric with respect to exchange of electrons. It may be written as

$$\begin{aligned}\mathcal{J} = \frac{S^2 e^2}{R} &+ \iint \frac{e^2}{r_{12}} \psi_a(1)\psi_b(2)\psi_b(1)\psi_a(2)\,d\tau_1 d\tau_2 \\ &- S\int \frac{e^2}{r_{a2}} \psi_a(2)\psi_b(2)\,d\tau_2 - S\int \frac{e^2}{r_{b1}} \psi_a(1)\psi_b(1)\,d\tau_1\end{aligned} \tag{17}$$

For small S the second of the four terms is the most important one. It has a positive energy and when substituted in Equation 15 stabilizes the triplet state relative to the singlet. It is this term that provides the energy of stabilization for parallel compared with antiparallel spins in orthogonal orbitals. The gap in energy, ΔE, between the singlet and triplet states may be obtained by substituting for Q and $\mathcal{J}$ in Equation 15 and subtracting E_+ from E_-. If we assume that S is small and drop terms in S, S^2, and S^4 we find

$$\Delta E \cong 2\mathcal{J} \cong 2\iint \frac{e^2}{r_{12}} \psi_a(1)\psi_b(2)\psi_b(1)\psi_a(2)\mathrm{d}\tau_1\mathrm{d}\tau_2 \tag{18}$$

If we introduce an additional pair of singlet states produced by placing both electrons on one nucleus, a process that costs an energy U, we find that Equation 18 is modified to

$$\Delta E \cong 2\iint \frac{e^2}{r_{12}} \psi_a(1)\psi_b(2)\psi_b(1)\psi_a(2)\mathrm{d}\tau_1\mathrm{d}\tau_2 + \frac{4t^2}{U} \tag{19}$$

where the **transfer integral** t is defined as

$$t = \int \psi_a(1)\psi_b(2)H\psi_a(1)\psi_a(2)\mathrm{d}\tau_2\mathrm{d}\tau_1 \tag{20}$$

This describes the perturbation introduced by transferring a singlet pair of electrons on separate atoms to a singlet pair on one atom.

To summarize this section, the balance between a singlet (antiferromagnetic) and a triplet (ferromagnetic) state depends on the balance between a term which corresponds to classical electrostatic repulsion (the Coulomb integral) and a term that has no equivalent in classical physics and is called the exchange integral. This latter effect is sometimes described as the additional stabilization provided when electrons with parallel spins are exchanged – though this does not actually explain the origin of the stabilization. The nett result of the competing influences is often expressed in general terms for exchange between general values of spin S_i and S_j as the **Heisenberg** or **Heisenberg–Dirac–van Vleck exchange Hamiltonian**:

$$H_{ij} = -2\mathcal{J}_{ij}S_i.S_j \tag{21}$$

where the **exchange constant** $\mathcal{J}_{ij}$ is defined as $-\frac{1}{2}(E_+ - E_-)$ and is positive for a ferromagnetic interaction.

3.2 Magnetic Exchange in Insulators[5,9]

The same principles may be carried through to materials in which the wavefunctions may not interact directly, but require an intermediate atom to support the coupling. Thus, in an insulating material such as NiO, the magnetic ions are too far apart for any significant **direct overlap** of their valence orbitals, and the wavefunctions mix through the covalent bond formed with the oxygen atoms. The term given to the various ways in which this form of indirect exchange operates is **superexchange**. The various contributions may be appreciated by considering the interactions between electron spins on two sigma-bonding orbitals on Ni^{2+} which overlap with a sigma-symmetry orbital on O^{2-}. We note that the principal interaction involves the half-filled e_g orbitals on Ni^{2+}.

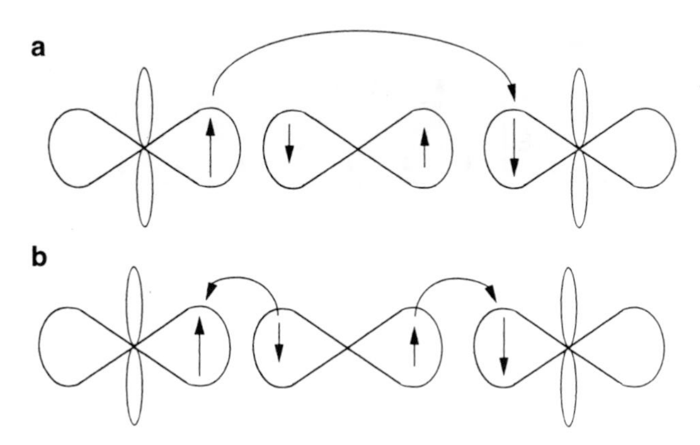

Figure 5 *Schematic representation of the different contributions to superexchange for e_g orbitals on Ni^{2+} interacting through sigma bonds to a bridging O^{2-} ion: (a) and (b) depict the kinetic and correlation terms respectively*

(i) Direct exchange. This corresponds to the exchange term in Equation 19 and involves the interaction between orthogonal orbitals: it always stabilizes a ferromagnetic interaction. Its strength increases with the degree to which the metal ion and ligand orbitals mix, and hence increases with the covalence of the M–X bonds for the more general case of direct exchange between M ions bound through X.

(ii) Kinetic exchange. This corresponds to the term t^2/U in Equation 19. It may be imagined as arising from the transfer of an electron from one e_g orbital on Ni^{2+} to the same orbital on a neighbouring Ni^{2+} ion [Figure 5(a)]: this interaction always favours antiferromagnetic exchange. Furthermore it is usually a stronger effect than the direct term.

(iii) Spin polarization terms. This may be envisaged as the transfer of one electron from a ligand orbital to an empty orbital on the metal ion. The lowest-energy transition of this sort corresponds to transfer of electrons from 2p orbitals on O^{2-} to 4s orbitals on Ni^{2+}. Hund's first rule indicates that this will produce ferromagnetic exchange.

(iv) Correlation effects. This arises when two electrons are simultaneously transferred from a filled orbital on the ligand to half-filled orbitals on the metal ions [Figure 5(b)]. It may be regarded as the simultaneous formation of bonds between the ligand and the metal ions. It produces antiferromagnetic exchange.

The superexchange processes we have described must be modified if one or both of the magnetic ions does not have a suitable partially-filled orbital into which electrons may be transferred. For example, if the e_g orbitals of a d^3 and a d^5 ion are mixed through sigma bonds to a bridging ligand, there are no electrons in the e_g orbital of the d^3 ion, and the favoured orientation of any transferred electron

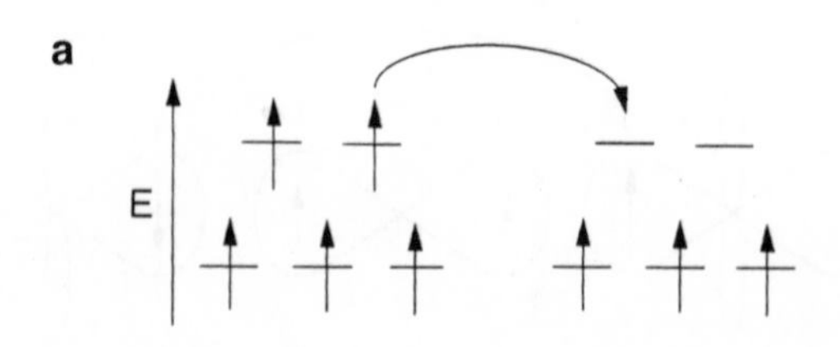

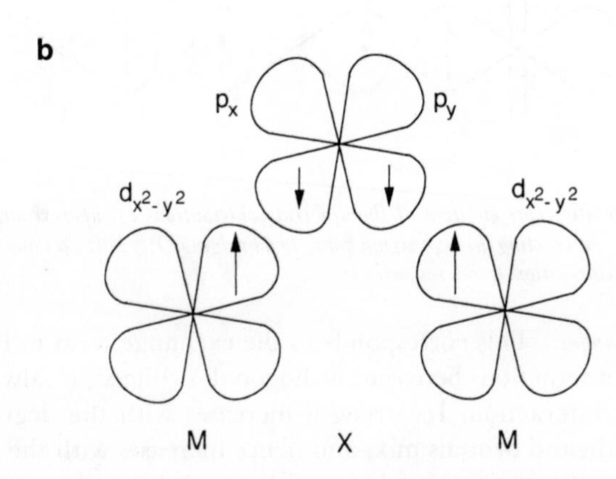

Figure 6 *Two different cases that lead to ferromagnetic exchange. (a) Kinetic exchange between d^3 and high-spin d^5 configurations. (d) direct exchange involving a 90° M–X–M bridge and sigma-bonding between the metal ion d-orbitals and the orthogonal p-orbitals on X*

would be parallel to the three electrons on the t_g orbitals according to Hund's first rule. This will favour an overall ferromagnetic interaction, though it will appear at a higher level of perturbation than the superexchange effects described above and will therefore be much weaker [Figure 6(a)].

It would appear that we need to perform a number of calculations just to predict the sign, let alone the magnitude of the magnetic exchange between ions in insulators. Such calculations require relatively sophisticated calculations and do not enjoy great success when compared with experimental data. Consequently, we resort to a series of simplified rules that allow us to discriminate between a few general cases.

We first consider the electron configuration of the participating magnetic ions, and the way in which their wavefunctions mix through the formation of sigma- and pi-bonds with the bridging ligands. In a few special cases ferromagnetic exchange may occur: these boil down to the cases where some form of orthogonality in the orbitals prevents transfer of electrons between the metal ion orbitals. The d^3–d^5 case we discussed above is one example of a general condition where the **orthogonality** is **within the magnetic ions**; alternatively, the **orthogonal link** may be **within the atom X** involved **in the M–X–M bridges**. A simple example of this is given in Figure 6(b) in which the covalent bonds between M and X make an angle of 90° and the principal contribution to the

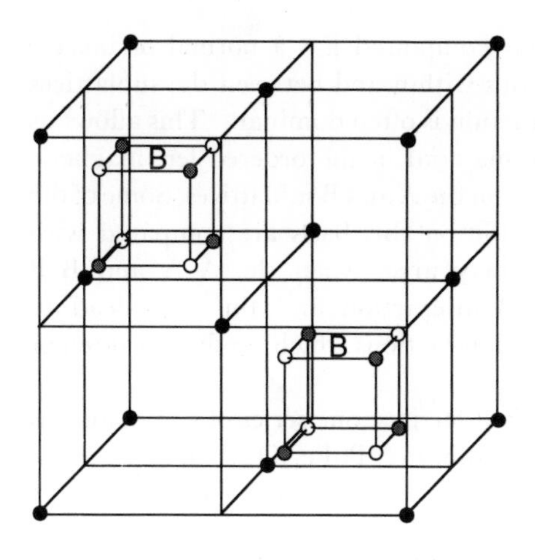

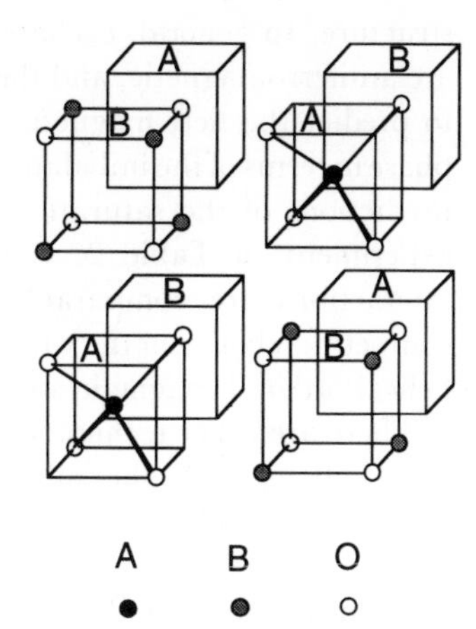

Figure 7 *Unit cell of the spinel AB_2O_4. On the right we show the relative dispositions of the tetrahedrally-coordinated A cations and the octahedrally-coordinated B cations; the eight cubes of this part of the Figure must be placed within the framework of the left-hand part to produce the complete unit cell*

sigma bonding from X is from orthogonal p-orbitals. As with the case of d^3–d^5 exchange, the interaction is weaker than would have been the case if such orthogonality had not existed. Unless there is no obvious way in which the relevant orbitals on the metal ions may mix, exchange between metal ions coupled through diamagnetic atoms is usually antiferromagnetic: its strength increases with the degree of covalence in the bonds, and is generally stronger for sigma compared with pi-bonds.

Although antiferromagnetic exchange is far more common than ferromagnetic exchange, we may synthesize a wide range of technologically significant magnetic materials by linking moments of different magnitude antiferromagnetically: such materials are called **ferrimagnets**. An example of such a material is provided by the compound with which we started this chapter – magnetite. This is one example of a class of compounds of general formula AB_2O_4 called **spinels**. The unit cell of the so-called 'normal' spinel is reproduced in Figure 7. It is based on a cubic close-packed array of oxygen atoms with the divalent ions A distributed over the tetrahedral sites and the trivalent ions B distributed over the octahedral sites. There is a so-called 'inverse' form too which may be written as $B(AB)O_4$ to indicate the distribution of ions over the A and B sites of the normal structure. Materials in which B is Al or Cr and A is Mg, Mn, Fe, Co, Ni, Cu, or Zn mainly have normal structures, while those in which B is Fe range from normal (A is Zn) to inverse (A is Fe, Co, Ni, and Cu) through materials with an intermediate occupancy of the tetrahedral sites by A and B ions (A is Mg, Mn).

Materials in which B is Fe are called spinel **ferrites**, and they find a wide variety of applications which are described in Section 5. Materials with different nett magnetizations may be produced by choosing ions with different moments and interactions with their crystal environment and placing them on the A and B sites, according to whether the particular compound has a normal or inverse structure. In general, exchange interactions within and between the sublattices are antiferromagnetic, and the A–B interaction is often dominant. This allows us to predict the nett magnetization of the material in the ordered ferrimagnetic phase in terms of the imbalance of moments on the A and B sublattices. Some of the predictions of the saturation moment made on this basis are compared with experiment in Table 2. Complications may arise when the A–A and B–B interactions are comparable to the A–B interaction and this may lead to competition between the various exchange interactions which produces canted or helical ordered magnetic structures.

There are other technologically significant ferrites: one series of compounds is isomorphous with the mineral magnetoplumbite, $PbFe_{12}O_{19}$, and the most common example is the barium salt which has the trade name Ferroxdure. A second set of materials is related to the minerals called **garnets**, which are silicates of general formula $A_3B_2(SiO_4)_3$, where A is Ca, Mg, or Fe and B is Al, Cr, or Fe. The silicon atoms may be replaced by other elements to produce a broader range of compounds whose formula may be written as $A_3B_2(CO_4)_3$ or $A_3B_2C_3O_{12}$, where A, B, and C have dodecahedral, octahedral, and tetrahedral coordination respectively. If we use the case of **yttrium iron garnet (YIG)**, $Y_3Fe_5O_{12}$, as an example of this type of material, and rewrite the formula as $Y_3Fe_2Fe_3O_{12}$ to indicate the occupancies of the different sublattices, we find that the various contributions to the bulk magnetization may be written as $(Fe^{3+}\uparrow)_2(Fe^{3+}\downarrow)_3$ (Y is diamagnetic). The net moment is therefore produced by the spin $(\frac{5}{2})$ on one Fe^{3+} ion. This may be controlled by substitution of the various ions. Thus, Y may be replaced by magnetic ions such as Gd or Ho, which reduces the bulk magnetization for antiferromagnetic coupling between A and C sublattices. Replacement of the B-site or C-site cations by non-magnetic ions will respectively lower or raise the magnetization. We return to these materials in Section 5.3.

3.3 Magnetic Exchange in Metals

The treatment of magnetism in metals differs from that in insulators in several ways. First, the electrons responsible for the magnetic behaviour may be regarded as localized – as would be the case for tightly-bound 4f electrons in lanthanide metals – or delocalized as is the case with s electrons in alkali metals. The way in which electrons of intermediate character should be treated – such as 3d electrons in transition metals – is less clear. What is certain is that there is a **connection between electronic conduction and magnetic interactions**: where there are significant exchange forces anomalies may be observed in the resistivity at the magnetic ordering temperature.

Magnetic exchange may be due to a direct term, as we have described for insulators. There is an additional interaction between localized moments which arises through the polarization of the conduction electrons: this is called the **Ruderman–Kittel–Kasuya–Yosida (RKKY) interaction** after those who contributed to its discovery. It has a long range and oscillates in sign with the separation of the atoms. Consequently, a particular moment may be subject to

Table 2 *Distribution of ions over A and B sites, and the saturated moments in a selection of ferrimagnetic spinel ferrites*

Compound	*Ions on sublattice A*	*Ions on sublattice B*	*Unpaired electrons on sublattice A*	*Unpaired electrons on sublattice B*	*Calculated moment (unpaired spins)*	*Experimental moment (unpaired spins)*	T_C (K)
$MgFe_2O_4$	Fe^{3+}	Mg^{2+}, Fe^{3+}	5	5	0	1.1	713
$MnFe_2O_4$	$Mn^{2+}_{0.8}Fe^{3+}_{0.2}$	$Mn^{2+}_{0.2}Fe^{3+}_{1.8}$	5	10	5	5.0	573
$FeFe_2O_4$	Fe^{3+}	Fe^{2+}, Fe^{3+}	5	9	4	4.2	858
$CoFe_2O_4$	Fe^{3+}	Co^{2+}, Fe^{3+}	5	8	3	3.3	793
$NiFe_2O_4$	Fe^{3+}	Ni^{2+}, Fe^{3+}	5	7	2	2.3	858
$CuFe_2O_4$	Fe^{3+}	Cu^{2+}, Fe^{3+}	5	6	1	1.3	728

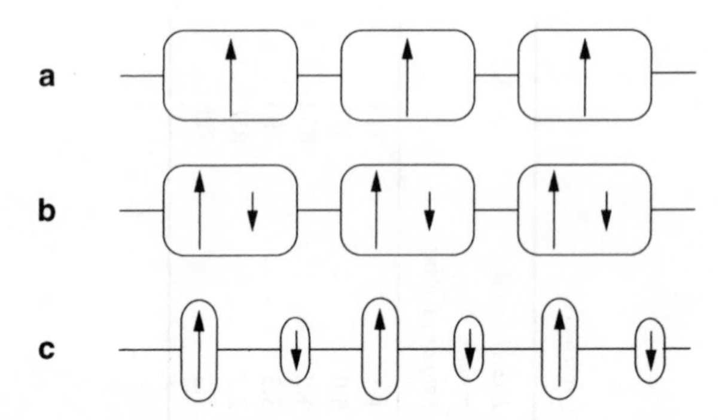

Figure 8 *Different arrangements of magnetic electron spin density that may provide a spontaneous moment*

exchange forces of different strengths and sign by a variety of moments at different separations. Conflict between the different influences may produce complex, non-collinear arrangements of moments below their ordering temperature as is the case with many lanthanide metals; where the spatial distribution of atoms has some degree of randomness, as is the case in an amorphous metal or a random substitutional alloy with a crystal lattice, the conflicting random exchange forces may produce a frozen random magnetic array at low temperatures which is called a **spin glass**.

3.4 Molecular Magnets[10,11]

So far we have discussed magnets in which the moments are located on atoms or ions or in bands. It is also possible to produce paramagnetic molecular species which, in principle, could be coupled to produce molecular solids with nett magnetic polarization. Such materials are likely to have a very low density and be very desirable for devices where low weight is an important factor. There are several strategies we might employ to achieve this ambition and these are depicted schematically in Figure 8. They are as follows:

(i) ferromagnetic coupling of isolated moments on atoms or molecules [Figure 8(a)]
(ii) antiferromagnetic coupling of molecules with regions of high and low spin density [Figure 8(b)]
(iii) ferrimagnetic coupling of two magnetic sublattices [Figure 8(c)]

It is difficult to arrange the formation of orthogonal exchange linkages in solids. A more common approach is to prepare crystals containing pairs of paramagnetic molecules that act as electron donors and acceptors. A donor–acceptor pair is denoted $D^{\bullet}A^{\bullet}$ before charge transfer, and $D^{\bullet+}A^{\bullet-}$ represents a possible charge-transfer state. Suppose that the acceptor molecule has orbital degeneracy, then Hund's first rule favours a spin triplet state for $D^{\bullet+}A^{\bullet-}$, which in turn stabilizes the triplet relative to the singlet ground state for the relative polarizations of the spins on $D^{\bullet}A^{\bullet}$. The study of **molecular charge-transfer salts** formed from

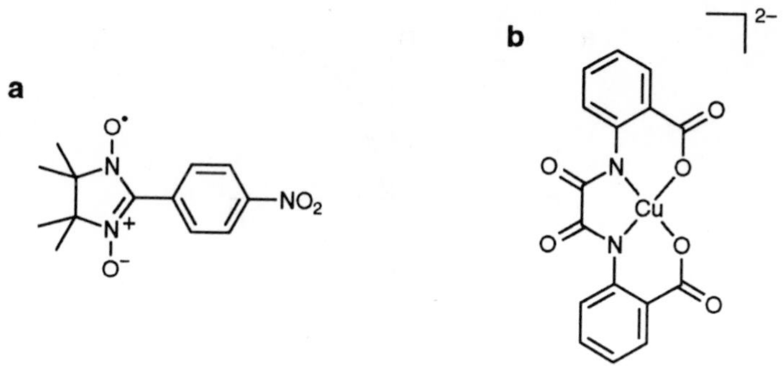

Figure 9 *(a) Nitronyl nitroxide radical whose crystals provided the first organic ferromagnet. (b) The ligand obbz [oxaminobis(*N,N'*-benzoato)] found in the ferrimagnet* $MnCu(obbz)(H_2O)_3$

paramagnetic molecules led to the preparation of $[Fe(C_5Me_5)_2][TCNE]$ (TCNE = tetracyanoethene) which shows a spontaneous magnetization below $T_C = 4.8$ K. The structure of this material consists of chains of alternating molecules of $Fe(C_5Me_5)_2$ and TCNE which have both intra- and interchain ferromagnetic coupling between moments arising from $S = \frac{1}{2}$, and bulk magnetization below $T_C = 4.8$ K. There are several other similar materials with different combinations of organometallic charge donor molecules, and molecular charge–acceptor species. More recently a spontaneous moment has been found in a compound containing no metal ions: crystals of the paramagnetic nitronyl nitroxide radical depicted in Figure 9(a) have a Curie temperatures of 0.65 K.

We might expect to have a greater variety of compounds to chose from if we include ferrimagnetic materials. Molecular entities may have different regions with different spin polarizations, as depicted in Figure 8(b). Antiferromagnetic exchange between these building blocks when they are aligned in the manner shown, leads to an overall magnetic polarization. Alternatively, we may try to produce ferrimagnets by linking two different metal complexes with different moments in the form of a polymeric solid (Figure 8(c)). One example of this type of material is the ferrimagnetic chain-like compound $MnCu(obbz)(H_2O)_3$ [obbz = oxaminobis(*N,N'*-benzoato), Figure 9(b)], in which the interchain coupling is ferromagnetic and $T_C = 14$ K.

4 HARD AND SOFT MAGNETS

4.1 Domains and Magnetic Hysteresis

Pure metallic iron has a Néel temperature of 1043 K, yet a piece picked at random from a chemical store will probably have a negligible magnetization at 300 K. The magnetization may be increased either by stroking it repeatedly in the same sense with a strong magnet, or by placing it in a unidirectional magnetic field $\boldsymbol{H}$. A typical graph of the way in which $\boldsymbol{B}$ changes with $\boldsymbol{H}$ as it is cycled from zero to a

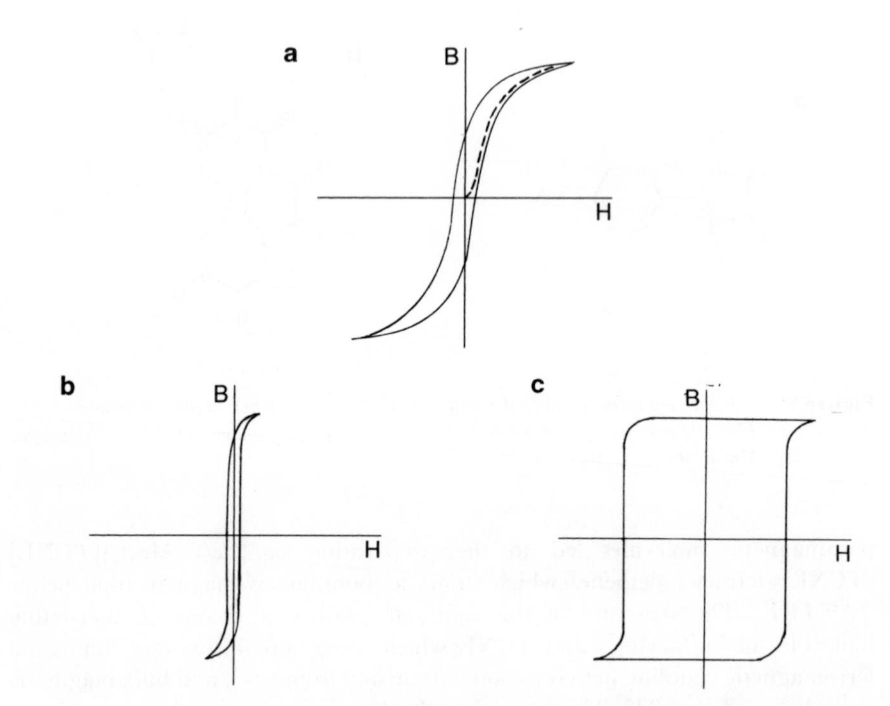

Figure 10 *Hysteresis curve for (a) a material that acts as a hard magnet. The dashed line shows the initial increase in* **B** *as* **H** *is applied to a demagnetized sample, and indicates the changing character of the domain distribution with positive* **H**. *Hysteresis curves for (b) a soft magnetic material suitable for use in a transfer core and (c) a ferrite magnetic switch of the sort that used to be used in computer memories*

large positive then to a large negative field and back to a large positive value of $\boldsymbol{H}$ is shown in Figure 10(a). This is called a **hysteresis curve**.

The initial state with $\boldsymbol{H}=0$ contains moments which are almost all aligned ferromagnetically, but grouped into regions called **domains** which adopt different polarizations relative to one another. The driving dorce for the formation of domains is the reduction of the unfavourable interaction between magnetic dipoles when the polarization is uniform. This may be visualized with the aid of the top part of Figure 11(a) which shows that for a bar magnet of finite length in which all the moments point in the same direction, there is a **demagnetizing field** $\boldsymbol{H}_\mathbf{d}$ which acts in opposition to **H** and reduces **B**. The strength of the demagnetizing field depends on **M** and the shape of the magnet: it is generally smaller for magnets which are long and thin, with the principal axis in the direction of **H**. It is greatly reduced if the moments rearrange themselves in the manner depicted in Figure 11(b).

The formation of a large number of domains is restrained by the unfavourable exchange energy associated with the non-parallel alignment of moments between them. This interface is called a **domain wall**. The change in the angle of the moments from one side of the wall to the other is commonly fixed at 90° or 180°

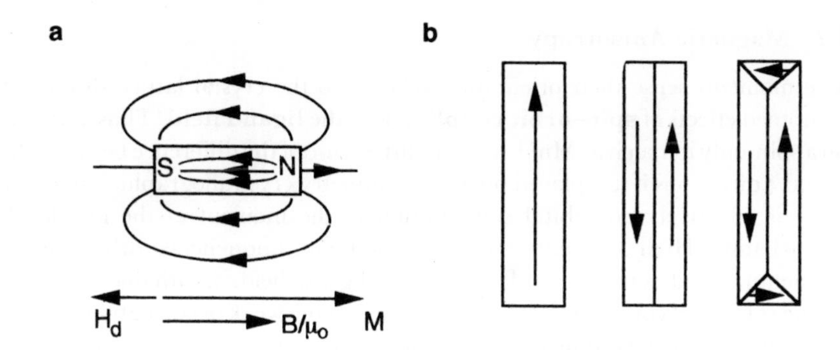

Figure 11 *(a) Production of an internal demagnetizing magnetic field* $\mathbf{H_d}$ *when a bar magnet is placed in a field* **H**, *reducing the magnetic flux,* **B**. *(b) Domain formation in a bar magnet*

[Figure 11(b)], but this may be spread over a large or small number of moments for wide or narrow walls with a small or large change in energy respectively.

If we return to the hysteresis curve of Figure 10(a), we observe that as **H** is increased, **B** rises to a limiting value as the magnetization saturates at a value that is equivalent to the perfect alignment of all the moments with **H**. The force required to produce **magnetic saturation** is small relative to the exchange energy per moment. If **H** is then reduced back to zero, a fraction of **B** called the **remanent induction** remains; this may only be eliminated by applying a reverse field called the **coercivity**.

This hysteresis curve is typical for a material used for permanent magnets: the typical working range is in the second quadrant where **B** is positive and **H** is negative, and the amount of energy that may be stored per unit volume of the material is the product **BH** which has units of J m^{-3} is called the **energy product**. The maximum value of this quantity, $(\mathbf{BH})_{max}$, is often used as a measure of the performance of a permanent magnet, together with the coercivity and the saturation induction. The amount of work done on the magnet by **H** is the sum of **H**d**B** around the hysteresis loop *i.e.* it is given by the integral $\oint \mathbf{H} d\mathbf{B}$. This is equal to the area of the curve, and is manifested as heat produced in the magnet.

In Figures 10(b) and 10(c) we show two quite different forms of hysteresis curves. That in Figure 10(b) shows low coercivity and remanence: this is characteristic of a ***soft* magnetic material** and is used where we require a material in which a large permeability may be produced in a small field. The area within the hysteresis loop is small and therefore the **energy losses** are low: this may be important in a device such as a transformer where energy efficiency is important. By contrast, Figure 10(c) shows a hysteresis curve for a material in which large fields are required to reverse **B**, and the transformation occurs very abruptly with **H**. This and the curve of Figure 10(a) are characteristic of ***hard* magnetic materials**; the squarer shape of 10(c) makes such materials more suitable for applications in which we wish to switch **B** very abruptly as **H** is reversed, such as a single element of a magnetic computer memory.

In order to design such materials, we need to know the origin of the saturation induction, the coercive force, and the remanence.

4.2 Magnetic Anisotropy

The moments sense their orientation relative to the crystal lattice through the combined effects of **spin–orbit coupling** and the **ligand field**. Thus, a moment on a spin-only ion such as Mn^{2+} has very little sense of the difference between the *x*, *y*, and *z* directions if it is placed on a low-symmetry crystrallographic site because there is effectively no orbital contribution to the moment at the first level of perturbation. High sensitivity requires an ion with unquenched orbital angular momentum and some form of anisotropic ligand field. Lanthanide ions have strong spin-orbit coupling which, combined with the weak perturbation produced by an anisotropic ligand field, strongly favours certain spin polarizations relative to the crystallographic axes if there is an orbital contribution to the moment. We call an axis or a plane that favours the polarization of the moment an **easy axis** or **plane**, and they have corresponding **hard planes** and **axes** respectively. This form of anisotropy is called the **magnetocrystalline anisotropy**.

The strength of the magnetic exchange is very sensitive to the distance between magnetically coupled atoms or ions: exchange constants increase if the relevant lattice constants are reduced. Conversely, magnetic exchange may induce small structural distortions. The relation between magnetic and elastic energy is called **magnetostriction** and it is quantified empirically in terms of the magnetostriction coefficient λ which is the fractional change in length experienced by the magnet when the magnetization is changed from zero to the saturation value. λ may take different values in different crystallographic directions. The different magnetic polarizations of adjacent magnetic domains usually give rise to an elastic strain at the boundary that may be bound to imperfections in the lattice. Thus, rapid movement of domain boundaries is hindered by defects, which produces an expenditure of energy and gives rise to irreversible magnetic phenomena when **H** is applied to a demagnetized sample for the first time. This form of barrier is eliminated if $\lambda = 0$.

As the size of the magnet is reduced, so the demagnetizing energy saved through domain formation is reduced relative to the unfavourable energy of domain wall formation until at a critical size only one domain is expected. This is strongly dependent on the shape of the sample, the tendency to remain as a single domain being highest for particles with the smallest magnetostatic energy, *i.e.* for needle-shaped as opposed to disc-shaped samples. The enhancement of the coercivity through the shape of the sample is called **shape anisotropy**.

5 APPLICATIONS OF MAGNETIC MATERIALS

We shall now explore the various requirements of technology and the way in which chemists, physicists, engineers, and metallurgists have responded by first dividing applications and materials into hard magnets, soft magnets and magnetic recording media. The division between hard and soft is somewhat arbitrary: it is usually accepted that a soft material has a coercivity of less than 10^3 A m^{-1}, and a hard material has a coercivity of greater than 10^4 A m^{-1}.

It will become apparent that the materials that are actually used are not always those with the best performance. A magnet that appears to be the most suitable in the laboratory may be much less appealing when the feasibility of building components from it, or its durability towards mechanical and chemical wear, or economic forces are taken into consideration.

5.1 Hard Magnetic Materials

Hard magnetic materials are used as **'permanent'** magnets in a wide variety of devices. We shall divide their functions according to the physical basis of their activity.

(i) Function based on the **Coulombic force law**, which quantifies the force of attraction or repulsion between magnetic poles: this converts magnetostatic energy into mechanical work and it will be familiar to anyone who has played with a permanent magnet. This simplest application is the magnetic compass. There is a wide range of devices used to attract or hold ferromagnetic materials, ranging from magnetic clamps to equipment used to sort ferromagnetic foreign matter from waste or mineral ores, to electrical switches which are activated by moving a magnet [Figure 12(a)]. Repulsion between permanent magnets is also exploited in low-friction bearings which reduce wear or drag.

(ii) Function based on the **Lorentz force law**, which quantifies the interaction between magnetic flux and moving charge. This establishes a coupling between electrical and mechanical energy. We illustrate the effect schematically in Figures 12(b) and (c) which show how the arrangement of magnetic flux and current may produce different types of force – a torque (twisting force), a transverse force and an axial force. The sense of the force may be determined in each case by the so-called right-hand rule: if you hold the thumb, forefinger and middle-finger of your right hand at right angles to one another in the most comfortable manner and take the **c**urrent direction to be along the middle (**c**entre) finger, and the **f**ield along the **f**orefinger, then the **m**otion induced by the force is in the direction of the thu**m**b. Production of a torque in this manner is the basis of a variety of AC electrical motors. Provision of an axial force produces the motion that allows a loudspeaker to work. Passage of current through the coil in Figure 12(c) forces the suspended magnet to move, and with it the diaphragm to which it is attached. This disturbs the fluid – usually air – in which the diaphragm is bathed, which in turn disturbs our eardrums.

When the Lorentz force is applied to free electrons, it changes their momentum; this provides a means of deflecting and focusing electron beams, as found in a mass spectrometer, or the magnetron at the heart of a microwave generator. We show a schematic diagram of a magnetron in Figure 12(d). Electrons leave the heated cathode and accelerate towards the anode in the electric field. A magnetic field is applied perpendicular to this trajectory, bending it. The electric and magnetic fields may be set so that electrons rotate in a circular orbit with an angular momentum that depends only on B and the ratio of the electrons charge to mass. The formation of a stable orbit, and the frequency of rotation (and hence the microwave frequency), depends on the production of a very stable magnetic field

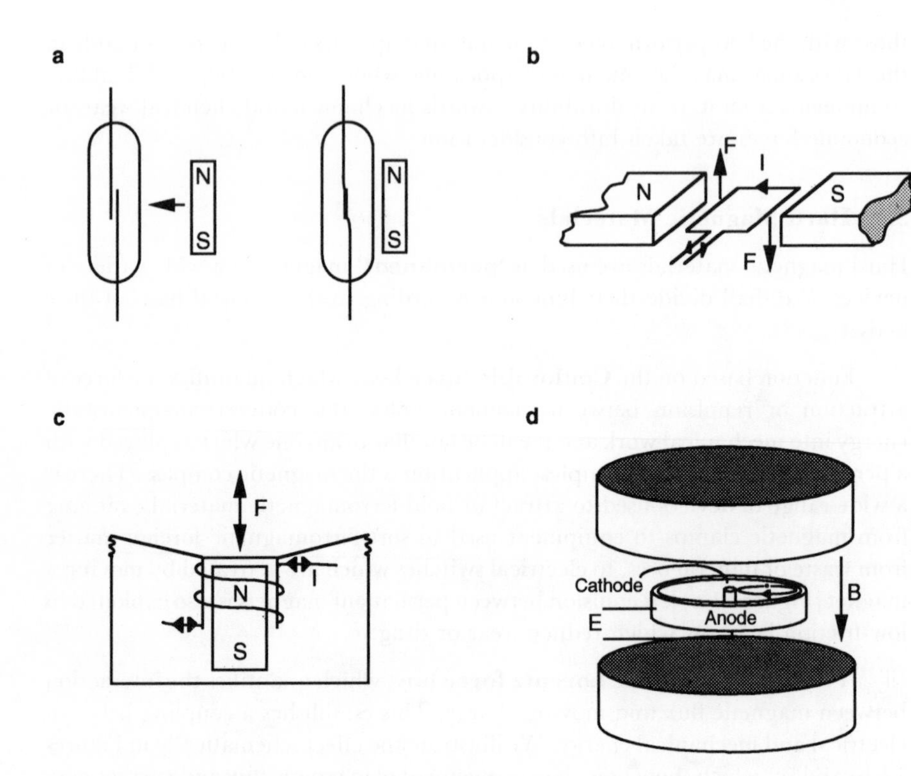

Figure 12 *Selected electromagnetic devices. (a) Reed switch activated by motion of a magnet relative to a pair of magnetically polarizable contacts. (b) and (c) show the sense of the torque and the linear force (F) produced when current (I) flows through a coil placed near the magnetic field produced by a permanent magnet; these form the basis of most electric motors and loudspeakers respectively. (d) The circulation of electrons in crossed electric and magnetic fields in a magnetron produces microwave radiation for a domestic microwave oven*

so there has been considerable research on both the materials and the way in which they may be formed into suitable shapes. Such devices may be run continuously, as in a domestic microwave oven, or in intense pulses for the transmission of radio, telephone, and television signals.

A similar sort of effect is used in some gas pressure gauges, most commonly called 'Penning' gauges. If a gas is introduced into a magnetron, the rotating electrons may ionize some of the molecules and produce a flow of charge between the electrodes. After calibration, the measurement of current provides an estimate of the gas pressure.

The earth's magnetic field focuses high-energy particles emitted by the sun towards its poles where their collisions with particles in the atmosphere results in the emission of the light known as the *aurora borealis* or northern lights in low northern latitudes, and the *aurora australis* or southern lights in low southern latitudes.

(iii) Function based on **Faraday's Law**, which quantifies the change in electrical current or potential when the magnetic flux passing through a conductor is changed. This provides a means of interconverting mechanical and electrical energy. Common applications are found in electrical generators and devices based on the production of Eddy currents such as the speedometer of a car. Microphones also operate according to this principle, as illustrated in Figure 12(c) which we first used to describe the action of a loudspeaker. However, we now take the sound waves to be the driving influence, and the induced current to be the result. This current may then be amplified and modified before being passed to a loudspeaker.

Until the 1930s most permanent magnets were based on **carbon-doped steel**. These were superseded by the **Alnico materials** – alloys of Fe with Co, Ni, Al, and small amounts of other metals – which had much higher coercivity and $(\mathbf{BH})_{max}$, but were very brittle. In order to make robust materials, the Alnico may be powdered and set in a rigid binder. Shape anisotropy may provide the basis for high coercivity in powdered samples of iron or iron–cobalt alloys. If the particles are sufficiently small they will form single domains in an applied magnetic field, which also serves as a means of aligning them. The nett polarization may be frozen by setting the aligned particles in a rigid binder, or sintering it.

High coercivity is found in **hard ferrites** with anisotropic crystal lattices and distinct hard axes or planes. Barium ferrite, $BaFe_{12}O_{19}$, is probably the most widely used example of this type of material, and is used in cathode ray tubes in domestic television sets. It would be prohibitively expensive, and not particularly useful, to try to make large devices out of single crystals of this material, so solids are fabricated from powders in the manner described above. Other examples of materials in which a high coercivity stems from large values of the magnetocrystalline anisotropy are given by metal alloys containing **lanthanide** metals: $SmCo_5$ has an exceptionally high anisotropy and saturation magnetization but suffers from the relative scarcity and therefore expense of cobalt. An alternative has been found in the class of compounds of general formula $R_2T_{14}X$, where is R is a lanthanide, T a transition metal, and X is B or C. The best known example is $Nd_2Fe_{14}B$. Such compounds are generally multiphasic and their activity is not fully understood. They are also more prone to chemical attack and have lower values of T_C than $SmCo_5$.

The properties of some hard magnetic materials are shown in Table 3.

5.2 Soft Magnetic Materials

Soft magnetic materials are used principally to amplify the **magnetic flux** created by an electrical current through the provision of a medium with very high permeability for a low coercivity. Their applications may be divided into dc and ac devices. The principles of operation of many of these devices is the same as those discussed in Section 5.1, with the permanent magnet replaced by an electromagnet.

Table 3 *Properties of iron and selected hard magnetic materials used as permanent magnets. Alnico 5 contains 50% Fe, 24% Co, 15% Ni, 8% Al, and 3% Cu by weight; while Alnico 8 contains 34% Fe, 35% Co, 15% Ni, 7% Al, 5% Ti, and 4% Cu by weight. Barium ferrite is* $BaFe_{12}O_{19}$

Material	*Ordering temperature* T_C (K)	*Coercivity* (A m^{-1})	*Energy product* $(BH)_{max}$ (kJ m^{-3})	*Remanent induction* (Tesla)
Iron	1043	88		
Alnico 5	1173	58 000	42	1.25
Alnico 8	1133	130 000	40	0.83
Barium ferrite	723	170 000	36	0.43
$SmCo_5$	973	670 000	160	0.91
$Nd_2Fe_{14}B$	585	1000 000	280	0.42

Common **dc devices** are electromagnets and electrical relays; the latter rely on the same principle as the electrical switches described at the start of the previous section.

ac applications include devices for the generation and conversion of electrical power. Most electrical generators are based on the device depicted in Figure 12(b) which we used to illustrate the action of an electrical motor; rather than convert electrical energy to mechanical energy, we may apply a mechanical force (in turn provided by a fossil fuel engine or steam pressure or falling water) to the coil and induce a current as it cuts lines of magnetic flux. The electrical potential generated may be increased or decreased using the transformer illustrated in Figure 13(a). A change in the electromotive force (emf) and hence the current through a primary coil produces magnetic flux which is amplified and fed to a second coil through the highly permeable core. This then induces an emf in the secondary coil such that the ratio of emf in the primary to the secondary is proportional to the ratio of turns in the two coils. The materials that are suitable for ac devices are subject to greater constraints than for dc devices because Eddy currents are easily generated in metals, and these lead to energy losses through resistive heating. At low frequencies such as the 50–60 Hz used in domestic power supplies, this is not too great a problem, and the effect may be suppressed adequately if the resistance of a metal magnet is increased through the addition of impurities. A common material which was designed with this principle in mind is iron containing a few percent of silicon.

The performance of a soft magnetic material may be improved if we reduce the magnetocrystalline or magnetostrictive effects. One way of doing this is to produce an alloy from two metals with different signs of the anisotropy such that at a particular composition the different contributions cancel. Such is the case for **Ni–Fe alloys** for which the composition close to Ni_3Fe provides a good static or low-frequency soft magnets which have trade names such as Permalloy, Supermalloy, and Mumetal. The properties of these and other soft magnetic materials are shown in Table 4.

Alternatively, strain between imperfections in the crystal lattice may be eliminated through the use of **non-crystalline materials**.[12,13] If a molten metal

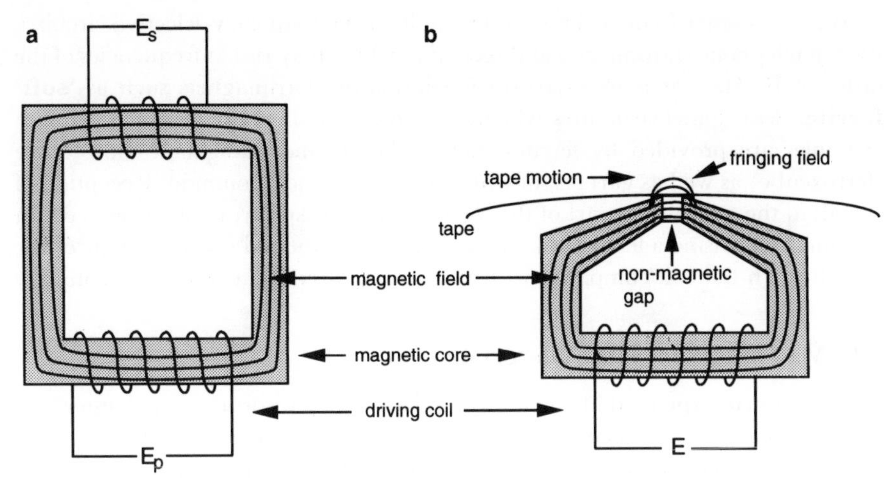

Figure 13 *(a) Illustration of a simple transformer in which an emf* E_p *in the lower primary coil is converted to a different emf* E_s *in the secondary coil with the aid of the enhanced flux in the high permeablility soft magnetic core. (b) Illustration of the action of a magnetic recording head in which a signal in the form of a current flowing in a coil controls the magnetic field experienced in a moving tape coated in a magnetically polarizable material. The top of the head is clearly visible in most audio cassette players*

is cooled sufficiently quickly, the atoms may freeze in positions that have no translational symmetry. This may be done by directing a jet of molten metal onto a cooled, rotating disc. The metal quenches at a rate of 10^6 K s^{-1} to form a ribbon which rapidly spins off the disc. The materials thus formed generally have lower saturation magnetizations and Curie temperatures than their crystalline counterparts, but have the advantage of higher resistivity, and hence lower energy losses in high-frequency applications, and may be produced with a wide range of compositions which may allow the magnetostriction to be tuned to very small values.

Table 4 *Properties of selected soft magnetic materials. 78-Permalloy is 78% Ni, 22% Fe; Supermalloy is 79% Ni, 16% Mo, and 5% Mo; and Mumetal is 77% Ni, 16% Fe, 5% Cu, and 2% Cr. MnZn ferrite is* $Mn_{(1-x)}Zn_xFe_2O_4$ *and Metglas is a tradename for an amorphous alloy of iron with B and Si. In this case we give figures for a glass with 4% B and 3% Si*

Material	*Ordering temperature* T_C (K)	*Coercivity* (A m^{-1})	*Relative permeability (max–min)*	*Saturation induction* (Tesla)
Pure Fe	1043	88		2.15
Fe−3.2%Si	1013	8	7500–55 000	2.01
78-Permalloy	873	4	8000–100 000	1.08
Supermalloy	673	4	100 000–1000 000	0.78
Mumetal	335	4	20 000–100 000	0.65
MnZn ferrite	403	16	1500–2500	0.34
Metglas	688	1.6	15 000–300 000	1.56

At much higher frequencies – for example in the frequency-selective circuitry used in telephone transmission and reception which may run at frequencies of the order of 10^9 Hz – it is necessary to use insulating ferrimagnets such as **'soft' ferrites** with spinel structures which have low crystalline anisotropy. Common examples are provided by ferrites such as Fe_3O_4 and manganese–zinc ferrite (ferroxcube) as well as garnets, of which YIG is the most common. Reception of signals in the microwave part of the electro-magnetic spectrum often relies on an antenna comprising a ferrite rod and a pick-up coil; the rod greatly enhances the flux through the coil compared with that experienced by a coil in a vacuum.

5.3 Magnetic Recording Materials

The **magnetic tape** used in an audio or video cassette recorder is a polyester film which is coated in a resin containing fine needle-shaped particles of ferrimagnetic γ-Fe_2O_3 or ferromagnetic CrO_2. Before the resin sets, the particles are subjected to a magnetic field so that the dried film contains an array of aligned single-domain particles, with the shape anisotropy playing an important role. γ-Fe_2O_3 has a defect inverse spinel structure in which there are ordered vacancies on the octahedral sites, and which is derived from Fe_3O_4 by oxidation. This results in a lower saturation magnetization for Fe_2O_3 (2.5β per iron atom, compared with 4.1β) but an ordering temperature that is about 85 K higher.

Information is written on the tape with a small electromagnet comprising a magnetic ring (typically of Permalloy) with a fine coil wrapped around it. There is a non-magnetic gap at the point where the tape passes over the head [Figure 13(b)]. Electrical impulses from a microphone pass through the coil, and activate the electromagnet, producing a 'fringing' field at the non-magnetic gap which passes through the tape and magnetizes the region that is passing at that point. The tape leaves the head with a remanent magnetization. The gap has been exaggerated for the sake of clarity in Figure 13(b), and is actually very small – of the order of 10 μm. The same magnet and coil may also be used to read the tape by measuring the current produced when magnetized region of the tape passes over it, and passing this via an amplifier to the electromagnets of a loudspeaker. A similar principle applies to computer memory disks. A hard disk is typically a stack of smooth aluminium plates coated in a magnetic oxide with a ferrite reading and recording head skimming over the surface. A floppy disk contains just one disk comprising a plastic base coated with a magnetic oxide.

Quite a different recording device is the **bubble memory** in which 'bits' of information are held in the form of small magnetic domains embedded in a ferrimagnetic film. Consider the pattern of domains in Figure 14 in which the shading denotes equal areas of uniaxial 'up' or 'down' polarization in the absence of any magnetic field. As a field is applied perpendicular to the plane of the film, one form of domain grows at the expense of the other and the serpentine arrangements of stripes is transformed into a collection of small cylindrical domains embedded in a film with the opposite magnetic polarization. These are called magnetic bubbles because the physical laws that govern their growth and stability are analogous to those that apply to soap bubbles. Both of these properties

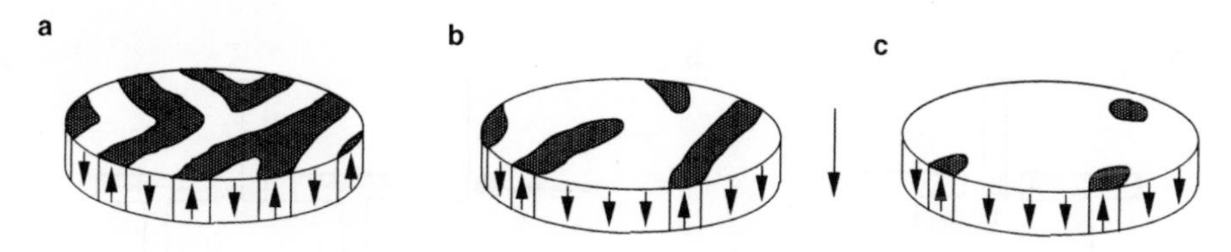

Figure 14 *Formation of magnetic bubbles in a ferrimagnet with easy-axis anisotropy in the direction of the applied magnetic field*

depend on the balance between the magnetocrystalline anisotropy and magnetostatic energy: the smaller the saturation magnetization, the smaller the bubbles that may be stable for a given thickness of film.

Suitable materials for bubble formation must satisfy several criteria. Clearly they must be ferro- or ferrimagnetic and it must be possible to prepare them as reproducible, high-quality thin films. In order to produce small, stable bubbles, the material must have a large magnetocrystalline anisotropy perpendicular to the plane of the film relative to the magnetization, so as to minimize the influence of the demagnetizing field.

The most successful materials to date are garnets. In Section 3.2 we described how the magnetic properties of such materials could be controlled by suitable substitution of the metal ions to produce materials with different saturation magnetization. At first glance, the garnets might not appear to be particularly suitable because their cubic structure indicates that uniaxial magnetocrystalline anisotropy will be negligible. However, this anisotropy may be controlled by substitution of the lanthanide ions to produce solid solutions containing two different lanthanide ions which order in the lattice perpendicular to the substrate on which the film is grown. It has been found empirically that the anisotropy is greatest when the difference in ionic radius of the lanthanide ions is greatest. One combination that works particularly well is the large Sm^{3+} ion combined with the small Lu^{3+} or Y^{3+} ion, and such garnets are used in most current commercial devices. Hexaferrites such as $BaFe_{12}O_{19}$ have large anisotropy energies but are disfavoured by difficulties in preparing films of acceptable quality. There has been some progress in the production of bubble memories based on amorphous metal alloys such as Gd–Co–Mo which offer advantages in the form of ease and expense of fabrication, but do not yet have the performance of the garnet films.

Bubbles are injected into one side of the film with a pulsed magnetic field produced by passing current through a small coil, and carried away from the injection point under the influence of a small bias field. The stream of bubbles is directed towards a series of magnetic sinks made out of a soft ferromagnet such as Permalloy. The bubbles latch onto these sites which act both as a means of directing the path the bubbles take, and also as a series of addresses: digital information is held as a series of locations which are either occupied or not occupied by bubbles. Bubbles are driven between these sites with a rotating magnetic field. This may be produced by placing a second coil perpendicular to that which injects the bubbles, and passing ac currents of equal frequency and 90° phase difference through the two coils.

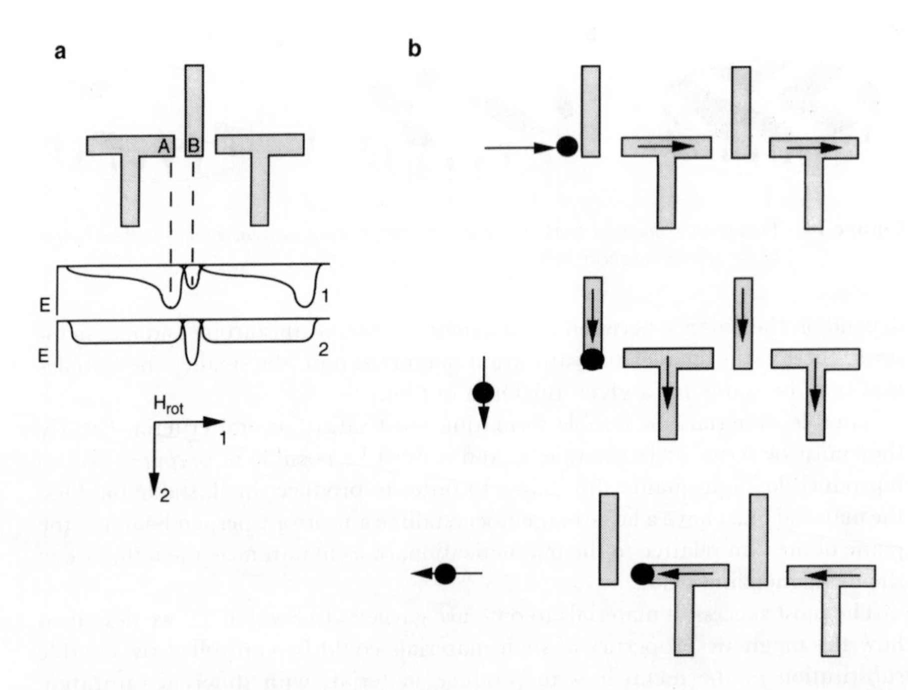

Figure 15 *(a) The favoured position of a magnetic bubble may be controlled by sweeping an applied magnetic field. In position 1, a bubble in the vicinity of sites A and B favours A where there is a local minimum in the potential energy, displayed on the upper of the two graphs below. In position 2, site B is favoured. (b) The effect of sweeping the rotating field through three positions at 90° to one another. A bubble, depicted by the black dot, is first injected then swept across the array of magnetic sinks*

In Figure 15(a) we indicate the effect of applying magnetic fields with different directions relative to a typical pattern of magnetic sinks – so-called 'T' and 'I' bars. As the field rotates the position of the potential well that traps each bubble changes. In position 1 the T bars are magnetized along the crossbar, providing a relatively deep well at the tip near the I bars, labelled A in the Figure. As the field is rotated to position 2, the favoured direction of magnetization is down the I bar, inducing the bubble to hop between addresses so that it is trapped at B. In Figure 15(b) we illustrate the effect of sweeping the rotating field through a full cycle, with the injection of a new bubble. Such a set-up would provide a constant stream of bubbles so long as the rotating field continued. In order to inject a vacancy in the stream, a counteracting magnetic field may be produced by passing current through a small loop of wire at the injection point.

The data may be retrieved without being damaged by driving the bubbles past a transducer which is composed of a magnetic conductor whose resistance is highly sensitive to changes in its magnetization, a phenomenon known as **magneto-resistance**. This converts the original series of electrical pulses back to electrical pulses. Research in the field of magnetic transducers received a fillip in 1988 when a particularly pronounced effect – dubbed **'giant magnetoresistance'** – was observed in thin films comprising layers of Fe and Cr. The origin of the effect lies in the scattering of conduction electrons by moments in the metals. Electrons passing through the magnetic layers, which are coupled antiferromagnetically, are scattering according to the relative polarizations of their spins. In zero magnetic field, there are equal numbers of either polarization of metal moments, and both polarizations of conduction electrons are scattered equally; when the material is magnetized, conduction electrons with one sense of polarization are scattered much less strongly than the other and the resistance drops by a factor of about 100. The effect is known to be sensitive to the composition, thickness, and magnetic coupling between the multilayers.

A magnetic bubble memory has the disadvantage of being relatively slow at retrieval of data compared with semiconductor memories, but it provides an extremely high storage density and there are no mechanically moving parts to wear out – all the motion involves bubbles driven by electrical impulses. Applications may be found in non-volatile computer memory whose stored data do not evaporate when the power supply is switched off.

More recent developments in magnetic recording are based on **magneto-optical phenomena**. When plane-polarized light passes through a transparent magnetized medium, the plane of polarization is rotated to a degree proportional to the magnetization in the direction of propagation of the light. This is known as the **Faraday Effect**; in reflection it is called the **Kerr Effect**. A common form of storage medium that operates using this principle is a disk coated with a thin film of an amorphous alloy of (Tb and Fe) or (Gd, Tb, and Fe) in which the easy-axis for magnetization is perpendicular to the film surface. The film is magnetized to its saturation point in this direction and then a fine laser beam is focused on a point on the surface so that it is heated locally above the Curie temperature. That small region will reverse its magnetization under the influence of the strong demagnetizing field, and a bit of data will be stored as a reversed domain. The information may then be read using the Kerr effect: a reflected laser beam and a series of polarization filters detect those regions whose Kerr rotation is different. Such magneto-optical storage systems are very fast and allow a much higher recording density (10 times faster and 1000 times more dense than a conventional floppy disk).

6 CONCLUSIONS AND FUTURE TRENDS

Advances in fundamental and applied magnetism enrich the lives of many people in many different ways, in the home and the workplace, and in medicine and science research. The design of new magnets is a very interdisciplinary activity, calling on expertise from chemists, physicists, materials scientists, and engineers. Indeed, many of those who work in this area would not recognize such divisions. However, if we use conventional divisions, it is chemists who use their chemical and crystallographic knowledge and intuition to place specific magnetic ions or molecules in specific sites in a solid and link them magnetically through bonds with particular chemical character and geometry. Physicists more commonly work on metallic elements and alloys, trying to understand the relationships between their electronic structure and magnetic properties. We have seen that bulk magnetic properties are also sensitive to the size and shape of a magnet, which is the domain of materials scientists and engineers who have been very active in preparing crystalline and amorphous solids in the form of lumps, powders, or films, and subjecting them to a variety of thermal and magnetic treatments to enhance their character as hard or soft magnets.

The well-established drive to produce magnets with larger or smaller coercivity, or with a larger value of $(BH)_{max}$ per unit mass, has been joined by initiatives aimed to produce thin films of magnetic materials or magnets with better magneto-optical activity or magneto-resistance. While there is research devoted to improving almost every aspect of the performance of magnetic materials, there are certain areas that are particularly prominent at present:

(1) **Molecular magnets with a spontaneous magnetization**. It is highly desirable to produce ferro- or ferrimagnets with low densities because this would greatly reduce the weight of many devices that require permanent magnets. While there has been little success in producing an organic ferrimagnet, with the record standing at less than 1 K for T_C for the nitronyl nitroxide described in Section 3.4, there has recently been remarkable progress in raising the ordering temperature in metal salts: a value for T_C of 240 K has been reported for the mixed-valence chromium(III)–chromium(II) cyanide $[Cr_5(CN)_{12}].10H_2O$.[14]

(2) **New magnetic films**. The production of thin films – and in particular of insulating thin films – is a developing field that presents many practical and theoretical problems. Layers composed of a single material may be used in magnetic and magneto-optical recording media, while multilayers comprising multi-decker sandwiches of different elemental metals or alloys may be of great importance in magneto-resistive devices.[15]

(3) There has recently been a great improvement in the performance of **hard magnets** containing lanthanide metals and iron when they are combined with nitrogen. The resulting alloys, with general formula $R_2Fe_{17}N_{3-\delta}$, and in particular with R = Sm, show a dramatic increase in T_C as the nitrogen content increases; the energy products of such materials may eventually exceed that of magnets such as $Nd_2Fe_{14}B$.[16] There have also been significant improvements in the performance of *these* magnets through the development of new processes for the preparation of fine powders. Usually the material is cast as an multiphasic ingot containing $Nd_2Fe_{14}B$ and a Nd-rich phase. When this is exposed to hydrogen and subjected to a number of heat treatments it breaks up to produce, after milling, a very fine powder of $Nd_2Fe_{14}B$. This not only aids the production of fine powders prior to binding to produce a composite magnet, but also reduces the extent to which the material is degraded through oxidation during processing and results in a very pure material. This sequence of treatments is known as hydrogen decrepitation.

(4) **Magnetic glasses**. The production of amorphous or nanocrystalline metals and alloys is growing as such materials are shown to possess useful – and often unique – magnetic properties. They may be used in bubble memories and magnetic recording heads as cheaper materials than crystalline magnets. They may also be produced with an extremely desirable combination of low coercivity

and high resistivity as an excellent material for the cores of transformers and motors and for magnetic shields.[17]

(5) **Fine particles**. The magnetic behaviour of fine particles is clearly relevant to conventional magnetic recording devices and the formation of composite magnets formed by sintering or binding the particles in a resin. An additional application that we have not touched on yet is the formation of magnetic colloids, such as Fe_3O_4 suspended in an oil with a surfactant to prevent aggregation.[18] Such liquids are called **ferrofluids** and may be manipulated with an external magnetic field, allowing fluid to be directed to places other than those favoured by gravity or hydrostatic pressure. They may be used as lubricants, sealants, or damping fluids as well as in medicine to stem the flow of blood when surgery would involve a higher risk.

7 REFERENCES

1. D. C. Mattis, 'The theory of magnetism I. Statics and dynamics', Springer, Berlin, 1988, Vol. 17.
2. P. W. Selwood, 'Magnetochemistry', Wiley, New York, 1956.
3. T. I. Quickenden and R. C. Marshall, *J. Chem. Educ.*, 1972, **49**, 114.
4. P. W. Atkins, 'Molecular Quantum Mechanics', Oxford University Press, 1983.
5. J. B. Goodenough, 'Magnetism and the Chemical Bond', Wiley, New York, 1963.
6. K. P. Sinha and N. Kumar, 'Interactions in Magnetically Ordered Solids', Oxford University Press, 1980.
7. C. A. Coulson, 'Valence', Oxford University Press, 1972.
8. J. N. Murrell, S. F. A. Kettle, and J. M. Tedder, 'Valence Theory', Wiley, London, 1969.
9. M. Kotani, *J. Phys. Soc. Jpn*, 1949, **4**, 293.
10. F. Palacio, in 'Magnetic Molecular Materials', ed. D. Gatteschi, O. Kahn, J. S. Miller, and F. Palacio, Kluwer Academic, Dordrecht, Netherlands, NATO ASI Series, 1991, vol. E198.
11. O. Kahn, Y. Pei, and Y. Journaux, in 'Inorganic Materials', ed. D. Bruce and D. M. O'Hare, Wiley, Chichester, 1992.
12. M. R. J. Gibbs, *Chem. Britain*, 1983, 837.
13. R. W. Cahn, *Contemp. Phys*, 1980, **21**, 43.
14. T. Mallah, S. Thiebaut, M. Verdaguer, and P. Veillet, *Science*, 1993, **262**, 1554.
15. J. A. C. Bland, R. D. Bateson, P. C. Riedi, R. G. Graham, H. J. Lauter, C. Shackleton, and J. Penfold, *J. Appl. Phys.*, 1991, **69**, 4989.
16. J. M. D. Coey, H. Sun, and D. P. F. Hurley, *J. Magnetism Magnetic Mater.*, 1991, **101**, 310.
17. K. Moorjani and J. M. D. Coey, 'Magnetic Glasses', Elsevier, Amsterdam, 1984, vol. 6.
18. R. E. Rosensweig, *Sci. Am.*, 1982, **247**, 124.

In addition to these specialized references, the following **general texts** are recommended:

J. H. Van Vleck, 'The Theory of Electric and Magnetic Susceptibilities', Oxford University Press, 1932.

B. N. Figgis, 'Introduction to Ligand Fields', Wiley, New York, 1966.

R. S. Drago, 'Physical Methods in Chemistry', Saunders, Philadelphia, 1973.

F. E. Mabbs and D. J. Machin, 'Magnetism and Transition Metal Complexes', Chapman and Hall, London, 1973.

R. L. Carlin, 'Magnetochemistry', Springer, Berlin, 1986.

R. J. Parker, 'Advances in Permanent Magnetism', Wiley, New York, 1990.

J. Crangle, 'Solid-state Magnetism', Edward Arnold, Sevenoaks, UK, 1991.

D. Jiles, 'Introduction to Magnetism and Magnetic Materials', Chapman and Hall, London, 1991.

The cross-disciplinary nature of magnetism means that advances and reviews are published in **journals** for general physics, chemistry, and engineering. In addition to reading the highlights presented in broad journals such as *Nature*, *Science*, *New Scientist*, *Scientific American*, *Physics Today*, *Physics World*, and *Chemistry in Britain*, details of most recent research may be found in the following journals:

Inorganic Chemistry
Journal of Magnetism and Magnetic Materials
Journal of Physics: Condensed Matter
Journal of the Physical Society of Japan
Journal of Solid State Chemistry
The Physical Review and Physical Review Letters

Every three years there is the International Conference on Magnetism, which concentrates on fundamental magnetism; the **proceedings** are published in the *Journal of Magnetism and Magnetic Materials*. The last such event, in Edinburgh in 1991, was reported in volumes 104–107 of that journal in 1992. Applied magnetism is the subject of the annual INTERMAG meeting which is reported in issues of the *IEEE Transactions on Magnetism*. Every year The Royal Society of Chemistry (Cambridge, UK) publishes an Annual Report which contains a **review** on magnetism. The last two reviews appeared in Section A (Inorganic Chemistry) as follows:

A. Harrison and S. J. Clarke, *Annual Reports on the Progress of Chemistry, Section A, Inorganic Chemistry*, 1992, **89**, 425 (published in 1993).

A. Harrison, *Annual Reports on the Progress of Chemistry, Section A, Inorganic Chemistry*, 1993, **90**, 407 (published in 1994).

Finally, there is a good overview of the status of research in fundamental and applied magnetism and an account of where advances are needed in a document produced by the UK Science and Engineering Research Council:

'Magnetism and Magnetic Materials Initiative', ed. J. Chapman, SERC, UK, 1990.

CHAPTER 11

Superconducting Materials

JOHN T. S. IRVINE

1 INTRODUCTION

The recent discovery of **high temperature superconductivity** in cuprates persisting to temperatures in excess of 77 K, the boiling point of nitrogen, has had a major impact on the solid state sciences and has brought the promise of exciting new technologies. The intense race in the late 1980s to find and exploit new, higher temperature superconductors engendered a high level of interdisciplinary co-operation between chemists, physicists, materials scientists, and engineers. This new emphasis upon interdisciplinary research has spread to many other areas of materials research and augers well for the effectiveness of future materials research.

The high public profile of superconductivity in the late 1980s, not to mention the high levels of research investment made by many companies, demanded that some devices based on high temperature superconductors should appear in the marketplace in the relatively short term. **Conventional superconductors** have found commercial success in the area of high field magnets, particularly in the area of **magnetic resonance imaging** (MRI). Some high temperature superconducting materials are already available in commercial **SQUID magnetometers**; however, it could be argued that novelty was a more important marketing factor than materials performance. Other applications, such as current leads for high field magnets and microwave devices, are likely to become commercially viable in the very near future.

The major impact of high temperature superconductivity is likely to be in the much longer term. Areas of application of great promise include power transmission, magnetic energy storage, propulsion, and superconducting electronics. Many visionaries have suggested the coming of a superconductivity economy, where superconductors would have an impact on much of everyday life. Such a prediction may be exaggerated, but industry should be able to harness the remarkable properties of high temperature superconductors in some major applications within the next few decades. Indeed, as the phenomenon is still only in its infancy, we should await the development of new, unforeseen applications to see the greatest impact of high temperature superconductivity.

The aim of this chapter is to provide a basic insight into superconductivity and its applications. To achieve this it is necessary to draw together concepts from **solid state physics, structural inorganic chemistry, and materials science**. This is a highly interdisciplinary subject: it is not really possible to fully understand the subject from the narrow confines of a single discipline.

The first topic to be addressed is the phenomenology of superconductivity. For most of the important applications of a superconductor it is not important how the electrons couple, rather it is the macroscopic physics of the 'Cooper pairs' that governs properties. From this we will go on to briefly examine the theoretical models of superconductivity. Next, the range of systems where superconductivity has been observed is reviewed and the crystal chemistry of the cuprates discussed in some detail. Stoichiometry is the most important factor in controlling properties of high temperature superconductors and so the general features of the defect chemistry and oxygen non-stoichiometry in the cuprates will be examined. It is obviously of key importance to accurately control and determine composition; thus the following sections address synthesis and characterization of the cuprates. The final topic to be addressed is the applications of high temperature superconductors, where a major emphasis is placed upon the role of solid state chemistry and materials processing in optimizing properties for technological application.

2 THE PHENOMENON

Frequently, superconductivity is defined in terms of the properties that a superconductor must process. A better definition is that a superconductor is a **charged superfluid**. The properties of a superconductor are a consequence of it being a superfluid. A superfluid is a condensed system of **bosons** (*e.g.* ^{4}He), whose common motion is described by a unique wavefunction. Superconductivity arises from a condensation of electrons (or holes), which are spin-$\frac{1}{2}$ particles; thus a superconductor is often described as a 'fermion' superfluid. The Pauli Exclusion Principle does not allow a system of fermions to be described by a unique wavefunction so the carriers must be paired to form bosons with integral spin. This idea can be modified to give the two fluid model, where the superconductor is seen to consist of a condensate of condensed pairs and excitations which are electron/hole-like.

Two properties are regarded as being hallmarks of superconductivity; they are **zero resistance** and **magnetic flux expulsion**.

On cooling, a transition from normal (*i.e.* 'metallic') to superconducting behaviour is observed at a certain **critical temperature, T_c**. The phenomenon was first discovered by Onnes in 1911 when he investigated the low temperature resistance of mercury. A sharp discontinuity in resistance was observed at 4.2 K, with the resistance dropping to below the limit of detection (10 μohms) at this critical temperature (Figure 1). When a dc current is passed through a superconductor, there is no potential drop across the superconductor: it exhibits zero electrical resistance.

A phase transition is also observed when the superconductor is cooled to below its critical temperature in a magnetic field. On cooling through the superconducting

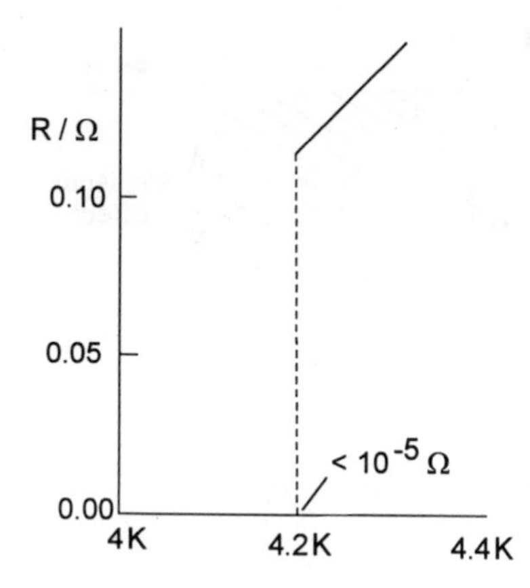

Figure 1 *The resistance of mercury as measured by Onnes in 1911*

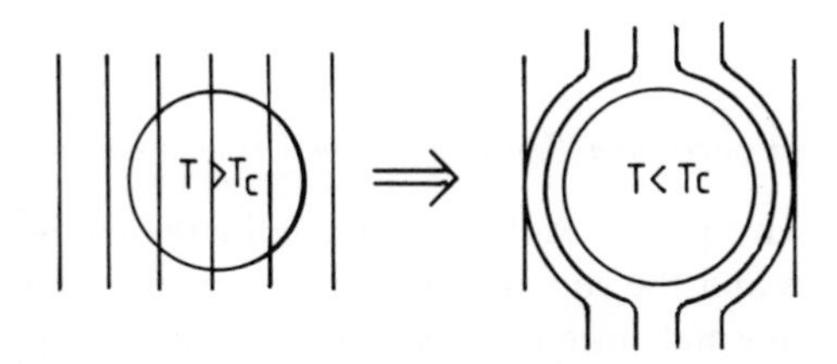

Figure 2 *Magnetic flux expulsion on cooling through the superconducting transition*

critical temperature supercurrents generated close to surface of the superconductor oppose the magnetic field and the flux is expelled from the sample (Figure 2). This is known as the **Meissner Effect**.

Superconductivity can be destroyed by increasing the applied magnetic field beyond a **critical magnetic field, H_c**. Two types of phase transition are observed (Figure 3). For a **Type I superconductor** at a temperature below its critical temperature, no magnetic flux enters at fields below the critical value. On increasing the field to the critical value, a first-order transition occurs and the superconducting state breaks down. If the field is now reversed, the flux is again expelled from the superconductor: this flux expulsion is another expression of the Meissner effect. Type I superconductivity is only observed in a limited number of superconductors, typically elemental superconductors.

Type II materials tend to exhibit poorer metallic properties in the normal state than type I materials; however, Type II superconductivity is more widely observed and it is the type of superconductivity exhibited by the technologically important materials, including the new high temperature cuprate superconductors. Type II superconductors exhibit two critical fields. Between zero magnetic field and the first critical field, H_{c1}, no magnetic flux enters the sample and **perfect diamagnetism** is observed as in Type I superconductors. At the first critical field, lines of magnetic flux start to penetrate the sample, but superconductivity is not

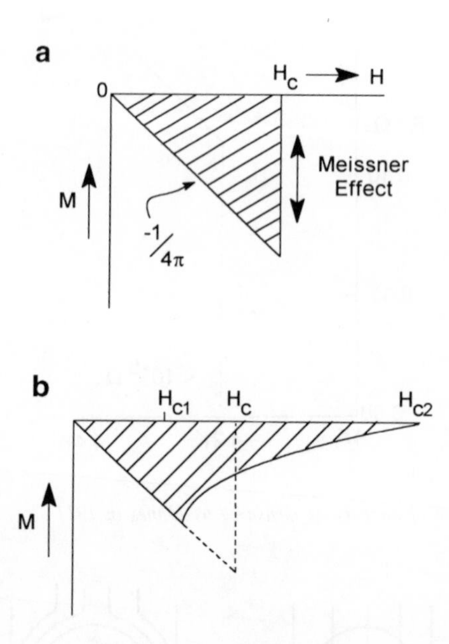

Figure 3 *Magnetization curves for a Type I superconductor (a) and a Type II superconductor (b)*

destroyed; supercurrents persist. As field is further increased the density of lines of flux further increases and the superconductor pair density falls until the system smoothly enters the normal state in a second-order phase transition at $\boldsymbol{H}_{c2}$.

Although a superconductor will expel a magnetic field below H_c or H_{c1}, the flux does in fact penetrate into the superconductor. The **penetration depth, λ**, is a characteristic length for a superconductor, dependent upon the pair density (Equation 1).

$$\lambda = \sqrt{(m_e/2\mu_0 e^2 n_p)} \qquad (1)$$

where m_e is the mass of the electron
μ_0 is the permeability of free space
n_p is the pair density

The pair density is a function of temperature, increasing from zero at the superconducting transition temperature to, ideally, encompass all the conduction electrons at 0 K.

The second characteristic length of a superconductor is the **coherence length,** ξ, which gives the distance over which superconducting order persists. Typically, the coherence length is of the order 0.1 μm in conventional superconductors; however, it is much shorter, *ca.* 1 nm, in high temperature superconductors.

3 BARDEEN, COOPER, AND SCHRIEFFER (BCS) THEORY

Most of the properties of a superconductor can be adequately described by assuming that condensed electron pairs exist; it is not necessary to understand the

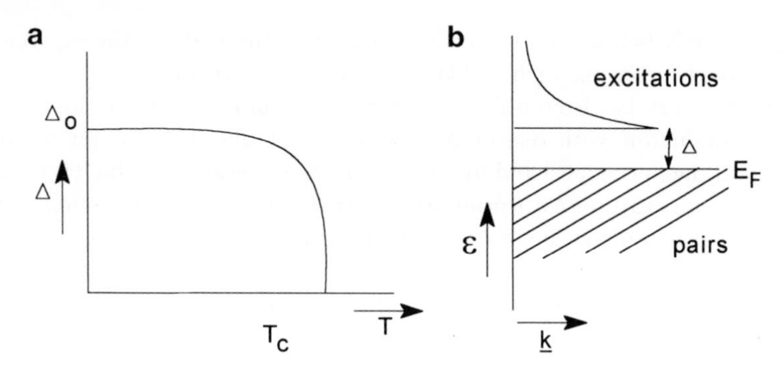

Figure 4 *Schematic density of states near the Fermi edge of a superconductor (b); and variation of the superconducting energy gap with temperature (a)*

nature of the pairing mechanism. It is, however, of considerable interest to understand the pairing mechanism and this is of key importance in the search for new improved superconductors and for further development of our understanding of the phenomenon.

Although superconductivity was first observed in 1911, it was not until 1957 that the fundamental mechanism was understood, when the BCS theory was published. The BCS theory has been highly successful, remaining essentially unchallenged until the advent of high temperature superconductivity in the late 1980s.

In the BCS model there is an attractive interaction between electrons close to the Fermi surface. The Fermi level is the energy of the highest occupied state in a metal at absolute zero, E_F. (At higher temperatures the probability of a state at the Fermi level being occupied is $\frac{1}{2}$.) The **Fermi surface** is the Fermi level in three dimensions. This attraction between electrons close to the Fermi surface is via an electron lattice interaction with electrons pairing in reciprocal space. In a conventional superconductor the coherence length is about 0.1 μm; thus the electrons interact over a large distance and should not be thought of as paired in real space. Pairs of electrons are linked via their momentum vectors, forming $\boldsymbol{k}\uparrow \ldots -\boldsymbol{k}\downarrow$ pairs. It is, perhaps, not obvious how an electron–phonon–electron interaction could be attractive. A simple illustration of an attractive electron–lattice interaction in real space terms is to think of one electron moving through the lattice, polarizing the lattice as it goes and so making it easier for a second electron to follow behind. In a superconductor, the interaction between electron wave vectors is effected by the polarization of the surrounding medium. Where resonance between this interaction and the lattice modes, or phonons, occurs there is an attractive interaction because the carriers are effectively overscreened by the lattice.

A consequence of electron pairing is the occurrence of a **superconducting energy gap** immediately above the Fermi edge (Figure 4b). In the normal, metallic state there is no discontinuity in the density of states at the Fermi edge; however, on cooling a gap arises at the superconducting transition, which increases on further cooling (Figure 4a). In the superconducting state, carriers immediately below the Fermi edge are paired; those above the superconducting gap are electron-like or hole-like excitations (Figure 4a).

BCS theory has been able to predict transition temperatures for conventional superconductors with reasonable accuracy, calculating these for the density of states at the Fermi level and the electron lattice interaction, which is related to the Debye temperature. **An upper limit for T_c of 35 K** seemed to follow from this model. Although modifications of the BCS model have been suggested to account for the much higher T_c values observed in the cuprates, the magnitude of the T_c values observed for the new superconductors strongly points to a deficiency in the BCS model.

Another feature often quoted as a test of the BCS model is the dependence of T_c upon the isotopes present in the superconductor. There have been conflicting reports as to whether an isotope effect is present or absent in the new superconductors; however it should be noted that some elemental superconductors, *e.g.* Zr, do not show an isotope effect.

4 MODELS FOR HIGH TEMPERATURE SUPERCONDUCTIVITY

Probably the most contentious area of modern solid state physics is the search for a theoretical model to describe the pairing interaction in the new superconductors. A large number of different models have been proposed, although these can easily be reduced to a small number of basic interactions. Although these models are usually considered independently, it is likely that different interaction mechanisms will occur simultaneously and it might be expected that it is a composite interaction mechanism that leads to such high transition temperatures. It is not the purpose of this chapter to describe these proposed mechanisms in detail; however it is worthwhile to briefly review these.

A compositionally induced **antiferromagnetic insulator** to anomalous metal transition is generally associated with high temperature superconductivity. The proximity of this antiferromagnetism has led many theorists to predict a magnetic or spin pairing interaction giving rise to high temperature superconductivity. Others suggest that a **polaronic interaction**, whereby the carrier is coupled to a localized distortion of the lattice, is important, whilst others suggest some sort of ferroelectric interaction. A modified BCS model, with stronger correlations, is also still widely thought to be a possible solution.

There are various approaches that are used to address the theoretical problem. One is to assume that the carriers remain as fermions, with superconductivity involving a BCS-like coupling of fermions close to the Fermi edge at T_c. A second approach is to assume the pre-existence of bosons, such as **bipolarons** above T_c, superconductivity arising from a Bose–Einstein condensation at T_c. A third approach, as introduced by Anderson in his resonating valence bond model, is to postulate the dissociation of the carrier into spin-only (spinon) and charge-only (holon) quasi-particles.

5 TYPES OF SUPERCONDUCTOR

Superconductivity is a very special phenomenon which can be regarded as a special state of matter; however, it is by no means limited in its occurrence. Superconductivity is frequently observed when metals, especially poor metals, are cooled to very low temperatures. In this section the range of superconductors is briefly reviewed, not only to illustrate how extensive the phenomenon is, but also to reinforce the belief that several further classes of superconductors are yet to be found.

5.1 Elemental Superconductors

At least 27 of the elements in the Periodic Table are known to superconduct at low temperatures, without the application of high pressures. The phenomenon is largely restricted to the transition and p-block elements. Interestingly the elements which exhibit the best metallic properties, *e.g.* Cu, Ag, and Au, do not superconduct, whereas those with the poorest metallic properties exhibit the highest T_c values, *i.e.* **Pb** (7.2 K) and **Nb** (9.5 K). The transition to superconductivity is also generally absent where magnetic ordering is present; thus few of the first row transition metals show superconducting transitions.

5.2 Intermetallic Alloys

A great many binary alloys have been found to exhibit superconductivity, although at least one of the component elements is usually a superconductor in the pure state. The most important of these have all exhibited the body centred cubic structure known as **A-15**, which is analogous to beta-tungsten (W_3O), see Figure 5. Critical temperatures of 18.3 and 23.2 K were observed for the alloys **Nb_3Sn** and **Nb_3Ge** in 1955 and 1973 respectively. These alloys are **Type II superconductors**, exhibiting high upper critical fields (*e.g.* $H_{c2} = 30$ T for Nb_3Sn) and so have important technological roles, being used for high field superconducting magnets. This value of **23.2 K** was the highest critical temperature known before the advent of high temperature superconductivity in 1986. The term high temperature superconductor is frequently used to describe any superconductor with a T_c higher than this 23.2 K value, although the definition of a T_c higher than the maximum predicted by BCS theory (35 K) is also used.

An interesting feature of intermetallic alloys is the optimization of the superconducting transition temperature at certain valence electron to atom ratios, with maxima being observed in T_c for 4.7 and 6.5 valence electrons per atom.

5.3 Non-oxide Inorganic Compound Superconductors

Several nitrides, borides, and carbides have been observed to exhibit superconducting transitions in the range 10–15 K (*e.g.* ZrN, NbC, MoC, NbN, La_2C_3).

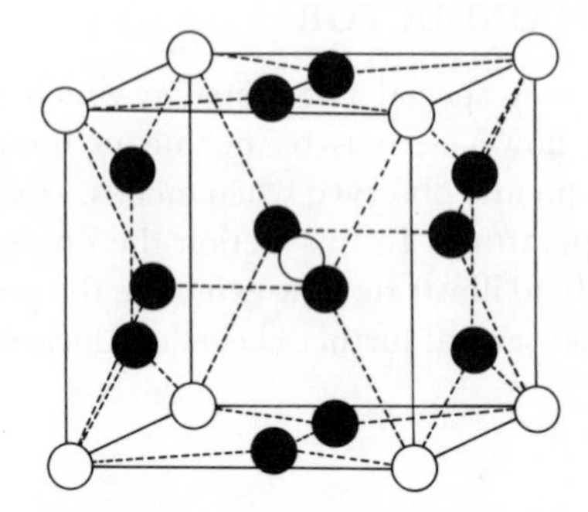

Figure 5 *The A-15 (Cr_3Si) crystal structure. Dark circles: Cr; white circles: Si*

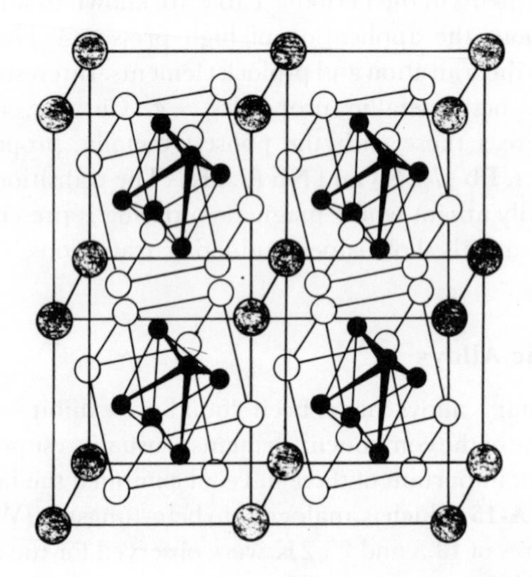

Figure 6 *The structure the Chevrel phase Pb(Mo_6S_8). Black circles: Mo; grey circles: Pb, and open circles: O*

$LuRh_4B_4$, which exhibits a T_c of 11.4 K is highly unusual as ferromagnetism and superconductivity coincide in the same system.

A particularly promising class of superconductor is the **Chevrel sulfides $M_xMo_6S_8$**, with T_c values of 15.2 K for M = Pb and 14.0 K for M = Sn. These materials exhibit very high upper critical fields, **50–60 T**, and appear to have possible high field applications. The structure of the Chevrel phase is based on Mo_6S_8 clusters, see Figure 6.

5.4 Carbon-based Superconductors

A large number of **organic charge transfer salts**, *e.g.* (BEDT–TTF)$Cu(SCN)_2$ have been found to superconduct at temperatures up to 13 K, although often only

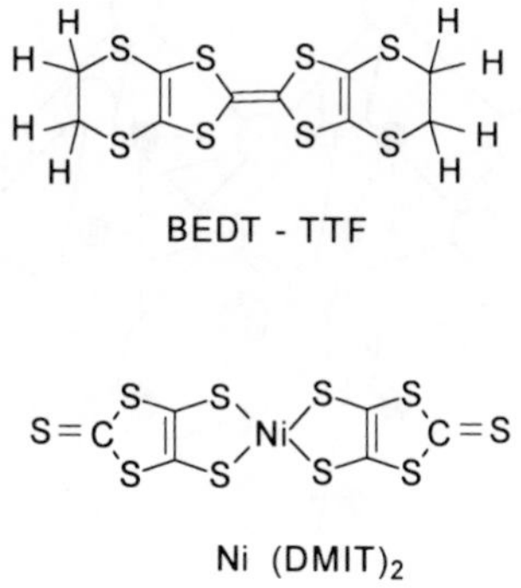

BEDT - TTF

Ni (DMIT)$_2$

Figure 7 *BEDT–TTF and Ni(DMIT)$_2$ ligands which form the conducting stacks in organic superconductors*

at high pressure. These are generally based on conducting cations such as BEDT–TTF or anions such as Ni(DMIT)$_2$, shown in Figure 7, which occur as homosoric stacks (*i.e.* the stacks all contain either the same cation or anion) in the crystal structure. These so-called **organic superconductors** are based on carbon sulfur complexes which are one- or two-dimensional conductors with the main conducting interaction via S–S intermolecular overlap.

There are also superconductors based solely on carbon. The first of these are two-dimensional graphite intercalation compounds, with T_c up to 2 K for NaC_8. More recently three-dimensional salts of the **fullerene C_{60}** have been found to superconduct. Mostly these have been **M_3C_{60} salts**, with structures based on cubic close packed arrays of C_{60} with all the octahedral and tetrahedral interstices filled. The first to be discovered was the potassium salt, K_3C_{60}, with a T_c of 18 K. The highest transition temperature, **32 K**, was observed for Rb_2CsC_{60}. Since these discoveries a further C_{60} salt, Ba_6C_{60}, with a body-centred cubic structure was found to superconduct at 7 K.

5.5 Oxide Superconductors

Superconductivity is far from uncommon in oxides; however, the phenomenon is associated with only a limited number of elements and structure types. In general, oxide superconductors exhibit a **rocksalt, spinel**, or **perovskite** structural type and contain at least one of the following elements: Cu, Bi, Pb, Ti, Nb, W, Mo, and Re.

Of the binary oxides, only **$Ti_{1-x}O$** and **NbO**, which have the rocksalt structure, are known to superconduct and even these have transition temperatures of less than a few Kelvin. It is particularly interesting to note that no copper oxide superconducts, despite the renowned importance of the CuO_2 layers in the high temperature superconductors. In the spinel structural family, **$LiTi_2O_4$** was found to superconduct up to **13 K** in 1973, which was the highest transition temperature to be reported for an oxide before the advent of high temperature superconductivity.

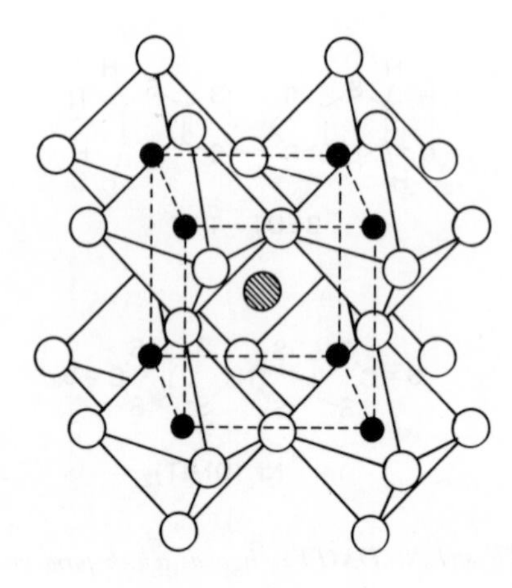

Figure 8 *The unit cell of the perovskite $CaTiO_3$, showing corner-sharing TiO_6 octahedral units*

5.6 Perovskites

The perovskite structure is by far the most prevalent structure amongst the oxide superconductors; the basic structure is illustrated in Figure 8. It can be seen that the perovskite structure is based on corner-shared **BO_6 octahedral units**. This is a very important feature because the shortest interatomic distance between nearest neighbour B atoms is significantly more than twice the atomic radius of B. As a consequence of this separation, it is necessary that the conduction process involves electrons moving from B atom to B atom via the oxide ion, thus the conduction band is of mixed metal and oxygen character. This is fundamentally different from the rocksalt or spinel structures where the MO_6 octahedra share edges and a conduction band arising almost entirely from a single species is quite feasible. In fact the perovskite family is the only clear example of **true compound superconductivity**. In all other systems the conduction process can be seen as being dominated by the orbitals of a single atomic species, whereas in the perovskites, geometry dictates that both metal and oxygen orbitals are involved in the conduction band.

The simplest perovskite to superconduct is the slightly oxygen deficient $SrTiO_{3-x}$, which has a transition temperature of 0.6 K. A large number of **tungsten bronze** ($A_{1-x}BO_3$) compounds of W, Mo, and Re have been found to superconduct at temperatures up to **6.5 K**. In this structural system superconductivity has been observed in cubic, tetragonal, and hexagonal variants, although the transition temperatures associated with the parent, cubic structural type are not as high.

The highest critical temperatures observed in non-cuprate perovskite oxides are for those with BiO_6 (and PbO_6) octahedral units. These systems are based on the end member **$BaBiO_3$** and **$BaPbO_3$** perovskites. In terms of simple band theory, an oxide with one unpaired electron per formula unit, such as $BaBiO_3$, would be expected to be metallic, whereas an oxide with no unpaired electrons per formula unit, such as $BaPbO_3$, would be expected to be an insulator. In fact the converse is

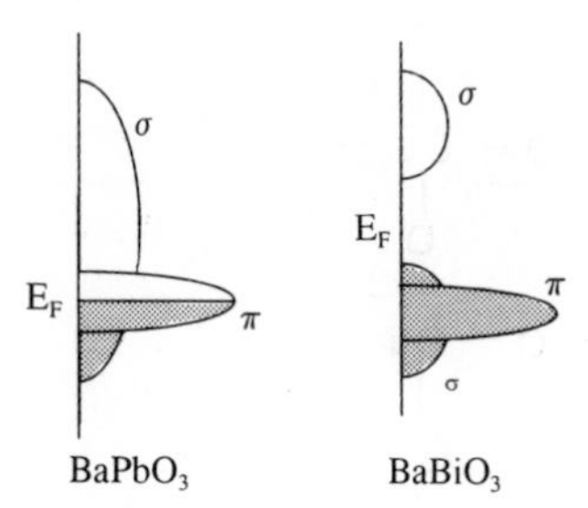

Figure 9 *Schematic band structures for $BaPbO_3$ and $BaBiO_3$*

true, $BaBiO_3$ is an insulator, $BaPbO_3$ is metallic. $BaBiO_3$ is what is known as a **charge transfer insulator**. Its structure is thought to consist of alternating Bi^{3+} and Bi^{5+} units; at room temperature it is a monoclinically distorted variant of the perovskite structure.

In band terms the charge disproportion is thought to split the band arising from the Bi 6s orbitals, rendering the lower half of this σ band fully occupied (Figure 9, right). $BaPbO_3$ cannot undergo charge disproportion as Pb^{IV} is a stable valence state; however, it does have a very slight distortion from the ideal perovskite unit cell, resulting in an orthorhombic unit cell. $BaPbO_3$ would be expected to be an insulator like $BaSnO_3$; surprisingly it is metallic at room temperature. The explanation for this is probably that the Pb 6s and the O 2p levels overlap to form a hybridized, partially occupied band at the Fermi Level, see Figure 9, left.

The two principal superconducting bismuthate perovskites are **$Ba(Bi_{1-x}Pb_x)O_3$** and **$(Ba_{1-x}K_x)BiO_3$**. $Ba(Bi_{1-x}Pb_x)O_3$ was discovered to superconduct up to 12 K in 1975. This was about the same time as the discovery of $LiTi_2O_4$, the only other oxide to superconduct to temperatures above 10 K before the advent of high temperature superconductivity. $Ba(Bi_{1-x}Pb_x)O_3$ exhibits a similar orthorhombic distortion to $BaPbO_3$, with its T_c maximum occurring at $x=0.75$. As $Ba(Bi_{1-x}Pb_x)O_3$ contains a Pb : Bi ratio of 3 : 1 in its optimum composition, it is perhaps better to regard it as a plumbate superconductor rather than a bismuthate; indeed $Ba(Sb_{1-x}Pb_x)O_3$ exhibits superconductivity up to 3.5 K for $x=0.75$.

$(Ba_{1-x}K_x)BiO_3$ undergoes a metal/insulator transition at $x=0.35$, transforming to a cubic structure. This cubic form of $(Ba_{1-x}K_x)BiO_3$ undergoes a metal to superconductor transition on cooling, exhibiting a T_c of 34 K for $x=0.4$. Importantly, this is the highest T_c exhibited by an isotropic material to date. In these systems it is thought suppression of the charge disproportionation in $BaBiO_3$ is the key to inducing superconductivity, by doping. Certainly this seems reasonable for $(Ba_{1-x}K_x)BiO_3$; however, $Ba(Bi_{1-x}Pb_x)O_3$ seems much closer to $BaPbO_3$ than $BaBiO_3$.

6 CUPRATE SUPERCONDUCTORS

6.1 Crystal Chemistry

Since the discovery of superconductivity at temperatures up to 35K in $(La,Ba)_2CuO_4$, by Bednorz and Muller in 1986, a large number of superconducting

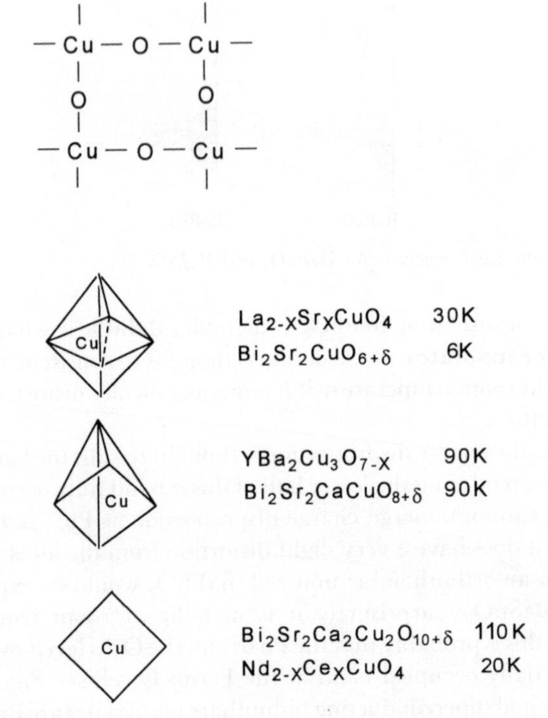

Figure 10 *Coordination environment of Cu in some High T_c oxides*

cuprates have been discovered. Although superconductivity has been observed in a broad range of structures, the phenomenon is only seen in structures which can be described as **oxygen deficient perovskites** or as **perovskite intergrowths**. The key structural feature in all of these systems is the presence of a **CuO_2 plane** or series of CuO_2 planes. The coordination environment of Cu in these planes can be octahedral, square pyramidal, or square (Figure 10). In general, only systems with Cu in 6-coordination exhibit the lowest transition temperatures.

The first material observed to exhibit a critical temperature **in excess of 77 K**, the boiling point of liquid nitrogen, was **$YBa_2Cu_3O_{7-x}$**, an oxygen deficient perovskite. The structure of $YBa_2Cu_3O_7$ can be simply derived from that of a triple perovskite unit, $3ABX_3$, by removing all of the oxygens from the yttrium layer and half of the O(5) oxygens from the basal CuO_2 plane (Figure 11). The only other non-intergrowth perovskite cuprates to superconduct are the infinite layer phases based on $SrCuO_2$ discovered in 1991. These phases which are highly defective are prepared at very high pressures. There are two important examples: **$Sr_{0.85}Nd_{0.15}CuO_2$** is an n-type superconductor with a T_c of **40 K**, and **$(Sr_{0.6}Ca_{0.4})_{0.9}CuO_2$** is a p-type superconductor with a T_c of **110 K**.

The vast majority of high T_c cuprates can be regarded as intergrowth phases between perovskite and either rocksalt or fluorite blocks. The most important of these are the **Ruddlesden–Popper phases** of formula **$(AO)_m[ABO_3]_n$**, which are rocksalt perovskite intergrowths. The two types of intergrowth are well

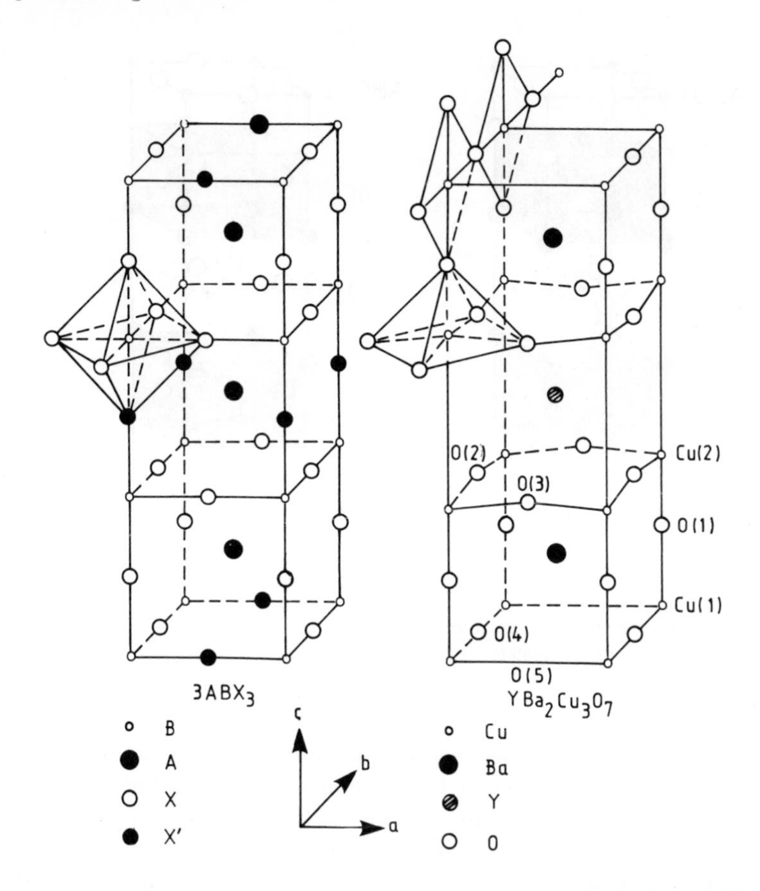

Figure 11 *Relationships between the $YBa_2Cu_3O_7$ and the triple perovskite ($3ABX_3$) unit cells*

illustrated by the Ln_2CuO_4 structures for Ln = La, Nd. **La_2CuO_4** exhibits the K_2NiF_4 structure, which is the Ruddlesden–Popper phase for $m = n = 1$ (Figure 12a). With a smaller lanthanide, *e.g.* Nd, a slightly different structure, which can be viewed as a fluorite–perovskite intergrowth, is found (Figure 12b). La_2CuO_4 doped with Sr is a **p-type** superconductor with $T_c = 38$ K, whereas **Nd_2CuO_4** is an **n-type** superconductor with $T_c = 24$ K.

The most important of the Ruddlesden–Popper phases are those based on single or double TlO, PbO, BiO, and HgO rocksalt layers. These belong to two ideal structural families where the indices relate to cation composition:

12[$n-1$][n], *e.g.* **$TlBa_2CaCu_2O_{7.5}$** is 1212 where $n = 2$
22[$n-1$][n], *e.g.* **$Bi_2Sr_2CaCu_2O_8$** is 2212 where $n = 2$

These series contain the highest T_c phases discovered to date, including $Bi_2Sr_2Ca_2Cu_3O_{10}$, $T_c = 110$ K; $Tl_2Ba_2Ca_2Cu_3O_{10}$, $T_c = 128$ K; and $Hg_1Ba_2Ca_3Cu_4O_{10}$, $T_c =$ **140 K** (156 K at 25 GPa). The members of the Bi and Tl double rocksalt layer series of phases are illustrated in Figure 13.

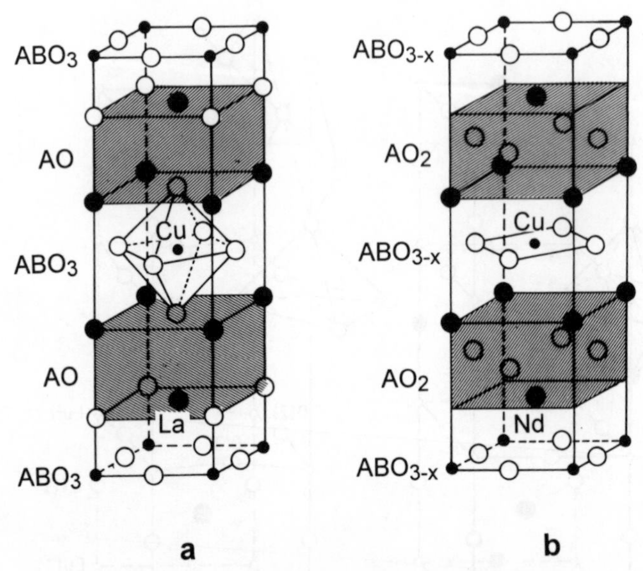

Figure 12 *Idealized structures of (a) La_2CuO_4 and (b) Nd_2CuO_4 showing perovskite/rocksalt and perovskite/fluorite intergrowths, respectively*

It should be emphasized that the ideal stoichiometries are rarely found; oxygen non-stoichiometry and cation intersite substitution are the norm. For example in the $Bi_2Sr_2Ca_{n-1}Cu_nO_{4+2n}$ series (Figure 13), the 2201 phase cannot be prepared in that ideal cation stoichiometry and the 2212 phase is much more difficult to prepare in 2212 cation stoichiometry than strontium-deficient phases with the same structure. Additional structural deviations arise from **supercell modulations**; these are particularly prevalent in Bi-based systems. It seems likely that this modulation arises from the lone pair on Bi, as almost identical modulation wavelengths are observed in other intergrowth structures containing Bi_2O_2 rocksalt layers such as $Bi_2O_2CO_3$.

6.2 Doping Mechanisms

The properties of high temperature superconductors largely depend upon carrier concentration, which is controlled by chemical doping. The mechanisms of doping cuprate superconductors are quite similar to those for a semiconductor such as Si (Equations 2 and 3, h^+ = positive hole):

$$Si^{4+} \rightleftharpoons Ga^{3+} + h^+ \quad (2)$$

$$Si^{4+} \rightleftharpoons P^{5+} + e^- \quad (3)$$

There are two principal methods of doping high temperature superconductors, **oxygen non-stoichiometry** and **aliovalent substitution**, although both

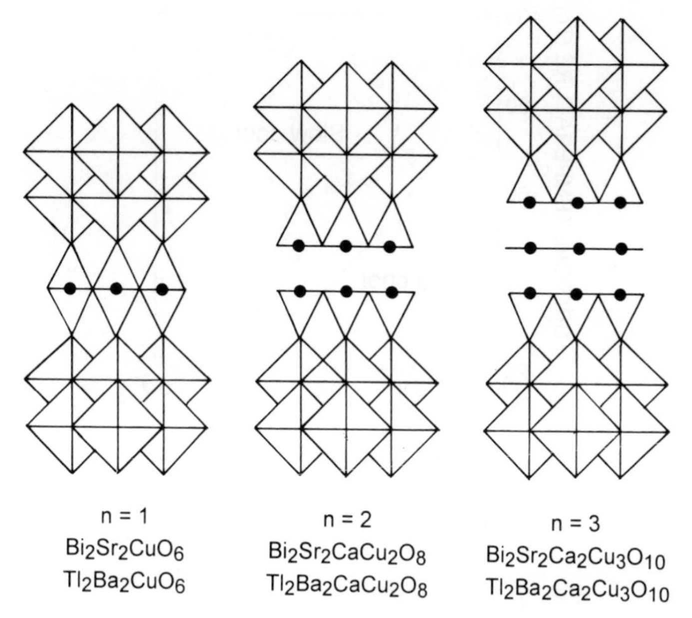

Figure 13 *Structures of phases in the ideal homologous series* $Bi_2Sr_2Ca_{n-1}Cu_nO_{4+2n}$ *and* $Tl_2Ba_2Ca_{n-1}Cu_nO_{4+2n}$*: the 22[n−1][n] family*

methods often operate in tandem. In $YBa_2Cu_3O_{7-x}$, carrier concentration is predominantly controlled by oxygen non-stoichiometry. The doping mechanism is a redox process involving uptake of atmospheric oxygen into the lattice with removal of electrons from the electron energy bands creating holes (Equations 4 and 5):

$$2\,Cu^{2+} + \tfrac{1}{2}O_2 \rightarrow 2\,Cu^{3+} + O^{2-} \quad (4)$$

$$Cu^{3+} \rightarrow Cu^{2+} + h^+ \quad (5)$$

It is usual for the chemist to consider doping in terms of nominal Cu valence; however, no assumption that the carriers reside on Cu is implied. In $YBa_2Cu_3O_{7-x}$ the critical temperature is found to decrease smoothly as the oxygen content is decreased for stoichiometries prepared at high temperature and rapidly cooled; for samples prepared at lower temperature, or slow cooled, plateaux are observed in plots of T_c versus composition (Figure 14). The plateaux behaviour, which is commonly encountered, is associated with oxygen ordering in the basal $YBa_2Cu_3O_{7-x}$ plane. From electron diffraction studies it appears that this ordering takes the form of alternating rows where the O(4) site (see Figure 11) is either fully occupied or empty.

In La_2CuO_4 oxygen non-stoichiometry generally is less important than aliovalent substitution. Here doping entails replacement of La by Sr with the creation of an electron hole (Equation 6). In this system, oxygen stoichiometry

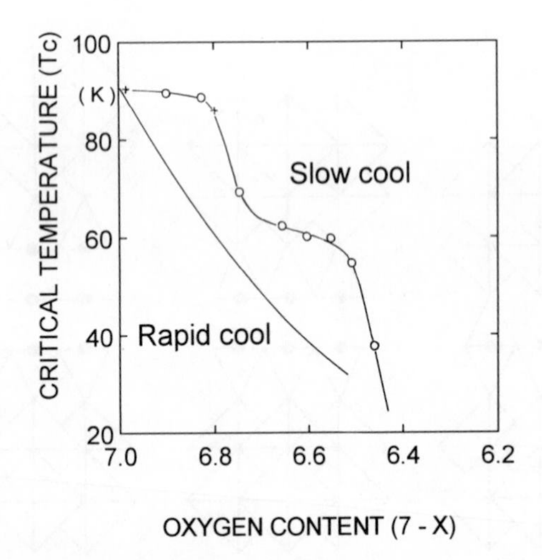

Figure 14 *Relationship between* T_c *and oxygen stoichiometry of* $YBa_2Cu_3O_{7-x}$

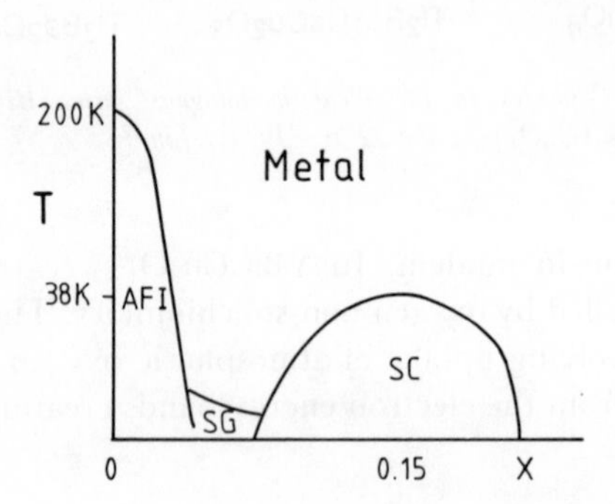

Figure 15 *Phase diagram of* $La_{2-x}Sr_xCuO_4$. *(AFI = antiferromagnetic insulator, SG = spin glass, SC = superconductor)*

remains, more or less, constant, thus the number of holes is proportional to the extent of substitution.

$$La^{3+} \rightleftharpoons Sr^{2+} + h^+ \qquad (6)$$

The effect of doping La_2CuO_4 with Sr is illustrated in Figure 15. On increasing hole concentration, La_2CuO_4 transforms from an antiferrogmagnetic insulator to a metal at high temperatures. At lower temperatures the transformation is from an antiferromagnetic insulator to a spin glass (a magnetically dilute state) to a superconductor to a metal.

Aliovalent substitution with a cation of lower charge does not always lead to increased carrier concentration; sometimes charge neutrality is maintained by loss of oxygen. For example, Sr vacancies in $Bi_2Sr_2Ca_{n-1}Cu_nO_{4+2n}$ are primarily compensated for by oxygen loss rather than hole creation.

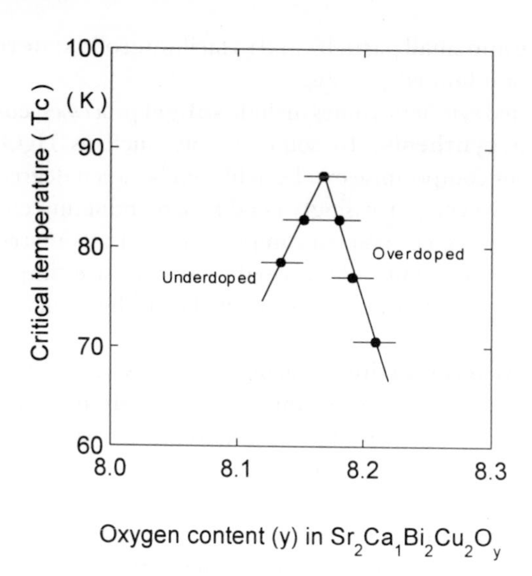

Figure 16 *Superconducting critical temperature as a function of oxygen content for $Bi_2Sr_2CaCu_2O_y$ (y=8+x)*

In Nd_2CuO_4, n-type doping is achieved by replacement of Nd by an ion of higher charge, Ce^{4+}. As no significant additional oxygen is taken into the structure, the charge is compensated by **electron creation** (Equation 7). In $Nd_{2-x}Ce_xCuO_4$, a maximum in critical temperature is observed at doping levels of $x=0.15$.

$$Nd^{3+} \rightleftharpoons Ce^{4+} + e^- \tag{7}$$

Superconducting critical temperature almost invariably shows a maximum at a particular level of doping, as is observed for $La_{2-x}Sr_xCuO_4$ in Figure 15. In $Bi_2Sr_2CaCu_2O_{8+x}$ this maximum is observed at an oxygen excess of about 0.17 (Figure 16). A number of workers have suggested that this maximum in T_c is actually a manifestation of a changeover in the conduction mechanism and refer to **overdoped** and **underdoped** regions, see Figure 16. In $YBa_2Cu_3O_{7-x}$, a maximum in T_c is not generally observed; however this seems to be because the maximum in critical temperature for unsubstituted $YBa_2Cu_3O_{7-x}$ occurs close to the highest oxygen stoichiometry that can be achieved at ambient pressure.

7 SYNTHESIS AND CHARACTERIZATION

The majority of the high T_c superconductors are ceramics; the basic method of synthesis is by solid state reaction. Appropriate amounts of oxides are heated together at temperatures high enough for reaction, but too low to suffer from volatility of reactants: typically 900–1000 °C. Frequently carbonates or nitrates are used instead of oxides, especially for oxides that readily carbonate or hydrate when handled in air. Precursors such as carbonate offer an additional advantage as they decompose into small particles and so facilitate solid state reaction, which is primarily a diffusion limited process.

More sophisticated synthetic routes include **sol-gel** processes, **co-precipitation**, and **combustion synthesis**. In some systems such as HgO or TlO based cuprates only a poor compromise can be achieved between degree of reaction and loss of reactants; however, surprisingly good superconducting characteristics can be achieved in incompletely reacted samples. As well as sintered ceramics, thin films, melt-textured ceramics, and single crystals are important for both fundamental and applied studies. More detailed consideration is given in the cited references.

Structural characterization is generally by X-ray diffraction, although neutron and electron diffraction techniques are widely used to supplement the information from X-ray diffraction. Electron probe microanalysis is a widely used technique for confirmation of cation homogeneity, or determination of the composition of phases in a phase mixture. Carrier contents can be determined either by thermogravimetry or chemical titration, although information from the two techniques is often complimentary. Basic characterization of superconducting properties is normally by critical temperature determination. Dual magnetic and electronic determination of transition temperature is instructive and is essential if superconductivity in a new system is to be proved. **Electrical** characterization is by four-terminal dc, or low frequency ac, whereas **magnetic** characterization is usually by either ac susceptibility or SQUID magnetometry.

8 APPLICATIONS

8.1 Introduction

Over one hundred new oxide superconductors with a T_c in excess of **20 K** have been discovered to date. Although the phenomenon is very new, it is surprising that the technological potential has only been assessed and optimized for a very small fraction of these. It is already very clear that no single material will be able to fulfil the wide-ranging demands of the various applications. Different physical and materials characteristics are required for different applications. To date, the most important materials for applications have been:

$YBa_2Cu_3O_7$ – **YBCO** or **'1,2,3-compound'**

$\left.\begin{array}{l} Bi_2Sr_2CaCu_2O_{8+x} \\ Bi_2Sr_2Ca_2Cu_3O_{10+x} \end{array}\right\}$ **BISCCO**

$(Ba, K)BiO_3$

Nb_3Sn and other A-15 alloys

The most important physical characteristics are critical current, J_c, and its field dependence. The requirements for various applications are summarized in Table 1 and illustrated in Figure 17, which also shows the best performance levels achieved before 1993. As can be seen the best critical currents obtained with thin

Table 1 *Target performance levels for superconductor applications*

Application	*Critical current* ($A\,cm^{-2}$)	*Critical Field* (Tesla)
Current transmission	10^8–10^9	not a limitation
Magnetic levitation	10^5	5
High field magnets	10^6	10
Superconducting magnetic energy storage (SMES)	10^6	20
Superconducting electronics	—	—

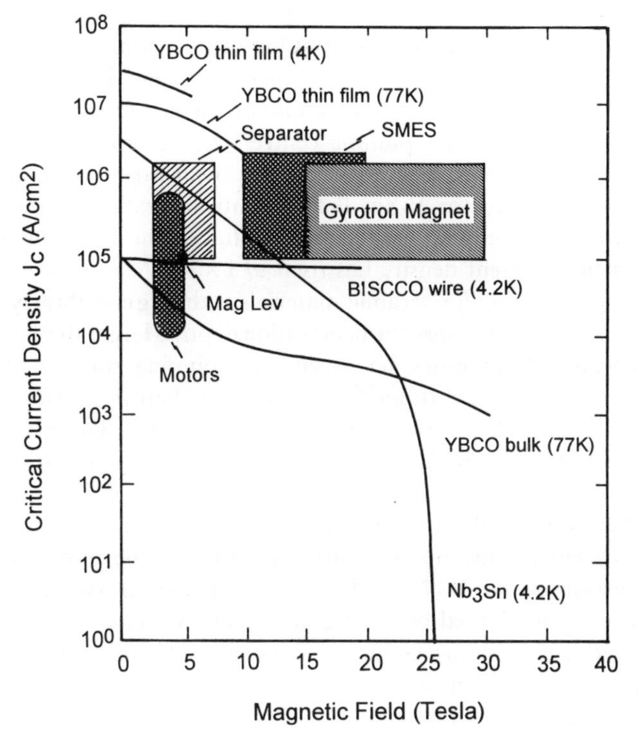

Figure 17 *Relationship between critical current and magnetic field, showing both requirements for specific applications and performance levels achieved*

films of YBCO are in excess of those for conventional superconductors at low fields and BISCCO shows little degradation in J_c with increasing magnetic field.

8.2 Critical Currents

The upper limit for critical current is where the carriers become depaired. Ginzberg–Landau theory shows that this limit depends on the inverse of the penetration depth, λ. If the particle size is similar to the penetration depth then the cross-sectional area of the superconductor is fully utilized at the critical current transition and the maximum possible value for critical current, 10^8–$10^9\ A\,cm^{-2}$, can be achieved. The best value achieved for YBCO is within two orders of magnitude of this maximum.

The materials that show the greatest potential for applications have tended to be Type II materials, because Type II materials tend to exhibit higher T_cs and generally have large penetration depths. When we consider the effect of magnetic field upon critical current, we are normally in the regime between lower (H_{c1}) and upper (H_{c2}); thus the superconductor will be penetrated by lines of flux. In a perfect lattice, the lines of flux would move in response to any applied electric field; thus a resistance would be observed. In a real lattice the lines of flux tend to be trapped at **defects** or **inclusions**; thus zero resistance can, and does, persist beyond H_{c1}.

Thus for high critical currents, the most important parameter is the **pinning density**, which tends to be high in oxide superconductors. In a material such as YBCO, $YBa_2Cu_3O_7$, there are often inclusions of $\mathbf{Y_2BaCuO_5}$ within the grains or crystals, which act as effective pinning centres. An example of how increasing pinning density can improve critical current is neutron irradiation of the conventional superconductor V_3Sn. Before neutron irradiation there is a low defect density and J_c is only 1 A cm^{-2}, after irradiation the defect density is much higher and critical current density has risen to 1×10^6 A cm^{-2}.

A further consideration for ceramic materials is their **granularity**. In Nb_3Sn granularity is beneficial, because the penetration depth is larger than the grain size and thus the grain boundaries act as effective pinning sites. In the ceramic cuprates, the penetration depth tends to be smaller than the grain size thus the grain boundaries act as *weak links* with low critical currents. These weak links are Josephson junctions: that is they involve tunnelling of superelectrons from one superconductor to another superconductor through an insulating barrier (see Section 8.4 for more detailed explanation). As a consequence of this granularity, there are two critical currents, one **inter-grain** and the other **intra-grain**. Magnetic measurements of J_c tend to give intra-grain values, whereas dc measurements are dominated by inter-grain effects. Weak link problems can be lessened by better ceramic processing to remove insulating impurities from the grain boundaries and to improve grain–grain alignment.

In YBCO ceramics weak links are the major obstacle to improving J_c; however, in BISCCO the problem lies in **flux creep**. The BISCCO materials are particularly two-dimensional; thus they tend to form flat platelike crystals. The short edge of the crystals, which is parallel to *c*, is about 100 times smaller than the larger crystal edges, which are in the *ab*-plane. The crystals are, thus, easy to process and grain alignment is good. Unfortunately the two-dimensionality of BISCCO also modifies its flux lattice; instead of lines of flux threaded through the crystal, or **flux vortices**, the BISCCO lattice comprises of **flux pancakes**, because there is poor coupling between superconducting layers. As these flux pancakes are restricted to a much smaller volume than the flux vortices, pinning is much poorer.

The improvement of critical currents is very much a problem of materials processing. The requirements are **high density ceramics**, **good grain alignment**, and fabrication into **wires** or **tapes**. These are very demanding targets, which are not necessarily compatible with one another. In BISCCO the most promising route involves mixing with silver powder, milling, extruding, rolling, and firing. The crystallite form facilitates grain alignment and the product produced at sintering temperatures, 800–850 °C, has close to optimal T_c. In YBCO the most promising fabrication route is via **polymer processing** of the green, unsintered, material. After initial sintering, density can be improved by **melt texturing** and well aligned, fully dense ceramics have been obtained. One particular problem for YBCO is that at sintering temperatures oxygen content is well below optimal. A post-anneal at 400 °C in oxygen is necessary to regain full oxygen stoichiometry; however, oxygen diffusion through a dense ceramic is difficult at these temperatures, so long term annealing is often required.

In addition to YBCO and BISCCO, there is a growing interest in a number of other materials for high critical current applications. Particular interest has centred on **Ruddlesden–Popper phases** similar to BISCCO, but which contain only a single rocksalt layer, especially those based on thallium.

8.3 High Current Applications

A large number of high current applications have been predicted for high temperature superconductors. Many of these may seem fanciful – certainly fabrication of suitable materials will stretch the ability of materials scientists and engineers – however, some of these high performance applications are already close at hand.

The most obvious application is the **transmission of electric power**. There are good reasons for optimism in this application as both a need and possible materials can be identified. The main power arteries of some large cities such as Tokyo are unlikely to be able to cope with future demands and will need to be replaced in the not too distant future. BISCCO wires of significant length, which can sustain higher critical currents at 20 K than conventional superconductors at 4 K, have been fabricated by Sumimoto. A related application is SMES, **Superconducting Magnetic Energy Storage**, which has been considered for **load levelling** of electric power. It is proposed to use large (km-scale) superconducting rings or helices to store energy.

It is also proposed to use superconductors in large motors to avoid large losses of energy to heat. Boats have been propelled using superconducting magnets and the film 'Hunt for Red October' was based on a futuristic submarine propelled in this manner. Similarly it has frequently been proposed to **levitate trains** using Meissner-type phenomena. Superconductors would be placed in the train and eddy currents induced in the aluminium track would provide the magnetic field. A more exotic proposal is to launch rockets into space using the Meissner effect.

The force exerted in levitating a superconductor can be quite large and it is has been suggested that superconductors could be used in **structural load bearing**. The possibility of using superconducting **frictionless bearings** is certainly attractive. High field magnets based on superconductors could be used for magnetic separation processes in refining (see Chapter 2). Such a process is unlikely to be commercially viable for terrestrial application, as the conventional technology is already installed and works adequately. If lunar mining does become a possibility, then a superconducting technology would be viable, not least because liquid hydrogen used for rocket fuel is stored at 20 K.

The one area where superconductivity is already highly commercially successful is high field magnets, particularly for **magnetic resonance imaging** (MRI). This technique has only recently become clinically available and is clearly a major medical technique of the future. Present-day magnets utilize Nb_3Sn and similar A-15 superconductors. These function well and are unlikely to be replaced by high T_c oxides, although there does seem to be an important niche for high T_c superconductors as **current leads** to the conventional superconductors. An opening may arise as the next generation of higher field magnetic resonance systems are developed. High T_c oxides such as **BISSCO** or materials such as the **Chevrel sulfides** have the potential to operate to much higher magnetic fields.

8.4 Microwave Devices, the Josephson Junction, and SQUIDs

In the short term the most important applications for high T_c oxides seems to be in microelectronic and electronic devices. There are two types of application: passive and active. In passive microwave devices the superconductor is used to form **microwave cavities, antennae, strip lines, and filters**. The most important property of a superconductor for passive devices is its surface impedance, which is related to skin depth in metals. High temperature superconductors have lower losses than copper at 77 K up to 10 GHz, although low T_c materials do perform better: *e.g.* epitaxial films of Nb_3Sn at 15 K.

The operation of most active devices is based on the **Josephson Junction**. The Josephson Effect relates to a current passing from one superconducting region to another superconducting region via an insulating barrier. If the barrier is sufficiently thin the superelectrons can tunnel through.

The current passing through the insulating layer is given by

$$I = I_0 \sin(\theta_2 - \theta_1) \tag{8}$$

where I_0 is the maximum supercurrent that can flow and $\theta_2 - \theta_1$ is the difference in phase of the macroscopic wave function across the junction. If we consider a Josephson junction in a superconducting ring (Figure 18), it is possible to show that the current through the junction is given by Equation 8, where

$$\theta_2 - \theta_1 = \frac{2\pi\Phi}{\Phi_0} \tag{9}$$

This relation shows that the current through the junction depends only on the magnetic flux threading the ring, Φ, and the value of the flux quantum, $\Phi_0 = h/2e$. An important aspect of this relationship is that the flux quantum depends on $2e$, the charge of the superconducting electron pair. Measurements on Josephson junctions made from high T_c superconductors have been used to prove carrier pairing in the high T_c cuprates.

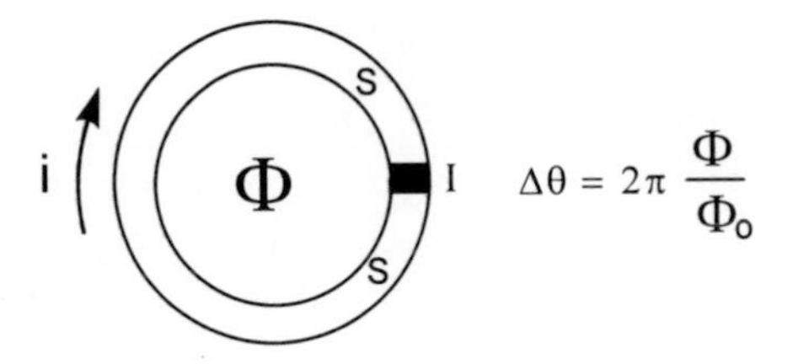

Figure 18 *Josephson Junction in a superconducting ring (S = superconductor, I = insulator)*

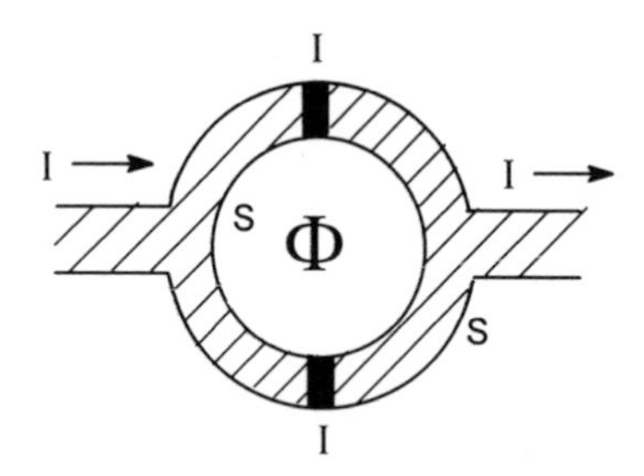

Figure 19 *Superconducting Quantum Interference Device (SQUID)*

The dc **SQUID**, Superconducting Quantum Interference Device, is made from a superconducting ring containing two Josephson junctions (Figure 19). This device can be fabricated from low or high T_c superconductors, although it is best as a thin film device. dc SQUIDs can be used to detect very small magnetic fields, down to 10^{-14} Tesla. As a consequence of this great sensitivity, SQUIDs show considerable potential for medical applications such as **electrocardiograms** and **brain wave studies**, as well as other applications such as submarine detection and corrosion monitoring. High T_c SQUIDs work well at low frequencies and some devices based on high temperature superconductor SQUIDs are already on the market.

Josephson effect devices may also be able to offer a basis for a **new microelectronics**. Such devices should be able to, at least, match conventional semiconductor devices. The prime example of this is the **Josephson computer** which was first developed by IBM, based on lead, in the 1970s. Such a computer would have faster switching speeds than one based on semiconductor technology and power dissipation would be lower; thus it would potentially be both smaller and faster. The computer memory consisted of an array of SQUIDs: if a flux vortex was present in a SQUID then the logic was 1, if no flux vortex then it was 0. The project was discontinued as the lead films were unstable to thermal cycling and the minimum dimension per flux quantum was 1 micron, which can now be achieved in conventional semiconductor devices. The small coherence length of high T_c superconductors would certainly facilitate smaller Josephson junctions; however, most high T_c materials give poor Josephson characteristics. This is largely due to their anisotropic nature and because in general the junction is SNS rather than SIS: the barrier is in the normal state rather an insulator. The recent

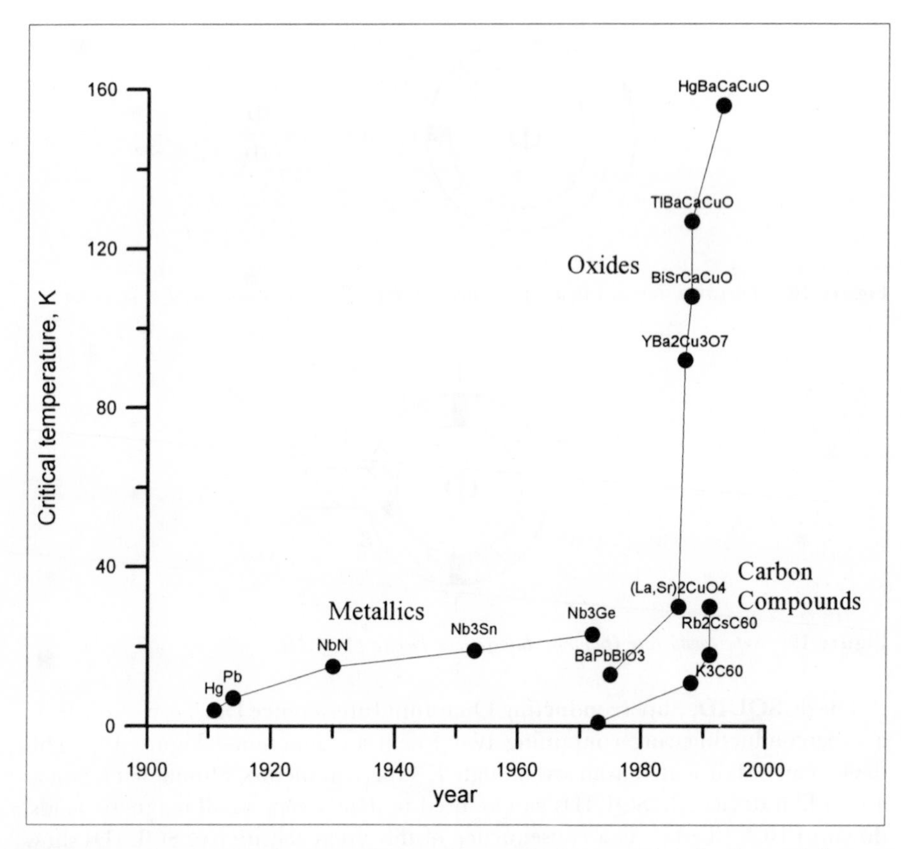

Figure 20 *Progress regarding the availability of superconducting materials during the twentieth century*

success in forming high quality Josephson junctions from thin films of the cubic perovskite (Ba,K)BiO_3 has rejuvenated work on the Josephson computer, however.

9 CONCLUSIONS AND FUTURE TRENDS

Superconductivity has been one of the most exciting areas for the development of new science and technology over the last seven or eight years, and this trend is likely to continue. An indication of progress over the years, and the extent to which recent developments have increased the rate of progress with respect to the availability of high temperature superconductors, is given in Figure 20. Although the size of the vast mountain of literature produced each year has subsided, the quality of output has greatly increased. Major developments still make the news, although much of the hype has subsided. There are still many important goals to achieve, and these include an **isotropic superconductor above 77 K**, higher critical currents, better wires, more powerful magnets, more sensitive SQUIDs, and even the achievement of **room temperature superconductivity**. An additional very important aspect is that a rational explanation for the **phenomenon of high temperature superconductivity** is still required.

If good progress is to be maintained, then the cross-disciplinary approach must continue. There are several areas where significant contributions are required from chemists, not only in synthesis and processing, but also in studying the relationships between composition, structure, and properties. Important targets are to optimize critical temperatures and currents in the cuprates and to develop a greater understanding of the exact relationships between stoichiometry, carrier concentration, and superconducting properties. The study of the non-cuprate systems will assist in the development of our understanding of the mechanism of high temperature superconductivity, and in the search for isotropic high temperature superconductors for superconducting microelectronic applications. Better synthetic routes to materials with improved J_cs are required, and these in turn will lead to better quality superconducting wires.

With regard to applications, as discussed above, there is likely to be a host of new uses demonstrated for high temperature superconductors. These developments will make a particular impact on methods for power transmission and storage, clinical medicine, and microwave technology. High temperature superconducting materials are also likely to make a major impact in the areas of space exploration and exploitation, transport, and information processing and storage. A 'superconducting economy' may not be too far distant!

10 REFERENCES

10.1 Specific References

2 The Phenomenon

2.1 C. Kittel, 'Introduction to Solid State Physics', Wiley, New York, 6th edn, 1986.
2.2 D. R. Tilley and J. Tilley, 'Superfluidity and Superconductivity', The Institute of Physics, Bristol, 3rd edn, 1990.
2.3 J. R. Waldram, in reference 1.1 (Section 10.2).

3 BCS Theory

3.1 J. Bardeen, L. N. Schrieffer, and J. R. Schrieffer, *Phys. Rev.*, 1957, **108**, 1175.

4 Models for High Temperature Superconductivity

4.1 See chapters by G. A. Sawatzky, N. F. Mott, V. J. Emery, T. M. Rice, and W. A. Goddard III, in reference 1.1 (Section 10.2).
4.2 M. F. Gunn, *Phys. World*, 1988, **1**, 31.

5 Types of Superconductor

5.1 B. L. Chamberland, in reference 1.2 (Section 10.2).

(Oxide Superconductors)

5.2 Sleight, in reference 1.1 (section 10.2).
5.3 Norton and Cava, in reference 1.2 (Section 10.2).
5.4 A. R. Armstrong and P. P. Edwards, 'Chemistry of Superconducting Oxides', in *Annual Reports on the Progress of Chemistry, Section C, Physical Chemistry*, 1991, **88**, 259 (published in 1993 by The Royal Society of Chemistry, UK).

6 Cuprates

6.1 J. G. Bednorz and K. A. Muller, *Z. Physik, B*, 1986, **64**, 189.
6.2 References 1.2 and 1.3.

7 Synthesis and Characterization

7.1 A. R. West, 'Solid State Chemistry and its Applications', Wiley, Chichester, 1984.
7.2 D. Segal, 'Chemical Synthesis of Advanced Ceramic Materials', Cambridge University Press, Cambridge, UK, 1989.

8 Applications

8.1 C. E. Gough and A. M. Campbell, in reference 1.1 (Section 10.2).
8.2 D. J. Connolly, R. R. Romanofsky, P. Aron, and M. Stan, *Appl. Supercond.*, 1993, **1**, 1231.
8.3 R. F. Giese, T. P. Sheahen, A. M. Wolsky, and D. K. Sharma, *IEEE Trans. Energy Conv.*, 1992, **7**, 589.
8.4 N. M. Alford, *Chem. Ind. (London)*, 1992, 809.

10.2 General References

1.1 'High Temperature Superconductivity: Materials, Mechanisms and Devices', ed. D. P. Tunstall and W. Barford, The Institute of Physics, Bristol, 1992.
1.2 'Chemistry of High Temperature Superconductors', ed. T. A. Vanderah, Noyes Publications, New Jersey, 1991.
1.3 'Chemistry of High Temperature Superconductors', ed. C. N. R. Rao, World Scientific, Singapore, 1991.
1.4 C. Gough, 'Challenges of High-T_c', *Phys. World*, 1991, **4(12)**, 26.
1.5 R. J. Cava, 'Superconductors beyond 1-2-3', *Sci. Am.*, 1990, (**8**), 42.

10.3 Core Journals

The chemistry associated with superconductivity is reported in the following journals:

Physica C
Journal of Solid State Chemistry
Physics Reviews B and Physics Reviews Letters
Japan Journal of Applied Physics Letters
Materials Chemistry
Journal of Materials Chemistry
Material Research Bulletin
Nature (very few but very important)
Science (very few but very important)

CHAPTER 12

Inorganic Colours and Decoration

CHRISTOPHER R. S. DEAN

1 INTRODUCTION

This chapter is aimed at outlining the use of inorganic chemicals for decorative purposes. In particular, inorganic pigments are described, and other decorative uses of inorganic chemicals are mentioned. In Section 2 of this chapter, the major physical and chemical properties required of inorganic pigments are discussed. In Section 3, some examples of inorganic pigments have been chosen to illustrate the range of materials, the chemical methods used for synthesis and manufacture, and the major uses and the important properties needed for commercially viable inorganic pigments.

Inorganic materials have been used for many centuries for decoration. Initially, naturally occurring ores were used, such as iron oxides. These were collected – yellow ochre, red earths, and brown siennas, such as Indian Ochre and Raw Sienna – and used by the early people to colour paintings and even their bodies. Although they also used colours obtained from plants, often extracted from berries and barks, the coloured earths or ores were far more durable and resistant to fading. In time, techniques to modify and improve the colours became known, often involving heating of the raw materials under controlled conditions to give new colours, such as Burnt Sienna. **Synthetic chemistry**, however, enables much better and more **reproducible** colours to be manufactured, principally because the purity of the chemicals can be higher, and the physical and crystallographic form of the product better controlled.

Most typically, inorganic colours are used as **pigments**. A definition of a pigment is given in the Pigment Handbook.[1] In essence a pigment is a particulate solid that may be black, white, or coloured, and which may be dispersed into a medium (usually a liquid) without significant solution or other interaction. In addition, a pigment will affect the appearance of an article by selective absorption of visible wavelengths of light, most commonly also scattering the light.

There is a fundamental difference between pigments and **dyes**, since the latter are essentially soluble in the medium in which they are used, and thus any particulate properties are lost in use. Inorganic chemicals may also be used as 'dyes' in the broadest sense, for example when metal oxides are added to molten glasses to colour the bulk of the glass. Cobalt and copper are well known, giving blue and blue-green glasses respectively. Non-pigmentary inorganic colours are not discussed further in this chapter.

Pigments are used very widely for decoration, in inks, paints, varnishes, and wood-stains, and also in plastics, tableware (pottery and glass items, *e.g.* plates and cups), sanitaryware (*e.g.* baths, toilets, and wash-basins), ceramic tiles, cement products, and cosmetics.

Decoration of ceramic ware is often achieved using **enamels**, a combination of a pigment or pigments with a ground glass or 'frit'. These may be applied as a layer on the surface of glazed ware (**'on-glaze'**) or glass, and be fixed in place by a firing cycle (temperatures up to about 900 °C), during which the glass component fuses to the glaze substrate. If higher temperatures are used, the decoration diffuses into the glaze, giving **'in-glaze'** decoration, which is typically more durable, and less prone to leaching out of soluble metal ions from the decoration. For even more durable decoration, the enamels may be applied to the body of the ware prior to glazing, thus protecting the decoration **'under-glaze'** after glaze-firing typically at temperatures well in excess of 1000 °C. Alternatively, ceramic decoration may be achieved by dispersing pigmentary colour throughout the glaze which is then fired on to the ware. This method is used widely for sanitary-ware and for tiles, and the highly temperature-stable pigments used are known as **'glaze-stains'**, even though they remain as particles rather than dissolving in the molten glaze.

Inorganic particulate chemicals are also used as phosphors, and finely divided metals can be used as metal pigments, for example in 'gold' and 'silver' inks for printing onto paper, but these applications lie outside the scope of this chapter and are not discussed further. Another form of decoration involving metals is **liquid gold** for tableware and jewellery. Liquid golds are inks for printing or brushing a layer onto ware, which then forms an adherent metallic layer on firing. **Bright golds** contain one or more gold compounds essentially dissolved in the medium, and this decomposes on firing to give a continuous 'mirror' of metallic gold of the order of 100 nm thick. **Burnish golds** contain gold particles, often in conjunction with a liquid bright gold component, and must be polished or burnished after firing on the decoration. The colour and adherence of these decorations are controlled by additions of other metals as soluble compounds, and white-metal decorations, **'liquid platinums'**, may be obtained either by using similar platinum compounds and metal particles, or by using white alloys of gold, such as with palladium.

2 PIGMENT SYSTEMS

2.1 Pigment Types

The **optical properties** of pigments are strongly dependant on the sizes of the particles. The majority of pigments interact with incident light, partially scattering and partially absorbing at all wavelengths. One of the first studies of the scattering of light by particles was by Mie,[2] even though this was much simplified by considering isolated spherical particles. In general, the capability of absorbing is increased with decreasing particle size (or mean particle size of the distribution). However, as the size of the particles decreases, the scattering capability increases until a maximum is reached. As the size decreases further, the scattering capability also decreases, as the particles become small compared with the wavelength of light.

Pigments with high **opacity** generally have a mean particle size of the order of 0.5 μm. The optimum particle size depends on the **refractive index** of the pigment and of the medium in which it is to be used. For some applications, very low scattering of the incident light is required and then smaller particle sizes are used, such that the scattering component is minimized. Examples of these are **'transparent'** coloured pigments for metallic car paints and for wood stains and varnishes, for which very low scattering of the incident light is essential in order to avoid a 'milky' appearance (see Section 3.1). For metallic car paints, the 'transparent' coloured pigment is used in conjunction with a metal (usually aluminium) flake pigment.

For uses at higher temperatures, such as for glass and pottery enamels and glazestains, larger particle sizes are normally used. This is partly because the refractive index of the medium (glass) in which they are to be used is higher, but also practical limits are set by the tendency for smaller particles to dissolve in the molten glass. In the case of glazestains, 'colour centres' are stabilized as inclusions in an inert matrix of a colourless or white host material. Glaze-stains based on the zircon lattice are described below in Section 3.6.

Pearlescent pigments are typically translucent materials in the form of smooth thin platelets, typically 5 to 20 microns in size (larger two dimensions) but much smaller in the thickness of the platelets. The decorative effect is achieved by interference between light scattered by the top and bottom surfaces of the platelets. Typical uses are in cosmetics and increasingly for automotive paints.

Almost all inorganic pigments have been included in a classification published by the DCMA (Dry Color Manufacturers' Association, now changed to the Color Pigments Manufacturers' Association).[3]

2.2 The Physical and Chemical Properties Required, and Their Assessment

2.2.1 Colour and Colour Strength. The colour of any article is determined not only by the pigment itself, but also by the medium in which it is dispersed, the geometry of the article, the nature of the incident lighting, and the way in which the article is viewed.[4,5] Human colour sight derives from three types of wavelength dependant sensors in the human eye. Each of three types of these sensors responds to a range of wavelengths of light, corresponding roughly to red, green, and blue. Signals from the eyes are interpreted by the optic nerve and the brain to provide information about colour as well as shape of objects viewed.

The most common way to assess or compare colours is by direct visual comparison. A pigment, or more usually a blend of pigments, is dispersed in an appropriate medium and compared with a standard, usually a dispersion in the same medium. As requirements for colour matching have become more stringent, and with increased need to match colours in different applications (such as

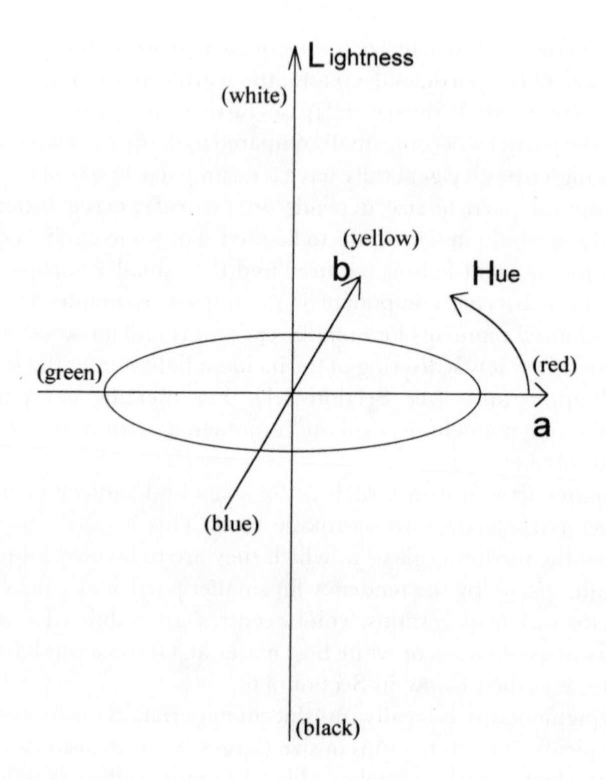

Figure 1 *CIELab colour space. Any colour may be represented by a point in this space, given by coordinates* **L**, **a**, *and* **b**. *More commonly coordinates* **L**, **C**, *and* **H** *are used, where* $\mathbf{C} = \sqrt{\mathbf{a}^2 + \mathbf{b}^2}$ *and* $\mathbf{H} = tan^{-1}(\mathbf{a}/\mathbf{b})$

matching glazed tiles to glazed sanitary ware and to a paint or wall-covering), there has been increased use of instrumental colour measurement. In practice, colour measuring instruments usually manipulate the data to give values of **lightness**, **chroma**, and **hue** (see Figure 1). The chroma is the strength of colouration, and the hue is an angle (0 to 360°; red at 0°, yellow at 90°, green at 180° and deep blue at 270°) representing the position in the spectrum.

Further complication is caused by **metamerism**; two specimens may give visually identical colours when viewed under one illumination, but give different colours under another illumination. This is caused by differences in the **reflectance spectra** of the two specimens, and demonstrates the limitations of assessing colour based only on three types of wavelength-dependant sensors.

A significant advantage of instrumental assessment of colour is the ability to compute blends of pigments to match a given colour in a medium, and also to assess possible metamerism.

2.2.2 Particle Size Distribution and Shape. Pigmentary properties are critically dependant on the mean particle size and the particle size distribution of the pigment powder (see Section 2.1 above). At large particle sizes (several microns), the number of particles in the medium is insufficient to afford good scattering by interaction with incident light. For example, for coloured plastics, pigment loadings of about 0.5% by weight are common (0.5 g pigment in 100 g pigmented polymer), typically giving a volume loading of about 0.1%. As the mean particle size decreases to about half of one micron, the colour intensity and opacity of pigmented decorations increase rapidly. However at particle sizes below about 0.1 micron, the ability of the particles to interact by scattering light decreases as colloidal sizes are approached, and in general the pigmentary strength decreases. However, the particles may still selectively absorb photons, and some pigments are designed to have particle sizes below 0.1 microns to give low opacity colouration ('transparent pigments'; see Section 3.1 below on iron oxide pigments and Section 3.5 on colloidal gold decoration). For paints, the objective is normally a completely opaque decorative layer, but for bulk colouration of plastics, satisfactory decoration is achieved at less than complete opacity.

The particle size, or more precisely the particle size distribution, of a pigment dispersion can be assessed in a number of ways.[9] Scanning and transmission **electron microscopy** are frequently used to assess directly the pigment particles, most especially when information about particle shape is needed. **Image analysis** methods allow statistical evaluation of large numbers of particles. Dispersions can be assessed by **light scattering instruments**[10] or, for distributions of finer particles, **Photon Correlation Spectroscopy**.[11] In addition, these methods give information about aggregation of the particles, which is strongly affected by the surface charge on the dispersed particles. This can be assessed by measuring the **zeta potential**.[12] Other important methods for assessing particle size distributions include sedimentation methods, timing the rates of passage of the particles through a viscous medium, frequently using centrifugal forces to speed up the measurements; and also measurement of the surface area of the powders by gas adsorption techniques.

2.2.3 Stability to Heat, Light, and Humidity. For some applications, *e.g.* inks for printing onto paper, the stability of the pigments in the application to heat, light, and humidity is not usually critical. For applications in plastics and ceramics, the heat stability is paramount, and for exterior decorations, **weatherfastness** giving decorative lifetimes of several years may be essential. In general, the degradation of materials with temperature is an innate material property and cannot be modified. The heat and weatherfastness of inorganic pigments tend to be superior to those of organic pigments. As a loose and general 'rule', the durability of any decoration is greater the higher the temperatures used during creation of the decoration; bulk colouration of plastics is more durable than surface decoration with a paint, and on-glaze enamel decoration is less durable than under-glaze decoration or colouration throughout the thickness of the glaze using glaze-stains. The temperature stabilities of pigments are normally assessed by dispersing in the medium under test using appropriate temperatures and times, and determining the effect on the colour of the decoration.

The reactivity of pigments in the presence of medium, moisture, and UV irradiation (*e.g.* titanium dioxide[20]) cannot easily be modified. If the instability is due to reaction involving interaction between the pigment and the medium, it may be affected by surface coating of the pigment to prevent close contact between the pigment material and the medium (see Section 3.3 below). **Lightfastness** or **weatherfastness** testing may often be by exposure of test items to outdoor conditions, Florida and the Arizona desert being particularly suitable testing locations. Alternatively, accelerated testing using artifical sources of ultraviolet light may be used, but great care must be taken to ensure appropriate wavelength distribution of the ultraviolet energy, and to provide moisture by immersion in or by spraying with water at intervals.

2.2.4 Interaction with the Medium or with Other Pigments. Any solubility of the pigment in the medium in which it is being used can result in migration of the pigment and **'bleeding'** of the colour. This is rare for inorganic pigments when used in organic or water-based media at ambient temperatures (*e.g.* for paints and inks) or even in plastics. This problem can be severe, however, when the inorganic colours are used in enamels. For example, cobalt blue enamels tend to show some solubility of the base colour in the glass frit component of the enamel, and this commonly leads to some migration of the colour during firing. Similarly cadmium enamels show some solubility of the cadmium sulfoselenide in the glass during firing, and this may result in reaction of the dissolved sulfoselenide if the cadmium enamel is used mixed with, or in close proximity with, other enamels.

Dispersions of particulate materials in plastics can affect the physical properties of the material. In some instances, such as long fibres, these effects can be very beneficial, but in other instances, the incorporation of pigments can detrimentally affect **tensile** or **impact strength**. Such detrimental effects are uncommon for inorganic pigments, but are well known for some organic pigments, such as the effect of phthalocyanine blue pigments on the impact strength of polypropylene.

2.2.5 Leaching or Extraction of Soluble Metals. Inorganic pigments containing heavy metals are subject to regulations and standards to ensure safety when used. The **Council of Europe Test**[6] evaluates the release into solution of metal ions from a pigment in the presence of 0.1 M hydrochloric acid, and sets limits for solubility of 0.005% for Hg; 0.01% for As, Ba, Cd, Pb, and Se; 0.05% for Sb; and 0.1% for Cr. This test simulates the tendency of the pigment to release soluble metal ions into the human body. The **Toys Test**[7] also evaluates the release of soluble ions, but from pigmented medium. Comminuted samples of pigmented paint or plastic are again contacted with hydrochloric acid (0.07 M), and the concentration of metal ions that are leached out of the pigmented samples is measured. The Food and Additives Consultative Committee Test assesses the leaching out of metal ions from pigmented plastics when in contact with foods. Typically, pigmented specimens are immersed in dilute acetic acid at elevated temperatures, and the concentration of metal ions going into solution is assessed.

2.2.6 Dispersion, and the Use of Surface Treatments. For maximum benefit from incorporation of a pigment in a system, the pigmentary particles must be dispersed at random throughout the medium, and must not remain clumped together as aggregates or agglomerates. In practice, some agglomeration of pigment particles is inevitable, but often major efforts are required to obtain sufficiently good dispersions of pigments for acceptable pigmentation. For paints, bead milling of pigments in a low viscosity component is frequently used. Commonly, surface coating agents are used, either as treatments of the pigment surface, or as components of the overall system, to improve the ease of dispersion, the extent of final dispersion, or the stability of dispersion of the pigments (see Section 3.3 below). The ease of dispersion is commonly assessed by determining the time of milling after which no further improvement in colour properties is achieved. Alternatively, microscopy may be used to assess the size and distribution of dispersed particles or agglomerates, or apparatus, such as a **Hegman Gauge**,[8] may be used to determine the maximum particle size in the dispersion.

2.3 Pigment Choice

The choice of a pigment, or often a blend of pigments, will depend critically on the application. A pigment that is ideal for a printing ink is unlikely to be ideal for ceramic applications. All the properties discussed above, together with others, most especially cost, have to be considered, and frequently a compromise must be accepted. A pigment system with borderline thermal stability may be used in an application, because the cost of pigments with fully acceptable thermal properties may be unacceptable for the chosen application. For ceramic applications, the choice of pigments, most especially in the yellow to red region of the spectrum, is very limited. Cadmium yellow and red enamels give strong bright colours, but cannot easily be mixed or used in combination with other types of enamels, and sometimes a purple or magenta enamel, such as a colloidal gold (see Section 3.5 below), has to be used as a compromise.

Table 1 includes examples of pigments commonly used for some applications, and the colour range for each.

3 EXAMPLES

3.1 Synthetic Iron Oxides

Although natural iron oxides[13] have been used since prehistoric times, only during this century has increasing demand for **higher chroma pigments** and for more controlled pigment properties resulted in synthetic routes to these products.[14] The three main colours are **haematite** (red; α-Fe_2O_3), **goethite** (yellow; FeO.OH or $Fe_2O_3.H_2O$) and **magnetite** (black; Fe_3O_4). Most other colours, especially browns, are mixtures or blends of the red, yellow, and black oxides, or are manufactured as mixtures. The exceptions are ferrites, notably zinc and magnesium, which are tan coloured, and maghemite (γ-Fe_2O_3) which is brown in colour but is chiefly important as a magnetic pigment for recording tapes and disks. Acicular (needle-shaped) particles are preferred for this application, and these magnetic pigments are generally prepared by controlled dehydration and reduction of yellow hydrated iron(III) oxide pigments to magnetite, followed by controlled oxidation to γ-Fe_2O_3.

Table 1 *Pigment colour ranges and uses*

	Purple/ Magenta	*Red*	*Orange*	*Yellow*	*Green*	*Blue*	*White*	*Black*
Paints and inks	------- *mono-azo- and di-azo organic pigments* ---------				---- *phthalocyanines* ---		titanium dioxide	carbon black
	------------ *quinacridone pigments* -----------------							
		-------- *isoindoline pigments* --------						
			Pb chrome/molybdate			ultramarines		
		------------- Fe oxides --------------				ferriferrocyanides		
			titanates (NiSb and CrSb)		Co chromite	Co aluminate		
	Mn violet			Bi vanadate/molybdate	Co titanate			
Plastics and rubber		----------- cadmium pigments -----------------------					titanium dioxide	carbon black
			Pb chrome/molybdate			ultramarines	zircon	Fe oxide black
		------------- Fe oxides --------------						
			titanates (NiSb and CrSb)		Co chromite	Co aluminate		Cu/Cr black
				Bi vanadate/molybdate	Co titanate			
	Mn violet				---- *phthalocyanines* ----			
Ceramic enamels		----------- cadmium enamels -----------------------				Co aluminate	titanium dioxide	Cu/Cr black
	Purple of Casius (Au)		Pb chrome/molybdate				zircon	
			titanates (NiSb and CrSb)					
					Cr/Cu/Co/Si/Al mixed oxides			
Glaze-stains		Fe zircon		Pr zircon		V zircon	zircon	Cu/Cr black
		--------- encapsulated CdS/Se in zircon ---------						Fe/Co(Cr) black

Items in *italics* are organic pigments.

The synthesis of the **yellow hydrated iron(III) oxide** was pioneered in the 1920s. In the 1930s, yellow and black iron oxides were prepared during reduction of nitrobenzene to aniline:

$$\underset{\text{nitrobenzene}}{Ph\text{–}NO_2} + 2\,Fe + 2\,H_2O \rightarrow \underset{\text{aniline}}{Ph\text{–}NH_2} + Fe_2O_3.H_2O \quad (1)$$

Classical aqueous precipitation processes gave the hydrated yellow iron(III) oxide. The red anhydrous oxide was prepared from the yellow oxide by calcination (heat treatment), causing undesirable sintering and aggregation at the temperatures required to drive off the water at acceptable rates. Modern precipitation methods were developed during the 1950s allowing effective control of particle growth and direct precipitation of the anhydrous red iron(III) oxide, resulting in a more easily dispersed pigment:[21]

$$4\,FeSO_4 + O_2 + 8\,NaOH \rightarrow 2\,Fe_2O_3 + 4\,Na_2SO_4 + 4\,H_2O \quad (2)$$

The preparation is generally carried out in two or more stages, the first being the precipitation or nucleation of seed material or very small crystallites. In subsequent stages, precipitated iron(III) oxide grows onto the seed particles, providing good control of particle growth and a much tighter particle size distribution. In the extreme case, the particle size of the precipitated yellow oxide can be maintained at 10–100 nm (0.01–0.10 μm), giving low-opacity or transparent grades of pigment for use in wood-stains and varnishes, and metallic car paints. Modern processes generally use iron(II) sulfate solution obtained as a by-product from steel pickling or from manufacture of titanium dioxide pigments (see Section 3.3).

The different processes used impart different properties to the pigments, most especially due to different particle size distributions, particle shapes, and extents of aggregation of the powders. Direct aqueous precipitation of the red anhydrous iron(III) oxide gives rhombohedral particle shapes, whereas dehydration of the hydrated yellow pigment preserves needle-shaped particle morphology. Controlling the particle size of the yellow pigment provides a range of colours from light yellow (smaller particles) through the much browner or darker colours for larger particle sizes.

Iron oxides are used in a wide range of applications, including paints, plastics, rubber, ceramics, and cement products, and also for cosmetics and food colouration (for pet foods). The pigments are generally of relatively low cost and have very low toxic hazard, but have much poorer colour characteristics than, for example, cadmium red and yellow pigments or typical red and yellow organic pigments. Also, iron oxides tend to dissolve in molten glasses, and thus are of only limited use for ceramic decoration.

3.2 Cadmium Sulfoselenides

Cadmium pigments[15] have been developed since the early twentieth century, and the growth in demand for these pigments has followed closely the increasing use of synthetic resins, and most especially the increased use of engineering plastics, such

Table 2 *Colours achievable by doping cadmium sulfide*

green-shade yellow	*yellow*	*orange*	*red*	*deep maroon*
(Zn,Cd)S	CdS	Cd(S,Se) (Cd,Hg)S	Cd(S,Se) (Cd,Hg)S	Cd(S,Se) (Cd,Hg)S

increasing Se or Hg content ⟶

as polycarbonate, polypropylene, nylon, and ABS (acrylonitrile–butadiene–styrene), since these polymers generally require higher processing temperatures (200–300 °C) for injection moulding.

Cadmium sulfoselenide pigments cover a major part of the spectrum, from green-shade yellow through oranges and reds to deep maroons. This is achieved by doping the cadmium sulfide lattice, *i.e.* replacing some of the cadmium atoms in the lattice with zinc or mercury, or some of the sulfur atoms by selenium.

The range of colours achieved is continuous (see Table 2), and depends on forming a true solid–solution lattice, that is with the dopant randomly replacing the Cd or the S atoms in the cadmium sulfide lattice. This modification of colour would not be achieved by mixing or blending of the cadmium sulfide with the dopant. Simple mixing of the very dark coloured cadmium selenide with yellow cadmium sulfide would give very dirty yellow, brown, or green colours. However, by mixing at the atomic level – randomly substituting Zn or Hg for Cd, or Se for S atoms in the crystal lattice of CdS – the semiconductor band gap of the cadmium sulfide is changed. This causes onset of absorption of light at different wavelength.

The reflectance spectra of cadmium pigments (Figure 2) show sharp transitions from absorption of light at the blue end of the visible spectrum to high reflectance of light in the red and green regions of the spectrum. Very similar reflectance spectra are observed for other semiconductor pigments, such as lead chromate, and for titanium dioxide (see Figure 3), for which the band gap is greater, giving a transition at about 400 nm, just at the blue edge of the visible spectrum.

Cadmium pigments are produced by a precipitation and calcination route. Cadmium sulfide is precipitated from an acidic solution of a soluble cadmium salt (typically the sulfate or nitrate) by the addition of sulfide ions. Sodium sulfide is commonly used (Equation 3), but if barium sulfide is used to precipitate cadmium sulfate, barium sulfate is co-precipitated to give a 'lithopone'. Selenium can be added co-dissolved with sodium sulfide, and is precipitated as very fine particles of elemental selenium, intimately mixed with the precipitated cadmium sulfide. When selenium is added, further cadmium is usually precipitated as carbonate or hydroxide. After washing and drying, the precipitate is calcined at temperatures above 500 °C. This causes a reaction between the selenium and the cadmium carbonate or hydroxide, forms the mixed crystal lattice, and converts the material from the precipitated cubic form to a hexagonal lattice.

$$CdSO_4 + Na_2S \rightarrow CdS + Na_2SO_4 \quad (3)$$

$$CdSO_4 + Na_2CO_3 \rightarrow CdCO_3 + Na_2SO_4 \quad (4)$$

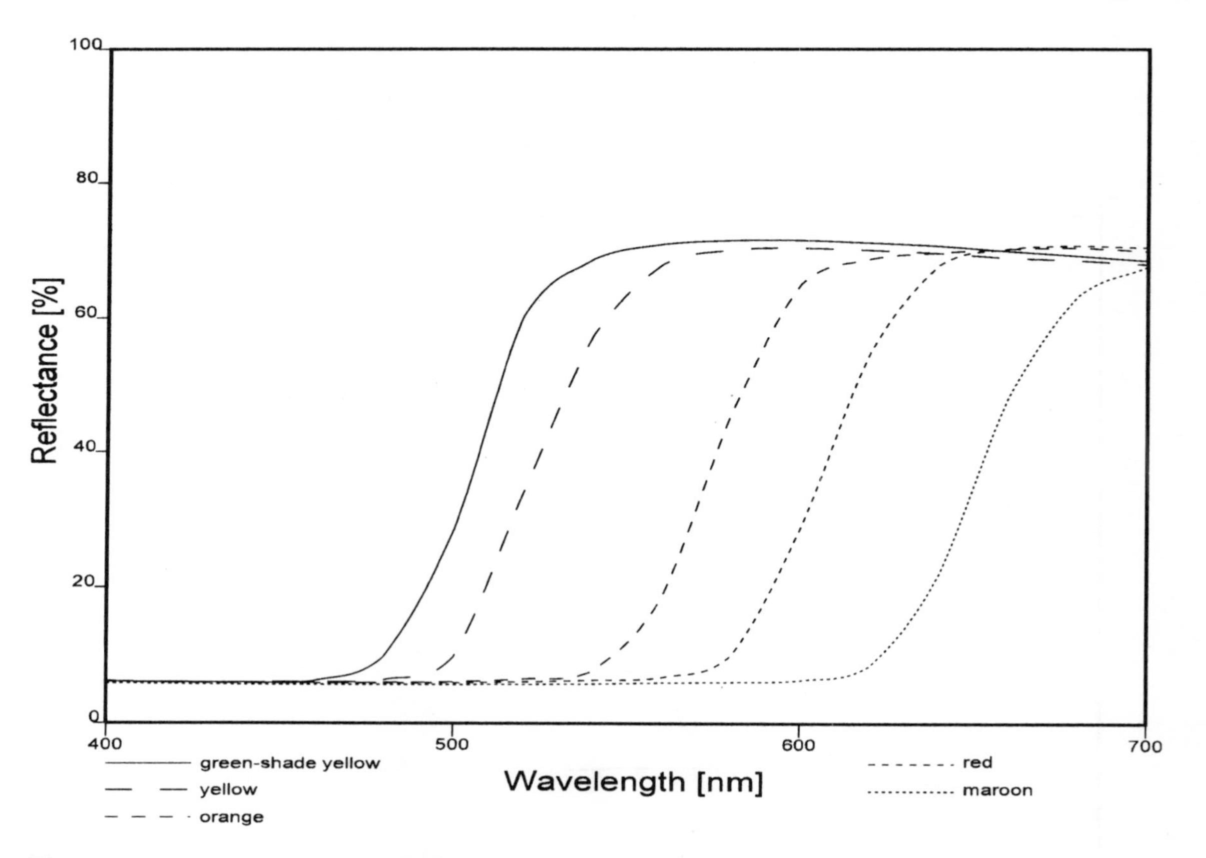

Figure 2 *Reflectance spectra of a range of cadmium pigments (0.5% in polystyrene)*

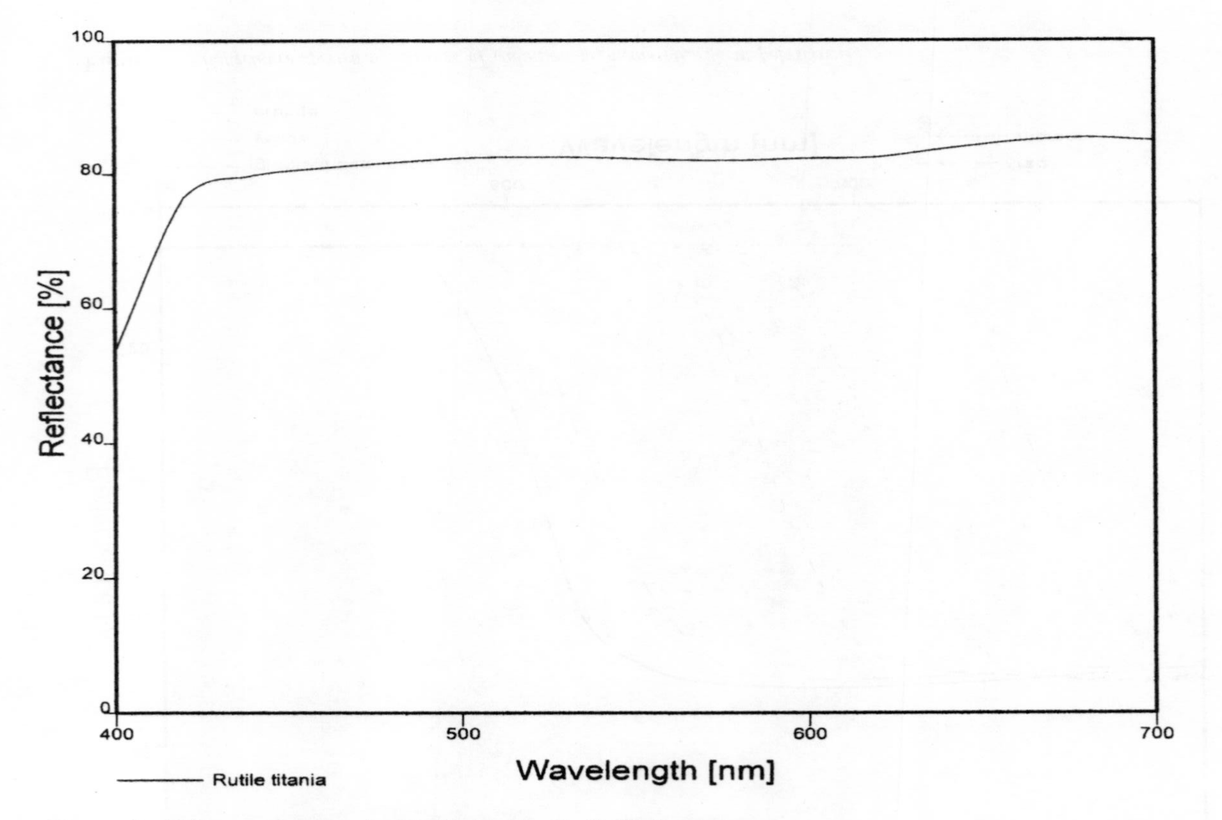

Figure 3 *Reflectance spectrum of rutile titanium dioxide (0.5% in polystyrene)*

The main uses of **cadmium pigments** are for colouration of **plastics**, especially the engineering polymers, and in specialist **paints**, including artists' colours, where durability of colour is essential. Cadmium pigments are also important for pottery and glass **enamels**, in which the sintering of the glass component of the enamel protects the cadmium pigments from oxidation by air at the elevated temperature needed for firing enamel decoration (typically from 600 to 900 °C). Cadmium colours may also be used at even higher temperatures as components of **glaze-stains**. To colour glazes that may be fired at temperatures in excess of 1200 °C, the cadmium colours are encapsulated in a zirconium silicate (zircon) lattice[22,23] (see also Section 3.6).

3.3 Titanium Dioxide (Titania) and Titanates

Titanium dioxide[16] is the most important **white pigment** currently in use. It is used widely on its own for pigmenting articles white, and in blends with other pigments to give paler colours and increased opacity. Titanium dioxide has almost completely displaced previously used white pigments, such as basic lead sulfate, antimony oxide, zinc oxide, and tin oxide, because it has superior optical properties and a very low toxic hazard. Two different crystal forms (rutile and anatase) are important as pigments, but although both possess tetragonal symmetry, the rutile lattice is more compact giving a higher refractive index and therefore greater pigmentary opacities for similar particle sizes.

Titanium dioxide is a **semiconductor** with a **large band gap**, giving a transition from absorption to reflectance of radiation at about 400 nm, just at the blue edge of the visible spectrum. The transition for rutile is at a slightly higher wavelength than for anatase, and falls just within visible wavelengths, leading to a cream overtone to the colour of rutile pigmented articles, whereas anatase offers a bluer overtone (more pure white). Overall, however, rutile is generally the preferred pigment form due to its higher refractive index. The colour and other pigment properties are significantly affected by impurities in the titanium dioxide, even at parts-per-million (ppm) levels, and manufacturing processes have evolved to ensure the required high purities.

The element Ti was discovered at the end of the 18th century, but it was not until the early 20th century that development of titanium dioxide as a pigment was started. Initially, only anatase pigments were synthesized, using a method that was the forerunner of the modern sulfate process (discussed below). Production processes for rutile pigments were developed because it was appreciated that rutile would have superior optical properties, and because anatase pigmented paint films tend to chalk on weathering. This progress was delayed by the second World War, but in the late 1940s, the chloride process was pioneered. Both processes give iron salts as a by-product.

3.3.1 The Sulfate Process. The source of Ti for the sulfate process is most commonly ilmenite ($FeTiO_3$), but the Fe content may be reduced by prior processing to give 'titanium slag'. The feedstock is ground and digested in sulfuric acid at high temperatures (typically reaching 100–200 °C). This converts the major components of the source into sulfates:

$$FeTiO_3 + 2\,H_2SO_4 \rightarrow TiOSO_4 + 2\,H_2O + FeSO_4 \tag{5}$$

Other impurities in the source are also converted into sulfates alongside the Fe. The soluble materials are filtered off, and the titanyl sulfate solution is reacted with water at high temperatures, being hydrolysed to precipitate titanium hydroxide. These conditions essentially leave the iron and other impurities in solution.

$$TiOSO_4 + 2\,H_2O \rightarrow TiO(OH)_2 + H_2SO_4 \tag{6}$$

The precipitated titanium hydroxide, or dioxide–hydrate, is washed, dried, and calcined to give the required titanium dioxide at high purity.

$$TiO(OH)_2 \rightarrow TiO_2 \tag{7}$$

3.3.2 The Chloride Process. The chloride process requires a higher-grade source than the sulfate process, and either 'titanium slag' or natural rutile is used. This is reacted with chlorine gas and coke at about 1000 °C to give titanium tetrachloride: see Equations 8 and 9. (This reaction is also used in the refining of titanium metal: see Chapter 2, Sections 4.2 and 5.2).

$$TiO_2 + 2\,Cl_2 + 2\,C \rightarrow TiCl_4 + 2\,CO \tag{8}$$

$$TiO_2 + 2\,Cl_2 + C \rightarrow TiCl_4 + CO_2 \tag{9}$$

The Fe and other impurities present are also converted to chloride, and the titanium tetrachloride is purified by distillation before being reacted with pure oxygen at high temperatures.

$$TiCl_4 + O_2 \rightarrow TiO_2 + 2\,Cl_2 \tag{10}$$

Titanium dioxide pigments from both processes are customarily coated to improve the properties. Organic coatings, such as silicones or amines, are used to improve dispersibility. However, inorganic coatings are also applied to improve the durability of pigmented resins. Titanium dioxide is very inert, but in the presence of moisture and ultraviolet radiation hydroxy and perhydroxy radicals can form at the surface of the titanium dioxide lattice. These may then react with an organic resin medium, causing degradation of the resin. Both anatase and rutile show this photochemical degradation of media, and most pigment grades limit this effect by coating the surface of the titanium dioxide particles with silica and alumina.

Titanium dioxide pigments are used in a **very wide range of applications**, including paints, inks and varnishes, thermoplastic resins and rubber, ceramics, paper, and textiles. Titanium dioxide has found other pigmentary uses, mainly resulting from its high refractive index and high reflectivity across the visible spectrum. Titanated mica pigments, in which a thin coating of titanium dioxide has been coated onto mica platelets, are very important pearlescent pigments, different effects being produced by using differing thicknesses of the titanium dioxide coating.

Titanate pigments which are based on the titanium dioxide lattice but use other transition metals in the lattice to produce non-white colours, are also important for durability and low toxic hazard. Nickel antimony titanate yellow and chrome antimony titanate buff are the most notable.

3.4 Cobalt Aluminate Blue

The organic phthalocyanine pigments[17] (mainly the blue copper phthalocyanine and the green chlorinated copper phthalocyanine) tend to dominate the market as blue pigments for paints and inks, and for many thermopolymers. However, for some applications, greater stability is essential. Ceramic enamels need greater heat stability, and coil coatings (premium coatings; pre-coated sheet metal, mainly used for industrial applications) need greater durability to weathering. For these applications the greater expense of inorganic cobalt pigments becomes justified.

Cobalt has been used to colour glazes and glasses since about 2000 BC, and in the 18th century, it was combined with alumina to obtain better colours. Cobalt aluminate blue pigments[18] are called **Thénard's blues** in honour of the chemist who pioneered the study of cobalt chemistry in the early 19th century.[24]

The chemical formation of these pigments is fairly straightforward, involving the calcination of an intimate blend of a cobalt and an aluminium compound, as illustrated by

$$\underset{}{CoCO_3} + \underset{\text{alumina hydrate}}{Al_2O_3} \rightarrow CoO.Al_2O_3 + CO_2 \qquad (11)$$

Temperatures up to about 1300 °C are used depending on the source materials, and mineralizers may be used to promote the reaction. The cobalt aluminate forms in a spinel structure, and must be pulverized and ground to achieve the necessary particle sizes for pigmentary applications.

For ceramic applications, a mean particle size greater than 1 μm and a relatively wide particle size distribution are commonly used. For applications in coatings and plastics, however, a lower mean particle size of about 0.4–0.8 μm is necessary, and a much tighter particle size distribution is preferred for good colour characteristics. While the chemical formation of these materials is relatively straightforward, much greater manufacturing skill is required to obtain the optimum particle size distribution needed for pigmentary applications. The raw materials used may also be very important in the production of these pigments, not only to control impurity levels in the product, but also to affect the ease of comminution.

A blue-green or turquoise pigment is also important, and is achieved by incorporation of chrome in the cobalt aluminate.

These pigments are considered **chemically inert**, with good resistance to heat and weathering, but have only medium pigmentary or tinting strength, and are relatively expensive. Besides ceramic enamels and premium coatings, cobalt blues are used for whitening ceramic bodies (offsetting the pink tint due to impurity iron), and for plastics, artists' colours, and cement products.

3.5 Purple of Cassius (Colloidal Gold)

The choice of pigment for red to purple colouration of plastics, inks, and paints is reasonably broad, including a very wide range of organic pigments, and inorganic pigments such as the cadmium sulfoselenides, iron oxide red and manganese violet. Of these, only cadmium sulfoselenides are sufficiently heat-stable for ceramic applications, and they are limited by the potential toxic hazard of leaching cadmium from the decorated ware and by chemical interaction of the cadmium enamels with other ceramic colours. For these reasons, 'Purple of Cassius' **enamels**, based on colloidal gold, are important for ceramic applications for colours in the range from **light pinks** to **deep purples**.

As with the transparent iron oxide discussed in Section 3.1 above, as the particle size of gold is reduced to less than the wavelength of visible light, so the scattering for the gold particles decreases. At sizes below 100 nm, the scattering or reflection from the gold particles becomes insignificant, but there is still interaction with the incident radiation causing absorption most especially in the green region of the visible spectrum. The transmitted light is seen as purple. Gold particles at these dimensions are not usually stable, and care has to be taken to stabilize the particles against growth or aggregation that would destroy the optical effect. To stabilize the gold for ceramic applications, the colloidal gold particles are supported on stannic oxide.

The preparation methods vary, but essentially a gold salt is reduced to gold metal, usually in the presence of stannic oxide, which may be produced simultaneously with the reduction of the gold salt. In practice, precautions must be taken to ensure that nucleation of the gold metallic particles is not followed by rapid particle growth, or by aggregation of the very small gold nuclei. These may be stabilized sterically using large organic molecules such as albumen, which adhere to the gold surface and prevent aggregation. Alternatively, or additionally, the particles may be stabilized by adsorbing on the gold surface materials that provide an overall charge, providing electrostatic stabilization.

To maintain the stability of the colloidal gold supported on stannic oxide, it is usually combined with an appropriate frit (ground glass) to form an enamel without being previously dried. In this respect, Purple of Cassius enamels are unusual in that most enamels are manufactured from a 'base colour' and a glass frit, allowing easy manufacture of a range of enamels for different applications (*e.g.* for earthenware, fine china, and porcelain) by the addition of different glass frits to one base colour. The colour of Purple of Cassius enamels may be controlled by adjusting the particle size distribution of the colloidal gold particles, by changes in the composition of the glass frit, and by additions of metals other than gold. Silver in particular tends to produce less blue fired colours.

3.6 Zircon Glaze-stains

The previous section on Purple of Cassius enamels briefly discussed the need for colours that have heat stability greater than that necessary for applications in paints and thermoplastics. In general, the more demanding the application, the more restricted the range of pigments from which a choice can be made. Thus for a non-toxic, intermixable pottery or glass enamel of purple colour, Purple of Cassius is almost unique.

The most demanding application is probably the colouration of glazes using glaze-stains (colour dispersed throughout the thickness of the glaze, rather than a coating on top of or under the glaze). Despite the term 'stain' that may be more normally associated with soluble materials and dyes, glaze-stains are true pigments, in that they remain discreet particles dispersed throughout the glaze, maintaining their particulate nature even in the molten glaze at the peak of the glaze-firing temperature cycle (up to 1300°C). Molten glasses are extremely efficient solvents for inorganic chemicals, and very few materials are therefore suitable for these applications. Very many of the colours used for these applications are based on zirconia (zirconium oxide) or on zircon (zirconium silicate).[25]

Zirconium silicate is used as a **white pigment** or **glaze-stain**,[19] providing good opacity for tile and sanitaryware glazes. Colours other than white may only be achieved by **doping** the zircon lattice with either transition metal ions (see Table 1) or with small discreet particles of coloured material. For example: vanadium ions impart a blue colour and praseodymium ions give a yellow; while inclusions of iron oxide, cadmium sulfoselenide, and of carbon can give pink, red to yellow, and grey colours respectively.

Figures 4 and 5 illustrate the colours achievable in sanitary and tableware respectively.

Unfortunately there are no known methods for effectively and directly incorporating dopant colour centres into the zircon lattice, and they must be incorporated during formation of the zircon lattice from zirconia and silica:

$$\underset{\text{zirconia}}{ZrO_2} + \underset{\text{silica}}{SiO_2} \longrightarrow \underset{\text{zircon}}{ZrSiO_4} \quad (12)$$

Temperatures in excess of 1000°C are normally required for this reaction, and most commonly the temperature and time required for the reaction are lowered by using mineralizers to promote the reaction. Alkali- or alkaline-earth metal fluorides are most often used, providing an intermediate volatile silicon–fluorine compound that then reacts with the solid zirconia phase. It is essential that the colour centres or their precursors are in the right location at the appropriate temperature and time so that they may be incorporated into the zircon lattice as it forms. For zircon-**encapsulated** cadmium sulfoselenides, **sol-gel** type methods may be used to lower as far as possible the temperature at which the zircon lattice is formed, since it is very difficult to maintain the cadmium sulfoselenide at the temperatures normally essential for zircon formation.

Zircon colours are used widely in glazes; and in enamels. In some cases they have even been recommended for plastics when non-toxic, non-interacting colouration is essential.

4 FUTURE TRENDS

The most likely trend for inorganic pigments will be a continuation to move away from potentially toxic elements, such as Hg, Pb, Cr(VI), and Cd. Increasingly stringent Health and Safety regulations are reducing the range of acceptable applications for existing pigments, and thus further limit the choice of pigments for any application. The **complete life** of the material will have to be considered; source, manufacture, use, and disposal. For example, the hazard associated with a pigment may be very low, such as for cadmium pigments due to their very low aqueous solubility. During disposal, degradation products from incinerated pigmented articles, however, can lead to soluble cadmium species. This may not automatically mean that organic pigments will become favoured, since the more complex organic pigments must be used for applications requiring heat stability or weatherfastness, and manufacture of some of these requires the use of potentially toxic intermediates, traces of which could remain in the product. Disposal of articles coloured with organic pigments may also start to pose problems as regulations tighten, potentially requiring that the decomposition products from polymers and pigments be controlled or collected following incineration.

As the number of traditional colours regarded as unsafe continues to increase, inorganic chemists will be required to be more ingenious in identifying **new colour systems**. Further compromises in the choice of decorative systems may otherwise mean colours that are less bright or strong, or which have reduced heat resistance or weatherfastness, have to be accepted.

Extrapolation from advances in recent years implies that **improvement of known materials**, making them more useful as pigments, is more likely than the identification of entirely new colour systems. Another likely trend will be towards new effects using existing materials, such as the use of ultra-fine titania, or of coated materials for lustrous effects. The development of ultra-fine particle size iron oxides as 'transparent' pigments (see Section 3.1 above) is another example where new technology has significantly increased the range of applications of chemicals that have been known and used as pigments for very many years.

Finally, another probable trend will be towards drastically **reducing the number** of coloured pigments that are widely used for less demanding applications. As discussed above in this section, general tightening of legislation and regulations will require a reduction in the range of choices of decorative systems. However, there are commercial advantages in being able to achieve a wide range of colours based only on a few pigments. **Trichromatic** or **process printing** can achieve a very wide colour range through combinations of only three coloured inks (generally cyan, magenta, and yellow) plus black. Pigments may be used in combinations to give new colours either by blending the pigments, by applying superimposed layers of colour, or by printing dots of colour in close proximity to each other. For bulk colouration, **blending** of a limited range of **very high chroma** (high colour intensity) pigments could give a wide range of colours,

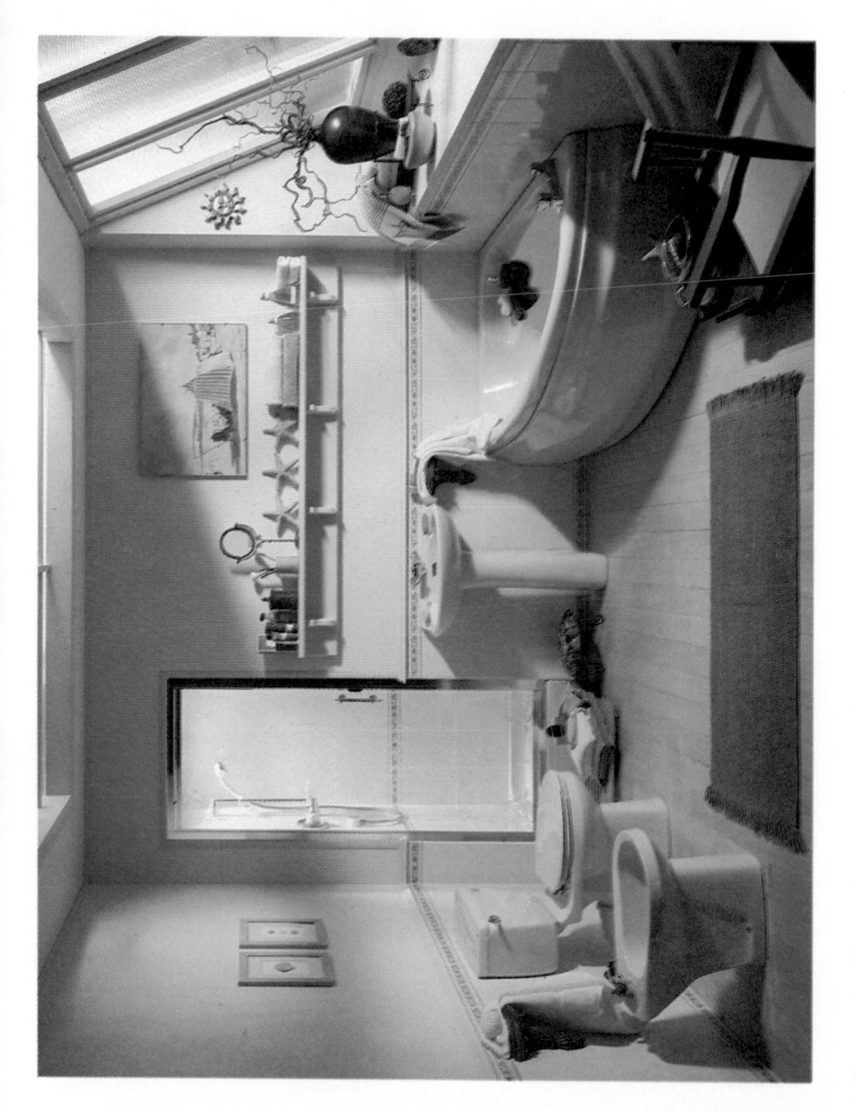

Figure 4 *Picture of bathroom, illustrating colour achievable via glaze-stains*
(picture kindly supplied by Trent Bathrooms, Hanley, Stoke-on-Trent, UK)

Figure 5 *'Haddon Hall' – Minton bone china tableware*
(picture kindly supplied by Royal Doulton plc, Stoke-on-Trent, UK)

allowing research, development, and production efforts to be concentrated on fewer pigments.

5 REFERENCES

5.1 General

1. 'Pigment Handbook', ed. P. A. Lewis, Wiley, New York, 2nd edn, 1987, vol. 1, p. vii.
2. G. Mie, *Ann. Phys.*, 1908, **25**, 377.
3. 'Classification and Chemical Description of the Complex Inorganic Color Pigments', Dry Color Manufacturers' Association, Alexandria, Virginia, 3rd edn, 1991.
4. D. B. Judd and G. Wyszecki, 'Color in Business, Science and Industry', Wiley, New York, 2nd edn, 1963.
5. G. Wyszecki and W. S. Stiles, 'Color Science', Wiley, New York, 1967.
6. Council of Europe Resolution AP(89) 1, 'On the Use of Colourants in Plastic Materials Coming into Contact with Food', adopted 13 September 1989.
7. British Standard BS 5665: Part 3: 1989 (equivalent to EN 71: Part 3: 1988), 'Specification for the migration of certain elements', The British Standard Institution, London.
8. '1985 Book of Standards', American Society of Testing Materials, Philadelphia, vol. 06.01, Standard D1210.
9. D. H. Everett, 'Basic Principles of Colloid Science', The Royal Society of Chemistry, Cambridge, UK, 1988, (243pp).
10. C. F. Bohren and D. R. Huffman, 'Absorption and Scattering of Light from Small Particles', Wiley, New York, 1983, (530pp).
11. B. J. Berne and R. Pecora, 'Dynamic Light Scattering', Wiley Interscience, New York, 1975, (400pp).
12. R. J. Hunter, 'Colloid Science: Zeta Potential in Colloid Science: Principles and Applications', Academic Press, London, 1981, (386pp).

5.2 More Specific

13. 'Pigment Handbook', ed. P.A. Lewis, Wiley, New York, 2nd edn, 1987, vol. 1, pp. 281–6.
14. 'Pigment Handbook', ed. P.A. Lewis, Wiley, New York, 2nd edn, 1987, vol. 1, pp. 287–307.
15. 'Pigment Handbook', ed. P.A. Lewis, Wiley, New York, 2nd edn, 1987, vol. 1, pp. 347–356.
16. 'Pigment Handbook', ed. P.A. Lewis, Wiley, New York, 2nd edn, 1987, vol. 1, pp. 1–42.
17. 'Pigment Handbook', ed. P.A. Lewis, Wiley, New York, 2nd edn, 1987, vol. 1, pp. 663–702.

18. 'Pigment Handbook', ed. P.A. Lewis, Wiley, New York, 2nd edn, 1987, vol. 1, pp. 389–94.
19. 'Pigment Handbook', ed. P.A. Lewis, Wiley, New York, 2nd edn, 1987, vol. 1, pp. 67–76.
20. G. Kaempf, W. Papenroth, and R. Holm, *J. Paint Tech.*, 1974, **46(598)**, 56.
21. '1984 Book of Standards', American Society of Testing Materials, Philadelphia, vol. 06.02, pp. 192–3, Standard D768–81.
22. A. Broll, H. Beyer, H. Mann, and E. Meyer-Simon, 'Ceramic Pigments based on Zirconium Silicate' and 'Ceramic Pigments', British Patents GB 1403 470 and GB 1419 166, 1972, assigned to Degussa, Frankfurt.
23. A. C. Airey and A. Spiller, 'Protected Pigments', European Patent EP 74 779, 1983, assigned to The British Ceramic Research Association, Stoke-on-Trent.
24. 'Cobalt: Its Chemistry, Metallurgy, and Uses', ed. R. S. Young, American Chemical Monograph Series no. 149, American Chemical Society, Reinhold, New York, 1960.
25. R. A. Eppler, 'Zircon-based colors for ceramic glazes', *Ceram. Inf.*, 1978, **13(151)**, 811; R. A. Eppler, 'Zirconia-based colors for ceramic glazes', *Indian Ceram.*, 1977, **20(3)**, 160.

5.3 Core Journals

Technical papers appear in a very broad range of publications, and new technology often only appears in patents. The best way to follow these is in the abstract journals:

World Ceramic Abstracts
World Surface Coatings Abstracts (Paint Research Association, London)
RAPRA Abstracts (RAPRA, Shrewsbury, UK)

Information about new developments in colour technology are published in the following journals:

British Plastics and Rubber
European Plastics News
European Polymers Paint Colours Journal
Ceramic Industry
Ceramic Industries International
Ceram Research
Global Ceramic Review

CHAPTER 13

Glasses, Ceramics, and Hard Metals

PHILIP G. HARRISON AND CAROLE C. HARRISON

1 INTRODUCTION

Ceramics, glasses, and hard metals form a very important class of materials. Applications of such materials include *inter alia* high temperature materials, ferroelectric materials, semiconducting materials, superconducting oxides, optical glasses, magnetic materials, porous materials, membranes, films, and abrasives. Although materials which exhibit these properties have been known in some cases for a very long time, in recent years demands have appeared for materials which have enhanced properties or new materials need to be invented which can fulfil novel requirements. Against this background the area of science and technology of ceramics, glasses, and hard metals has flourished. The empirical techniques by which many existing materials have been manufactured and processed into commercial products have been replaced by an understanding of the fundamental chemistry involved, and this in turn has lead to the ability to design novel materials with superior properties. Of course, there is also the essential effect of serendipity, as long as the scientist can recognize that he has accidentally found something which is new and revolutionary. Mimicking the ways of nature too has found a very important place in the search for new materials. All of these are very important to the way which this area of chemistry is now approached. Demands for new materials to meet new applications as well as old applications which need new, higher performance materials have both served to promote a vast interest worldwide, an interest which is growing at an enormous pace. New synthetic methods are emerging, and new techniques for the investigation of the properties of materials are continually being devised. Not surprisingly, the materials involved cover a wide range of chemistry, and in this chapter an attempt has been made to gather together some of the more important aspects in this area.

2 OXIDE CERAMICS

2.1 Simple Ceramic Oxides

Binary ceramic oxides fall into two classes: those which are predominantly **ionic** in nature and those which are predominantly **covalent**. In the former category are magnesia, MgO, alumina, Al_2O_3, titania, TiO_2, and zirconia, ZrO_2, whilst the only binary oxide of importance in the second category is silica, SiO_2, although related to this is a vast array of metal silicates and zeolites. In ionic ceramics metal cations and oxide anions form close-packed arrays, whereas silica, silicates, and zeolites comprise covalently-bonded networks with tetrahedral silicon. These oxides are generally **hard** with **high melting points** (*e.g.* MgO, m.p. 2800 °C) and are used as abrasives, refractories in furnaces, and for ceramic monoliths and fibres. The cubic form of alumina occurs naturally as the mineral corundum which is extremely hard (9 on the Mho scale). Titania is used widely as a pigment due to its high refractive index, mostly in paints but also in paper making and plastics. Both modifications, rutile and anatase, are used although the rutile form has the better optical properties.

The simplest ionic ceramic oxide is magnesia, MgO, which has the cubic NaCl structure. An alternative way of viewing this structure, however, is as face-centred cubic (fcc) close-packing of the large oxide anions with the small magnesium cations occupying the octahedral interstices. This method is a useful way of viewing ceramic oxides. For example, the structure of the cubic modification of zirconia, ZrO_2, is quite simply the same fcc close-packing of the oxide anions with zirconium cations occupying the tetrahedral interstices in the lattice. Alumina, Al_2O_3, may be described in a similar fashion although in this case the oxide anions are hexagonally close-packed with the aluminium cations again in two-thirds of the octahedral interstices, although this leads to some distortion of the hexagonal close packing. In all the forms of silica, SiO_2, (*e.g.* quartz, cristabolite, tridymite) the silicon is bonded covalently to four oxygen atoms leading to a three-dimensional network. The form of silica which is stable at low temperature is α-quartz, and the change to β-quartz occurs at 573 °C. This transformation involves a slight distortion of the lattice known as a **displacive transformation** and similar displacive transformations occur between the different forms of cristabolite and tridymite. The transformation between β-quartz and β-tridymite (at 867 °C) and β-tridymite and β-cristabolite (at 1470 °C) involve bond rupture and bond forming and are termed **reconstructive transformations**. Obviously, reconstructive transformations require substantially more energy than displacive transformations.

2.2 Electroactive Ceramic Oxides

Crystalline coating films of many binary and ternary heavy metal oxides are of great importance for their **electronic** and **opto-electronic** properties. A non-exhaustive list is shown in Table 1. Ternary oxide systems are particularly interesting. Much attention has been paid to ferroelectric thin films for opto-electronic applications, capacitor and non-volatile IC-memories in micro-

Table 1 *Thin film applications and properties of some metal oxide materials*

Application	*Materials*	*Properties*
Packaging/IC substrates	Al_2O_3, BeO	high insulation, high thermal conductivity, low permittivity
Capacitors	$BaTiO_3$, rare earth oxides	high permittivity, low dielectric loss, controllable coefficients of thermal expansion, high breakdown voltages
Piezoelectric transducers and surface acoustic wave devices	$Pb(Zr,Ti)O_3$, $BaTiO_3$, $LiNbO_3$	high piezoelectric coefficients, high coupling coefficients
Thermistors	$BaTiO_3$	change of resistance with temperature
Varistors	ZnO	change of resistivity with applied field
Pyroelectric detectors	$Pb(Zr,Ti)O_3$	change of polarization with temperature
Electro-optic components	$LiNbO_3$, $Pb(La,Ti)O_3$	change of birefringence with electric field
Gas sensors	ZrO_2, SnO_2	ionic conductivity, surface controlled conductivity
Display devices	WO_3, Pb(La,Ti)O, indium tin oxide (ITO)	high contrast images, relatively good fatigue characteristics
Transparent electrodes	SnO_2	good transparency (380–650 nm), high conductivity
Electromagnetic signature control	$Na_4Zr_2Si_3O_{12}/Y_3Fe_5O_{12}$	low coefficients of thermal expansion, high temperature stability, high electromagnetic damping characteristics
Solar battery electrodes	Cd_2SnO_4	high current density
Insulating glass coatings	SnO_2, F-doped SnO_2	good infrared reflectivity, hard

electronics (see also Chapter 9). Important examples are $BaTiO_3$ (BT), $Ba_{1-x}Sr_xTiO_3$ (BST), $PbTiO_3$ (PT), $PbZr_{1-x}Ti_xO_3$ (PZT), La-doped PZT (PLZT), $Pb(Fe_{1/2}Nb_{1/2})O_3$ (PFN), and $Pb(Mg_{2/3}Nb_{2/3})O_3$ (PMN). These materials are characterized by **large dielectric constants**, PFN having a magnitude of 2000 at 60 Hz. All exhibit the perovskite-type crystal structure as does the antiferroelectric material $PbZrO_3$, and $LiNbO_3$ which forms **piezoelectric films** used in surface acoustic wave devices. Highly efficient **ferroelectric** and **ferromagnetic** materials can result if there is crystalline orientation in the coating films. Coating films of **electronic conductor oxides** are also important as electrodes and for other purposes. Here indium tin oxide (ITO) and Cd_2SnO_4 (CTO) have found tremendous application in display devices and solar battery electrodes, respectively.

Several methods are employed for the synthesis of electroactive ceramic materials either as powders or as films. For ceramic powders, a useful and attractive method is **hydrothermal synthesis**. This involves reactions between heterogeneous phases, and the physical characteristics of the powder are determined by the interaction between the solid and fluid phases. An illustrative example of this is the formation of crystalline tetragonal $BaTiO_3$ from $Ba(OH)_2$ and TiO_2 at 240° in the presence of chloride ions whereas without chloride the metastable cubic modification is obtained. Non-stoichiometric solid solutions may be obtained either by melting together the appropriate amounts of the individual components, *e.g.* $La_{1-x}Sr_xTiO_3$ $(0<x<1)$, or by dc arc-melting the reactants, *e.g.* $Ln_xBa_{1-x}TiO_{3-\delta}$ (Ln = La, Nd, Gd, Er) from $BaTiO_3$, Ln_2O_3, anatase, and Ti.

In recent years **sol-gel methodology** (see Chapter 9 and Section 4 below) has been applied extensively in the preparation of ceramic powders. Although giving good quality products, the procedures are invariably empirical in nature, and rely on the determination of suitable processing conditions by trial-and-error techniques. Many variables including solvent, concentration, order of addition of reactants, reaction times, ageing times, *etc.*, can exert a profound and largely unknown influence on the production of the desired product. A good example is provided by the formation of MVO_5 (Nb, Ta) mixed oxides from MCl_5 and $VO(OBu^t)_3$ in Pr^iOH. This is a complex process involving several steps via an empirically determined procedure leading to a sol of undetermined structure which is aged to a gel, dried to a xerogel, and finally crystallized to give MVO_5. Apart from the initial reactants and the final product essentially nothing is known about the intermediate steps.

For some oxides systems, notably ITO, the preferred method of film formation is by **sputtering** or **chemical vapour deposition** (see Chapter 9, Section 1.3.3). Films of ITO prepared by dip- or spin-coating from solutions show conductivities lower by an order of magnitude than those prepared by CVD. This inferior property is due to porosity from the insufficient connection of constitutent conducting particles – a direct result of the method of preparation. Films of many other oxide materials are formed from solutions containing a suitable mixture of precursor reagents in a similar fashion as for sol-gel procedures. Generally, the procedure comprises making a solution of a suitable metal alkoxide or carboxylate precursor, adding the second metal reagents followed by 'modifiers' such as β-diketones or organic bases added to control some of the hydrolysis/condensation steps, and allowing the mixture to form a sol. At an appropriate time this mixture is used to dip- or spin-coat the substrate, which is then processed thermally to give the ceramic film. Examples of perovskite-type ferroelectric oxide ceramic films which have been obtained by this method are $SrTiO_3$, $BaTiO_3$, and $Pb(Mg_{1/3}Nb_{2/3})O_3$ (PMN).

In order to optimize the **electrical properties** of films, it is extremely important to control particle size and crystallite orientation within the film. The heterogeneity which develops in a multi-component material is kept to a minimum if the particle sizes are restricted to nano- or even sub-nanometre dimensions. Compared to conventional ceramic processing, the homogeneity gain using nanosized sols is up to 3 orders of magnitude.

Processing temperature can have a profound effect. Processing at very high temperatures leads to poor film properties arising from crystallite disorder. One of the benefits of colloidal sol processing is a substantial reduction of sintering temperatures, slow crystal growth, and the production of extremely small crystallites. In addition to improved electrical properties, other benefits accrue from the use of lower processing temperatures including the ability to apply films to a greater variety of substrates and devices.

The development of crystalline microstructures in films is a complicated process, where both nucleation and crystal growth play important roles. Conventional processing methods lead to relatively large domain sizes, although it known that **improved electrical properties** arise from materials containing well ordered small crystallites. It should be possible to obtain microstructures by designing molecular precursors in which the different components have positions close to those required in the final structure. This leads to crystalline films with domains on the nanometre scale which exhibit good electrical properties otherwise observed only for the much larger domain dimensions obtained under conventional processing conditions.

An elegant example of what can be termed the molecular structural unit synthetic approach is provided by the formation of $CdNb_2O_6$ from the bimetallic molecular precursor $CdNb(\mu\text{-}OAc)_2(\mu\text{-}OPr^i)_4(OPr^i)_6$. In this molecule the two metals are linked by oxygen atoms and possess the $\{MO_6\}$ stereochemistry required in the final columbite-type structure. The ceramic is obtained by hydrolysis and thermal treatment at a fairly low temperature of *ca.* 600 °C. The synthesis has two other features worthy of note with respect to the programme proposed here. The bimetallic precursor is obtained by the reaction of $Cd(O_2CMe)_2$ with $Nb(OPr^i)_5$ in toluene or hexane. However, no reaction occurs with $Nb(OEt)_5$. Secondly and quite remarkably, $Cd(O_2CMe)_2$ is insoluble in hydrocarbon solvents, but does dissolve in these solvents in the presence of $Nb(OPr^i)_5$! These two observations serve to illustrate the profound influence of the choice of alkyl group on such syntheses, and the quite unexpected behaviour of some compounds which may otherwise be excluded from selection as potential precursors.

3 GLASSES

Glasses are amongst the oldest of man-made materials, having been employed for both their **cutting edge** as well as for their **optical properties**. The nature of glass melting changed very little between the earliest known artifacts until the end of the eighteenth century when physicists began to understand the nature of heat and chemists started to isolate and identify many individual elements. Many inventions have been crucial in the development of glass engineering, for example the continuous tank furnace in the mid-nineteenth century. Since then, and particularly in the last few decades, glass technology has advanced greatly. New applications for glasses are continually being found, and new technologies are driving the search for specialist, high performance glasses.

3.1 What is a Glass?

The most frequently used **definition of a glass** is that proposed by the American Society for Testing Materials in 1945: 'Glass is an inorganic product of fusion which has cooled to a rigid condition without crystallizing'. Subsequent developments in processing technology have permitted materials to be produced by methods other than cooling from a melt, *e.g.* non-crystalline solids obtained by sol-gel methods, chemical vapour deposition, or sputtering, and such materials have an equally strong claim to be included in any definition of a glass.

Structurally, glasses have amorphous, *i.e.* **liquid-like structures**, but behave as **solids** at room temperature. They also undergo a **glass transition** at a particular temperature, T_g, or, more typically, over a range of temperatures. This transformation range is the range of temperatures at which structural relaxation times are measurable. Within this range different physical properties such as density, viscosity, heat capacity, *etc.* can be changed reversibly by appropriate heat treatment. However, as different properties involve different mechanisms of relaxation, the value of T_g depends on the property being used to measure it.

Volume changes which occur on cooling of a melt are illustrated diagrammatically in Figure 1. If crystallization takes place on cooling a sharp decrease in volume (increase in density) occurs at the melting point (T_f) due to the greater order and compactness of the crystal structure (point *c*) compared to the melt (point *b*). However, if the melt supercools, no sudden decrease in volume is observed at the melting point. Instead, the volume continues to decrease steadily

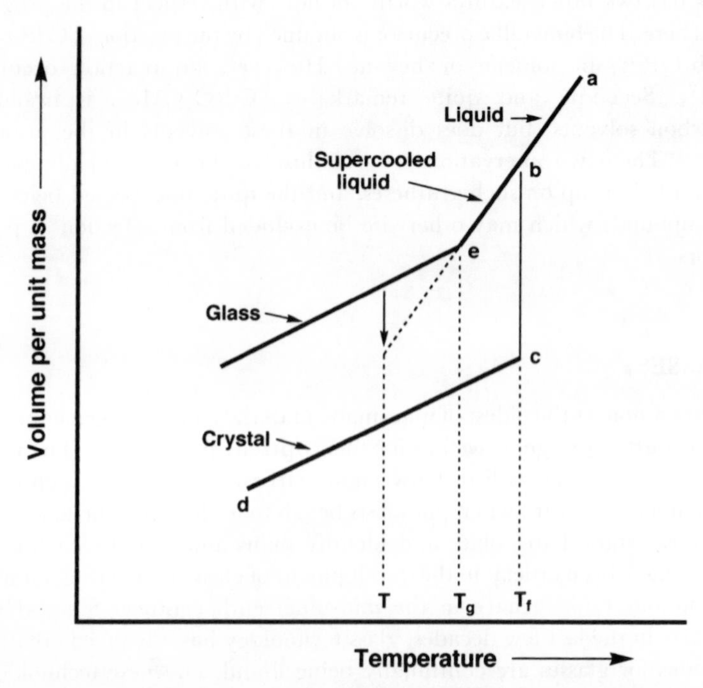

Figure 1 *Plot showing the change in volume with time during glass formation*

along the same line until the glass transition temperature, T_g, is reached (point *e*). This decrease is partly due to the decreasing magnitude of atomic vibrations, but is also due to the melt changing structure to become more compact with falling temperature. As the temperature falls these structural changes occur more slowly due to increasing viscosity until the viscosity becomes so high that no further rearrangement is possible. At the glass transition temperature (point *e*), the gradient of the volume/temperature plot decreases indicating the transition between the supercooled liquid and a glass. The change is not sudden, and the temperature at which the glass transition occurs depends on the rate of cooling of the melt. If the melt is cooled more slowly the point *e* would move further along the dashed line beyond point *e* causing the formation of a more dense glass with a correspondingly lower transition temperature.

The structure of a glass may be described in terms of its **fictive temperature**, that is the temperature at which the existing structure would be at equilibrium. Clearly this must lie in the transformation rate and will be lower for a more slowly cooled glass than for a glass of identical composition but cooled at a faster rate. This can be illustrated by imagining a glass at a temperature T' above T_g, being cooled infinitely quickly to a temperature T well below T_g. The glass formed will have the same structural configuration as the glass at T', *i.e.* the equilibrium configuration. The fictive temperature of the glass at T is T'. Unfortunately, since the fictive temperature is determined by comparison of some property of the glass with that of a glass cooled rapidly from a known temperature, the fictive temperature varies with the property chosen to observe it.

Although glasses are, as implicitly stated in the idea of a fictive temperature, not in the equilibrium state for the material at a temperature below the glass transformation temperature, it is wrong to assume that glasses become more crystalline with time. Indeed, glass artifacts produced four thousand years ago are no more crystalline now than when they were manufactured. This is because at temperatures significantly below T_g, the changes in the glass structure are infinitely slow. Glasses, therefore, may be considered to be kinetically stable or in a state of metastable equilibrium.

3.2 Types of Silicate Glass

There are many different types of glass, although by far the majority of glass produced is based on silica. The earliest glasses were based on silica, but it is only relatively recently that pure silica glass has been produced. Silica glass is the only single component oxide glass to be produced, but is difficult to work with because of its high working temperatures and high viscosity which limits the shapes of objects which can be formed. Nevertheless, it has very good chemical durability and is transpartent to visible and ultraviolet radiation.

The most common silicate glass is **soda lime–silica glass**, with a typical composition of $15Na_2O–10CaO–70SiO_2$. This has much greater chemical durability than $Na_2O–SiO_2$ and is used in the most common applications such as window glass and bottles. In applications where chemical durability and heat resistivity are very important borosilicate glasses with a typical composition of

$5Na_2O$–$15B_2O_3$–$80SiO_2$ are used, but the high cost of these prohibits very wide usage except when absolutely necessary. Many different types of glass may be made in which different metal ions are incorporated into the glass via the metal oxides. **Crystal glass** contains lead ions, introduced into the glass as PbO. This has the effect of joining the glass network rather than disrupting it, and produces a glass with a high dielectric constant, high refractive index, and high resistivity. Addition of alumina to an alkali silica melt produces a glass with a viscosity about 1000 times greater than that of a typical soda lime–silica glass and has extremely good resistance to devitrification. Such glasses are employed in situations where other glasses might devitrify.

3.3 The Structure of Glasses

According to the **Zachariasen–Warren hypotheses**, glasses maintain short range order but not long range order. In silicate glasses this is via the tetrahedral unit, and in the glass phase these tetrahedra are not uniformly and symmetrically interconnected as is the case for crystalline silicates. In the simplest case, that of silica itself, the crystalline forms (*e.g.* quartz, cristobalite, tridymite) the tetrahedra are arranged in a very regular way (Figure 2). In contrast, in amorphous silica slight variations in bond angles of the bridging oxygen atoms cause an irregular interconnected spatial network of tetrahedra to be formed (Figure 3).

Zachariasen's theory formulates the following rules for the formation of glasses from simple compounds such as silica:

(a) If the smallest building unit of the oxide easily forms polyhedral building groups then it will tend to form a glass.

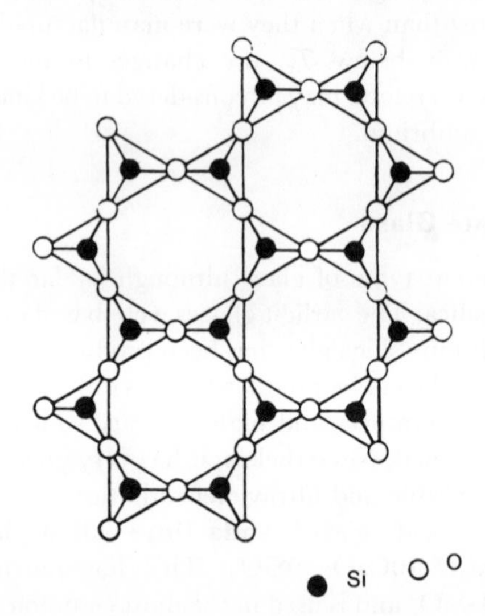

Figure 2 *Illustration of the regular structure of crystalline silica*

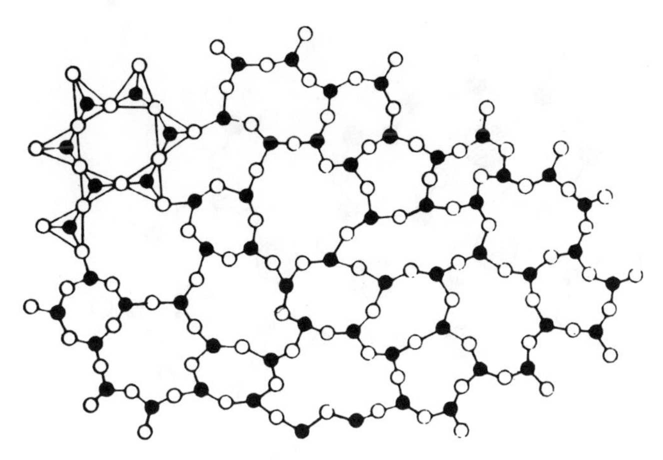

Figure 3 *The structure of amorphous silica*

(b) Any two such polyhedra must not have more than one vertex in common.
(c) Anions must not be linked to more than two central atoms of a polyhedron, *i.e.* they form bridges between polyhedra.
(d) The polyhedron must have less than six vertices.
(e) At least three corners of a polyhedron must be linked to adjoining polyhedra.

If large cations are incorporated into the structure, for example by fusing sodium oxide with silica, bridge failures occur caused by the anion (*i.e.* oxide) introduced with the cation occupying the vertex of one tetrahedron while the cation fills the vacancy produced in the open lattice as illustrated in Figure 4. Glasses may be considered to have a structure containing both amorphous and ordered zones with the ordering being maintained over only a few unit cells.

Clearly, if too much sodium oxide is incorporated into the glass, then it is not possible to form a network of $\{SiO_4\}$ tetrahedra. Because of this it is not normally possible to make a glass with a sodium:silicon ratio greater than 1:1. If this ratio is exceeded an average of two of the four oxygen atoms surrounding each tetrahedron become non-bridging and a glass does not form.

Zachariasen categorizes cations which participate in glass formation into three groups:

(a) **network formers**: these predominantly have coordination numbers of 3 or 4, and include Si, B, P, and Ge
(b) **intermediate oxides**: these predominantly have coordination numbers in the range 4–6, and include Al, Zn, Mg, and Pb
(c) **network modifiers**: these predominantly have a coordination number of 6, and include Na, K, and Ca

As their name suggests, intermediate oxides exhibit behaviour which lies between that of network formers and network modifiers. In a silicate glass, for

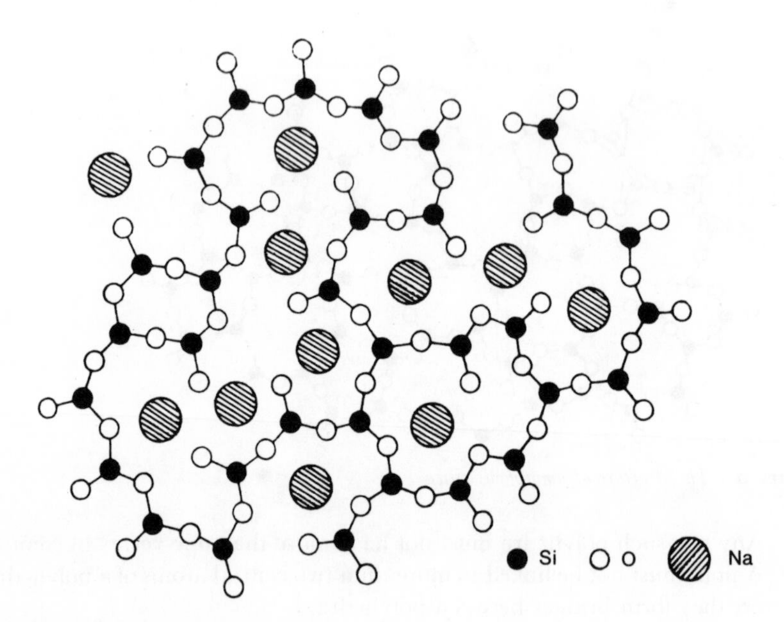

Figure 4 *The structure of sodium silicate glass*

example, Al^{3+} can replace Si^{4+} and form tetrahedra if a single Na^{+} cation is present for each $\{AlO_4\}$ tetrahedron formed so that electrical neutrality is maintained. Since this produces no non-bridging oxygen ions, the glass structure is reinforced. Alternatively, if an insufficient number of sodium cations are present, the aluminium may function as a modifier of coordination number 6 and be surrounded octahedrally by oxygen ions, and so disrupt the continuous network. Clearly, any change in the structure of a glass will alter its properties.

3.4 The Preparation of Glasses

In 1945 all glasses were produced by cooling a melt. Since then, however, alternative preparations have been devised, but these tend only to be used for the production of specialist high quality glasses and the majority of glass is still produced by a similar process to that used fifty years ago. The important steps in glass manufacture are the melting of the batch, fining, homogenizing, and heat conditioning. All these processes take place concurrently in different parts of a large glass tank furnace capable of producing large quantities of silicate glasses for most everyday applications such as windows, bottles, and car windscreens. The end use of the glass determines the properties required and hence also determines the composition used.

The components required in the glass are mixed together in a batch and the conversion of this batch into a useful glass occurs as it travels through the furnace. This process is brought about by a variety of physical and chemical processes which occur in the furnace. These are:

(a) evaporation
(b) formation and loss of gaseous constituents formed by decomposition of batch constituents (*e.g.* CO_2, SO_2, SO_3, *etc.*)
(c) formation of liquid phases from melting of some of the batch constituents and also from multicomponent mixtures
(d) solution of liquid phases formed to yield a homogeneous melt free from crystalline material
(e) volatilization of components from glass and batch (*e.g.* Na_2O, K_2O, F_2)
(f) solution of gas phases in the melt leading to either a metastable equilibrium or a discharge of gas

A commercial silicate glass batch comprises a mixture of approximately seven to twelve individual components. The bulk of the batch needs to contain **silica**; hence the principal ingredient is **sand**. Large quantities of limestone, dolomite, soda ash, and feldspar are frequently present as well as compounds of lead and barium. These other oxides are added to the melt to produce more specialist glasses. The remainder of the batch is made up of several additional minor ingredients present in only small quantities ($<1\%$), and are added to fulfil specific functions in the glass preparation process. These include **glass fining agents** ($NaNO_3$, NaCl, various sulfates), **glass decolorizing agents** (selenium, cobalt oxide, rare earth compounds), **glass colouring agents** (oxides or salts of iron, cobalt, nickel, manganese, copper, cadmium, chromium), **oxidizing agents** (various sulfates), reducing agents (carbon, sulfur, sulfides), melting aids (sulfates, fluospar), **descumming agents** (Na_2SO_4, other sulfates), **opacifiers**, and **nucleating agents**. Also included in most batches (between 15–30%) is **scrap glass** ('cullet'), recovered from trimming the produced glass or from poorly finished glass which does not meet the intended specification. The cullet is useful in decreasing the time taken for the batch to melt.

The melting phase begins when the batch is introduced into the furnace and is generally considered to be complete when the melt is free from crystalline materials which were present in the batch. It is desirable to reduce the time taken to complete melting since this allows the fining and homogenizing processes to reach completion under better conditions and therefore increase the quality of the glass produced. The most obvious way of increasing the melting rate is to increase the temperature of the furnace. This, however, does introduce some problems. Excessively high temperatures reduce the refractory life, and there is the associated cost increase of heating the furnace to these increased temperatures. There is also some evidence that raising the temperature above a certain level may actually promote segregation during melting. These problems make it useful to be aware of other factors which may influence the melting rate. These include:

(a) Glass composition. Reducing the silica content of the glass generally increases the melting rate, *i.e.* reduces the temperature required to obtain batch free glass in *ca.* one hour. This effect can be attributed to the continuous network of silica being interrupted by cations.
(b) Grain size of the batch constituents. When the grain sizes of all raw materials

are the same, the melting rate increases as the grain size decreases. In contrast, if the grain sizes are different the melting rate is decreased due to batch separation prior to melting.

(c) Quantity and grain size of the cullet. When the grain sizes of the raw materials differ, the addition of a cullet increases the melting rate. When the grain sizes of the raw materials are similar, the cullet has very little effect unless present in very large quantities.

(d) The melting rate is also affected by batch homogeneity and formulation, but in a very complex way.

Fining is the part of the process which removes bubbles from the melt and commences when the batch is introduced into the furnace and continues into the refining chamber. One of the principal limits on the fining time is the viscosity of the glass. Clearly, the greater the viscosity the slower will be the rate of rise of bubbles within the melt.

Generally, the fining action of a particular agent reaches a maximum after a small amount of reagent has been added. Exactly how fining agents function is not fully understood. Both physical and chemical processes contribute in varying degrees, and the composition of the glass itself is sometimes a factor in the effectiveness of a fining agent. One of the most frequently used types of fining agents are metal sulfates, especially **sodium sulfate**. As sulfates are generally insoluble in silicate melts, they tend to be concentrated at phase boundaries. At a temperature of 1350 °C the sulfate starts to decompose liberating sodium ions which diffuse through the phase boundary. This migration disrupts the interface violently and so stirs the melt, and if the boundary is the liquid–gas interface bubbles reach the surface. At 1450 °C the decomposition pressure of the sulfate reaches one atmosphere and large bubbles of sulfur oxides are formed. These rapidly rise to the surface of the melt sweeping away any smaller bubbles in their path.

When applied to a glass, **homogeneity** is a relative term, and complete homogeneity probably never exists in a glass. Practically, however, any glass in which no variation in properties can be determined is said to be homogeneous. The degree of homogeneity achieved is a reflection of the entire preparation process of the glass. In addition to processes taking place in the furnace such as convection and stirring, which themselves are affected by the temperature and time spent in the furnace, the homogeneity is also dependent on the batch materials and on the degree of mixing of the batch on entering the furnace. Better control over each stage of the glass-making process can produce increased homogeneity in the final glass.

Heat conditioning is the process which brings the glass to the uniform temperature required for successful forming. As the viscosity of a glass changes rapidly with the temperature, it is necessary to ensure that the temperture of the glass during the forming is consistent. The heat transmission properties of a glass are affected by the absorption of the glass in the near infrared. This depends on the composition of the glass and, if a uniform temperature is to be reached easily, it is important that the Fe^{2+} and OH^{-} content of the glass are controlled. Therefore, the rate at which the glass reaches a uniform temperature depends on batch composition as well as furnace design.

3.5 Properties of Silicate Glasses

Glasses are rarely employed as structural load bearing materials. The only common application of a glass for its high strength is in the use of **glass fibres** for the **reinforcement** of **plastics** and **cement**. Rather, glasses are brittle materials and, like most brittle materials, are weak in tension but surprisingly stronger in compression. Fracture stresses for glasses are typically in the range 50–100 MN m^{-2} (*cf.* high strength steels can be as strong as to fail at 3000 MN m^{-2}).

Glasses with greatly improved strength can be produced in several ways, all of which depend on either ensuring that no damage is done to the glass surface or by removing the damaged surface layer. The surface layer can be removed by drawing the glass through a flame at temperatures in excess of 1000 °C. This procedure volatilizes the surface material to give strengths as high as 1400 MN m^{-2}. Alternatively, the glass surface may be removed using an etchant containing hydrofluoric and sulfuric acids. Unfortunately, such high strengths are of little practical use since there is no method of completely protecting the surface. Flaws are quickly reintroduced and most of the high strength is soon lost.

In practice, glass is **strengthened** by preventing existing cracks (Griffith's cracks) from growing. As crack growth only occurs if there is a tensile stress at the tip; strengthening can be achieved by introducing compressive stresses parallel to the glass surface and to a depth greater than the maximum crack depth. If this is done then it is necessary to apply a load large enough to neutralize the compressive stress before the crack tip experiences a tensile stress. By so doing, the strength of a glass may be increased by a factor of five.

Such compressive stresses can be introduced by either chemical or thermal processing. Thermal strengthening (or **toughening**) is employed to produce car windscreens, oven doors, and tumblers, but it is more difficult to toughen more intricate shapes. The process involves heating the glass to 750–800 °C, after which the surface of the glass is cooled rapidly by air jets. This introduces a compressive stress in the surface of the glass and a corresponding stress in the centre. When the glass fractures, the strain energy in the glass is rapidly released, mainly in the form of the large surface area of the resulting fragments. This is the cause of the small size of the fragments formed when a windscreen shatters. These small pieces are less dangerous than the large pieces which would be formed otherwise, and this is the main reason why windscreen glass is toughened, rather than for its increased strength.

Chemical strengthening may be used to cast more complicated or thinner shapes than thermal toughening, but takes more time. The glass to be strengthened is immersed in a salt solution containing potassium ions whereupon exchange takes place. Sodium ions diffuse into the solution from the glass and are replaced by the larger potassium ions. This size difference results in compressive stresses being set up at the glass surface. Since this method depends upon diffusion whereas thermal toughening depends upon heat flow, different stress distributions are produced. As the stresses in the chemically strengthened glass are lower, the stored strain energy is less, and larger more dangerous fragments are produced when the glass fractures.

4 THE SOL-GEL PROCESS AND CERAMICS

4.1 The Basic Chemistry of the Sol-gel Process

Sol-gel technology enables the production of glasses and ceramics through **chemical reactions** at near **ambient temperatures**. Materials can be prepared in a variety of forms (*i.e.* bulk, powder, wire, thin film, aerogel, *etc.*) suitable for a wide range of optical, structural, and electroceramic applications (see Chapter 9). The use of low temperature routes also makes it possible to extend the process to ORganically MOdified CERamics (ormocers) and other materials suitable for applications in areas such as nonlinear optics and biosystems. Exciting prospects exist for technological exploitation as new inorganic and inorganic/organic composites can be produced which exhibit novel properties and improved performance.

Sol-gel processing* is extremely complex and comprises many steps, and many factors are important in the formation of the final product including the precursor chemistry, hydrolysis and condensation reactions, gelation and drying. A summary of the sol-gel process is shown in Figure 5. It should be noted here that the route to the preparation of a sol is common to all final products. It is at the gelation stage and subsequent treatment stages that materials with grossly different physical appearance and often physical properties result.

Sol-gel materials can be produced by/from colloidal or alkoxide routes, utilizing largely aqueous or non-aqueous environments for the reaction chemistry. The most flexible route is the alkoxide (or organic) based route as it enables homogeneous mixing of reactants at the molecular level (important for multicomponent systems), although there are problems associated with shrinkage and cracking during drying.

The bulk of the work carried out in sol-gel research has concentrated on **silica** and silica based systems. This is because of the **glass forming** ability of silica and the desire to densify materials prior to crystallizing them, as well as the traditional dominance of silica based materials in the application of ceramics. However, a large number of other elements have also been studied (see Section 4.3.3 below).

The chemical reactions taking place in the early stages of the process are the following (M is any electropositive element and R any organic group):

* Terms commonly employed in sol-gel processing:

Sol: A sol is a colloidal system where particles or large molecules (1–500 nm diameter) are dispersed in liquid.
Gel: A gel is a colloidal system of solid character in which the dispersed substance (minimum 1–3% by weight) forms a continuous, ramifying coherent network that is interpenetrated by a system (usually liquid). Thence, the sol-gel process is the transition of a system of colloidal particles in a solution into a disordered, branched, continuous network, which is interprenetrated by liquid. Gels are usually classified according to the dispersion medium used, *e.g.* hydrogel or aquagel, alcogel, and aerogel (for water, alcohol, and air respectively).
Xerogel: These materials are produced when the liquid within the gel is removed by simple evaporation.
Aerogel: Aerogels are obtained when the liquid within the gel is removed above its critical temperature and pressure (*i.e.* under supercritical or hypercritical conditions).
Cryogel: Cryogels are produced from gels in which the liquid phase is first frozen into a solid and then sublimed. The vapour is removed by vacuum pumping.

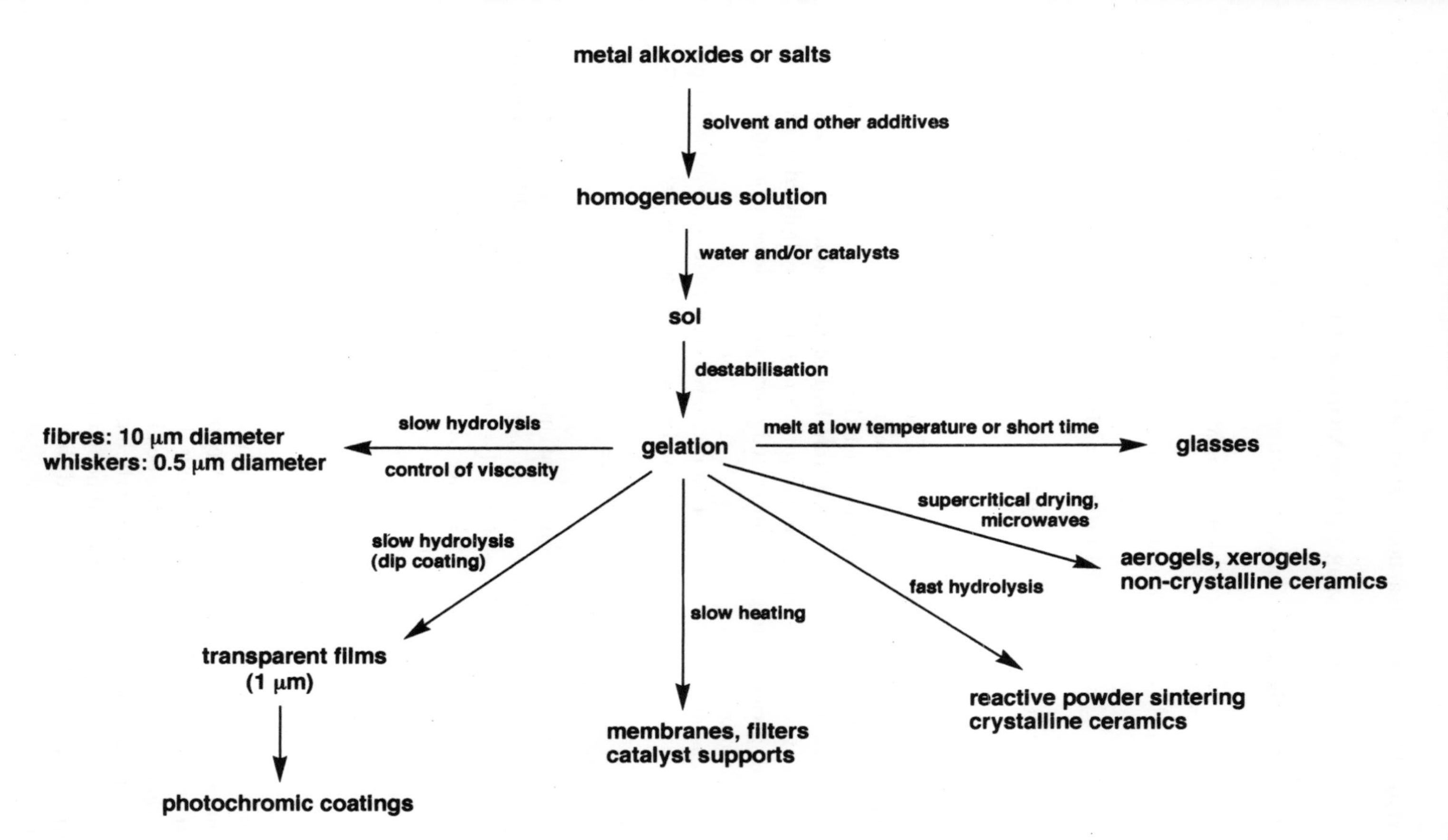

Figure 5 *Important stages of the sol-gel process*

(a) Hydrolysis (forward reaction), esterification (reverse reaction):

$$MOR + H_2O \rightleftharpoons MOH + ROH \quad (1)$$

(b) Alcohol condensation (forward reaction), alcoholysis (reverse reaction):

$$MOR + HOM \rightleftharpoons MOH + ROH \quad (2)$$

(c) Water condensation (forward reaction), hydrolysis (reverse reaction):

$$MOH + HOM \rightleftharpoons MOM + H_2O \quad (3)$$

The overall reaction for the formation of a silica gel is:

$$Si(OR)_4 + 2\,H_2O \rightleftharpoons SiO_2 + 4\,ROH \quad (4)$$

The silica produced according to Equation 4 also contains many residual hydroxyl/alkoxy groups depending on the method of formation.

In essence, the polymeric gel (from which the dried material monolith, powder, film, fibre, *etc.* is produced) is formed from hydrolysis products of the alkoxide, which react with each other liberating water, or from the hydrolysed or partially hydrolysed species reacting with unreacted precursor to release an alcohol as the condensation by-product.

Other reactions in which water is not required to initiate the polymerization pathway, and which give ethers as the condensation by-products, may also occur:

$$MOR + MOR \rightleftharpoons MOM + ROR \quad (5)$$

In the synthesis of multi-component materials by the sol-gel route (*e.g.* electroactive ceramics) it is necessary to take into account the different reaction rates of the precursors. The rate of hydrolysis of silicon alkoxides is slow compared with those of other alkoxides, and hence in order to obtain an homogeneous gel, the preferential hydrolysis of one precursor must be prevented.

Hydrolysis and **condensation** reactions (Equations 1–3) occur concurrently but at different rates, and so once hydrolysis has occurred, it is unlikely that the two reactions can be separated. These reactions have important consequences for the prehydrolysis of slowly reacting precursors as the formation of partially hydrolysed species can lead to self-polymerization of the precursor, which in turn can lead to a decrease in the overall homogeneity of the final material.

The order and the relative rates of the hydrolysis and condensation reactions are the most important parameters defining the initial nature of the gel. Reaction conditions for the polymerization may be divided into two general types: (a) low water concentration and low pH, and (b) high water concentrations and high pH. It should be noted, however, that although there is much information available relating to the mechanisms concerned there is not total agreement as to which operates.

(a) Low water/acid catalysis. Under these conditions it is generally accepted that hydrolysis takes place via substitution of an alkoxy group for an hydroxyl group. Both three- and five-coordinate silicon species have been proposed as the transition state intermediates. The chemically most sensible intermediate (and the one for which there is a greater consensus of opinion) is the five-coordinate intermediate which results from a biomolecular S_N2 nucleophilic substitution reaction. In this reaction sequence (Figure 6), the alkoxide is protonated by the acid, thus increasing the acidity of the group and allowing the central silicon atom to be attacked from the rear by a water molecule. The water acquires a partial positive charge and consequently partially reduces the charge on the alkoxide thus making it a better leaving group.

In all cases where the transition state acquires a partial charge (whether it be positive or negative) inductive effects will be important. For hydrolysis of precursors already containing Si–O–Si bonds it is important to maintain the structural integrity of the polymerized units and this happens if the water to alkoxide ratio is kept below *ca.* 15. Under these conditions, water attacks preferentially the silicon alkoxide bonds in preference to the Si–O–Si bonds.

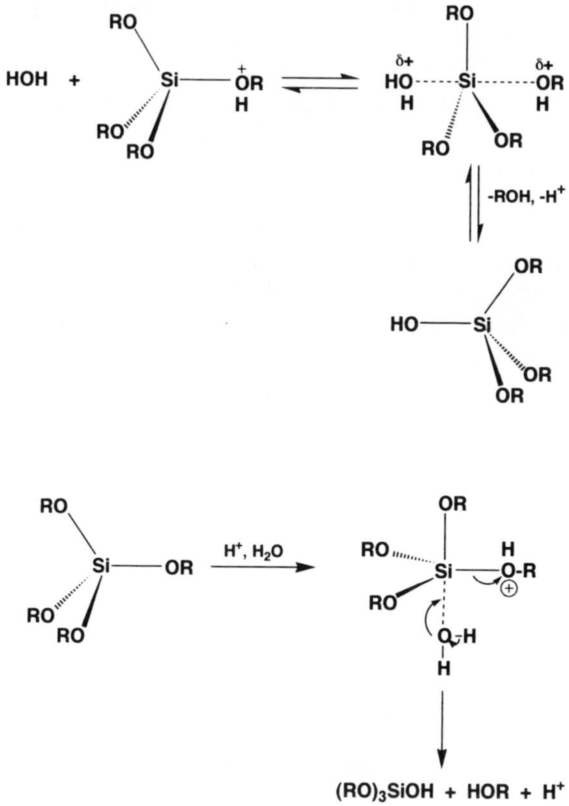

HOH + (RO)₃Si–$\overset{+}{O}$R(H) ⇌ [HO(H)$^{\delta+}$···Si(RO)(RO)(OR)···$^{\delta+}$OR(H)] —(-ROH, -H$^+$)→ HO–Si(OR)₃

(RO)₃Si–OR —(H^+, H_2O)→ [(RO)₃Si–$\overset{\oplus}{O}$(H)-R ··· O(H)-H] → $(RO)_3SiOH$ + HOR + H^+

Figure 6 *Mechanistic stages in the low water, acid catalysed hydrolysis of tetraalkoxysilanes*

(b) High water/base catalyses. A pentacoordinate transition state is proposed for this type of reaction with the hydroxyl anion directly attacking the electropositive silicon resulting in a partial negative charge developing on the silicon atom (Figure 7). Redistribution of charge occurs with the partial negative charge being accommodated by an alkoxy group. The alkoxy group is then able to leave the reactive intermediate, a proton being abstracted from a readily available water molecule, with additional hydroxide ions being generated in the process.

Alcohol exchange (transalcoholysis) may also occur, although the precise mechanism of this process is not known. Not surprisingly, the exchange mechanism is dominated by steric effects, although not exclusively so, and is also favoured by acid catalysis.

$$Si(OR)_4 + x\,R'OH \rightleftharpoons Si(OR)_{4-x}(OR')_x + x\,ROH \tag{6}$$

The process is thought to occur via an S_N2 type reaction with reaction rates in the order primary alcohols > secondary alcohols > tertiary alcohols.

The effect of alcohol exchange is to produce a variety of species for hydrolysis, and hence modifications of the hydrolysis rates of the precursor molecules. The presence of sterically demanding alcohols in solution or alkoxy groups attached to the central silicon atom leads to changes in the proximity of reactive species and hence to changes in condensation rates. The major consequence of this is to produce higher concentrations of partially hydrolysed species and thereby decrease the gelation time, so affecting the physical nature of the resulting materials. **Gelation** occurs fastest in methanol and the pore volume decreases with increasing ramification of the alkoxy group, ($Me > Et > Pr^n > Pr^i$).

Condensation occurs via a **nucleophilic condensation mechanism** and water or alcohols may be released (Figure 8). Water release is favoured unless water concentrations are very low when alcohol liberation becomes significant. At low pH, the most basic silanols are protonated first (monomeric singly hydrolysed species rather than oligomers). Protonation of the silanol increases the acidity of

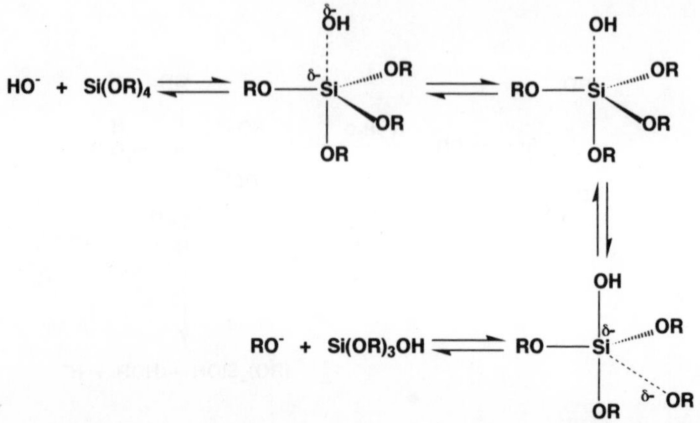

$$HO^- + Si(OR)_4 \rightleftharpoons RO\text{—}\overset{\delta-}{Si}(OR)_3\cdots\overset{\delta-}{O}H \rightleftharpoons RO\text{—}\overset{-}{Si}(OR)_3\cdots OH$$

$$RO^- + Si(OR)_3OH \rightleftharpoons RO\text{—}\overset{\delta-}{Si}(OH)(OR)_2\cdots\overset{\delta-}{O}R$$

Figure 7 *Mechanistic stages in the high water, base catalysed hydrolysis of tetraalkoxysilanes*

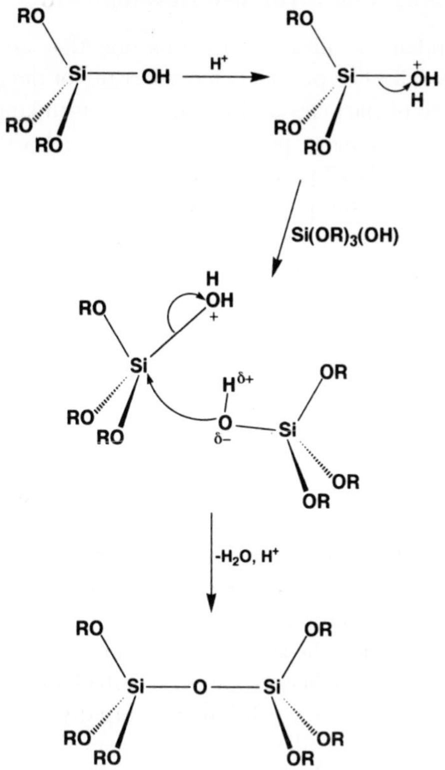

Figure 8 *Mechanistic stages of condensation in the hydrolysis of tetraalkylsilanes*

the group and renders the silicon atom more susceptible to attack by a nucleophile.

Under basic conditions condensation is expected to proceed via a penta- or hexa-coordinated transition state involving the deprotonation of a silanol, the oxygen acquiring a formal negative charge and becoming the nucleophile for attack on another silicon centre. As the acidity of the groups attached to the silicon increases, the greater will be the stabilizing effect on the expanded coordination transition states. Note that these reactions are subject to the same steric and inductive effects as for the hydrolysis reactions.

Hydrolysis and **condensation** reactions do not occur in isolation from one another and the final material which results derives from a combination of the two reactions. There are a wide range of parameters which affect the nature of the final material including the nature of the precursor(s), pH, type of catalyst used, the solvent, the water/alkoxide ratio, the addition sequence, the atmosphere above reaction vessel, the time of mixing, and the relative rates of hydrolysis for multi-component systems. *All* of these factors will affect the final structure of the gel, and they also have effects on the mechanism and rate of reaction of both hydrolysis and condensation reactions. How these two reactions interrelate is of extreme importance in determining the final outcome of the sol-gel procedure used.

4.2 Gelation, Ageing, and Structure Development

With time the condensed silica species link together to become a **three-dimensional network**. The physical characteristics of the gel network depend greatly upon the size of particles and extent of cross-linking prior to gelation. Acid catalysis leads to a more polymeric form of gel with linear chains as intermediates. Base catalysis yields colloidal gels where gelation occurs by cross-linking of the colloidal particles. At gelation, the **viscosity** increases sharply, and a solid object results in the shape of the mold. With appropriate control of the time-dependent change of viscosity of the sol, fibres can be pulled or spun as gelation occurs.

The gelation point of any system, including sol-gel silica is easy to observe qualitatively and easy to define in abstract terms but extremely difficult to measure analytically. The sol becomes a gel when it can support a stress elastically. It is extremely difficult to define the point at which a distinct transition from sol to gel occurs. However, the sharp increase in viscosity that accompanies gelation essentially freezes in a particular polymer structure at the gel point. At this point gelation may be considered a rapid solidification process.

Gelation is nearly instantaneous for alkoxides under strongly acidic or strongly basic pH conditions. The amount of water has a dramatic effect on the gelation times. In the silica system, gelation may occur in as short as 10 minutes but can be as long as 7 hours. Increasing the size of the solvent molecule and increasing the size of the alkoxide also lead to increases in the length of time to gelation (*cf.* effects on hydrolysis rates). Under basic conditions a gel rapidly forms but the majority of the precursors are at that point still unreacted. Under acidic conditions most of the precursors have been partially hydrolysed and have taken part in condensation, but still have M–OR bonds.

Ageing is the process which occurs when a gel is left in its mother liquor after gelation and reactions continue such that the structure and properties of the material change significantly. During this time, polycondensation continues along with localized solution and reprecipitation of the gel network which increases the thickness of interparticle necks and decreases porosity. The strength of the gel thereby increases with drying. The reason for doing this is to produce a gel of sufficient strength to withstand the stresses encountered during drying.

During the ageing process four reactions can occur, often simultaneously: (a) polycondensation, (b) syneresis, (c) coarsening, and (d) phase transformation.*

* *(a) polycondensation:* Provided that silanol groups are close enough to react they will condense to increase the connectivity of the network. This is particularly true of acid catalysed gels. The process can be accelerated by hydrothermal treatment.

(b) syneresis: This arises from expulsion of liquid from the pores and results in gel shrinkage. It is thought to be associated with bridge bond formation and can be controlled by the addition of electrolyte.

(c) coarsening: This is the irreversible decrease in surface area due to dissolution and reprecipitation events. The pore diameters usually increase during this phase of the reaction. Strong acids (HCl, HNO_3, and H_2SO_4) increase the pore volume of the gel but without lowering the surface area, by promotion of coalescence between particles without particle growth and coalescence. Addition of acids at lower concentrations (*e.g.* 8 M sulfuric) results in a dramatic reduction in surface area. Ageing in alkali leads to dissolution of silica and formation of networks with larger pore sizes.

Different methods of ageing lead to materials with different structures. Gels shrink on drying to give small pore volume and diameter. If they are wet-aged they increase in coalescence and shrink little on drying, yielding materials with larger pore sizes than materials that have not been wet-aged. Further heat treatment stabilizes these materials and structure coarsening leads to **reductions in surface area**, an increase in pore diameter but no change in pore volume.

As well as changes to **surface area** and **porosity**, the **mechanical properties** of the gel are also modified during ageing. Due to the increase in bridging bonds the stiffness of the gel network increases, as does the elastic modulus, the viscosity, and the modulus of rupture. This is important when considering the drying stage since the greater the stiffness of the aged gel the less shrinkage during drying.

During drying liquid is removed from the interconnected pore network. If the pores are small (<20 nm) large capillary stresses develop during drying. These stresses cause the gels to crack catastrophically unless the drying process is controlled by decreasing the liquid surface chemistry by addition of surfactants or elimination of very small pores, by hypercritical evaporation, which avoids the solid–liquid interface, or by obtaining monodisperse pore sizes by controlling the rates of hydrolysis and condensation.

There are three **stages of drying**:

Stage 1: the decrease in volume of the gel is equal to the volume of liquid lost by evaporation. The compliant gel network is deformed by the large capillary forces which causes shrinkage of the object. This behaviour may be modified when the system contains small <20 nm diameter pores. During this first stage, the **rate of loss of solvent** from the material is **constant**. Most of the water present in the pores is 'free' and its removal does not affect the bound layer surrounding the silica network.

Stage 2: this stage begins when the strength of the network has increased, due to the greater packing density of the solid phase, sufficient to resist further shrinkage. At this stage liquid is driven to the surface by gradients in capillary pressure, where it evaporates due to the ambient vapour pressure being lower than inside the pores. At this point in the drying process, the materials turn opaque, starting at the edges and progressing linearly towards the centre. The most plausible explanation for this phenomenon is that the pores (isolated or in groups) are of such a dimen- sion that they are able to scatter light. There is an associated increase in open porosity and a reduction in moisture content of the sample from *ca.* 15 to 6%. Most failures (cracks) occur in the early part of this stage.

Stage 3: This stage is reached when the pores have substantially emptied and **surface films along the pores cannot be sustained**. Under these conditions, the rate of loss of solvent is reduced dramatically and materials can be treated thermally at temperatures of *ca.* 180 °C without sample cracking.

Drying an alcogel requires the removal of the interstitial liquid and its replacement with a dry gas. The method used to remove the liquid influences the characteristics of the dried gel. There are many problems with the development of internal tensile forces. These may be eliminated by preventing the liquid–vapour boundary from developing. This can be achieved by supercritical drying in an autoclave. If the solvent in the pore is suddenly put under pressure greater than its critical pressure and then slowly heated to a temperature 20–25 °C above the critical temperature, the liquid will expand but not boil at such a high pressure. At the **critical point**, the liquid ceases to be a liquid in the thermodynamic sense and is termed a fluid. During this process, no interfacial tension is created. Both ethanol and carbon dioxide can be used for supercritical drying although carbon dioxide is the system of choice because of the lower temperatures involved. The conditions for supercritical drying are as follows – MeOH: 240 °C/81 atm; CO_2: 35 °C/70 atm pressure.

Dehydration, **dilation**, and **contraction** of the silica network are all necessary in the production of fully dense materials or in the production of porous gel silicas. In all cases, surface dehydration is essential prior to full densification in order to prevent water molecules from being trapped in the glass. Dehydration can be achieved using a variety of chlorinated compounds such as carbon tetrachloride which can react completely with surface hydroxyl groups to form hydrochloric acid. This then desorbs from the gel in the temperature range 400–800 °C when the pores are still interconnected. Dehydration at temperatures of up to a few hundred degrees centrigrade is reversible and physisorbed water is readily removed. Chemisorbed water (directly associated with silanol groups) is more difficult to remove and does not take place to any significant extent until temperatures significantly higher than 400 °C are used.

Heating the porous gel at high temperatures causes densification to occur. The pores are eliminated and the density ultimately becomes equivalent to fused quartz or fused silica. At temperatures above *ca.* 400 °C, removal of water occurs with associated Si–O–Si bond formation from adjacent hydroxyl groups. This process is irreversible but only occurs to any major extent at temperatures above 700 °C. Above 800 °C only isolated silanol groups exist and these can also be removed by extended thermal treatment.

For **optical applications**, complete transmission over a broad range of wavelengths is essential and all water must be removed from the sample. There are very few examples where this has been achieved for materials produced in air. The use of chlorine here has proved useful although again it is essential that chlorine groups are removed from the system before full densification occurs.

The fully dense gel silicas have excellent transmission from 160 to 4200 nm with no OH adsorption peaks. The UV cutoff is shifted to lower wavenumbers by removal of OH from the glass. The advantages of these materials over high-grade fused silica are net-shape casting, and lower coefficients of thermal expansion (0.2×10^{-6} cm/cm compared with 0.55×10^{-6} cm/cm).

4.3 Sol-gel and Aqueous Chemistry of Metal Oxides

The sol-gel chemistry of metal alkoxides incorporates several nucleophilic substitution reactions of the general type:

$$-\mathrm{M{-}OR} + \mathrm{XOH} \rightarrow -\mathrm{M{-}OX} + \mathrm{ROH} \tag{7}$$

and includes hydrolysis (X=H), condensation (X=M), or transalcoholysis (X=R′).

Most industrial processes are based on inorganic precursors since they are much cheaper and easier to handle than metal alkoxides in organic solvents. However, the **aqueous chemistry of metal salts** appears to be much more complicated than that of alkoxide systems owing to the occurrence of spontaneous hydrolysis and condensation reactions in the aqueous medium. These reactions depend on many parameters such as pH, concentration, or temperature.

The general reaction in the hydrolysis of metal cations is

$$[\mathrm{M(OH_2)}_N]^{z+} + h\,\mathrm{H_2O} \rightarrow [\mathrm{M(OH)}_h\mathrm{(OH_2)}_{N-h}]^{(z-h)+} + h\,\mathrm{H_3O^+} \tag{8}$$

Hydrolysis leads to a set of deprotonated monomeric species ranging from aquo-ions, {M–(OH_2)}, to hydroxy species, {M–OH}, and oxo-anions, {M=O}. The equilibria which are established depend on many parameters such as the pH of the solution, temperature, and concentration. The goal is, therefore, to see if it is possible to predict the value of the hydrolysis ratio h and which species can act as precursors for condensation at a given pH. Proton exchange occurs between the precursor and the solution until the mean electronegativity of the hydrolysed precursor χ_p becomes equal to the mean electronegativity χ_w of the aqueous solution. The relationship between χ_w and pH is given by:

$$\chi_w = 2.732 - 0.035\mathrm{pH}$$

Thus, proton exchange between the precursors and the medium occurs until χ_p is equal to χ_w. The hydrolysis ratio h may then be calculated at any pH value, and that of a given precursor depends on both the pH of the aqueous solution and the oxidation state $z+$ of the cation. Condensation reactions are expected to occur when the system is brought from the aquo (low valent cations) or oxo (high valent cations) into the hydroxo one. Therefore pH conditions for condensation can be predicted.

Two main mechanisms are known for condensation:

(i) *olation:* where the nucleophilic attack of a negatively charged OH onto a positively charged metal cation leads to the departure of an aquo ligand from the coordination sphere of the metal. This leads to the formation of "ol" bridges:

$$\mathrm{M{-}OH} + \mathrm{M{-}OH_2} \rightarrow \mathrm{M{-}OH{-}M} + \mathrm{H_2O} \tag{9}$$

This process only occurs when aquo-precursors have their maximum coordination number.

(ii) *oxolation:* where nucleophilic addition of OH groups onto metal ions is followed by 1,3-proton transfer within the transition state (M–OH...M–OH), forming an 'oxo' bridge with removal of a water molecule.

The stability of the M–OH bond in aqueous media can be analysed by the Partial Charge Model:

(a) If the metal ion is more electronegative than hydrogen then hydroxyl groups exhibit acidic behaviour and dissociate in water. Examples include electronegative elements in the top right hand corner of the Periodic Table and highly charged transition elements such as Mn(VII).
(b) If the metal ion is less electronegative than the hydrogen ion then charge is transferred towards the hydrogen atom of the hydroxyl bond. At one extreme, base dissociation of ions such as Na^+ and Ba^{2+} can occur. An electronegativity range can be defined where the M–OH bond remains stable in an aqueous medium. Condensation can proceed as long as the partial charge on oxygen δ_O is less than zero. Five groups of cations can be distinguished:

  (i) those which undergo acid ionization leading to inorganic acids which do not condense in aqueous solutions
  (ii) those which undergo basic ionization leading to inorganic bases which do not condense in aqueous solutions
  (iii) Those which condense only through olation leading to hydroxide precipitation
  (iv) those which condense through olation and oxalation leading to hydrous oxide ($MO_{z/2}.xH_2O$) precipitation
  (v) those which condense only through oxolation leading to molecular polyacids.

4.3.1 Oxolation and Polyanions. High valent cations ($z > 4$) give rise to oxo-hydroxo anions $[MO_x(OH)_{m-x}]^{(m+x-z)-}$ in aqueous solutions. Condensation is only possible via oxolation as no water molecules are coordinated to the metal ion. Protonation of anionic species is usually performed by acid. Two mechanisms for oxolation are possible depending on whether coordination expansion of the metal ion is possible or not:

(i) **Nucleophilic addition** via M–OH or M=O groups occurs if the coordination state can be increased. No water molecules are removed and chains and rings are formed very rapidly. The mechanism leads to edge or face sharing $[MO_n]$ polyhedra.
(ii) If the metal atom already exhibits the maximum coordination state then

reaction occurs by **nucleophilic substitution** and leads to corner sharing polyhedra. Nucleophilic addition of an OH group is quickly followed by proton transfer towards the leaving ligand.

$$\text{M–OH} + \text{M–OH} \rightarrow \text{M–O–M–OH} \rightarrow \text{M–O–M–OH}_2 \qquad (10)$$

The formation of water is catalysed by base (first step) or acid (second step) and occurs over a wide range of pH. At room temperature, highly condensed species are obtained around the point of zero charge but the partial charge on the hydroxyl groups must be negative in order for the oxolation to occur.

For Cr(VI) aqueous solutions, the fate of both $CrO_2(OH)_2$ and $[CrO_3(OH)]^-$ species must be considered.

$$h=7 \qquad 2\,[CrO_3(OH)]^- \rightarrow [Cr_2O_7]^{2-} + H_2O \qquad (11)$$

Condensation occurs by oxolation but stops at this stage as no more hydroxyl groups are available in the dimer.

$$h=6 \qquad 2\,[CrO_2(OH)_2] \rightarrow [(HO)O_2Cr\text{–}O\text{–}CrO_2(OH)] + H_2O \qquad (12)$$

OH groups are still available in the dimer but they become positively charged meaning that condensation stops at this stage. Owing to the small size of Cr(VI), coordination expansion is not possible and only dimeric species are found in solution. Coordination expansion is possible for both Mo(VI) and W(VI) species leading to more complex behaviour in aqueous solutions. The effect of this is to extend the numbers of polynuclear species which are possible, and cages based on Mo_6, Mo_7, Mo_8, W_{10}, and W_{12} are all known.

4.3.2 Olation and Polycations. Olation is observed with aquo-hydroxo precursors $[M(OH)_h(OH)_2)_{N-h}]^{(z-h)+}$ where $h<n$ and $z<5$, and it leads to the formation of OH bridges via the formation of bridging $H_3O_2^-$ species.

$$\text{M–OH} + \text{M–OH}_2 \rightarrow \text{M–OH–M} + H_2O \qquad (13)$$

$$\text{or M–OH} + H_2O\text{–M} \rightarrow \text{M–O}\cdots\text{H–O–M} \rightarrow \text{M–OH–M} + H_2O \qquad (14)$$

As water ligands have to be removed from the metal coordination sphere the olation rates depend upon the lability of the $M–OH_2$ bonds. Rates increase with increase in ionic radius, decrease in oxidation state, and for transition metals. Crystal field stabilization effects have to be taken into account.

The early stages of condensation can be readily followed for Cr(III) complexes (d^3). The initial condensation reaction leads to a corner sharing dimer (green) then to an edge sharing dimer (blue). At this stage the hydroxo ligands have a small positive partial charge and react no further. Addition of base leads to hydrolysis of the complex (the dimer is more acidic than the monomer) and further addition of monomeric species gives rise to trimeric and tetrameric species.

Figure 9 shows an example from the chemistry of titanium, illustrating the importance of the timing of the olation and oxolation reactions in the production of different crystallographic forms of titania.

4.3.3 Solution Chemistry of Alkoxide Precursors. Figure 10 indicates elements for which alkoxides have been characterized. Many of these species have also been used for sol-gel preparations although their cost is often prohibitive for industrial usage. Alkoxides of the d^0 transition metals (*e.g.* Ti, Zr), are widely used as molecular precursors to glasses and ceramics. The differences between these and the silicon alkoxides are as follows:

(i) Their lower electronegativity causes them to be less stable towards hydrolysis, condensation, and other nucleophilic reactions.
(ii) Transition metals often exhibit several stable coordinations and can expand their coordination, and oligomers formed by alkoxy bridges are common.
(iii) The greater reactivity of the transition metal alkoxides requires stricter controls of processing conditions such as moisture, *etc.* in order to get homogenous materials.
(iv) The rapid kinetics of nucleophilic substitution make fundamental studies of the reactions and the species present difficult.

4.3.4 Mechanisms of Hydrolysis and Condensation. For coordinatively saturated metals in the absence of catalysts, reactions occur by nucleophilic substitution (S_N) mechanisms involving nucleophilic addition followed by proton transfer from the attacking molecule to an alkoxide or hydroxo-ligand within the transition state and removal of the protonated species as either alcohol (alcoxolation) or water (oxolation):

$$H_2O: + M{-}OR \longrightarrow H_2O{:}\rightarrow M{-}OR \longrightarrow HO{-}M\leftarrow{:}O(H)R \longrightarrow M{-}OH + ROH \qquad \text{hydrolysis}$$

$$M{-}O(H){:} + M{-}OR \longrightarrow M{-}O(H){:}\rightarrow M{-}OR \longrightarrow M{-}O{-}M\leftarrow{:}O(H)R \longrightarrow M{-}O{-}M + ROH \qquad \text{alcoxolation}$$

$$M{-}O(H){:} + M{-}OH \longrightarrow M{-}O(H){:}\rightarrow M{-}OH \longrightarrow M{-}O{-}M\leftarrow{:}OH_2 \longrightarrow M{-}O{-}M + H_2O \qquad \text{oxolation}$$

When $N - z > 0$, condensation can occur by olation:

$$\mathrm{M{-}O{-}H} + \mathrm{M \leftarrow :O(H)R} \longrightarrow \mathrm{M{-}O(H){-}M} + \mathrm{ROH}$$

$$\mathrm{M{-}O{-}H} + \mathrm{M \leftarrow :OH_2} \longrightarrow \mathrm{M{-}O(H){-}M} + \mathrm{H_2O}$$

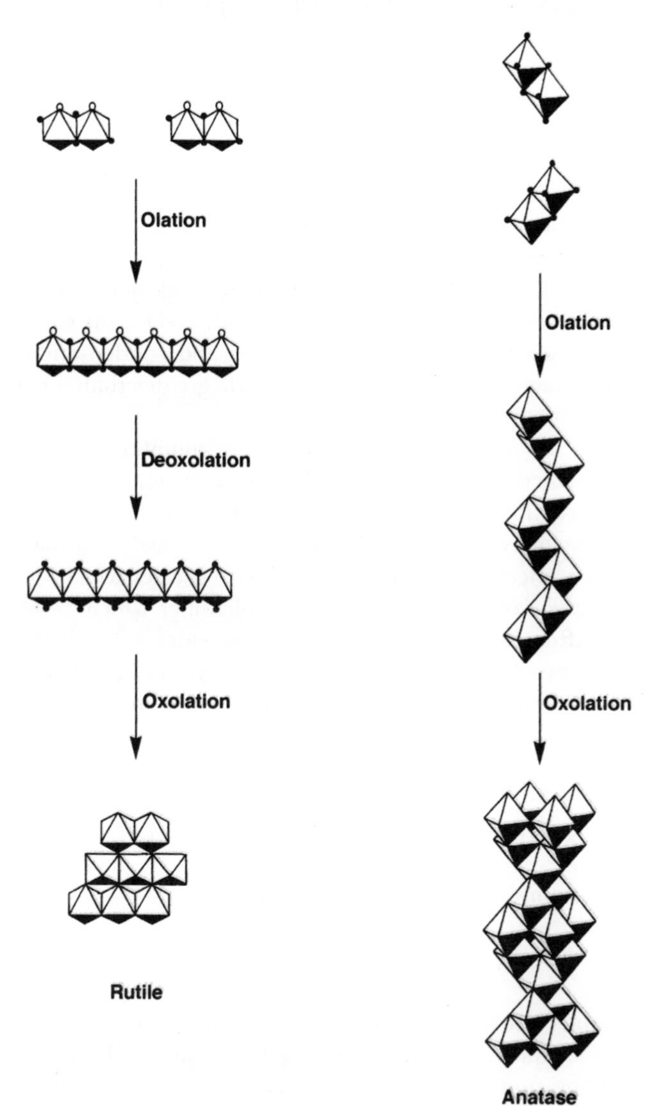

Figure 9 *Condensation pathways for the evolution of (left) the rutile and (right) the anatase modifications of* TiO_2

H																	He
Li	**Be**											**B**	C	**N**	**O**	**F**	Ne
Na	**Mg**											**Al**	**Si**	**P**	**S**	Cl	Ar
K	**Ca**	**Sc**	**Ti**	**V**	**Cr**	**Mn**	**Fe**	**Co**	**Ni**	**Cu**	**Zn**	**Ga**	**Ge**	**As**	**Se**	Br	Kr
Rb	**Sr**	**Y**	**Zr**	**Nb**	Mo	Tc	Ru	**Rh**	**Pd**	Ag	**Cd**	**In**	**Sn**	**Sb**	**Te**	I	Xe
Cs	**Ba**	**La**	**Hf**	**Ta**	W	Re	Os	Ir	**Pt**	**Au**	Hg	Tl	**Pb**	**Bi**	Po	At	Rn
Fr	Ra	Ac															

Ce	**Pr**	**Nd**	Pm	Sm	**Eu**	**Gd**	**Tb**	Dy	**Ho**	Er	Tm	**Yb**	**Lu**
Th	Pa	**U**											

Figure 10 *Periodic Table showing which element alkoxides are known (elements in bold)*

Information obtained for the **titanate system**, for example, suggests that alcoxolation rather than oxolation is the favoured condensation reaction between partially hydrolysed, coordinatively saturated titanate precursors. This is in contrast to that found for the silicon system where oxolation is the preferred reaction. The rates of reaction are somewhat different, and for the corresponding titanium and silicon alkoxides the hydrolysis rates and condensation rates for titanium are in the range of 4–5 orders of magnitude greater than for the silicon system.

Reaction kinetics are also affected by molecular complexity (*e.g.* $Ti(OEt)_4$ exhibits oligomeric structures whereas $Ti(OPr^i)_4$ remains monomeric) and also by correct choice of reaction conditions (*e.g.* $Zr(OPr^n)_4$ in Pr^nOH produces a precipitate, but in cyclohexane a homogeneous gel is obtained). The nature of the R group also has an effect on the size of oligomeric species which develop. In general, the larger the size of R, the smaller the oligomer which develops.

Chemical modification of transition metal alkoxides with alcohols, acids, and bases, and chelating ligands is commonly employed to retard hydrolysis and condensation reaction rates in order to control the condensation pathway of the evolving polymer. Reaction may occur by nucleophilic substitution (coordinatively saturated metal):

$$x\,XOH + M(OR)_z \rightarrow M(OR)_{z-x}(OX)_x + x\,ROH \tag{15}$$

or nucleophilic addition (coordinatively unsaturated metal):

$$x\,XOH + M(OR)_z \rightarrow M(OR)_z(XOH)_{N-z} \tag{16}$$

The modifying ligands (usually more electronegative) are removed more slowly (if at all) during the condensation reaction thereby reducing the functionalities available for condensation leading to the production of less highly condensed products and promoting gelation. Common modifying ligands include alcohols, chloride, acetate, acetylacetonate, *etc.* The use of these ligands is particularly

Table 2 *Optical applications of sol-gel derived materials*

Optical effect	*Application*	*Compositions*
light transmission	optical wave guide	$PbO–SiO_2$, $SiO_2–TiO_2$
		$Na_2O–B_2O_3–SiO_2$, SiO_2
optical absorbtion	UV shielding	$TiO_2–SiO_2$
	optical absorption	Fe_2O_3, CoO, NiO
	colouration of glass	SiO_2 + transition metal oxides
		$CeO_2–TiO_2$, $GeO_2–V_2O_5$
interference	colouration	TiO_2
reflection	radiation shield	$In_2O_3–SnO_2$, $VO_2–SiO_2$
	solar energy reflecting	$PbO–TiO_2$, $Bi_2O_3–TiO_2$
anti-reflection	increase of damage threshold of fusion laser glass, clear vision	$Na_2O–B_2O_3–SiO_2$, SiO_2 $TiO_2–SiO_2$
fluorescence	optical shutter	CdS, $CdS_{1-x}Se_x–Al_2O_3$
	switch, optical IC	SiO_2, $Sr_{1-x}Ba_xNb_2O_6$
		$LiNbO_3$
fluorescence	luminescent solar	SiO_2 doped with organics
	concentrator lasers	$Al_2O_3$with organic dyes
electrochromism	displays	WO_3
	transparent electrode for ion passing	$CeO_2–TiO_2$
patterning	optical ROM disk	B_2O_3, TiO_2, $ZrO_2–SiO_2$

important when multicomponent materials are required as they promote the reduction of hydrolysis and condensation rates, and also the numbers of positions available for these reactions, such that in mixed silica/metal oxide systems the formation of mixed M–O–Si bonds can be favoured.

4.4 Optical Materials

A wide range of sol-gel materials (with very diverse compositions) show optical effects and have applications in areas as diverse as solar energy reflection, solar concentrators, lasers, and optical ROM (Read Only Memory) disks (Table 2).

Problems in the preparation of silica glass monoliths arise because of residual impurities and silanol groups within the glass which can be trapped within the pores during the drying/thermal treatment process and leads to disintegration of glass samples. The use of **drying chemical control additives** (DCCAs) such as formamide greatly increases the ability to produce pieces up to 20 cm in diameter. The **formamide** (added at the mixing stage) appears to function by reducing the evaporation rate of solvent and producing a more uniform pore size distribution in the gel network thereby minimizing differential drying stresses and producing more uniform growth of the network during ageing which increases the strength of the gel. The glasses which result can either be fully dense or porous, and for the latter have surface functionality dependent upon the treatment regime which also determines their optical quality. The lowest UV cutoff known for a bulk silica is found for these materials and ranges from 250 to 300 nm.

The porous glasses (15–35% interconnected porosity) are chemically and thermally stable and can serve as a host matrix for organic polymers/dyes and metal ions. **Laser densified waveguides** have been produced from these materials. As these materials are porous they can be impregnated with 30–40% by volume of a second-phase such as an optically active organic or inorganic compound.

The greatest market for optical glasses (beyond silica) lies in the production of compositionally complex multicomponent materials whose optical and other properties can be tailored by variations of composition and thermal history. Most success in this area lies in the area of **thin films** where the problems of drying are vastly reduced relative to those encountered in the preparation of monoliths.

4.4.1 Graded Refractive Index Materials. These are generally fabricated by the ion exchange of glass rods using molten salts. Larger pieces of material can, in principle, be prepared via the sol-gel route. Two methods have been employed:

(a) Rod shaped wet gels (SiO_2–GeO_2 or SiO_2–TiO_2 from alkoxides) are immersed in neutral or acidic solutions to leach out part of the dopant under diffusion controlled conditions. Materials are subsequently rinsed, dried, and heated in order to produce the fully densified body.
(b) Gels of composition $26PbO.7B_2O_3.67SiO_2$ [prepared from TMOS, TEOS, $B(OEt)_3$, and $Pb(CH_3COO)_2$] are put into acetone in order to increase the mechanical strength of the material. The compositional gradient of lead is produced by soaking the gel in an ethanolic solution of potassium acetate. The gel is subsequently washed, dried, and then heated as before in order to produce a robust material.

4.4.2 Nonlinear Optical Materials. Silica glasses doped with PbS, CdS, CdTe, ZnS, CuCl, and CuBr can be produced and have χ^3 values of the order of 10^{-4}–10^{-6} esu. Such materials can be prepared by several different routes.

(a) preparation of SiO_2–MO phases, for instance by addition of the acetate to the silicon alkoxide, with subsequent treatment with H_2S.
(b) addition of cadmium and sulfur [$Cd(NO_3)_2$ and $(NH_2)_2CS$] to the starting solution. Thermal treatment of the bulk gel results in a CdS crystallite doped silica glass which is yellow in colour (see Chapter 12, Section 3.2). Further control of the crystallite sizes is possible by subsequent treatment with hydrogen sulfide gas at elevated temperatures. If small (of the order of 1 nm) crystals are produced these show quantum-confinement effects and it is not clear whether they are behaving as a cluster or as a small piece of the bulk material. Emission bands from 470 to 490 nm are observed for particles of diameter *ca.* 1 nm and 640–670 nm for particles *ca.* 10 nm in diameter.

Electrochromic layers on glass can be produced using the chemistry of mixed valence compounds. Specific optical properties arise from the optically activated hopping of electrons from low to high valence states. This can be effected for thin films of WO_3 in an electrochemical cell.

$$WO_3 + x\,Li^+ + e^- \rightleftharpoons Li_xWO_3 \tag{17}$$

The pure tungsten oxide films are colourless and the lithium tungstate blue. These layers can be prepared from tungstic acid solutions or tungsten alkoxides (from tungsten oxychloride). For the above materials, the switching time and the long term memory decrease with the amount of water added due to faster ion diffusion through the gel network. Thus, the properties of the layer can be tailored to suit specific applications. Other oxides which have been studied include MoO_3, V_2O_5, TiO_2, Nb_2O_5, IrO_2, and NiO. The colour of TiO_2 layers can be modified (grey to blue) by changing the coordination environment around the titanium by modification of the alkoxide with acetic acid. Molecular bronzes can also be made by the *in situ* oxidative polymerization of organic monomers such as aniline or pyrrole into a $V_2O_5.nH_2O$ matrix. The gel layer then exhibits metallic properties with a high optical reflectivity.

5 CARBIDES

Carbides are compounds of carbon in combination with elements of approximately the same or lower electronegativity (*i.e.* other inorganic compounds of carbon such as CO, CO_2, CS_2, or CCl_4 are not carbides). Preparation is usually by one of three methods: (i) direct combination of the elements at high (>2300 K) temperature, (ii) reaction of the metal oxide with carbon at high temperature, and (iii) reaction of the heated metal with the vapour of a suitable hydrocarbon. In some cases the carbide may be obtained by reaction of acetylene with an aqueous solution of the metal salt (*e.g.* silver carbide).

Classification of carbides is not straightforward, but there are three general categories, ionic carbides, covalent carbides, and interstitial carbides. However, several carbides exist which exhibit intermediate properties.

Ionic carbides are formed by the more electropositive metals and the lanthanides and actinides, and exist as colourless transparent crystals comprising close packed metal cations and anions containing one, two, or three carbon atoms. Ionic carbides undergo facile hydrolysis with evolution of hydrocarbons. Beryllium carbide, Be_2C, and aluminium carbide, Al_4C_3, are examples of ionic carbides containing single carbon anions (sometimes referred to as methanides) and give methane on hydrolysis.

The most numerous type of ionic carbide are those containing the C_2^{2-} anion (acetylides), and are typified by calcium carbide, CaC_2. The structure is very similar to that of NaCl with C_2^{2-} anions replacing chloride anions. Hydrolysis gives acetylene, a reaction which has been used as the basis for the storage of acetylene. Carbides of the lanthanides and actinides have the general formulae MC_2 or M_2C_3 [$M_4(C_2)_3$] and are formally similar to the alkaline earth carbides. All have the CaC_2 structure, but the C≡C bond distance shows a distinct increase over that in CaC_2 due to donation of electron density from metal to the antibonding orbital of the $[C{\equiv}C]^{2-}$ anion. Metal–anion interaction also modifies the nature of the hydrolysis reaction where acetylene is no longer the major product and a mixture of hydrocarbons and hydrogen are produced. In contrast to the insulating character of CaC_2, the lanthanide carbides are good conductors of electricity (*e.g.* the conductivity of $LaC_2 \approx$ that of lanthanum metal) due to the availability of free electrons ($La^{3+} + [C{\equiv}C]^{2-} + e^-$). Only a single example is known of an ionic compound with three carbon atoms in the anion, Mg_2C_3, which gives either methyl acetylene or allene on hydrolysis depending on the pH.

The two principal examples of **covalent carbides** are silicon carbide and boron carbide. **Silicon carbide**, SiC, forms a lattice comprising a three-dimensional array of silicon and carbon atoms analogous to the structures of diamond and elemental silicon. In the ideal form each silicon atom is surrounded tetrahedrally by four carbon atoms and each carbon atom is surrounded by four silicon atoms. Structures are based on the hexagonal α-SiC (Wurtzite ZnS form) and the cubic β-SiC form (zinc blende form). However, imperfections lead to a large number of polymorphs. Industrially SiC is obtained by the reaction of high-grade quartz with excess coke or anthracite in an electric furnace at $>2000\,°C$. Silicon carbide is extremely hard (9.15 Mohs) (hence its trivial name 'carborundum' from *carbo*n and co*rundum*) and chemically inert, and finds extensive application as an abrasive. It exhibits an extremely low thermal coefficient of expansion and outstanding resistance at temperatures even above the melting point of steel, and is often used as a coating for metals, carbon–carbon composites, and other ceramics to provide protection at these extreme temperatures. Other uses are as a particulate and fibrous reinforcement in both metal and ceramic matrix composites.

Boron carbide is quite often described as having the composition B_4C, although the structure shows it to actually be $B_{13}C_2$. The compound is closely related to β-rhombohedral boron and contains icosahedral B_{12}units connected by linking linear C–B–C units. That the bonding is not purely covalent is indicated by the significant electrical conductivity of the crystals. It is chemically very unreactive but reacts with molten alkali giving the borate and carbonate. It is very hard yet unusually lightweight and finds uses in applications requiring excellent abrasion resistance and as a portion of bulletproof armour plate, although it has poor properties at high temperatures.

Interstitial carbides arise from the filling by carbon atoms of octahedral interstices in the close-packed metal structure. Since the number of octahedral interstices equals the number of metal atoms, the limiting stoichiometry in MC, and the resulting structure corresponds to the NaCl structure in which metal atoms occupy the Cl^- positions and the carbon atoms the Na^+ positions. Occupation of half the octahedral interstices leads to the stoichiometry M_2C. These metal carbides exhibit properties characteristic of metal alloys (*e.g.* opacity, conductivity, metallic lustre) indicating that the free electron structure of the metal is not drastically modified by the inclusion of carbon. In addition, the carbide is often harder and has a higher melting point than that of the metal itself, demonstrating the presence of metal–carbon bonding. Often carbides adopt the cubic NaCl structure even though the metal does not itself possess a close-packed structure. The size of the metal is important in determining the properties of the carbide. Large metals (those with radii >135 pm, including Ti, Zr, Hf, V, Nb, Ta, Mo, and W) have interstices large enough to accommodate carbon atoms without any distortion of the cubic lattice. The interstices in smaller metals (those with radii <135 pm, *e.g.* Cr, Mn, Fe, Co, and Ni) are not large enough to accommodate carbon atoms without significant distortion of the lattice, and these metals do not form carbides with the ideal MC composition. Rather, these metals form several carbides of different stoichiometries usually with complex structures. Several (*e.g.* Fe_3C or cementite, a constituent of steel) are also of great importance industrially.

Unlike ionic or covalent carbides, interstitial carbides are resistant to water and generally chemically inert, only being attacked under extreme conditions such as concentrated nitric acid. This relative inertness is most probably due kinetic factors (*e.g.* the inaccessibility of the carbon for attack) since reactions with oxygen and water are favoured thermodynamically. The carbides of the smaller metals such as Fe_3C and Cr_3C_2 are much more reactive due to the severe distortion of the lattice. All react with dilute acids and in some cases water to give hydrogen and hydrocarbons.

A number of interstitial carbides, particularly those of W, Ti, Ta, Co, and Cr, have properties which make them eminently suited for use as cutting tools. Most carbide tools in use today are WC-based (either straight WC or multicarbides of W–Ti or W–Ti–Ta), depending on the material to be machined, with cobalt as the binder. These tools are much harder and chemically more stable; they have better hot hardness, high stiffness, lower friction, and operate at higher cutting speeds than high-speed steels and cobalt alloys. Cemented carbide tool materials based on TiC have been developed primarily for auto industry applications using predominantly Ni and Mo as a binder. These are used for high speed (>1000 ft/min) finish machining of steels and some malleable cast irons. Earlier versions, still widely used, have WC as the major constituent, with a cobalt binder in amounts of 3–13%. Often only a coating of the abrasion-resistant, hard, and chemically inert metal carbide, usually TiC, is applied to the surface of the cutting tool, most commonly by chemical vapour deposition. Typically a mixture of hydrogen, methane, and titanium tetrachloride gases is presented to the blank cutting tools at a temperature of *ca.* 1000 °C. The deposited TiC also catalyses the reaction which produces the TiC coating.

6 BORIDES

Metal borides are notable for the diverse range of stoichiometries and structural types. Stoichiometries range from metal-rich (M_5B) to extremely boron-rich (MB_{100}), although the most common stoichiometries are M_2B, MB, MB_2, MB_4, and MB_6. Apart for a group of metals found in the middle of the Periodic Table (Zn, Cd, Hg, Ga, In, Tl, Ge, Sn, Pb, and Bi), all metals form binary borides. Particular metals often form several borides of different stoichiometries; for example, chromium and nickel form at least eight and most metals form four or five. Many borides possess properties which allow their application in numerous areas. Metal-rich borides are extremely hard, involatile, chemically inert, refractory materials. Melting points are often much higher than the metals themselves, for example TiB_2, ZrB_2, HfB_2, NbB_2, and TaB_2 all have melting points ≥ 3000 °C. Further, many borides have interesting electrical properties which again are often much higher than the metals (that of TiB_2 is *ca.* five times that of titanium metal). Although normally obtained as powders, metal borides used for high performance engineering purposes (*e.g.* TiB_2, ZrB_2, and CrB_2) may be fabricated into the form desired by usual powder metallurgical techniques.

Metal borides are obtained by a number of different methods including

(a) direct combination of the elements. Here an intimate mixture of finely divided metal and boron are heated to temperatures in the absence of air, *e.g.*:

$$Cr + n\,B \rightarrow CrB_n \qquad (18)$$

Although this method is capable of giving well-defined products free from impurities such as the oxide, care needs to be exercised in the weighing and mixing stage to ensure that the desired stoichiometry is obtained.

(b) reduction of the metal oxide with elemental boron, or boron and carbon, or boron carbide, *e.g.*:

$$Sc_2O_3 + 7\,B \rightarrow 2\,ScB_2 + 3\,BO \qquad (19)$$

When boron alone is used, a substantial amount of the boron is lost as volatile BO. The use of carbon as the coreductant reduces the loss of expensive boron, but introduces the possibility of contamination of the boride by metal carbide. Boron carbide is a useful and cheap source of boron and reacts with most metals and their oxides.

(c) coreduction of a boron halide and metal halide with hydrogen, *e.g.*:

$$2\,TiCl_4 + 4\,BCl_3 + 10\,H_2 \rightarrow 2\,TiB_2 + 20\,HCl \qquad (20)$$

(d) reduction of a boron halide with a metal, *e.g.*:

$$BCl_3 + W \rightarrow WB + Cl_2 + HCl \qquad (21)$$

(e) Coreduction of the metal and boron oxides using carbon, *e.g.*:

$$V_2O_5 + B_2O_3 + 8\,C \rightarrow 2\,VB + 8\,CO \qquad (22)$$

(f) Electrolytic deposition from fused salt mixtures containing B_2O_3 or borax and the metal oxide in a molten alkali metal halide or fluoroborate.

All the methods require high temperatures, usually in the range 1000–2000 °C.

The structures of metal borides vary enormously as the metal:boron ratio changes. Those at the metal-rich end have structures which are dominated by the nature of the metal lattice. In these the small boron atoms occupy octahedral interstices in the close packed metal lattice either as isolated boron atoms (*e.g.* in

Mn_4B, Cr_2B, Co_3B, Re_3B, Pd_5B_2, Re_7B_3, Ru_7B_3, MoB_2, MnB_2, FeB_2, PtB) or pairs of directly bonded boron atoms (*e.g.* in V_3B_2, Nb_3B_2, Ta_3B_2, Cr_5B_2). As the boron:metal ratio increases so does the possibility of direct boron–boron bonding, and in the stoichiometry MB (*e.g.* for FeB, CrB, and MoB) the boron is present as linear zig-zag chains of boron atoms. Nevertheless, the overall structure is still dominated by the close packing of metal atoms. Branched chain (*e.g.* $Ru_{11}B_8$) and ribbon or double chain structures (*e.g.* M_3B_4 (where M = V, Nb, Ta, or Cr) are also formed. Further increase of the boron:metal ratio leads to the formation of two-dimensional and three-dimensional structures.

Most borides exhibiting two-dimensional structures have stoichiometries which approximate to the composition MB_2. The predominant structural type is that of AlB_2 and is also found for the diborides of Mg, Sc, Y, Ti, Zr, Hf, V, Nb, Ta, Cr, Mo, Mn, the lanthanides, U, and Pu, although other structural types are known. The structure comprises alternate planes of metal atoms and hexagonal puckered nets of boron atoms and represents a vast distortion from the close packed structure of the metal. Nevertheless, the trigonal prismatic coordination of boron by metal atoms as in more metal rich borides is still retained.

Borides of stoichiometry MB_4 (M = Ca, Y, Mo, W, lanthanides, and actinides) form a transition between the two-dimensional structures adopted by the diborides and the truly three-dimensional structures of more boron-rich hexaborides. In these the boron atoms form a regular mesh of B_4 squares and B_7 heptagons, although closer examination shows that the B_4 squares are actually fragments of B_6 octahedra.

Metal hexaborides, MB_6, are very numerous and are formed by the alkali metals, alkaline earths, the lanthanides and actinides, Sc, Y, Zr, Cr, and As. The structure consists of a three-dimensional network of B_6 octahedra interconnected in all three directions to afford a robust but open framework which exhibits a very low coefficient of thermal expansion usually in the range $6–8 \times 10^{-6}\ K^{-1}$. The metal atoms occupy the cavities in the structure. Molecular orbital (MO) bonding theory requires the transfer of only one electron to the boron framework (*cf.* the two electrons which are required for $B_6B_6^{2-}$), and hence alkali metal hexaborides are formed readily, those of Ca, Sr, Ba, Eu, and Yb are semiconducting, whilst those of tripositive lanthanide cations and Th^{IV} have high metallic-type electrical conductivity. The structures and properties of these boron-rich borides are now dominated by the bonding demands of boron, and are similar in many respects to the allotropic forms of elemental boron. Indeed, borides with even higher contents contain structural units characteristic of elemental boron. Thus, the dodecaborides MB_{12} (M = Al, Sc, Y, lanthanides, actinides), AlB_{10}, and YB_{66} all contain the icosahedral B_{12} unit.

7 NITRIDES

Like metal carbides, metal nitrides may be classified into three general types: **ionic** nitrides, **covalent** nitrides, and **interstitial** nitrides. Ionic nitrides are formed by metals of Groups 1 and 2 and have low melting points. Not unexpectedly they are easily hydrolysed and are unimportant as ceramics. In contrast, the covalent nitrides formed by the Group 13 and 14 elements and interstitial nitrides formed by the transition metals and the lanthanide and actinide elements have extremely useful properties and find application in an increasing number of areas. In particular, the nitrides of silicon, titanium, boron, and aluminium are important ceramic materials with diverse applications. In the interstitial nitrides, the band structure of the metal is retained, and these nitrides exhibit similar electronic, electrical, and magnetic properties to the metals as well as excellent mechanical behaviour. Indeed, the properties exhibited by some nitride materials, *e.g.* the ferromagnetism exhibited by iron samarium nitride, suggest that nitride materials will replace more common materials in the future and that there is in nitrides an as yet relatively untapped wealth of new materials. More esoteric potential applications include the use of uranium and plutonium nitrides in nuclear reactor fuels. In addition, mixed oxynitrides of some elements are important as ceramic materials.

Quite generally, in thermodynamic terms metal nitrides are more stable than the corresponding carbides but less stable than the corresponding oxides. The most stable nitrides are those of aluminium, cerium, titanium, zirconium, hafnium, and the actinides thorium, uranium, and plutonium. Others, including those of boron, silicon, vanadium, niobium, tantalum, and gallium, are less stable but can be formed. Gallium nitride is becoming more important as the favoured material for electronic display screens. In some instances, as for example with tin, the binary nitride previously thought to be too unstable to be prepared has recently been obtained as synthetic procedures have improved.

7.1 Silicon Nitride

Although known for a long time, the utility of silicon nitride, Si_3N_4, as an **engineering ceramic** was first demonstrated in 1957 and is now produced in large scale as powders for refractory and advanced engineering applications. Although first prepared by the ammonolysis of $SiCl_4$ followed by thermal decomposition of the intermediate imide:

$$3\,SiCl_4 \xrightarrow[-NH_4Cl]{+NH_3} \frac{1}{n}\,[\text{-}Si(NH_2)_2\text{-}]_n \qquad (23)$$

$$\frac{1}{n}\,[\text{-}Si(NH_2)_2\text{-}]_n \rightarrow \tfrac{1}{3}\,Si_3N_4 + \tfrac{1}{3}\,N_2 + 2\,H_2 \qquad (24)$$

the most common commercial method of preparation is by simple heating of the elements:

$$3\ Si(s) + 2\ N_2(g) \rightarrow Si_3N_4(s) \qquad (25)$$

Reaction commences at a temperature of 950 °C, but since the reaction is very exothermic releasing 2080 kJ mol^{-1} care is necessary to prevent the temperature rising sufficiently to melt and agglomerate the silicon (m.p. 1420 °C) reducing the surface area and the conversion rate. The temperature of the reaction is usually kept in the range 1250–1350 °C with iron and alkaline metal fluorides being added to accelerate the reaction.

An alternative method of preparation which has great potential as a cheap source of Si_3N_4 powders is by the carbothermal reduction of silica:

$$SiO_2 + 6\,C + 2\,N_2 \rightarrow Si_3N_4 + 6\,CO \qquad (26)$$

Reaction temperature is 1450 °C, and one per cent of iron nitrate is essential for the stoichiometric reaction with carbon to occur. The reason for this appears to be the formation of a ferrosilicate liquid from which Si_3N_4 precipitates.

Two methods, reaction sintering and liquid phase sintering, are used for the densification and formation of shaped bodies of Si_3N_4. In the reaction sintering process, Si_3N_4 synthesis and consolidation of the product takes place in one step. So-called reaction-bonded silicon nitride (RBSN) is made by nitriding a preform of elemental silicon in nitrogen or ammonia at a temperature below the melting point of silicon, forming Si_3N_4 as a wool in the pores. In spite of the considerable weight increase (66.6%), no increase in dimensions occurs allowing the manufacture of very high precision products.

In liquid phase sintering of Si_3N_4, 4–8% of an oxide such as magnesium, yttrium, cerium, or zirconium is added to the nitride,which is then heated to a temperature in excess of 1650 °C. At this temperature the oxide forms a silicate liquid by reaction with the surface layer of SiO_2 on the Si_3N_4 particle. The Si_3N_4 then dissolves in the silicate liquid forming an oxynitride liquid which when supersaturated with nitrogen precipitates out Si_3N_4 in an elongated form. On cooling the oxynitride liquid forms a glass which remains at the Si_3N_4 particle boundary. Unfortunately, since glasses have a softening point which is significantly below the glass melting point, liquid phase sintering is deleterious for the thermomechanical properties of the material. Sintered Si_3N_4 containing grain boundary glass may, however, be heat treated at *ca.* 1300 °C to crystallize the grain boundary liquid thereby improving the thermomechanical properties. High temperature sintering may also lead to material loss by either decomposition into the constituent elements or by reaction of Si_3N_4 grains with SiO_2.

Crystalline Si_3N_4 occurs in two modifications, α-Si_3N_4 and β-Si_3N_4. Both have very similar structures although the β-form is somewhat more stable by 30 kJ mol^{-1}. In each form the silicon atom is surrounded tetrahedrally by four nitrogens, and the polymeric structure of the β-form, which is that of the phenakite ($BeSiO_4$) type. The principal difference between the two forms is in the stacking, which in the α-form gives rise to long channels in the structure into which may be incorporated heterometal atoms such as yttrium, praseodymium, and neodymium.

The properties of Si_3N_4 are similar to SiC although its oxidation resistance and high-temperature strength are somewhat lower. Both silicon nitride and silicon carbide are potential candidates for components for automotive and gas turbine engines, permitting higher operating temperatures and better fuel efficiencies with less weight than traditional metals and alloys.

7.2 Silicon Oxynitrides

Replacement of silicon by aluminium and nitrogen by oxygen affords a family of mixed silicon aluminium oxynitride ceramics known commonly by the acronym SiAlONs. The structures of **sialons** are based on that of β-Si_3N_4 and the general formula is $Si_{6-n}Al_nO_nN_{8-n}$ where n is the number of aluminium atoms replacing silicon. Sialon phases and solid solutions are quite numerous, and in addition it is possible to obtain metal-substituted sialons of the type $M_n(Si,Al)_{12}(N,O)_{16}$, where M can be Li, Mg, Ca, Y, or a lanthanide (excluding La or Ce).

A typical sialon composition is $Si_3Al_3O_3N_5$ which can be prepared by the carbothermal reduction of meta-kaolin:

$$3\,(Al_2O_3.2SiO_2) + 15\,C + 5\,N_2 \rightarrow 2\,Si_3Al_3O_3N_5 + 15\,CO \tag{27}$$

Other sialon phases may be obtained similarly from other luminosilicates. Sialon crystals are typically embedded in a glassy phase based on Y_2O_3. The resulting ceramic is relatively lightweight with a low coefficient of thermal expansion, good fracture toughness, and a higher strength than many of the other common advanced ceramics. Sialons may find applications in engine components and other applications involving both high temperatures and wear conditions.

Silicon oxynitride, Si_2N_2, is also refractory although its use as a structural material has not been explored. Preparation can be either by the reaction of elemental silicon with silica under an atmosphere of nitrogen at 1450 °C, or by the high temperature reaction of Si_3N_4 and SiO_2.

7.3 Aluminium Nitride and Oxynitride

Unlike Si_3N_4, aluminium nitride, AlN, is moisture sensitive and therefore has not found the same level of application. Its major use is as the substrate material for high density integrated circuits. Several methods have been used for its preparation including the direct nitridation of the metal, carbothermal reduction of Al_2O_3, the reaction of aluminium trihalide with nitrogen and hydrogen, and the thermal decomposition of $(NH_4)_3AlF_6$, all of which need very high temperatures usually well in excess of 1000 °C. Structurally, AlN adopts an hexagonal pseudo-Wurtzite structure, but like all covalent ceramics cannot be self-sintered. Sintering to the full density is achieved by the addition of oxides of yttrium, rare earths, and alkaline earths. The advantages exhibited by AlN are its high thermal conductivity, high electrical resistivity, and low temperature coefficient of thermal expansion.

Powders of various stoichiometry in the Al_2O_3–AlN system can be obtained by reacting the two components or by reduction and nitridation of alumina, and the products adopt the spinel structure. The incorporation of dopants such as boron, yttrium, or lanthanum during sintering imparts useful optical transparency [transparent between 5.12 and 0.27 μm (UV cut-off)].

7.4 Boron Nitride

Several routes to boron nitride have been devised but the principal commercial methods for its manufacture are

(i) the direct reduction of borax:

$$Na_2B_4O_7 + 7\,C + 2\,N_2 \rightarrow 4\,BN + 7\,CO + 2\,Na \tag{28}$$

(ii) reduction of B_2O_3:

$$B_2O_3 + 2\,NH_3 \rightarrow 2\,BN + 3\,H_2O \tag{29}$$

(iii) carbothermal reduction and nitridation:

$$B_2O_3 + 3\,C + N_2 \rightarrow 2\,BN + 3\,CO \tag{30}$$

(iv) by reaction of volatile boron compounds with ammonia:

$$BCl_3 + NH_3 \rightarrow BNH_3Cl_3 \tag{31}$$

$$BNH_3Cl_3 \rightarrow BN + 3\,HCl \tag{32}$$

Purification of the BN is usually by heating at 1900 °C in nitrogen when impurities volatilize. The last method is used to obtain high purity material.

The structure of boron nitride is very similar to the layer structure adopted by graphite, and also exists in two polymorphic forms, hexagonal and cubic. Cubic boron nitride (CBN), developed by General Electric (called Borazon), retains its hardness at elevated temperatures and has low chemical reactivity. Though not as hard as diamond, CBN can be used to machine efficiently and economically difficult materials at higher speeds and a higher removal rate than cemented carbides such as tungsten carbide, and with superior accuracy, finish, and surface integrity. As a covalent nitride, boron nitride does not self-sinter, and is consolidated by addition of B_2O_3 and hot pressing.

7.5 Nitrides of Groups 4 and 5

Both the Group 4 elements (Ti, Zr, and Hf) and the Group 5 elements (V, Nb, and Ta) for nitrides of composition MN. All are high melting, are extremely hard, and exhibit metallic type electrical conductivity. Unfortunately they have poor resistance to oxidation, titanium nitride being the most resistant withstanding oxidation up to *ca.* 700 °C. For this reason only **titanium nitride** has found major application as a gold-coloured coating on high speed cutting tools. The advantages of TiN coatings include reduced tool wear, higher hardness, relative inertness, and a low coefficient of friction.

Three methods have been used for the preparation of these nitrides:

(i) direct reaction of the elements at temperatures above 1000 °C, in which the stoichiometry is controlled by the pressure of nitrogen
(ii) dissolution of the transition metal in molten copper, nickel, or copper–nickel, the solution being treated with a high pressure of nitrogen when the nitride precipitates out (the Menstruum technique)
(iii) reaction of the metal halide with nitrogen and hydrogen at high temperature or in a plasma

Physical vapour deposition (PVD) is the most viable process for TiN-coating high speed steel (HSS), primarily because it is a relatively low temperature process which does not exceed the tempering point of HSS, thereby eliminating the necessity for subsequent heat treatment of the steel. In PVD, sputtering is used to eject atoms, ions, and/or neutral or charged clusters of atoms from a titanium source in a vacuum chamber. The titanium particles travel in more or less straight lines and condense on the parts to be coated. Because the PVD process is primarily line-of-sight, the parts are rotated during the coating process to obtain a uniform coating. Substrate heating enhances coating adhesion and film structure; therefore PVD processes are carried out with the workpieces heated to a temperature in the range 400–900 °C. Because surface pretreatment is critical, tools to be coated are subjected to vigorous cleaning processes typically involving degreasing, ultrasonic cleaning, and freon drying.

8 BIOSOLIDS

Biosolids are materials which are **recognized** and **assimilated by the body**. A high degree of 'bioactivity' is desirable and is measured by 'the ability of a material to form surface hydroxyapatite in a controlled *in vitro* environment'. Tissue response to diverse bioceramic compositions leads to four primary categories of behaviour:

(1) For toxic materials, the surrounding tissue dies.
(2) For non-toxic materials and those which dissolve, the surrounding tissue replaces the implant.
(3) For biologically non-toxic and inactive materials, a fibrous tissue capsule of variable thickness forms.
(4) For non-toxic and biologically active materials, an interfacial bond forms.

In reality, the materials of most interest are those in category (4) and are known as **bioactive ceramics**.

Four major categories of bioactive ceramics have been developed: (1) dense hydroxyapatite (HA) ceramics, (2) bioactive glasses, (3) bioactive glass-ceramics, and (4) bioactive composites. Each of these materials develops a bond to living bone with bond thickness varying from 0.01 to 200 μm. The physical properties and biomechanics at the implant–tissue interface vary such that all four types of material have found use in a variety of clinical applications.

The general theory of bone bonding of bioactive glasses is due to the establishment of a competitive advantage for osteogenic precursor cells over fibroblasts. It is suggested that this is due to the presence of biologically active silanol, calcium, and phosphate sites on the surface. When bioglass is exposed to water or body fluids several key reactions occur. Cation exchange of sodium and calcium ions in the glass for protons from the surrounding solution results in hydrolysis of surface silica groups and leads to interfacial dissolution. As the solution becomes more alkaline, repolymerization of the silanol groups occurs producing a silica rich surface layer. Another direct consequence of the high pH at the glass solution interface is that calcium and phosphorous oxides which have been released into solution during network dissolution crystallize into a mixed hydroxy–carbonate apatite (HCA) on the surface. It is proposed that crystallites of HCA phases bond to interfacial metabolites such as mucopolysaccharides and collagen. It is hypothesized that this incorporation of organic biological constituents within the growing HCA- and SiO_2-rich layers appears to be the initial step in establishing bioactivity and bonding to tissues.

A common characteristic of the materials which have been used is the chemical components which include CaO, P_2O_5, Na_2O, and SiO_2. The bonding to bone has been associated with the formation of **hydroxyapatite** (HA) on the surface of the implant. Although a range of compositions can be used (up to 60% silica), an even narrower range of compositions can bond to soft tissues. A characteristic of the soft-tissue bonding compositions is a very rapid rate of HA formation. This has previously been attributed to the presence of Na_2O or other alkali cations in the glass composition which increases the solution pH at the implant–tissue interface and thereby enhances the precipitation and crystallization of HA. The rate of HA formation has also been shown to be strongly dependent on the ratio of SiO_2 (glass network former) to Na_2O (network modifier) in the glass. When the glass contains over 60% SiO_2 or more, bonding to tissues is no longer observed.

Bioactive glasses are conventionally prepared by the traditional methods of mixing particles of the oxides or carbonates and then melting and homogenizing at temperatures of 1250–1400 °C. The molten glass is cast into steel or graphite molds to make bulk implants. A final grind and polish are often required. If powders are required these are produced by grinding and then sieving to achieve the desired particle size characteristics.

Several problems arise which have been associated with this method:

(1) Highly charged impurities such as Al^{3+}, Zr^{4+}, Sb^{3+}, Ti^{4+}, Ta^{5+}, *etc.* can be picked up at any stage in the preparation process. This phenomenon is also related to the low silica and high alkali content of the traditional bioactive glass compositions. The incorporation of impurity ions leads to dramatic reductions in bioactivity.
(2) Processing steps such as grinding, polishing, *etc.* all expose the bioactive powder to potential contaminants.
(3) There is a compositional limitation on materials prepared by the conventional

high temperature methods due to the extremely high equilibrium liquidus temperature of SiO_2 (1713 °C) and the extremely high viscosity of silicate melts with high SiO_2 content.

(4) High temperature processing leads to increased processing costs.

Low temperature **sol-gel processing** offers an alternative to the conventional methods. Advantages include ease of powder production, a potentially broader range of bioactivity due to changes in composition and/or microstructure through manipulation of the processing parameters. The materials so produced can be sintered at relatively low (600–700 °C) temperatures thus giving a much better control over the purity of the resultant materials. It has been possible to reduce the number of chemical elements required to produce bioactivity from four to three by this route. $CaO–P_2O_5–SiO_2$ gels have been produced with silica content from 50–90% and calcium oxide from 46–1%, all with 4% phosphorous pentoxide.

In vitro testing of bioactivity requires the study of the development of an hydroxyapatite layer on the surface of the particles. This can be carried out by immersing powders in a Tris buffered solution [tris(hydroxymethyl)-aminomethane, $pH = 7.2 \pm 0.1$] at 37 °C and subjecting the reaction vessel to a controlled orbital motion. The as prepared powders and reacted powders can be investigated by reflectance Fourier transform infrared spectroscopy, X-ray diffraction, and surface area analysis. In general, in accordance with the conventionally prepared bioglasses, the sol-gel derived bioactive gels show no crystallinity for the as prepared materials after heating to 600–700 °C.

Comparisons between materials produced by the sol-gel method and those produced by conventional high temperature methods show that the bioactivity exhibited by a Bioglass material 45S5 (containing 24.5% Na_2O, 24.5% CaO (in weight %), and a Ca/P ratio of 5.2) is matched by extremely high silica containing gel glasses containing of the order of 90 mol% silica. This extension of bioactivity is associated with materials of high surface area and total pore volume of 0.3–0.6 $cm^3 g^{-1}$. It is thought that these ultrastructural features may give rise to an increased density of potential **nucleation sites** for the formation of the hydroxyapatite surface layer. This is in accord with data derived from *in vivo* studies on bone which found that under the conditions of the test, if the materials developed **surface areas of at least 100 $m^2 g^{-1}$** then they would bind to bone. Materials with surface areas below 1 $m^2 g^{-1}$ are not bioactive. It is clear, therefore, that the high surface areas and pore volumes lead to highly bioactive materials.

9 CONCLUSIONS AND FUTURE TRENDS

In order for materials to satisfy the demands of existing or new applications, the one essential need is an understanding of how **properties of materials** are affected by the **microstructure** of the material, *i.e.* at the **nanoscale level**. Two factors are important: the preparative methodology and the subsequent processing into the final material. Thus, emphasis needs (and nowadays certainly is) to be placed on the fine microscopic detail of these two stages in the formation of materials, and what is emerging are numerous examples of advanced technology materials.

Until relatively recently, in many areas (perhaps the majority) the emphasis lay with metal-based materials. That situation has changed drastically, and now new materials based on **ceramics, glass, and hard metals are replacing metals** in many traditional applications. Two such areas are **motor vehicles** where composite materials and ceramics are finding increasing use, and **ferromagnets** where materials such as iron neodymium boride and iron samarium nitride are replacing metal alloys. There is almost no area of advanced technology where these types of material are not used or are being considered for use. In communications, energy, and transportation technologies these materials are making a tremendous impact and will continue to do so as new materials with **high performance properties** are discovered. The future in this area is indeed full of rich potential.

10 REFERENCES

Oxide Materials

C. N. R. Rao and J. Gopalakrishnan, in 'New Directions in Solid State Chemistry', Cambridge University Press, Cambridge, UK, 1989.

J. Gopalakrishnan, in 'Chemistry of Advanced Materials', ed. C. N. R. Rao, Blackwell Scientific Publications, Oxford, 1993, p. 41.

Glasses

W. Vogel, 'The Structure and Crystallization of Glass', Pergamon Press, Oxford, 1971.

M. Cable and J. M. Parker, 'High Performance Glasses', Blackie, Glasgow, 1992.

The Sol-gel Process and Glasses

C. J. Brinker and G. W. Scherer, 'Sol-gel Science', Academic Press, San Diego, 1990.

'Chemistry, Spectroscopy and Applications of Sol-gel Glasses', *Structure and Bonding*, Springer-Verlag, Berlin, 1993, vol. 77.

L. L. Hench and J. West, *Chem. Rev.*, 1990, **90**, 33.

H. D. Gesser and P. C. Goswami, *Chem. Rev.*, 1989, **89**, 765.

Carbides

'Kirk-Othmer Encyclopedia of Chemical Technology', Interscience, New York, 3rd edn, vol. 4, 1978.

H. H. Johansen, *Surv. Prog. Chem.*, 1977, **8**, 57.

Borides

'Boron and Refractory Borides', ed. V. I. Matkovich, Springer-Verlag, Berlin, 1977.

R. T. Paul and C. K. Narula, *Chem. Rev.*, 1990, **90**, 73.

Nitrides

J. Mukerji, in 'Chemistry of Advanced Materials', ed. C. N. R. Rao, Blackwell Scientific Publications, Oxford, 1993, p. 169.

H. H. Johansen, *Surv. Prog. Chem.*, 1977, **8**, 57.

Biosolids

L. L. Hench and J. Wilson, 'Introduction to Bioceramics', World Scientific Publishers, London and Singapore, 1993.

General

'Better ceramics through chemistry' series; Materials Research Society (MRS) Symposia series published biannually by MRS, Pittsburgh, Pennsylvania.

'Ultrastructure processing of advanced materials', ed. L. L. Hench and J. West, Wiley, New York, 1992.

D. L. Segal, 'Chemical Synthesis of Advanced Ceramic Materials', Cambridge University Press, Cambridge, UK, 1989.

'Comprehensive Inorganic Chemistry', ed. A. F. Trotman-Dickenson, Pergamon Press, Oxford, 1973.

'Encyclopedia of Inorganic Chemistry', ed. R. B. King, Wiley, Chichester, 1994.

Journals

Chemistry of Materials
Journal of Materials Chemistry
Inorganic Chemistry
Journal of Non-crystalline Solids
Journal of Sol-gel Science and Technology
Journal of Materials Science
Journal of the American Ceramic Society
Journal of the American Chemical Society
Journal of Materials Research
Journal of Solid State Chemistry

CHAPTER 14

Advanced Cementitious Materials

FRED P. GLASSER

1 INTRODUCTION

One characteristic of civilizations is the development of an infrastructure. Most of us depend on a **built environment** – roads, bridges, houses, factories, schools, offices, *etc*. Some aspects of the infrastructure are highly visible, but beneath our feet is an invisible but essential network of pipes, drains, tunnels, conduits, *etc*.

Throughout historic time, the sophistication required to process building materials has tended to increase. Stone and wood require mainly transportation and dressing, whereas brick, tile, and cement require more sophisticated processing. Even stone, however, requires to be cemented into place. From these beginnings there has been a steady evolution in the sophistication of cementitious materials and with it, an expansion of their application areas.

Historically, **gypsum and lime cements** were well established 5000 years ago. Natural gypsum, $CaSO_4.2H_2O$, is carefully dehydroxylated by removing $1.5H_2O$, to yield the 'hemihydrate', $CaSO_4.0.5H_2O$. When mixed with water, hemihydrate rehydrates rapidly and sets solid. The process is still in use today, mainly for interior use in the form of gypsum plaster, wallboard, *etc*. Lime plaster has also been in use since antiquity. Limestone, $CaCO_3$, is burnt at 850–900 °C to yield CaO. The resulting CaO is rehydrated, or 'slaked' to yield $Ca(OH)_2$; the process is strongly exothermic. $Ca(OH)_2$, mixed with sand, forms a weak mortar which bonds to brick, tile, stone, *etc*. It is somewhat less soluble and more water resistant than gypsum plaster.

Development of modern cements began early in the 19th century. The modern product, known as **Portland cement**, is a precisely formulated material consisting principally of four oxides: CaO, Al_2O_3, Fe_2O_3, and SiO_2. Various raw materials including limestone, clay, and sand are used to achieve the correct composition. These are fired together until partial fusion occurs at 1375–1475 °C. The final product consists of four phases whose idealized formulae are Ca_3SiO_5, Ca_2SiO_4, $Ca_3Al_2O_6$, and $Ca_2(Al,Fe)_2O_5$ in proportions dependent upon the final application. The fired product, termed **'clinker'** from its appearance, is finely ground to specific surfaces in the range 3000–5000 $cm^2\,g^{-1}$ prior to use.

This review furnishes background to this chapter, which will deal with cementitious materials and composites. One consideration which the reader should constantly bear in mind is that cementitious materials are used in **two** distinct applications: **structural** (*e.g.* as slabs and columns) and **non-structural** (panels, cladding, *etc.*). Structural uses are governed by strict codes of practice. Given the long lifetime expected for structural components, changes tend to be evolutionary rather than revolutionary. Non-structural applications tend to be less inhibited by specification and it is easier to introduce novel processes and products. Compared to other materials, *e.g.* electronics, the pace of innovative technological changes in building materials appears very slow. But there are sound legal and practical reasons why this should be so especially with regard to structural uses. Yet even here, demands for reliable, high strength, durable, but economic construction, force changes to existing practice.

There are also many novel uses for cement: in mine stabilization, in oil production, and in environmental controls of toxic and radioactive wastes. The relevant chemistry will be explored.

2 IMPROVEMENTS IN THE QUALITY OF CEMENT

2.1 Portland Cement

Economic and environmental considerations have led to a number of **process improvements**. In Western countries, the **rotary kilns** used to clinker raw materials have been extensively modified in two ways: by adding suspension preheaters, which use offgas to heat incoming ground materials prior to their introduction into the rotary kiln section; and by using dry or semi-wet feed, rather than wet feed. Figure 1 shows a simplified schematic flow diagram. Both measures improve the fuel economy.[1] The clinker grinding technology has also improved: although grinding uses only *ca.* 10% of the total energy, electrical energy is more expensive than thermal energy. Offgas is generally passed through **electrostatic precipitators** to remove a high proportion of fine particulate matter which would otherwise be emitted.

Considerable interest has been shown in the use of supplementary fuel sources such as industrial solvents, waste oil, scrap rubber tyres, *etc.* Because of the relatively long gas residence time, high temperature, and normally oxidizing conditions in the kiln, these materials can be incinerated cleanly, without excessive stack emissions. Nevertheless, the chemistry of the waste may have an impact on the cement quality which must be anticipated: for example, tyres contain steel reinforcement which is oxidized in the kiln, adding Fe_2O_3 to the batch. Moreover, tyres contain appreciable zinc which is known to retard somewhat the set of cement. However, in practice this and other potentially adverse effects can be compensated by regulating other process variables.

In the course of cement clinkering, the high temperatures achieved in the presence of excess oxygen result in volatilization of some of the alkali metals and sulfur in the kiln. Circulation of alkali metals (Na, K) and S, as SO_2, are connected. Measurement of vapour pressures confirms the high volatility of these elements at clinkering temperatures.[2] The dew point of alkali metal sulfates may

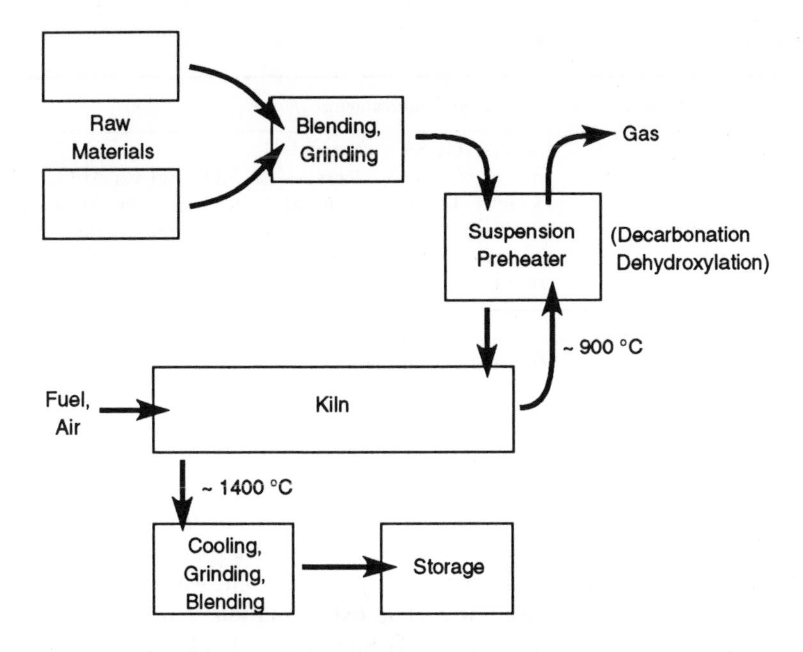

Figure 1 *Schematic flow diagram of the cement-making process. Typical kiln designs produce 800–4000 tonnes per day*

be exceeded at some points in the kiln, with the result that alkali sulfates condense, typically on cooling clinker, or in electrostatic precipitator dusts, or both. Alkali metal sulfate condensed on clinker greatly modifies the early hydration properties of cement. Disposal of alkali metal sulfate-rich condensates is difficult. While dusts can be reintroduced into the kiln, it is desirable to minimize the total alkali content of the clinker. Thus disposal of dust is a problem: other potential uses will be explored subsequently.

Kilns also emit nitrogen oxides. Attention to burner design has resulted in significant decrease in the potential for nitrogen oxide emission.

The mineralogy of the clinker has also undergone refinement. Table 1 shows some of the variations which are used commercially. The intention is to tailor the clinker mineralogy so as to create cementitious material for **specific applications**. Because Ca_3SiO_5 reacts more rapidly with water than Ca_2SiO_4 to give a space-filling, strong gel, the content of the former is enhanced in high early strength cements. These are deliberately more finely ground than normal to facilitate reaction. On the other hand, massive constructions such as dams, thick slabs, *etc.* may require special low-heat cements in order to prevent excessive temperature rise and subsequent thermal cracking of fresh concrete. Low heat cements are proportioned so as to contain less $Ca_3Al_2O_6$ and Ca_3SiO_5 and rather more Ca_2SiO_4 than normal. The concentrated burst of heat evolution in the first few days, so characteristic of ordinary cements, is thereby spread out over a much longer time. Such cements may also be more coarsely ground. Strength development occurs more slowly than normal but by 6 to 12 months, is little

Table 1 *Clinker mineralogy*

Cement type and purpose	*Chemical and mineralogical change*
High early strength	Increase Ca_3SiO_5 content
Low heat of hydration	Increase Ca_2SiO_4; decrease Ca_3SiO_5 and $Ca_3Al_2O_6$
Sulfate resistant	Decrease $Ca_3Al_2O_6$ content or decrease total Al_2O_3 or both
Oil well cement*	As above, for sulfate resistance; also, coarse grind
White cement	Decrease Fe_2O_3 content and/or burn under reducing conditions

* Note. Oil well cements embrace a wide range of formulations tailored to specific applications. For example, cements intended for service at high temperatures (90–200 °C) normally contain quartz interground with clinker.

different than in comparable formulations made with ordinary cement. Sulfate resistant and oil well cements are normally proportioned so as to contain low $Ca_3Al_2O_6$ contents: either restrictions are placed on their total Al_2O_3 content or else iron oxide is added to tie up the Al_2O_3 as **'ferrite'**, $Ca_2(Fe,Al)_2O_5$.

The relationship between chemical and phase content was worked out many years ago from the phase equilibria. The oxide analysis is recast into a phase formula using principally four phases – Ca_3SiO_5, Ca_2SiO_4, $Ca_3Al_2O_6$, and $Ca_2(Al, Fe)O_5$ – by sequentially assigning first the iron to ferrite, then the remaining Al_2O_3 to $Ca_3Al_2O_6$ and finally, solving a set of simultaneous equations to proportion the remaining components between the two calcium silicates. Bogue[3] also gives variants of the method; for example, if the Al/Fe ratio of the ferrite is not equal to 1.0, or if some unreacted free lime is present. Taylor has recently suggested improvements to the accuracy of the calculation by including corrections for the magnesium, alkalis, and other minor components present in clinker.[4] The revised calculation is said to significantly improve the correlation between calculated and observed phase contents; the latter can be independently determined by optical microscopy, with point counting, or by quantitative X-ray diffraction.

Mention should also be made of **colour control** in cements. Normal Portland cements are grey, the depth of colour varying with the chemistry. Iron and manganese are the principal colourants and architectural white cements usually impose restrictions on the contents of **iron** and **manganese**. Where suitable raw materials are unavailable, clinkering under reducing conditions may be used. This reduces iron and manganese to Fe(II) and Mn(II) which have less colourant properties than the equivalent and more usual Fe(III) and Mn(III) components.

Fluxes are occasionally used in cement burning. **Fluxes** assist reaction and therefore differ from **mineralizers** because the latter also affect the equilibria amongst the solid phases. The most effective flux is **fluorine**. Acting in conjunction with Al, it is also a mineralizer and stabilizes Ca_3SiO_5 – which itself has a lower limit of stability at 1250 °C, below which it decomposes to CaO and Ca_2SiO_4 – to as low as 1050 °C. The pair (Al, F) form substitutional solid solutions in which Al replaces Si and F replaces O.[5]

The incorporation of fluorine in low concentrations, *ca.* 0.2–1.0 wt%, has no significant effect on the cementing properties of the final product. The same is not

Table 2 *Blending agents used in cements*

Generic type	*Natural (N) or Artifical (A)*	*Constitution*	*Normal replacement level (wt%)*
blast furnace slag (iron making)	A	>90–95% glass. $CaO–MgO–Al_2O_3–SiO_2$; some sulfide and sulfur	10–70
Coal combustion fly ash	A	Very variable. Glass is reactive but ash may also contain reactive crystalline phases	10–40
Silica fume (ferrosilicon by-product)	A	Mainly glassy SiO_2. Very fine mean particle size	2–12
Clay-derived pozzolans	A	Metakaolin (kaolin heated to 700–800°C)	2–10
Zeolite and glassy pozzolans	N	Very variable. Both zeolite and hydrous glass are potentially reactive	0–100 [but high concentrations require an activator, *e.g.* $Ca(OH)_2$]

true of **chloride** which has a very deleterious and specific action of **enhancing corrosion** of steel embedded in concrete. For that reason, cement-making raw materials must be kept low in chlorine.

Advances in cement making thus depend on **process engineering** as well as **chemistry**. The basic nature of Portland cement remains relatively static, but improved chemical processing contributes to energy efficiency and an improved product. Novel cements and novel ways of using Portland cement more efficiently will be discussed subsequently.

2.2 Blended Cements

All Portland cements are blended in the sense that calcium sulfate is interground with the clinker to control the initial set characteristics. **Gypsum** is normally used as the sulfate source although it is partly converted to hemihydrate by the heat and chemical energy of grinding. However, the blending agents more usually envisaged include **slag**, coal combustion **fly ash**, and natural and artificial **pozzolans**. These materials are added, often for economic reasons, but when properly used they may also significantly improve the properties of the final product. Table 2 describes briefly the constitution of these materials. Technically, all are 'pozzolanic': that is, they are not normally cementitious in their own right but when suitably activated, as by reaction with cement constituents, they produce substances which enhance the cementitious potential of the system.

Slag, typically iron blast furnace slag, has perhaps the simplest constitution because it is nearly homogeneous **glass**. The molten slag is water or air quenched and granulated to ensure high glass content. It may be ground separately or else interground with clinker.

Various indices have been proposed to evaluate the potential suitability of glassy slags. Generally, these employ **acid–base theory**, an acid in this context being an oxygen acceptor, *e.g.* $[SiO_4]^{4-}$, and a base being an oxygen donor, *e.g.* CaO. The reactivity is usually expressed in terms of a ratio: Σ(acidic constituents) / Σ(basic constituents). Moir and Glasser have reviewed various empirical formulae for determining the reactivity of slags from their analysis.[6] Unfortunately 'reactivity' is a concept which is difficult to measure and which cannot be expressed by a single specific quantity; partly for this reason, none of the proposed correlations are adequate except for screening purposes – to distinguish suitable from unsuitable glasses. In general, the highest lime content consistent with preservation of high glass content gives the most reactive slags.

Modern coal combustion produces micron-size fly ash particles. **Ashes with low residual carbon contents** are preferred for cement blending. The composition of the ash particles is highly variable, reflecting the mineral content of the coal. Many ashes contain an aluminosilicate glass together with inert substances (hematite, mullite, quartz, *etc.*). But ashes range widely in CaO content; higher lime ashes contain less glass, but more reactive crystalline phases, *e.g.* Ca_2SiO_4 and $Ca_3Al_2O_6$, which contribute to the cementitious potential. Because of their small particle size relative to cement, often 2–10 μm relative to 10–40 μm for ground cement, fly ash facilitates better packing of grains with the result that hydration products are better able to fill space. The consequence is a denser, less permeable cement matrix. Some **aluminosilicate fly ashes** have a high content of hollow glassy spheres. These so-called **cenospheres** can be concentrated by flotation and are useful to produce **low-density cement materials** without undue sacrifice of strength.

Silica fume furnishes perhaps the best example of how particle sizing influences packing which, in turn, benefits the properties of the hardened product. Although silica fume is not easily dispersed in cement, a good dispersion containing only 5–10% fume gives **higher strength** and affords greater **chemical resistance** to aggressive environments relative to the equivalent Portland cement formulation.[7] Silica fume adversely affects the rheology of the wet mix and on that account, is usually used in conjunction with a plasticizer to reduce water demand.

Many other pozzolanic materials are potentially worth exploring. **Kaolin**, $Al_2Si_2O_5(OH)_4$, is a commonly occurring clay mineral, often as micron-sized platelets. It contains strongly-bonded hydroxyls which can only be removed above $\sim$450 °C. Ignition in the range 500–750 °C removes water but preserves in degraded form some parts of the structure of the precursor clay. This product is an **active pozzolan**. Not surprisingly, crushed and ground **common brick** is also somewhat pozzolanic, thus opening the possibility of recycling, albeit in different form, fired clay building materials.

Naturally-occurring pozzolans are widely distributed. They are mainly of volcanic origin and derive their intrinsic rectivity from the presence of glassy shards or of alteration products, typically zeolites, or mixtures thereof. As might be expected, the materials are highly variable and require selective extraction.

Table 3 *Some special cements*

Type designation	*Principal phases*	*Minor phase(s)*	*Uses*
High alumina cement (HCA)	$CaAl_2O_4$, $Ca_{12}Al_{14}O_{33}$	Ca_2SiO_4	High early strength
Calcium sulfoaluminate	$4CaO.Al_2O_3.SO_3$	Ca_2SiO_4	Bricklaying mortar, expansive agent*
Phosphate cements			Dental and medical uses. Repairs of ordinary concrete

* with added lime, or when mixed with Portland cement

2.3 Non-Portland Cements

Portland cement is not the only type which is used in construction industries. Table 3 shows other cement types. **High alumina cement** (HAC), based on calcium monoaluminate, is widely available. Its advantages are **rapid strength gain**, often achieving nearly full strength within 16–24, and compatibility with hardened Portland cement, to which it adheres strongly, and good refractoriness. Disadvantages include higher production costs. This cement type is difficult to sinter and is often made by total fusion; the melt is cast into ingots which are allowed to freeze with crytallization.

This process has historically been more difficult to understand and control than the clinkering cycle in Portland cements. During fusion, iron is partly reduced and is present as a mixture of Fe^{2+} and Fe^{3+}: thus, we have a five component system rather than one of four components, as in Portland cement. Nevertheless great strides have been made in determining the phase equilibria.[8] A principal problem has been that the clinker mineralogy is abnormally sensitive to the amount of SiO_2 in the batch. Increasing SiO_2 could lead to formation of gehlenite, $Ca_2Al_2SiO_7$, and/or pleochroite, both of which are inert, or to Ca_2SiO_4, which will hydrate and contribute to the strength at longer ages, or mixtures of the three. Pleochroite has been assumed to be thermodynamically metastable since it decomposes readily upon heating in air. However, it requires essential Fe(II) and is stable so long as reducing conditions are maintained. The crystal chemistry and thermodynamic stability of this phase have been explored.[9]

The **'conversion'** reactions of high alumina cement have led to problems. The normal hydration products obtained at 0–25 °C include $CaO.Al_2O_3.10H_2O$. This phase is, however, unstable and 'converts' to the more stable phases, mixtures of **hydrogarnet**, $3CaO.Al_2O_3.6H_2O$, and **gibbsite**, $Al(OH)_3$. In a moist atmosphere the rate of conversion is temperature dependent, increasing rapidly with rising temperature ≥ 20 °C. The products of the conversion reaction have a smaller molar volume than the reactants. The volume decrease manifests itself as an increase in porosity rather than as shrinkage. This increased porosity leads to a decrease in strength and increased susceptibility to attack by aggressive salts, *etc.* Very slow conversion, as happens in HAC used in cool or warm but dry environments results in less strength decrease than if conversion is more rapid. In any event, conversion problems are lessened if mixes are allowed to set and hydrate warm, at $\geq 40\,^{\circ}C$; conversion occurring in the still-plastic mass does not decrease the strength. Alternatively, initial formation of $CaO.Al_2O_3.10H_2O$ can be avoided by using HAC–slag mixes.

Refractory cements are normally based on high alumina compositions. While ordinary Portland cements are refractory in the sense of having high melting points, heating lowers their strength drastically: true ceramic bonds do not readily replace the hydrogen-bonded network of ordinary Portland cement. High alumina cements, on the other hand, retain much water to higher temperatures, 300–400 °C, with the result that dehydroxylation and ceramic sintering overlap upon continued heating. Moreover, and unlike Portland cement, the resulting ceramic product will not readily rehydrate in ambient atmospheres. Normal commercial high alumina cements contain too much iron oxide to achieve the best refractory potential. Therefore special low-iron refractory grades are prepared, usually made by sintering with partial fusion. Typical products are low in iron and may contain 50–60% (or more) Al_2O_3. The clinker consists of mixtures of $CaAl_2O_4$ and $CaAl_4O_7$. The cement is usually mixed with aggregate, preferably **calcined Al_2O_3** to enhance the refractory potential. In this way mortars and concretes can be made, as is done with Portland cement. Moist mixes can be tamped in place or else wetter mixtures can be poured. Set and strength gain occur within a few hours. The set shapes are dried slowly and are thereafter ready for use. Quite complex shapes and/or large, joint-free monoliths are readily produced without need for complex ceramic forming processes. The upper service temperature depends on the grade of cement and aggregate used, but service temperatures of ~1600 °C are easily achieved.

Other special cements are described in Table 3. **Calcium sulfoaluminate**, $4CaO.3Al_2O_3.SO_3$ (or $3CaO.3Al_2O_3.CaSO_4$) forms the basis of a newly-developed range of cements. The sulfoaluminate compound is closely related in structure to sodalite, a naturally-occurring sodium aluminosilicate. However whereas sodalite is inert towards water, calcium aluminosulfate is very reactive. It is formed, like Portland cement, by calcination in a rotary kiln of $CaSO_4$, $CaCO_3$, and an alumina source. If aluminosilicate clays are used as the alumina source, the batch is generally proportioned so as to produce clinker which consists of two phases: Ca_2SiO_4 and $3CaO.3Al_2O_3.CaSO_4$. Both phases contribute to strength gain.[10] Clinker firing temperatures are generally on the order of 1250 °C, resulting in significant energy saving relative to Portland cements. The resulting product is handled and used like Portland cement, with which it is compatible. Experience thus far suggests that its durability, including compatibility with embedded steel, is comparable with that of conventional Portland cement formulations.

3 CHEMICALS IN CEMENT HYDRATION

3.1 The Hydration Process

Cements are normally mixed with more water than is required by the chemical hydration process. The weight ratio of water:cement, or w/c, for complete hydration is somewhat variable, but typically lies in the range w/c = 0.24–0.26. However, to ensure **plasticity** and **flow**, an excess of water is required. Mixtures

Table 4 *Main hydrate phases in Portland cement*

Amorphous
C-S-H gel (*c*alcium *s*ilicate *h*ydrate)
Ca/Si mole ratio ~1.8–2.0 in ordinary cements, but may decrease to ~1.0 in the presence of reactive siliceous components, *e.g.* fly ash (OPC, HAC)*

Crystalline
$Ca(OH)_2$, Portlandite (OPC)*
$3CaO.Al_2O_3.3CaSO_4.31\text{–}32H_2O$, Ettringite (OPC)
$3CaO.Al_2O_3.CaSO_4.12H_2O$, 'Monosulfoaluminate' (OPC)
$3CaO.Al_2O_3.6H_2O$, Hydrogarnet (HAC, OPC)
$CaO.Al_2O_3.10H_2O$ (HAC)

*(HAC) Characteristically present in high alumina cement
*(OPC) Characteristically present in ordinary Portland cement

of cement with sand (termed **mortars**) require ratios of w/c 0.4–0.5, while **concretes** (mixtures containing sand and coarser rock) may range up to w/c ~0.6. A typical hydrated cementitious material thus contains three physically distinct regions: aggregate, cement hydrate products, including unhydrated cement in the early stages of reaction, and entrapped water. The latter is known as pore water. Of course the cement hydrates are not homogeneous. But the fine-grained nature and poor crystallinity of the hydration products have only recently been elucidated. Table 4, after Taylor,[11] lists the main hydration products.

The various solids react with water at very different rates. Soluble sulfate quickly reacts with calcium aluminate phases to product **ettringite**, $3CaO.Al_2O_3.3CaSO_4.32H_2O$. This and $Ca(OH)_2$ are often the first two hydrate phases detected, often with a few minutes of mixing. The main bonding phase is a calcium silicate gel, shorthand C-S-H (C for calcium, S for silicon, H for water). Its C/S ratio is typically 1.8–2.0 in normal cements, although in cements blended with more siliceous additives, it may decrease to 1.5 or less after $Ca(OH)_2$ is consumed as a consequence of pozzolanic activity. C-S-H begins to form in quantity after about 4–6 hours. The system also becomes depleted in available sulfate and in the ettringite phase; the latter is partially replaced by calcium aluminate hydrates and calcium aluminate monosulfate hydrate.

The literature abounds in hydration models. Often these concentrate on explaining a particular aspect of hydration. The models which have the most appeal in present context are primarily physical. This is because the early behavior of cements, especially within the first few hours or days, is much influenced by the presence of added chemicals. Thus a basic formulation can be modified or tailored so as to achieve particular objectives. Table 5 lists some of the purposes for which additives are used.

3.2 Set Retarders and Accelerators

Normally, cements take on an initial set within 2–3 hours of mixing and a final set within 3–6 hours. Occasionally, it is desirable to reduce or extend this period. For

Table 5 *Chemical additives to cement and concrete*

Purpose	*Comment*
Set accelerators and retarders	Useful to enhance early strength gain or, for retarders, to delay set to permit transport or holding prior to emplacement
Plasticizers and superplasticizers	Enhance workability and permit use of lower water:cement ratios
Shrinkage control additives	Minimize shrinkage of still-plastic cement and reduce cracking
Anti-freeze	Permit low temperature use
Air entrainers	Increase freeze–thaw performance

example, if high alumina cement is mixed with an ordinary Portland cement grout or mortar, set occurs almost instantly, a phenomena known as **'flash set'**. Such mixtures are employed, for example, in sealing water leaks in foundations, tunnels, *etc*. Less drastic acceleration is obtained by adding various salts. Some of these are shown in Table 6.

$CaCl_2$ added to the batch somewhat shortens set time and substantially increases strength gain within the first few days or weeks. However $CaCl_2$ is not now permitted in steel-reinforced concrete because of its potential to enhance steel corrosion. Other non-corrosive **accelerators**, *e.g.* **calcium formate** are preferred even though they are less effective. Retarders are also used. The set time of cement is not very sensitive to temperature but in unusual situations, such as in oil well cementing, where downhole temperatures may exceed 100 °C, retarders are necessary in view of thermal acceleration as well as long pumping times.

A wide range of **retarders** are known; *e.g.* **sugar** or **EDTA**. In these examples the retarder acts by sequestering calcium which is required for reaction. Other retarders, *e.g.* $CaSO_4.2H_2O$, apparently retard by forming ettringite very rapidly. It forms a protective coating over unhydrated or partially hydrated grains. But the list of potential retarders is very long and in many instances the retarding mechanism is either speculative or not well-established.

Table 6 *Calcium salts which accelerate strength gain*

Salt	*Notes*
Chloride	Most effective accelerator but use now restricted
Thiocyanate Thiosulfate Propionate	Give modest degree of benefit
Perchlorate	Strong oxidizer: safety problems
Formate Nitrite	Either accelerate: mixtures of the two are sometimes used

3.3 Plasticizers

Cement formulations require a certain consistency for placing, pumping, *etc.* The necessary fluidity can be achieved by adding more water but the ultimate strength is reduced as the w/c ratio increases. Moreover, it is almost always necessary to use low w/c ratios to achieve low permeability in well-cured products. Low permeability, in turn, enhances the resistance of the concrete to ingress of substances corrosive to cement or those which increase corrosion of embedded steel.

Many substances have been used as **plasticizers** or **superplasticizers** (there is no technical distinction to be made between the two). Their constitution is often obscure, partly in the desire to promote proprietary formulations but partly because they are industrial by-products and comprise mixtures of substances. Typically, plasticizers are water soluble organic polymers containing functional ionic groupings. Figure 2 shows a typical polymer backbone unit. Many are sulfonates, *e.g.* sulfonated melamine formaldehydes or napthalene formaldehydes, or lignosulfonates. Several more recent types include sulfonic acid esters and carbohydrate esters. The dosage depends on the workability and degree of water reduction desired; typically, it lies in the range 0.5–2.0% plasticizer.

The mode of action of plasticizers is understood, at least in general. Cement particles which are placed in water develop a surface charge and normally attract each other, but only weakly. The polar groups of the plasticizer become bonded to the surfaces of hydrating cement particles, giving rise to a relatively non-polar film, which may well be monomolecular; typical plasticizer molecules have molecular weights in the range 20 000–50 000. The resulting non-polar sheath further reduces attractive forces between adjacent grains, allowing them to move past each other more freely in response to shearing. Hence the plasticizing action. Since the rheology of cement–water systems is very complex – they comprise a non-Newtonian system – specialist monographs should be consulted for a more detailed mathematical treatment on the rheological consequences.[13] Typical superplasticizers retard set, the amount of retardation depending on dosage and temperature. Under warm conditions, however, superplasticizers which retard in cold batches may actually accelerate set.[14] Suffice it to note that the majority of high strength concretes and of high quality concretes intended for service in severe environments employ plasticizing agents.

The presence of plasticizer cannot generally be detected on the basis of sulfur analysis; the S in cement interferes. Some plasticizers are supplied as either sodium

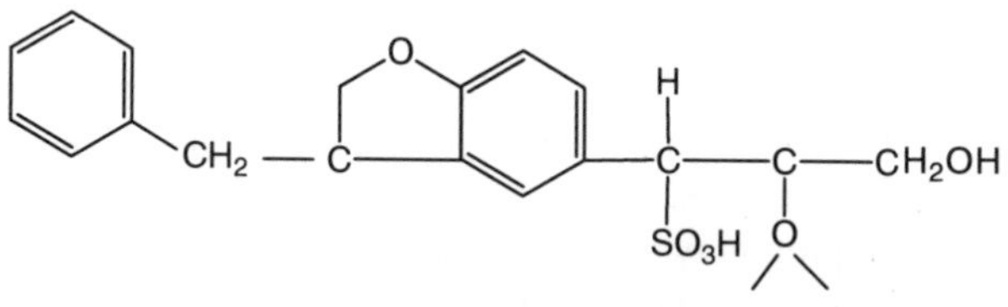

Figure 2 *Backbone configuration of a typical water-soluble polymer used to plasticize cement–water mixes*

or calcium salts; the calcium contribution of the latter is of course overwhelmed by Ca from cement, but in conjunction with low sodium cements, it may be possible to determine Na if the sodium salt was used. Infrared methods are most successful when applied to fresh mixes. **Lignosulfonates** have characteristic absorptions at 2950 and 1250 cm^{-1}, both sharp and strong, with a broader band at 1200 cm^{-1} (all arising from sulfonate). The **sulfonated melamine** and **formaldehyde plasticizers** have a band at 1200 cm^{-1} with a characteristic group of three in the range 1330–1540 cm^{-1}. The triazine group of melamine gives rise to a band at 820 cm^{-1}. **Carboxylic acid** derivatives lack the band at 1200 cm^{-1} but have a characteristic narrow band at 1600 cm^{-1}. Infrared bands of OH^- may interfere and care should be taken in applying methods developed for superplasticizer to its mixtures with alkaline cement. It is not practicable to recover plasticizer from mature concretes as the initially soluble material gradually undergoes irreversible hydrolysis and is thereby insolubilized.

3.4 Other Organic Additives

The constitution of commercial polymer–cement systems is an area of considerable trade secrecy, and only the general outlines of the many proprietary formulations will be given. These materials are popular with the industry because the handling properties of polymer mortars and concretes are much like those of ordinary mixes.

Two methods of making polymer concretes and mortars are in use. One is a two-step process: mortars are impregnated with a monomer solution, usually **methyl methacrylate**, which is then subjected to ***in situ*** **polymerization**. In the second method, the polymer or its precursor, is an essential part of the mix. Amongst the polymer systems compatible with cement are rubbery compounds, *e.g.* **styrene–butadiene** or latex, and harder polymers, *e.g.* **ethylene–vinyl acetate, polyacrylic ester**, and **epoxies**. Bonding develops between the cement and polymer. The resulting mixes find widespread and diverse applications as repair materials, floor screeds, decorative coatings, *etc.* Further practical details are given in a review.[15]

3.5 Oil Well Cements

Oil well cements differ substantially in composition, mineralogy, and particle size from ordinary constructional cements. Service conditions are different and, typically, much harsher. Elevated temperatures, up to 200 °C, may be encountered as well as exposure to aggressive formation water rich in sulfide, sulfate, chloride, *etc.* Moreover, there are physical restrictions placed on the cementitious material. Like normal Portland cement, it must bond to and protect steel. It must be pumpable over long distances: the well may be several km deep and, moreover, interruptions to pumping may occur which require the grout to be held, without set occurring, for several hours or more.

These considerations are catered for by taking a few basic cement types, and adjusting grout characteristics by means of additives to suit particular applications.

The basic cement types used generally have **sulfate resistance** built in by virtue of restrictions in their content of alumina or aluminates. For progressively warmer service conditions the cement is more coarsely ground and, for service in **hydrothermal conditions** in the range $\geq$90–120 °C, silica flour as quartz is also added in sufficient quantities to bring the mean Ca:Si ratio close to unity. Normal cements have a Ca:Si ratio of ~2.5. These give poor high temperature strength but, by adding silica as ground quartz, formation of **tobermorite**, (ideally $Ca_5(Si_6O_{18}H_2).8H_2O$: Ca/Si 0.9) is promoted. Tobermorite formation leads to better preservation of strength and is preferentially developed in most autoclaved cement products.

However, much of the other property adjustment is achieved by additives. These are chosen on a performance basis: if they work, they are generally permissible. Much secrecy governs exact formulations and it is therefore only possible to give a general survey of widely-used additives.

Calcium chloride is widely used as an accelerator. At elevated temperatures it causes set cements to gain strength rapidly. The chloride does not significantly affect corrosion since formation waters are typically saline. Sea water, often used as mix water, also accelerates strength gain. Sodium silicate is occasionally used as an accelerator but organic accelerators are preferred. However, given the degree of thermal acceleration typically encountered, the demand for retarders usually outstrips that for accelerators.

The specific retardation of a typical **retarder**, e.g. lignosulfonate, tends to decrease with increasing temperature. For that reason, higher temperature cementing conditions usually demand more powerful retarders. These are **mainly organics**: for example saccharides, hydroxycarboxylic acids, or cellulose derivatives such as carboxymethylhydroxyethylcellulose (CMHEC in shorthand). The retarder may also thicken slurries and prevent bleed. Alkylene phosphoric acid salts are also used; despite cost, their performance is said to be less affected by small variations in cement quality.

Inorganic salts are occasionally used as retarders. High sodium borates and phosphates, *e.g.* Na_3PO_4, are effective. Lead oxide, zinc oxide, and certain acids, *e.g.* chromic or hydrofluoric are also retarders but their use is increasingly discouraged on safety grounds.

Fillers, known as **'extenders'** are often added to decrease slurry densities, discourage segregation in slurries, and increase grout yields. **Bentonite** is a commonly-used additive. The active component is a clay of the **montmorillonite** family. Its incorporation reduces fluid loss from grouts and inhibits segregation. Sodium silicates are very effective in these applications, especially when grouts need to be mixed with highly saline water, *e.g.* sea water. In well applications, corrosive aqueous phases may be encountered. Aluminosilicate fly ash is used to enhance resistance to chemical attack. If the fly ash contains a high proportion of cenospheres, these are also effective in lowering slurry densities. If the cement needs to be very lightweight, foaming agents may also be added together with gas under pressure.

Occasionally, slurry densities need to be *increased* above the normal range, 1600–2200 kg m^{-3}. Commercially available inert powders with densities in the range 4300–5000 kg m^{-3} are used; these include haematite (Fe_2O_3), ilmenite (iron titanium oxide), and barite ($BaSO_4$).

As indicated, siliceous fly ash can improve the chemical resistance of cement. A number of other reactive, fine particulate materials can be used. The requirements for additives are less strict than for constructional concretes, partly because **oil well cements** need only attain moderate strengths and partly because the generally higher service temperatures, relative to surface conditions, facilitate rapid reaction. Amongst the special materials incorporated in oil well formulations are expanded **perlite** (crushed volcanic glass which, upon prior heating, expands), an asphaltic material known as **gilsonite**, and powdered coal. Silica fume, a by-product of ferrosilicon smelting, is occasionally used.

During the long holding and pumping times associated with oil well grouts, it may be necessary to supplement the action of bentonite, sodium silicate, *etc*. Some other materials used to control fluid loss are given in Table 7. The action of these materials is often very sensitive to salinity of the mix water, cement type, temperature, and the presence or absence of other admixtures. This again highlights the need for expert knowledge and experience with formulations for a specific set of conditions. Nelson[16] has given perhaps the best detailed technical survey.

3.6 Other Mineral Additives

Of the other additives used in cement, the most important is **calcium carbonate**. If $CaCO_3$ is interground with clinker the grinding process results in a very fine mean grain size for $CaCO_3$ which, on account of its excellent cleavage, is not as hard as clinker. The $CaCO_3$ product is not, as might be supposed, inert but reacts readily with cement components. The principal reactions are with the alumina phases, to form the **monocarboaluminate** and **hemicarboaluminates**. Figure 3 taken from reference 17 shows the stability range of the two phases with respect to the other calcium aluminate hydrates. Formation of carboaluminate has been suggested as a method for averting the loss of strength experienced when high alumina cements are used in warm, moist service conditions.[18,19]

Table 7 *Some water-soluble materials used to control fluid loss in oil well cements*

Type	*Example*
Cellulose Derivatives	ethylenediaminecarboxymethylcellulose carboxymethylhydroxyethylcellulose hydroxypropylcellulose
Non ionic-synthetic polymers	polyvinylpyrrolidone polyvinyl alcohol
Anionic synthetic polymers	polyacrylamide (partially hydrolysed) 2-acrylamido-2-methylpropane sulfonic acid
Cationic synthetic polymers	poly(ethyleneimine) polyallylamine methacrylamidopropyltrimethylammonium chloride

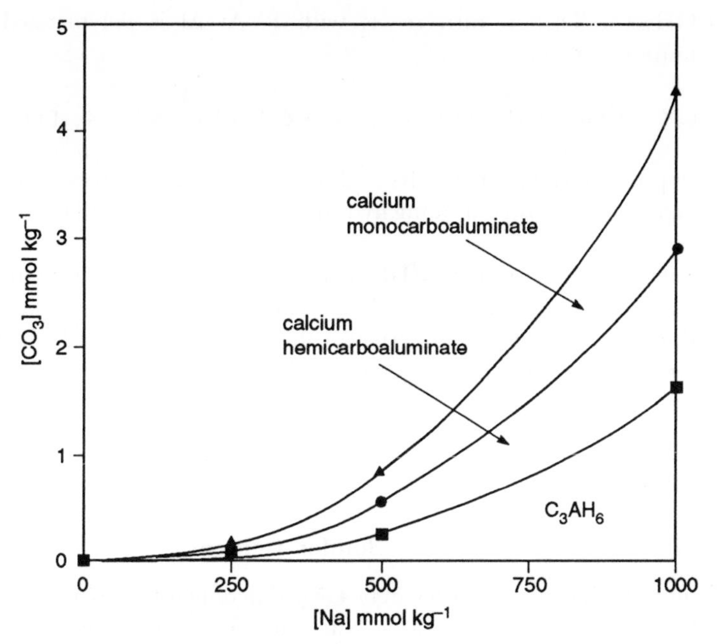

Figure 3 *Reaction of calcium aluminate hydrates, $3CaO.Al_2O_3.6H_2O$ (abbreviated C_3AH_6) with CO_2. The aqueous CO_2 concentration and the alkali content affect the reaction. At low soluble alkali contents, $3CaO.Al_2O_3.6H_2O$ is unstable at very low CO_2 concentrations, but its stability range is enhanced at higher alkali contents. At any fixed alkali content up to 1 M, the decomposition products are encountered in a constant sequence. At high CO_2 contents, above the curve marked by triangles, the only stable solids are $CaCO_3$ and $Al(OH)_3$.*
The diagram is meant also to indicate how thermodynamics can be applied to the hydration of cement and to predict the reactions occurring in the environment
(Adapted from reference 17)

3.7 Macrodefect-free Cements

Ordinary cements incompletely fill space, with the result that they contain pores. In the mechanical sense these pores act as flaws. The publications of Birchall and collegues[20] on the original macrodefect-free (MDF) cements have stimulated much interest, given the substantial gain in **compressive stength** and the even greater gain in **flexural strength**, from 10 MPa in conventional pastes to perhaps 200 MPa in the best MDF products. The original system, which is still one of the best, utilizes mixtures of high alumina cement, a water soluble polymer such as **PVA** (polyvinyl alcohol acetate), and a minimal amount of water. These components are typically premixed at low shear, then subject to intensive high shear mixing and calendaring to form sheets. The product is pressure cured at ~0.5 MPa and 80°C. High alumina cement shortens the cure period; its Al component is apparently the key to a chemical interaction between the inorganic hydration products and PVA. Since the composite contains insufficient water to hydrate all the cement, some clinker grains remain as microaggregate.

The chemical reactions are initiated by dissolution of Al during hydration to form $[Al(OH)_4]^-$. This in turn reacts with PVA: Al is polymerized in a polycondensation reaction:

$$4\,(CH_2\text{–}CHOH)_n + n\,[Al(OH)_4]^-_{(aq)} \rightarrow n\,[(CH_2\text{–}CHO)_4Al]^- + 4n\,H_2O \quad (1)$$

The water produced is free to dissolve and transport Al, although eventually the cycle is brought to a halt by continuing hydration, which converts all free water to bound water.

One problem with the original MDF cements was their loss of strength upon continued wetting or in wet–dry cycling. This problem has prevented full commercial exploitation of the properties of this unique system. It is, however, likely that research currently in progress will overcome these drawbacks, perhaps by developing new combinations.[21]

4 SPECIAL CEMENTS

4.1 Alkali Activated, Chemically Bonded Cements

Inorganic aluminosilicate polymers can be formed in dimensionally-stable masses using NaOH (or KOH) as the activator. Numerous sources of alumina and silica are potentially available. Extensive use has been made of glassy slags in the former Soviet Union, Finland, and elsewhere. Blast furnace slag is usually granulated and ground, typically to 5000 $cm^2\,g^{-1}$, but the mix water is replaced by an equivalent volume of 10–30% NaOH solution. Sand and mineral aggregates may be incorporated to make mortars and concretes, and the mixes are said to give good corrosion protection for steel. The mixes characteristically have relatively low temperature rise and rapid strength gain during hydration relative to Portland cements; the ultimate strengths are similar.

However, it is not always necessary to use highly alkaline activators. Flue dust from cement plants can be a useful activator as well as $CaSO_4.2H_2O$ (gypsum) and the so-called soluble anhydrite, $CaSO_4$. De Silva and Glasser[23] found that when muscovite (a potassium-bearing mica mineral) was activated by firing and subsequently incorporated into $Ca(OH)_2$–metakaolin activated mixes, the K became mobile and served as an important secondary activator.

Other sources of alumina and silica have been used. **Metakaolin**, derived from the commonly-occurring clay mineral kaolin $[Al_2Si_4(OH)_2]$ is an example. Kaolin is dehydroxylated by heating to 45–750 °C: the resulting product retains in degraded form the structure of the parent kaolin. This makes it especially reactive. The necessary activators include $Ca(OH)_2$, to furnish Ca, and NaOH or KOH. Thus the overall chemistry is similar to that of slag-based systems.

At low hydration temperatures, in the range 10–40 °C, the principal bonding agent in these systems is a **gel-like aluminosilicate polymer** modified by inclusion of Na, K, and Ca ions. However, the higher temperature cure, 55 °C or more, the amorphous gel-like product begins to **crystallize**. Several studies[22,23] have identified sodalite and various zeolites amongst the crystalline products. Other investigations have confirmed that a range of **zeolitic phases** can be developed in warm cures. These have the general formula $R_2O.Al_2O_3.xSiO_2.nH_2O$. With cures at or near ambient temperature, x tends to lie in the range 2.0 to 3.0 but with steam cure x approaches 4. The formulations patented by Davidovitz[24] and marketed under the 'Pyrament' trademark are in this class. Typically, a drop in compressive strength occurs as the gel bond is replaced by crystalline bonding. Nevertheless, products often retain adequate strength for most purposes. The development of zeolitic phases, with their capacity for ion exchange and catalysis, suggests new areas of application for cementitious formulations.

4.2 Expansive and Self-stressing Cements

It is desirable in many instances to develop limited and controlled expansion during strength gain. Most methods to achieve this involve chemical controls. Such cements are classified according to the nature of the additive used to modify Portland cements: Type K cements use calcium sulfoaluminate, $4CaO.3Al_2O_3.SO_3$; Type M cements use high alumina cement with additional gypsum; and S type cements use enhanced clinker levels of $3CaO.Al_2O_3$, together with added gypsum.

Application of all these types has been handicapped by the ability to secure reproducible, controlled expansivity. What is certain is that the **expansion is caused by ettringite formation**. Ettringite, $3CaO.Al_2O_3.3CaSO_4.32H_2O$, has a physically low density, $\sim 1720\,kg\,m^{-3}$. On that account, its formation from denser solids is expansive.

However, there is less agreement about the precise mechanisms whereby expansive forces developed. Mehta[25] proposed that an ettringite gel is first formed (this is a contradiction in terms: ettringite refers to a crystalline substance). In a second stage this gel imbibed or absorbed water, resulting in a physical expansion. Others have supposed that mechanical pressures, generated by physical growth of ettringite crystals, caused expansion. Both theories have adherents. But whatever the mechanism, there is no doubt that large expansions can be obtained; some formulations are sufficiently expansive that when poured into holes drilled in rock, they replace conventional explosive as a means of splitting or breaking even the hardest rocks. However, in practice, most applications demand less expansion, 0.05–0.1% in length, in order to ensure shrinkage compensation or modest expansions necessary to stress steel. But these have proven difficult to obtain reproducibly.

4.3 Other Inorganic Cements

Amongst the non-silicate cements, perhaps the most successful have been the **phosphates**. The magnesia–phosphate system is typical. It is essential to use a fine grained source of magnesia: either MgO or $Mg(OH)_2$. This is mixed with an aqueous solution containing dissolved P_2O_5. The phosphorus component is furnished by a variety of salts: by $NH_4H_2PO_4$, $(NH_4)_2HPO_4$, or H_3PO_4 itself, or by $Al(H_2PO_4)_3$. The hardened product made with ammonium phosphates contains a mixture of phases, both crystalline and amorphous.[26] Some unreacted magnesia almost invariably persists.

Part of the ammonium phosphate can be replaced by sodium phosphate. But most investigators indicate that the preferred bonding phase is **struvite**, $NH_4MgPO_4.6H_2O$. The products reach high strength rapidly and are apparently mechanically compatible with Portland cement: as such, they find widespread use as **repair materials** for conventional cement and concrete.

There is also much interest in **calcium phosphate cements**, which have been reviewed by Brown and Chou.[27] The main mix components are $Ca_4(PO_4)_2O$ (tetracalcium phosphate), water, and an acid: either phosphoric acid or polycarboxylic acids or mixtures. An obvious target of these researchers is to produce **hydroxyapatite-bonded cement** with the intention of developing **biocompatible materials**.

5 ENVIRONMENTAL APPLICATIONS

Cements are widely used in the fixation of liquids, sludges, particulates, *etc.* containing **toxic or radioactive wastes**. Several specialist reviews have been published.[28,29] The conversion of mobile particulates and liquids to solids greatly reduces dispersion and lowers the surface area available for leaching. In addition to the physical action, cements have a definite potential for chemical fixation. There is a definite sorption of high surface area gel for many toxic and radioactive species; the high internal pH tends to precipitate many metals as hydrous oxides or hydroxides: and the cement components themselves may react with waste species to form low solubility substances.

Moreover, the **oxidation–reduction potential** can be manipulated; for example, by adding iron blast furnace slag which characteristically contains ~ 1% sulfur in its most reduced (sulfide) form. Cr(VI), if present, remains relatively soluble. But it can be reduced to less-soluble Cr(III) by adding slag, iron(II) sulfate, iron scrap, *etc.* Cr(III) is precipitated initially as $Cr(OH)_3$ but in the longer term forms dilute solid solution with the aluminate phases of the matrix in which Cr(III) substitutes for Al(III). Chromium solubilities, as determined from analysis of expressed pore fluid from well-cured compositions, are 0.5–1.0 ppm.[30] A survey of the behaviour of other elements in cement has been given.[31] On account of the stability and persistence of cement in natural disposal environments, cements find widespread acceptance as a preferred matrix.

6 CONCLUSIONS AND FUTURE TRENDS

Several types of cement are available and these are tailored to suit a wide range of applications by adding various chemicals. In some cases, the additives comprise a major part of the system. Blended cements, using slag and fly ash are examples. In other applications, the set and strength gain can be regulated to suit a range of applications including oil well cements, which may experience high temperatures and aggressive environments. Not all additives are inorganic: polymer cement blends are widely used and defect-free cements, which are polymer modified and comprise a new, high-technology class of materials.

Cements are relatively cheap the raw materials essentially inexhaustible. Hence they show great promise for the future. Nevertheless, researchers face a number of challenges. The first of these is in production. World production exceeds 10^9 tonnes yr^{-1} and is rising. Production is energy intensive, and will come under increasing scrutiny and regulation. There is need to make a **more consistent product**; this, in turn, demands a better understanding of how clinker microstructures and phase development affect reactivity and strength gain. Demand for fossil fuel can be reduced by burning intractable and non-recyclable materials *e.g.* tyres, sump oils, used organic solvents. But these may also introduce inorganic contaminants: tyres, for example, introduce major amounts of iron from steel reinforcement as well as zinc. These must not be allowed to reduce clinker quality.

Another and very energy-effective way of using cements is to achieve **better lifetime** of concrete products. This demands better fundamental understanding of the adverse chemical reactions which can occur between cements and mineral aggregate, and of improving the resistance of concrete to penetration by aggressive environments. In this respect, sea water is notably aggressive as it contains three components: magnesium, chloride, and sulfate, all of which are aggressive. We have already shown how low value organics have valuable plasticizing action which helps produce more durable materials: other substances remain to be found.

Damaged concrete requires repair, and the fundamentals of adhesion of **repair materials** (and also their durability) require further work. The performance of cements in **composites** also requires more study. New, high performance **organic fibres** are being developed, but research on low cost renewable fibres, mainly of plant origin, is also required. **Glass fibres** work well and are used to reinforce cement, but they degrade at high pH. Less alkaline cements remain a possible area to be developed.

Cement will also be developed for special purposes. Research on **biocompatible cements**, intended for bone and tooth replacement, is one such example (*cf.* Chapter 13, Section 8).

Many of the developments envisaged do not demand expensive equipment or extremely sophisticated laboratory facilities. But they are less and less likely to depend on empirical discoveries. Instead, they will increasingly be linked to materials chemistry and an improved understanding of the fundamental physical and chemical processes occurring in cement systems.

7 REFERENCES

7.1 Specific References

1. K. Peray, 'The Rotary Cement Kiln', Edward Arnold, London, 2nd edn, 1986.
2. C. Gang-Soon and F. P. Glasser, *Cem. Concr. Res.*, 1988, **18**, 367.
3. R. G. Bogue, *Ind. Eng. Chem. (Analytical)*, 1929, **1**, 192.

4. H. F. W. Taylor, *Adv. Cem. Res.* 1989, **2**, 73.
5. E. G. Shame and F. P. Glasser, *Brit. Ceram. Trans. J.*, 1987, **86**, 13.
6. G. K. Moir and F. P. Glasser, Proceedings of the 9th International Congress on the Chemistry of Cement, National Council for Cement and Building Materials (New Delhi, November 1992), 1992, vol. 1, p. 125.
7. P.-C. Aitcin, S. L. Sarkar, and Y. Diatta, *Mater. Res. Soc. Symp. Proc.*, 1987, **85**, 261.
8. (a) F. P. Sorrentino and F. P. Glasser, *Trans. J. Brit. Ceram. Soc.*, 1975, **74**, 253; (b) *ibid.*, 1976, **75**, 95.
9. A. Sourie, F. P. Glasser, and E. E. Lachowski, *Inst. of Materials Ceram. Trans. J.*, 1994, **93(4)**, 41.
10. A. K. Chatterjee, 'Special and New Cements', in Proceedings of the 9th International Congress on the Chemistry of Cements, National Council for Cement and Building Materials (New Delhi, November 1992), 1992, vol. 1, p. 177.
11. H. F. W. Taylor, 'Cement Chemistry', Academic Press, London, 1990.
12. V. S. Ramachandran, 'Concrete Admixtures Handbook', Noyes Publications, Park Ridge, New Jersey, 1984.
13. G. H. Tatersall and P. F. G. Banfill, 'The Rheology of Fresh Concrete', Pitman, London, 1983.
14. 'Superplasticizers and Other Admixtures in Concrete', SP-119, ed. V. M. Malhotra, American Concrete Institute, Detroit, Michigan, 1989.
15. Y. Ohsama, in reference 12, p. 341.
16. E. B. Nelson, 'Well Cementing', Schumberger Educational Services, Houston, Texas, 1990.
17. D. Damidot, S. Stronach, A. Kindness, M. Atkins, and F. P. Glasser, *Cem. Concr. Res.*, 1994, **24**, 563.
18. 'Calcium Aluminate Cements', R. J. Mangabhai, Spon, London, 1990.
19. C. H. Fentiman, *Cem. Concr. Res.*, 1985, **15**, 622.
20. S. R. Tan, A. J. Howard, and J. D. Birchall, 'Advanced Materials From Hydraulic Cement', in 'Technology in the 1990's: The Promise of Advanced Materials', The Royal Society, London, 1987.
21. J. A. Lewis and W. M. Kriven, MRS Bulletin XVIII no. 3, 1993, p. 72.
22. A Palomo and F. P. Glasser, *Brit. Ceram. Trans. J.*, 1992, **91**, 107.
23. P. S. De Silva and F. P. Glasser, 'The Hydration Behaviour of Metakaolin – $Ca(OH)_2$ – Sulphate binders'. Proceedings of the 9th International Congress on the Chemistry of Cements, National Council for Cement and Building Materials (New Delhi, November 1992), 1992, vol. 4, p. 671.
24. D. Davidovitz, US Patents 4 340 486 (September 1982) and 45 009 985 (April 1985).
25. P. K. Mehta, *Cem. Concr. Res.*1976, **6**, 169.
26. J. H. Sharp and D. H. Winbow, 'Magnesia–Phosphate Cements', in 'Cements Research Progress' , ed. P. Brown, American Ceramic Society, Columbus, Ohio, 1986, p. 233.
27. W. E. Brown and L. C. Chou, 'A New Calcium Phosphate Water-Setting Cement', in reference 26, p. 351.

28. 'Chemistry of Cements for Nuclear Applications', ed. P. Barret and F. P. Glasser, European Materials Research Society Proceedings, vol. 27, Elsevier–Pergamon, Oxford, 1992.
29. 'Chemistry and Microstructure of Solidified Waste Forms', ed. R. D. Spence, Lewis Publishers, Boca Raton, Florida, 1993.
30. A. Kindness, A. Macias, and F. P. Glasser, *Waste Manag.*, 1994, **14(1)**, 3.
31. D. E. Macphee and F. P. Glasser, MRS Bulletin XVIII no. 3, 1993, p. 66.

7.2 List of Journals

Cement and Concrete Research
Advances in Cement Research
Magazine of Concrete Research
Il Cimento
Journal of the American Ceramic Society

7.3 Other Serial Publications

International Congress on the Chemistry of Cement (Proceedings)
- 9th Congress, New Delhi, 1992
- 8th Congress, Rio de Janiero, 1986
- 7th Congress, Paris, 1980

Cement Research Progress (1976–); issued annually by the American Ceramic Society, Columbus, Ohio.

7.4 Books

F. M. Lea, 'The Chemistry of Cement and Concrete', Edward Arnold, London, 1970; new edition, 1995.

H. F. W. Taylor, 'Cement Chemistry', Academic Press, New York and London, 1990.

G. Bye, 'Portland Cement: Composition, Production and Properties', Pergamon Press, Oxford, 1983.

CHAPTER 15

Inorganic Chemicals as Metallic Corrosion Inhibitors

PETER BODEN AND DON KINGERLEY

1 INTRODUCTION

Aqueous metallic corrosion may be described as the interaction between the metal and an electrolyte. Adverse effects upon the metal resulting from this interaction may be reduced or avoided by the use of **corrosion inhibitors**. This procedure is one of the main methods of corrosion prevention and is useful when a barrier coating (such as paint) cannot be used, such as on sliding surfaces and when a paint or metallic coating would be difficult to apply, such as down a long pipeline, and the debonding of the coating could cause blockages as in a high pressure boiler tube condenser or heat exchanger. Typical concentrations of inhibitors are from milligrams to grams per kilogram of water.

A corrosion inhibitor is a chemical that is added in small amounts to a corrosive water environment to reduce the rate of attack on the metal; and can be used directly by dissolving in the water, or added to paints, waxes, greases, and oils to protect metals when the environment becomes humid or wet, as in the many oil/water emulsions used in industry. Inorganic corrosion inhibitors, as opposed to organic inhibitors, are mainly used in near-neutral solution to form thin insoluble films by precipitation of an insoluble anion with the metallic cation, or to raise the electrode potential to more positive values where passive oxide film formation is possible.

Many inhibitors have been developed from organic chemicals and the majority are used on bare metal surfaces, where inhibition mainly occurs by adsorption of a few monolayers by electrostatic attraction. This requires an acid environment, *e.g.* pH 1–5 to give a film free surface. However, some anions such as benzoate and benzotriazole also form barrier films which reduce corrosion. The organic inhibitions are generally not used at high temperatures.

Inorganic inhibitors have been used successfully to reduce or eliminate the many forms of corrosion including uniform attack, pitting, crevice, bimetallic, and stress-assisted corrosion, including **corrosion fatigue** and **stress corrosion cracking**, although the effectiveness of the treatment with inhibitor depends on the conditions of aggressive ion concentration, temperature, flow, and stress levels.

Many types of corrosion inhibitor are commercially available mainly with either an inorganic or organic chemical base. The choice of which inhibitor to use (and whether to use inhibitors at all!) depends on the economic factors; for instance in cooling water systems, it is usually cheaper to add corrosion inhibitors to protect pipelines rather than to use a more corrosion-resistant material or internal coatings.

Even low rates of corrosion can lead to **loss of efficiency** in some machines such as pumps, where small amounts of corrosive wear (the most common form of wear), can lead to loss of fit of bearings, and wear of pump impellers and seals.

In all cases, the use of inhibitors must be considered against the needs of health, safety, and environmental contamination. Thus the use of inhibitors in the food industry and in drinking water systems, or in systems that may contact potable water supplies, have to be carefully considered for toxicity and possible disposal problems.

The use of corrosion inhibitors (including **vapour phase inhibitors**) is important, especially where water can condense in inaccessible places from saturated water vapour or from steam. This is especially important in packaging, where corrosion inhibitors either impregnated in the packaging, or loose in sachets or other slow release systems, assist moisture absorbing dessiccants or even replace them when the condensate cannot be removed, *e.g.* aircraft bilge spaces (human perspiration).

Corrosion mechanisms and especially the action of corrosion inhibitors can be better understood when interpreted on the basis of the **electrochemical processes** that occur when a metal is in contact with an aqueous environment. The electrochemical basis of corrosion is described and the effect on the thermodynamics and kinetic parameters of the corrosion inhibitors are indicated.

2 PRINCIPLES OF CORROSION INHIBITION

Chemicals may be added to the environment to inhibit corrosion with three objectives:

(1) to **react** with the agents in the environment responsible for corrosion thus effectively eliminating the cause.
(2) to **alter the pH** of the system so as to make the environment less aggressive to a particular metal.
(3) to **interact with the metal surface** in some manner not always clearly understood, but without affecting the composition or properties of the environment.

3 CORROSION AS AN ELECTROCHEMICAL PROCESS

3.1 Principles

Corrosion involves the passage of electrons between anodes and cathodes on the same or, in the case of bi-metallic corrosion, between two different metal surfaces.

At the **anode**, the electrochemical process is the formation of metal ions from the atoms at the metal surface which then move into the aqueous environment. The overall anodic reaction is

$$M \rightarrow M^{z+} + z\,e^- \quad (1)$$

This process results in the release of electrons which must be consumed by some other electrochemical process (**cathode reaction**) if the corrosion of the metal is to continue.

For many corrosion processes these electrons may be removed by either **reaction with hydrogen ions** in acid; or in neutral solutions (when hydrogen ions are in low concentration) by the **reduction of dissolved oxygen** molecules. For hydrogen ion reduction by electrons:

$$2\,H^+ + 2\,e^- \rightarrow H_2(g) \quad (2)$$

For reduction of dissolved oxygen:

$$O_2 + 2\,H_2O + 4\,e^- \rightarrow 4\,OH^- \quad (3)$$

(Note that when the concentration of hydrogen ions is low and when dissolved oxygen is absent, very little corrosion can occur of the structural materials based on iron, aluminium, copper, *etc.*)

Corrosion inhibition is consequently the prevention or slowing down of the release of metal ions, *i.e.* **anodic inhibitors**, or the prevention or slowing down of reactions at the cathodes, *i.e.* **cathodic inhibitors**. However, most corrosion inhibitors provide a **thin layer across the metal surface** effectively preventing anodic and cathodic reactions by restricting access from the environment, especially hydrogen ions and dissolved oxygen.

Corrosion products often **slow down** the initial corrosion rate but their effectiveness depends on coverage, adhesion, porosity, and insolubility in the environment.

Since at the anode, metal ions are produced (see Equation 1), then insoluble salts will provide a barrier by adding an appropriate soluble anion, often the sodium salt. The production of good adhesion, low porosity, and good coverage depends on many other factors.

At the cathode where the dissolved oxygen, necessary for corrosion to occur, is reduced, the solution next to the surface becomes alkaline due to the formation of hydroxyl ions (see Equation 3). This effectively changes the pH; and if cations are added, insoluble metal hydroxides which are produced provide a barrier layer over the cathode area, see Figure 1.

If the metal hydroxides have **good adhesion**, **low porosity**, and **low solubility** at neutral or lower pH values then corrosion inhibition will occur since the corrosion rate depends on the rate of diffusion of the oxygen molecules to the surface. The pH of the surface now changes from the alkaline pH to the bulk solution, say pH 7, but the solid oxide is slow to dissolve. Low corrosion rates can be achieved by removing or lowering the concentration of the dissolved oxygen via

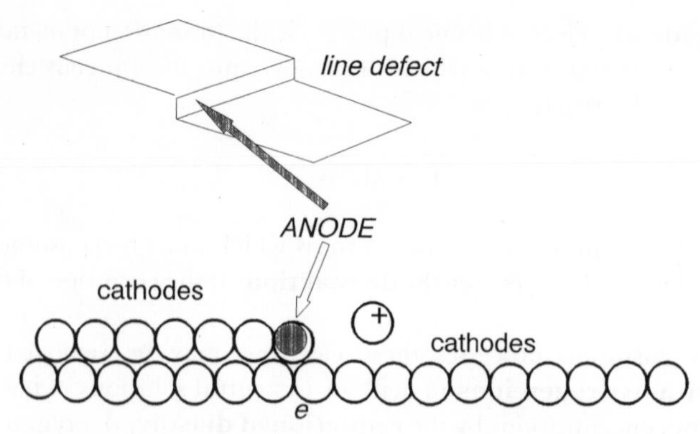

Figure 1 *A simplified view of cathode/anode areas on a metal surface, showing anodes as relatively small areas in the form of atom ledges and line defects arising from emergent dislocations, slip planes, kink sites, and crystal boundaries*

the use of chemicals which react with oxygen. This approach can only be considered for closed systems however, owing to the expense of replenishing the oxygen scavenger in an open system.

There is an important class of inorganic inhibitors which provides an electrochemical reaction which makes the electrode potential more positive and encourages the growth of a protective metal oxide film. These are called **passivators** rather than inhibitors and rely on a **thin oxide film**, sometimes only a few atoms thick and formed by a solid state growth process rather than by precipitation from solution. Passivators are often regarded as dangerous because if sufficient concentration of the inhibitor is not maintained then deep pitting corrosion occurs at the exposed areas because of the more positive electrode potential. These types of corrosion inhibitor are often used, however, because of their stability, especially at elevated temperatures.

In order to gain a better understanding of inhibitor action by insoluble coatings and by passivation then reference may be made to electrode potential–pH diagrams. These are often referred to by the name of their creator, *i.e.* **'Pourbaix Diagrams'**.

3.2 Electrode Potential–pH Diagrams

The diagrams are calculated from thermodynamic data and from solubilities of the various oxides and hydroxides as a function of pH. For many metals there are oxides which have low solubility over a wide pH range around the neutrality point of pH 7. It is necessary to decide on an arbitrary value for solubility which could be regarded as stable; Pourbaix made the choice that an oxide was stable when the solubility of the metallic cation was less than 10^{-6} gram ions/kilogram water. The pH at which the metal oxide had this concentration of metal ions in equilibrium with the solution was considered to be the **boundary** between **corrosion by dissolving the oxide** and **passivity**, that is the formation of an insoluble oxide film on the metal surface.

There is now much experimental evidence for this concept to be acceptable as a practical criterion for the **protection of metals** by the adjustment of the pH.

An example of these diagrams is given in Figure 2 for the case of zinc/water. Here the formation of a passive, insoluble barrier layer on the zinc surface occurs in the pH range pH 8.5–11.9; at a lower pH than 8.5, dissolution of the zinc oxide is possible to give $Zn^{2+} + H_2O$, and at a pH above pH 11.9, the zinc oxide becomes soluble as the zincate (ZnO_2^{2-}) anion. The solubility of the zinc oxide and zincate ion can be expressed as follows:

$$\log Zn^{2+} = 10.96 - 2pH \tag{4}$$

$$\log ZnO_2^{2-} = -29.78 - 2pH \tag{5}$$

The pH when the oxide solubility yields 10^{-6} g ions/kg water (the Pourbaix criterion for no corrosion) can be calculated from these equations to give the boundary between the passivating oxide and the corrosion as zinc ions on the potential–pH diagram, see line 1 (pH = 8.5) in Figure 3, and the boundary between passivating oxide and corrosion as zincate ions, line 2 (pH = 11.9) on Figure 3.

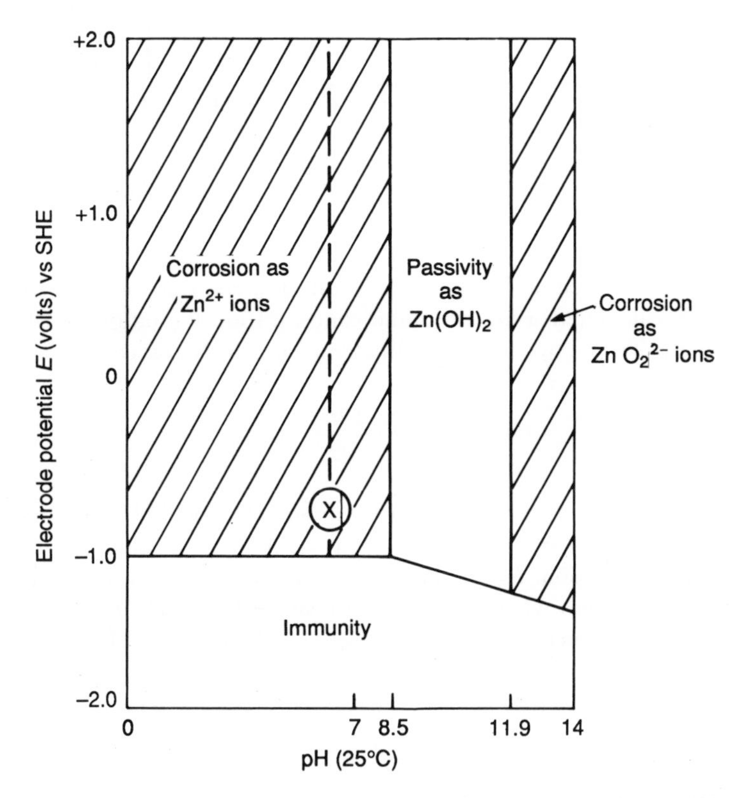

Figure 2 *Potential–pH diagram for zinc/water system (after Pourbaix[3]). The dotted lines refer to the zinc/water/CO_2 system showing the more protective carbonate film which extends the passive region*

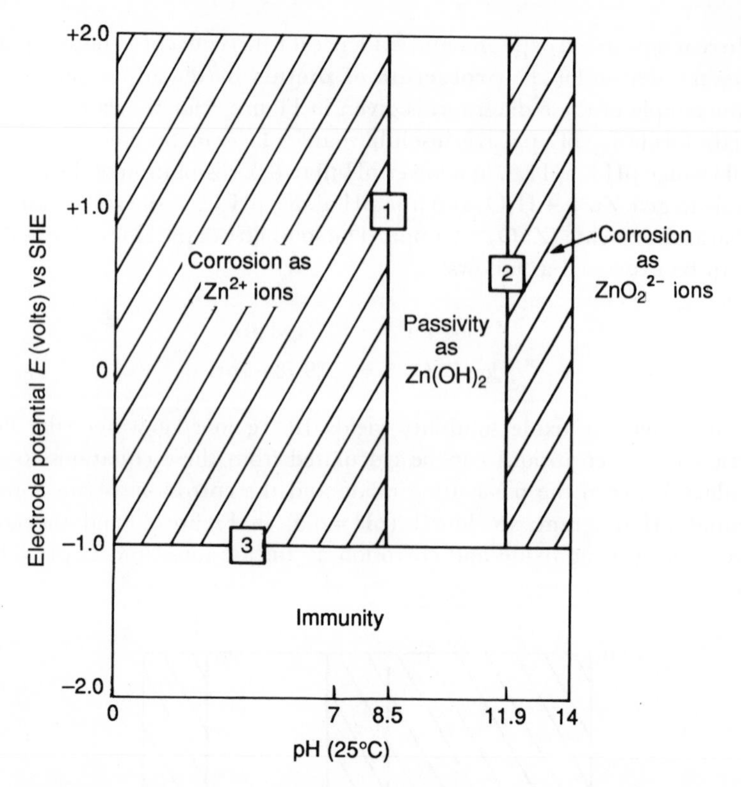

Figure 3 *Construction of the zinc/water potential–pH diagram from the Nernst Equation (line 3) and the Solubility of zinc hydroxide (lines 1 and 2)*

The electrode potential for the equilibrium between zinc ions and zinc metal can also be calculated using the same concentration limitation of 10^{-6} gram ions/kilogram of water by substituting this figure into the Nernst equation as follows:

$$E \text{ (volts)} = E^{\circ} + \frac{2.3RT}{zF} \log [M^{z+}] \tag{6}$$

where E = electrode potential vs standard hydrogen electrode (SHE), E° = standard electrode potential, T = the temperature in K, R = gas constant, z = number of electrons in the reaction, F = the Faraday constant, and M^{z+} = concentration of metal ions. (Note that when the activity of $M^{z+} = 1$ then $E = E^{\circ}$.)

For zinc at the concentration where corrosion is considered to be low:

$$E \text{ (volts)} = -0.76\text{ V} + 0.06\log(10^{-6}) = -1.12\text{ V} \tag{7}$$

Where all the electrode potentials refer to the standard hydrogen electrode (SHE).

Line 3 on Figure 3 can now be added and is parallel to the pH axis as no pH effects can occur under the conditions represented by Equation 7.

The line calculated from the **Nernst equation** represents the potential where

corrosion or immunity may be possible. At potentials more negative than this the metal can be regarded as immune and at more positive potentials then corrosion can occur. This depends on pH, for if the solubility of the oxide is low, then passive films form on the metal surface.

Pourbaix and his collaborators have calculated diagrams for all the metals and some anions. Some of the anions are inhibitors and can be used on a combined diagram to show where the area of passivation can be extended. See Figure 4 for the effect of the molybdate ion on extending the diagram for iron, and Figure 5 for the phosphate/iron diagram. In Figure 2, the area has been outlined for the Zn/water diagram to show the effect of CO_2 and the carbonate ion CO_3^{2-} ion which are much less soluble than the zinc hydroxide and are generally produced in natural waters which contain these species. There is a considerable extension of the passive zone with these 'natural inhibitors' for zinc.

Calculations can also take into account variable valency and the more complex reactions between a metal and water where hydrogen ions (pH) and electrode potentials are in the same equation, for example Equation 8:

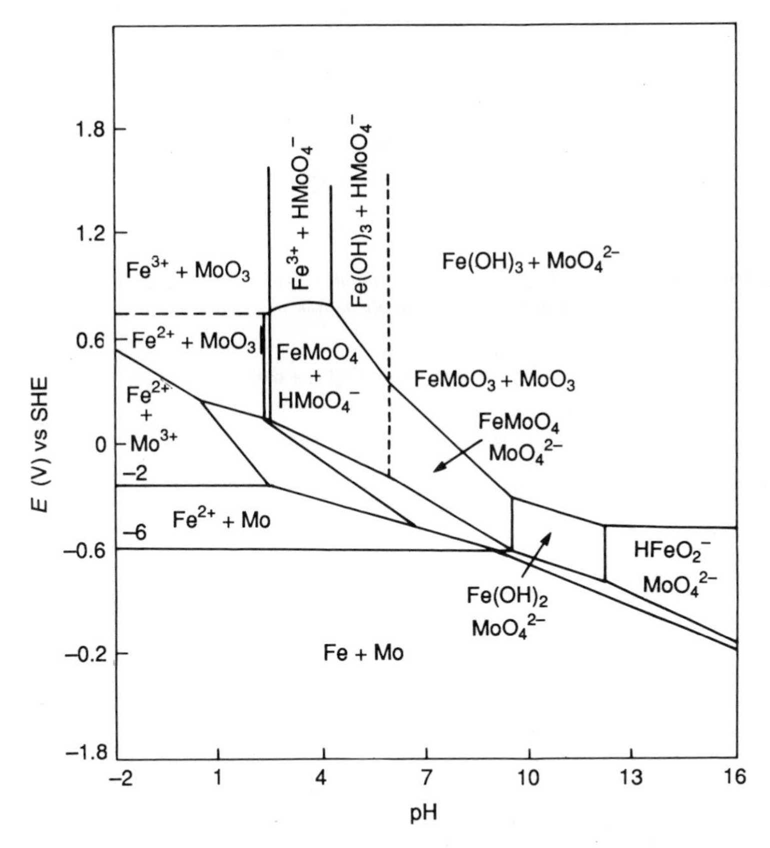

Figure 4 *Potential–pH diagram for the iron/molybdate/water system, showing the extended area of the passive zone due to insoluble iron molybdates. (Mo concentration $10^{-2}M$; Fe 10^{-6} M; 298K)*

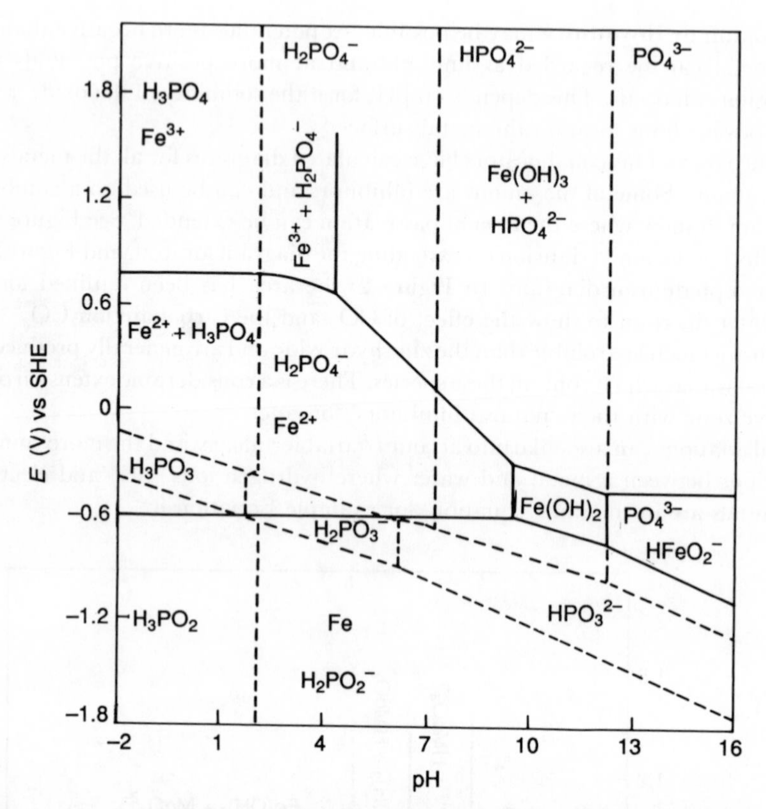

Figure 5 *Potential–pH diagram for the iron/phosphate/water system, at 298 K, showing the extended area of the passive zone due to insoluble iron phosphates*

$$2\,Fe^{2+} + 3\,H_2O \rightarrow Fe_2O_3 + 6\,H^+ + 2\,e^- \tag{8}$$

This is shown in the potential/pH diagram for the iron/water system (Figure 6) where this reaction is affected by both pH and electrode potential and so gives a sloped line. Also this oxide γ-Fe_2O_3 is very insoluble and gives a very effective passive film.

The diagrams form a guide to the corrosion behaviour but have certain limitations. They do not take into account the presence of various anions such chlorides, sulfates, and nitrates which make it more difficult to produce continuous passive films.

3.3 Competitive Adsorption

Inorganic anion inhibition is mainly through the **anion** forming **insoluble films** at the metal surface and their efficiency depends on the **competition** with other **non-protective anions** that form soluble salts with the metal cations from the corroding metal. Soluble anions include the sulfates, chlorides, and nitrates that occur in natural waters and can be considered as aggressive anions.

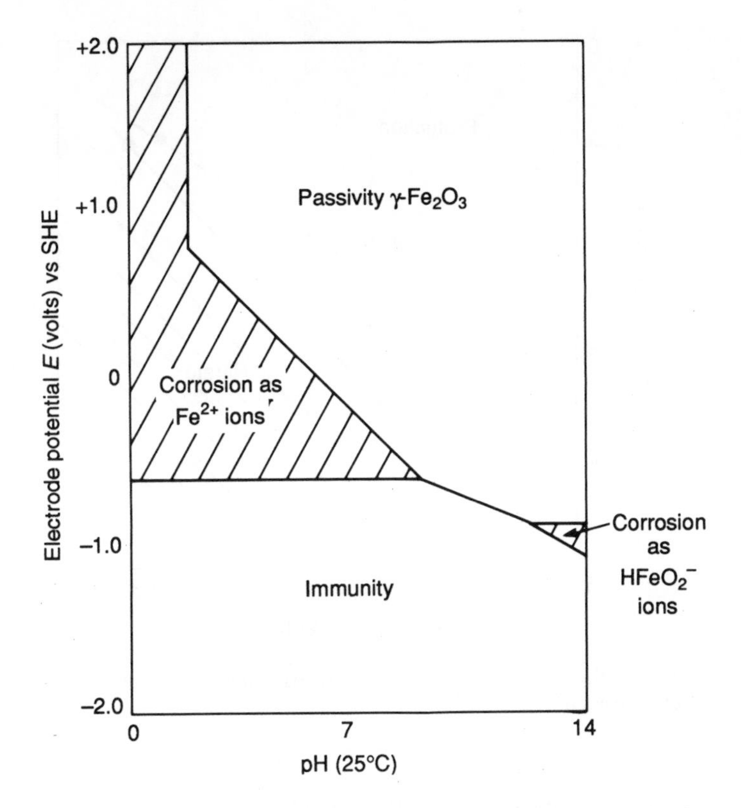

Figure 6 *Potential–pH diagram for the iron/water System (after Pourbaix[3]) showing region of passivity extending to the acid regions at high anodic potentials*

Many studies have shown that there is a **critical ratio** of the inhibitor anion to aggressive anion which determines whether corrosion or protection occurs. Thus for a low concentration of say chloride, a low concentration of inhibitor is required. As the chloride concentration increases the amount of inhibitor required for corrosion protection has to be increased to maintain the critical ratio for protection (see Figure 7).

It has been found[4] that there is a logarithmic dependence of the inhibitor concentrations on the concentrations of the aggressive anions:

$$\log c_{\text{inhibitor ion}} = \log c_{\text{aggressive ion}} + \text{constant} \quad (9)$$

where $n \simeq 1$ for monovalent ions such as chloride and nitrate, and $n \simeq 0.6$ for sulfate ions.

3.4 Formation of Insoluble Films

The addition of zinc salts greatly reduces the corrosion of iron in aerated neutral water because of precipitation of low solubility zinc hydroxide in the alkali

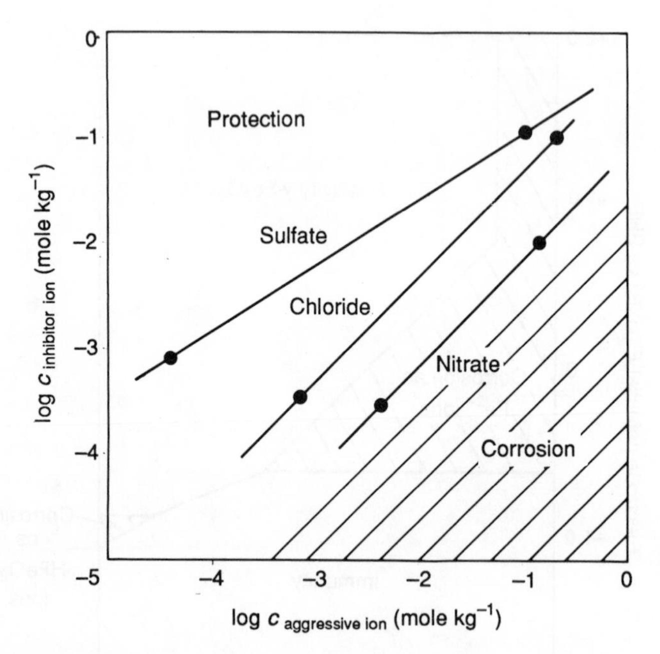

Figure 7 *Relationship between concentration of aggressive/inhibitive anions, showing competitive absorbtion (after Rozenfeld[4])*

generated at the cathode surfaces of the corroding iron. Since the cathode area is geometrically greater than the anode areas (emergent dislocations, grain boundaries) (see Figure 1) this physical barrier film prevents access of the dissolved oxygen to the cathode sites and corrosion is geatly reduced. The success of zinc salts as an inhibitor also depends on the low porosity of the zinc hydroxide layer and its adherence to the steel which appears to be good.

The formation of passive films depends on the type of Potential–pH diagram. Iron falls into the category of a mixed valency oxide formation where the oxides of different valency have different solubilities. Such an example is iron, see Figure 6, where at certain electrode potentials the higher valent oxide is formed which has lower solubility in acid solution (lower pH) than the lower valent oxide, so that above a certain potential the higher oxide forms on the metal surface and reduces considerably the corrosion rate by this form of **passive oxide layer**.

Whilst the anode potential can be raised by electrochemical polarization, it is also possible to achieve a suitable potential difference by allowing a redox reaction to take place which raises the electrode potential of the surface above the potential for the formation of the passive oxide film as shown in the potential–pH diagrams.

The two **most important passivators** which allow the protection of **iron** by their redox reactions are due to the **chromium(III)/chromium (II)** and **nitrite/ammonia** reactions. This effect is achieved merely by adding the soluble sodium or potassium salts of chromate or nitrite.

The level of concentration required varies with the ratio of aggressive/inhibitor ions in solution. In the case of anions, the ratio of the inhibitor ions to the aggressive ions is important (see Figure 7). In the case of the cations this will depend on the formation of insoluble compounds over a range of pH values, particularly in the acid region; this could assist in the formation of the barrier film but has to occur on the metal surface to be effective, *e.g.* iron phosphate.

Passive film formation by a solid film diffusion process always forms on the surface next to the metal with good adherence although some 'salt' precipitation films are found to be non-porous and have good adhesion.

3.5 Typical Inorganic Inhibitors and their Mechanism of Protection

Inorganic inhibitors are invariably **anions** and added to neutral, aerated water systems as the sodium or potassium salt. Cations such as arsenic and antimony have also been used, particularly for inhibition of acid solutions but their high toxicity now precludes them from industrial application.

The anion forms an insoluble species, usually with the metallic cation from the corroding metal, or from calcium ions often present in natural waters or from added cations such as zinc. The films formed must be of low porosity and of good adherence to the metal surface to reduce the diffusion of oxygen to the cathodic sites on the metal surface.

True passive films are formed from a solid state reaction occurring above a certain protection electrochemical potential as a result of an oxidizing anion undergoing a redox reaction which raises the metal surface electrochemical potential. In this case the barrier film is a metal oxide: *e.g.* for iron, γ-Fe_2O_3.

4 PRACTICAL ASPECTS OF CORROSION INHIBITION

4.1 Effects of Inhibitors

Anion inhibitors that are effective for a particular metal or alloy may not be protective for other metals. Often, corrosion inhibition is specific for a particular metal or alloy.

Since a surface reaction is involved, then the surface roughness and cleanliness is important. The surface may be very reactive because of the crevices, but the formation of a continuous oxide film may be more difficult. If the surface is greasy, barrier film formation may not occur and if it does then there will be poor adhesion and the corrosion process may continue, as water and oxygen can diffuse through oil and grease.

Bimetallic corrosion can often be successfully treated by anion inhibitors providing the inhibitor is protective to one but not aggressive to the other. For many industrial systems, involving multi-metal joints where bimetallic corrosion is possible, the use of inhibitors becomes more complicated and 'cocktails', *i.e.* **mixtures of several inhibitors**, are used.

Because an air or gas/water interface is present, there is the danger of corrosion in the area of condensation, in the wet gas area, where the inhibitor can only protect the fully immersed area. **Vapour phase inhibitors** have however been developed where the anion is carried in an **organic chemical 'parachute'**, *e.g.* cyclohexylamine carbonate and dicyclohexylamine nitrite.

Many anionic inhibitors only protect in a narrow pH range and some inhibitors are transformed by a pH change, *e.g.* polyphosphates to less protective orthophosphates. In water treatment practice, corrosion is restrained by adjustment of pH and the dissolved oxygen concentration. Strictly, it may be said that this is not inhibition but these changes encourage formation of a stable passive film and/or reduce the corrosion reaction by removing the cathodic process stimulant.

4.2 Synergism of Corrosion Inhibitors

Synergism is observed by the **use of two or more inhibitors together**, and in some cases this restrains the corrosion rate more effectively than the simple sum of the two individual substances. This phenomenon enables considerable savings to be made and is extensively employed in commercial inhibitor formulation.

It has been shown[3] that in aerated, near neutral solution where the corrosion of iron and steel is at a high rate, synergism can occur by the joint action of

(a) Dissolved oxygen and anions such as nitrites and chromates. These anions will reduce the corrosion of iron but more so in the presence also of dissolved oxygen.

(b) Mixed oxidizing and non-oxidizing inhibitors, *e.g.* nitrite and borate. The nitrite supports the formation of a passive film on iron of γ-iron(III) oxide and the borate forms insoluble compounds which effectively plug weaknesses in this barrier film.

(c) There are many successful combinations with organic corrosion inhibitors, *e.g.* phosphate (for iron) and benzotriazole (for copper in a steel/copper system such as that found in central heating systems).

5 ANION INHIBITOR PROPERTIES IN NEUTRAL ELECTROLYTES

5.1 Sodium Phosphates

These anions are commonly used either as the mono-orthophosphates or as polyphosphates formed from the dehydration of the hydrogen orthophosphates:

$$NaH_2PO_4 \rightarrow NaPO_3 + H_2O \tag{10}$$

The **monophosphates** give good inhibition in neutral solution but the passive film appears to be a mixture of γ-$Fe_2O_3 + FePO_4.2H_2O$ (see also Figure 5).

The efficiency of inhibition depends on competitive adsorption with sulfates, nitrates, and chlorides, and this depends on the ratio of ions, *e.g.* the **polyphosphates** have been used with the naturally occurring calcium ions to form thin films of calcium tripolyphosphate $Ca(P_3O_{10})_2$ as the barrier film but zinc ions are often added and are more effective than calcium ions. The addition of these inhibitors, in the presence of a divalent ion such as calcium or zinc, lowers the electrode potential which is indicative of a cathodic inhibition, *i.e.* coverage of these areas preventing further diffusion of oxygen which then lowers the corrosion rate.

Polyphosphates are more effective inhibitors when oxygen is present owing to the higher pH generated at the cathode areas as a result of the corrosion process, which causes precipitation of the calcium and/or zinc phosphate at the surface. The presence of a film on the surface is confirmed because corrosion does not occur when the inhibitor concentration is removed. The film has been identified as $NaFeCa(PO_3)_5.8H_2O$.

The stability of the polyphosphates is reduced at pH values less than about 5; the usual range for their use is therefore pH 5–7. They are most effective at ambient temperature because hydrolysis takes place more readily at higher temperatures to give orthophosphates. This process takes place even at 30 °C over several days and in 24 hours at near the boiling point of water. These inhibitors are non-toxic.

5.2 Sodium Silicates

Soluble silicates are of variable composition having a general formula $n Na_2O.m SiO_2$ where the ratio m/n is known as the module and is variable and depends on the pH. The **mechanism** of inhibition appears to be the formation of complex, **colloidal systems** which react with the heavy metal ions such as Fe^{2+} and Al^{3+} to give barrier coatings of metal silicates.

The sodium silicates can contain Na^+, OH^-, SiO_3^{2-}, $HSiO_3{}^-$, molecules Na_2SiO_3, $NaSiO_5$, H_2SiO_3, SiO_2, and other complexes such as $(SiO_2)_x H_2O.y SiO_3$. The actual colloids produced depend on concentrations, pH, temperature, and other anions in solution.

The **stability** of the colloids in solution is important and under certain conditions of the above parameter can lead to crystallinity, a coarsening, and gel precipitation. This can lead to non-uniform coverage and the possibility of local corrosion, *i.e.* pitting. When silicates are used the conditions for coagulation are avoided.

When the soluble ions such as sulfates and chlorides are in high concentration (as in sea water) the silicates are ineffective because non-protective silicates are produced at the surface. Silicates are most effective when sulfate and chloride anions are less than about 500 ppm.

Silicate inhibitors are attractive because they are non-toxic and have a **mixed inhibitor effect**, *i.e.* they form insoluble inhibitors with heavy metal ions and slow down the diffusion of oxygen. The silicates are good inhibitors for Pb, Sn, and Cu and are effective in concrete for the protection of steel reinforcing rods and structural components embedded in concrete.

5.3 Sodium Molybdates

As with other anions, there is an increase in corrosion until a critical concentration is reached, when inhibition occurs. Sodium molybdate is subject to selective ion competition so that for aerated and neutral potable waters containing 20 ppm of

Na_2SO_4 and 30 ppm NaCl, 200 ppm of molybdate can suppress corrosion of steel at ambient temperatures. However, with 200 ppm chloride, 1000–2000 ppm molybdate is required to suppress corrosion. There is a change in electrode potential to more positive values and this may be interpreted as anode inhibition.

The mechanism of protection appears to be the **formation of a passive film**, (see Figure 4, page 397) but it is not yet clear whether a molybdate is formed at the anode sites or whether the passive film of the metal, *e.g.* iron, is strengthened. Synergism with phosphates occurs when phosphomolybdates are formed. Phosphomolybdates are notable for being insoluble in acid solution, thus they are valuable in the case of **pitting corrosion**. Here, when the pits become more acidic than the bulk solution, phosphomolybdates precipitate within the pits, giving barrier coatings and slowing down the pitting rate.

5.4 Sodium Nitrite

Sodium nitrites can be used alone or in conjunction with other anion inhibitors. The nitrite does not form insoluble salts with other metals so that the mechanism of inhibition is thought to be associated with the **redox reaction** which occurs on the metal surface. The nitrite inhibition is mainly used for iron and the potential change caused by the redox potential for the conversion of nitrite to ammonium ions:

$$NO_2 + 8\,H^+ + 6\,e^- \rightarrow NH_4^+ + 2\,H_2O \tag{11}$$

$$6\,Fe(OH)_2 + NO_2 \rightarrow 3\,(\gamma\text{-}Fe_2O_3) + NH_4^+ + 3\,H_2O + 2\,OH^- \tag{12}$$

These reactions take place at a more positive electrode potential and the **iron is passivated** by the formation of γ-Fe_2O_3. The anion concentration is important in that a critical concentration is required in order to raise the potential to the required level. Lower concentrations than this critical value can lead to acceleration of corrosion. In a system it would be important to keep up the required level of concentration throughout the life of the equipment. The passive film on iron can be broken down by chloride ions to give pitting, in which case the concentration of inhibitor must be increased. In the absence of aggressive ions, *e.g.* distilled water, only 10–100 mg/kg is required.

This inhibition shows a logarithmic relationship within aggressive ions so that a minimum critical ratio is required. For nitrites the slope log [inhibitor ion]/log [aggressive ion] = 1.0 for nitrates and chlorides and about 0.5 to 0.6 for sulfates (see Figure 7, page 400). It has been noted that these ratios are only reliable on new surfaces and that for previously corroded surfaces, say in chloride, then higher ratios are required. The nitrite anion inhibitor, usually sodium nitrite, is often used in conjunction with other inhibitors.

Nitrite is frequently used together with sodium benzoate and the combination appears to be more effective, especially at higher temperatures (*e.g.* in central heating system). This appears to be the effect of benzoate inhibition of pit formation in the passive film formed in the presence of the nitrite anion. However, there is a pH range over which this combination is most effective, *i.e.* above pH 6.0.

5.5 Dissolved Oxygen Scavengers

These are not inhibitors in the sense defined above but are included because these scavengers represent an alternative approach to corrosion control.

As discussed before (Equation 3, page 393), an important cathodic reaction in corrosion processes is the reduction of dissolved oxygen to form hydroxyl ions. If **dissolved oxygen** can be **removed** from solution the **extent of corrosion** will be **decreased**. This may be done by thermal, vacuum-extraction, or chemical methods. In many situations two methods are used successively, the thermal means first, as this is cheaper, and then the chemical method to remove the residual oxygen. Hence the term 'scavenger' for the reagents employed.

Sodium sulfite was widely used in high pressure boiler plant and still finds a place in low pressure plant. It finds extensive use in primary oil extraction, when water is pumped down steel pipes in order to extract more oil. It reacts with oxygen to form sodium sulfate, *i.e.*

$$Na_2SO_3 + \tfrac{1}{2}\,O_2 \rightarrow Na_2SO_4 \quad (13)$$

A catalyst, *e.g.* the cupric or cobalt ion, is added when used in cold water systems.

A disadvantage of the use of sodium sulfite for higher pressure boiler plant, however, is that it leads to a substantial increase in the dissolved solids content. This is undesirable as the concentration of dissolved material in the boiler water has to be limited to avoid carry-over of solutes in steam.

Hydrazine is used in the power generation industry to remove traces of oxygen, after vacuum extraction:

$$N_2H_4 + O_2 \rightarrow N_2 + 2\,H_2O \quad (14)$$

This is accompanied by some thermal decomposition:

$$3\,N_2H_4 \rightarrow 4\,NH_3 + N_2 \quad (15)$$

As the products from these reactions are gaseous, they do not increase the solute concentration. As well as acting directly with oxygen, hydrazine reacts with iron(III) oxide which has been formed on the internal surfaces of high pressure plant when oxidizing conditions have been prevalent, and reconverts it to the normal magnetite. Concern about the toxicity of hydrazine has led to the search for alternatives. As might be expected, hydroxylamine, NH_2OH, will also react with oxygen, but the diethyl derivative is preferred.

6 SOME APPLICATIONS OF CORROSION INHIBITORS

6.1 Cooling Water Circulation – Once-through Systems

In practice, corrosion occurs in conjunction with other problems or effects, *e.g.* **deposition of mud, silt, scale, and biofouling**; it is artificial to discuss any one of them in isolation so there is a need for a multi-pronged attack. In once-through systems as found for example in large power stations, where the volume of water is so large as to rule out, on economic grounds, any chemical treatment (except chlorination to control algae) the approach has been to 'engineer-out' problems. This involves the following aspects:

(a) choose appropriate corrosion resistant materials
(b) limit fluid velocity to avoid corrosion–erosion
(c) prevent the generation of particularly aggressive environments
(d) follow as far as is practicable appropriate operation/maintenance/design philosophy to minimize corrosion, *e.g.* to avoid stagnant water in heat exchangers which may induce deposit attack. Iron(II) sulfate may sometimes be dosed in these operations. With this method, iron(II) sulfate when added to the slightly alkaline cooling water forms iron(II) hydroxide which is precipitated in colloidal form and the charged particles travel preferentially to flaws in the protective oxide film on the condenser tube. The additions are made in the immediate vicinity of the tank, and small quantities of iron(II) sulfate are a temporary expedient pending repair.

The system may still be subject to scaling so that **inhibited acid washing** from time to time may be used, but otherwise the system will not be treated with inhibitors.

6.2 Cooling Water Circulation – Open Systems

Here, volume does not necessarily prohibit chemical treatment. This method depends upon cooling achieved by evaporation in a tower so that the solutes will eventually become increasingly concentrated and tend to precipitate and form scale. Where this is so, it may be desirable to practice scale control as well as inhibition by any of a number of means, *e.g.*

(a) **Water softening**, *i.e.* remove calcium and magnesium ions from the cooling water, partially or wholly, and then add inhibitors to control corrosion in scale-free conditions.
(b) **Acid dosing**, to a pH at which scaling is tolerable, may be an option where the volume is too large to permit any degree of softening, but no inhibitor is used in such operations.
(c) Occasional **off-load chemical cleaning** may be applied to remove scale and then inhibitors are necessary.
(d) **Mechanical cleaning** – both on-load, and off-load.

For conditions outlined in (a), chemical additives (excluding acid dosing) include specific inorganic inhibitors such as zinc, orthophosphate, polyphosphate, phosphonate, molybdate, and nitrite. Mixtures of inhibitors include: zinc ions plus chromate; zinc ions plus polyphosphate; zinc ions plus phosphonate; polyphosphate plus orthophosphate; polyphosphate plus phosphonate; and mixtures of phosphates, molybdate, and nitrites.

An interesting variant is **inhibition by scale control**. Scale is formed when certain solutes come out of solution. Many natural waters contain dissolved salts of calcium and to a lesser extent magnesium which cause scaling problems ranging in size from the familiar domestic kettle 'fur' to deposits of the order of a tonne in industrial heat exchangers. Calcium bicarbonate is far more soluble than the carbonate in water. Either heating the solution, which expels some carbon dioxide, or addition of alkali which reacts with the bicarbonate will precipitate calcium carbonate. In order to control the thickness of scale to give protection but *not* to block tubes or reduce heat transfer significantly it is necessary to apply a chemical equilibrium technique. The equilibrium between calcium and carbonate, bicarbonate, *etc.* depends upon the pH. Various relationships amongst these contributory factors to scaling have been derived and may be used to facilitate control of the scaling/corrosive properties of a water, notably those equations proposed by Langelier[7] and by Ryznar.[8]

In Langelier's approach, the actual pH (pH_a) of the cooling water is compared with the pH if saturated with calcium carbonate (pH_s). The difference ($pH_a - pH_s$) is Langelier's Index. According to whether this is positive or negative the **water is designated as scaling or corrosive**. Then pH adjustment via the addition of either acid or alkali is made to control the scale to the appropriate thickness. Ryznar uses a similar approach but his Index is $2pH_s - pH_a$. Both Indices are subject to the effect of other factors, *e.g.* chloride concentration in solution. Indices analogous to these but for phosphate systems have also been proposed.

At relatively **low volumes** of circulating water (below 100 m^3), some degree of **water softening** may be economical for make-up water to the system when the concentration of salts is reduced by dumping and diluting with softened water.

6.3 Engine Radiation and Cooling Systems

The engine coolants used in cars and compressors need to have antifreeze additives for use in cold climates. These are usually polyhydric alcohols, normally ethylene glycol. The resultant **water–alcohol mixture needs inhibiting against corrosion** caused by both developing acidity, as the glycol decomposes, and the periodic presence of dissolved oxygen, which is more soluble in glycol than water. The precise nature of the inhibitors used depends upon the materials of construction of the engine.

The main constituents of engines have steel, copper-based alloys, and aluminium. In the UK corrosion inhibitors to use are given in the British Standards BSS 3150, 3151, and 3152, respectively, which recommend sodium benzoate–nitrite, triethanolamine–sodium mercaptobenzothiazole, and sodium borate. These mixtures also form the basis of many commercial formulations, but some variations may be found.

To ensure reliability, inhibitor concentration must be maintained by periodically checking; and as and when necessary, possibly annually flushing out the circuit and replacing with new coolant. Whilst antifreeze is not needed in the summer in some countries, the metals of the circuit will still be vulnerable to corrosion if not protected and this is probably most conveniently achieved by using an **inhibited antifreeze** formulation throughout the year.

Diesel cooling systems in locomotives have been protected by sodium borate–nitrite mixtures, sodium nitrite additions having been made where aluminium is present. Sodium benzoate–sodium nitrite formulation has also been used. Cavitation can also be a problem in Diesel engines, this being combatted by the use of an oil–water emulsion added to the coolant. Marine diesels have been protected by borate–nitrite, and by phosphate–nitrite mixtures. Large marine diesel engines do not normally need an antifreeze and for many years sodium chromate was successfully used as an inhibitor, but is now being phased out because of toxicological objections.

6.4 Central Heating Systems

Many central heating systems function satisfactorily without inhibitors, but treatment is so easy to achieve in comparison with the costs of a failure due to corrosion that it is probably prudent to **add an inhibitor formulation** in any case. Most town's waters are relatively benign and on filling the system there may be some initial deposition of calcium carbonate together with formation of magnetite from the dissolved oxygen present in the water. Thereafter, provided losses of water from the system are negligible, there should be **virtually no corrosion**. In many systems this is achieved, but this does depend on operation and design.

Efforts should be made to prevent the ingress of air into the system, *e.g.* persistent leaks. Otherwise, oxygen transported into the system with the water used to make-up the losses will corrode the steel radiators to form iron(II) hydroxide. This, in turn, will decompose via the **Schikorr reaction** to form magnetite and hydrogen:

$$3\,Fe(OH)_2 + H_2O \rightarrow Fe_3O_4 + H_2 \quad (16)$$

The hydrogen along with oxygen, nitrogen, *etc.* will accumulate in radiators and require periodic 'bleeding'. The magnetite can lead to blockages in the system, particularly in the circulating pump after the summer 'off-load' period, and can lead to costly repairs. In addition, pitting occurs beneath deposits of *e.g.* magnetite, with possible penetration of the tube wall and consequent leakage.

Usually this corrosion will require some alkalinity control as well as deoxygenation. A specific inhibitor for copper (organic, *e.g.* benzotriazole) to protect the pipes is usually also included, together with possibily some specific additives, *e.g.* phosphate and molybdate which is one of several constituents of the old Gas Council formulation. Where nitrite is used, this is a nutrient for certain bacteria so that a biocide is also included.

In the case of central heating systems where the water can return to the drinking supply, care must be taken to avoid the use of toxic chemicals for treatment.

6.5 Refrigeration Plant and High Chloride Systems

Many refrigeration plants use strong solutions of **brine** (sodium or calcium chloride) at about 20% concentration. Such solutions, when oxygen-free, are not substantially more corrosive than water and may be used without trouble for many years. However with oxygen ingress, *e.g.* leaks and when replenishing during maintenance, then there is a danger of corrosion being initiated. Chromate was formerly used as an inhibitor, but toxicity considerations now forbid this. Other inhibitors used include nitrite and phosphate.

6.6 The Pickling of Metals

Various metals and alloys may be pickled to remove coatings such as oxides, formed during manufacture and/or service, as well as unwanted deposits such as hardness salts. The **pickling agents** are frequently, although not invariably, **acidic** solutions. **These must be inhibited** to restrain undue attack on the metal substrate after the oxide or scale has been removed.

In plant practice, **organic inhibitors** are now used virtually exclusively, *e.g.* substituted thioureas and quinolines. However, **arsenic** is capable of inhibiting acidic attack on metals, as is **antimony chloride** which is used in hydrochloric acid for removing tin, zinc, and chromium from steel. Clark's solution [tin(II) chloride in hydrochloric acid] is used for derusting steel in the laboratory.

Although organic substances provide the main inhibitor for acid cleaning in practice, some synergism is obtained by additions of inorganic substances. Thus sodium chloride, bromide, or iodide is used in the sulfuric acid employed for steel pickling and one of these together with thiocyanate for treating titanium in sulfuric acid. Steel in nitric acid is protected by certain sulfur compounds, *e.g.* sulfide, sulfite, and thiosulfate.

A novel method[8] of oxide removal has been developed in the **nuclear industry**, where certain power generation plant had such low tolerance to metal loss that the inhibited organic acid system formerly used for cleaning was unacceptable. The novel approach was to emphasize the reductive dissolution aspects of descaling by using a strong redox system together with a powerful complexing agent to take into solution the reduced metallic oxide. Suitable formulations for this purpose include vanadium(II) trispicolinate and cobalt(III) bipyridyl. It should be noted that these processes, known as 'LOMI' for **low oxidation state metallic ions**, are not inhibited in the sense that an added substance decreases the corrosion rate. In this case, the potential is such that on the Pourbaix Diagram the system is in the region of **immunity**.

6.7 Water for Steam Raising

The water conditions and the types of treatment used for steam raising plant are different from those employed in domestic central heating systems owing to the higher temperatures and pressures obtaining. Whilst some of the reactions remain the same, they occur at higher rates so that corrosive attack with the risk of consequent failure may be more severe.

An example is the Schikorr reaction, the decomposition of the iron(II) hydroxide initially formed to magnetite and hydrogen. As seen above in central heating systems (Equation 16, page 408), this can lead to blockage with magnetite and the need to 'bleed' gas from radiators. In high pressure plant the increasingly thick magnetite may induce severe corrosion due to the concentration effect under the influence of heat flux (values of 10^2–10^4 times the concentration in the bulk solution), and the **hydrogen gas can embrittle the metal** and so lead to explosive and costly failure. There are three aims of water treatment for steam raising, *etc.* These are to minimize

(a) scaling
(b) corrosion
(c) carry-over of boiler water solutes in the steam – this is only a significant risk at higher boiler pressures, *e.g.* above about 60 bar

The water used for boiler feed in steam raising usually requires some prior form of purification. This may be **base exchange**, in which the potential scale forming salts of calcium and magnesium are replaced by sodium; **evaporation**, where most of the solutes (whether solid, liquid, or gaseous) are separated from the water proper; or **ion exchange**, which within the efficiency of the equipment removes both cations and anions. There are many variants on these themes.[6]

When water of an appropriate purity is produced, the first of the above objectives (a) has been achieved and no scale should form. Corrosion, however, is still possible as a result of both the pH and the dissolved presence of oxygen, and to combat such corrosion, adjustments have to be made to both of these parameters. pH may be altered by additions of alkali whilst, after convenient thermal deaeration, residual dissolved oxygen is 'scavenged' by chemical means. For this purpose, sodium sulfite and hydrazine have been extensively used. Other substances investigated for removing dissolved oxygen include diethylhydroxylamine and iso-ascorbic acid.

A number of other factors of importance in boiler water practice are not relevant in the context of inhibition, *e.g.* the tendency to carry-over boiler water solutes. Various bodies issue standards *etc.* for boiler waters.[5]

A need for the addition of specific substances to restrain corrosive attack of a particular type arises with rivetted-seam boilers, now less common than formerly. These seams were subject to caustic cracking, sometimes accompanied by fatal consequences. Two recommended palliatives were sulfate and nitrite treatment. In the first of these, a ratio of sodium nitrite to sodium hydroxide of 0.4 is maintained. In the case of sulfate treatment, the sodium sulfate/sodium hydroxide ratio required is 2.

Treatment for hot water systems is essentially the same as that for domestic central heating systems.

6.8 Corrosion Inhibitors for Paint Coatings

Most paint polymers are permeable to moisture and gases so that corrosion can occur beneath the film. In some types of paint, this is minimized by the high ionic resistance of the film but in others, inhibitors are incorporated into the paint formulation to combat any corrosion. For steel, paint primers are compounds of low solubility such as lead oxide (red lead), basic lead salts, or calcium plumbate.

These react with the linseed oil to form a **soap**, *e.g.* lead linoleate, which acts as an anodic inhibitor to preserve the iron(III) oxide film on the steel. On aluminum, chromates of low solubility are used, *e.g.* zinc or strontium chromate. These function in the same way as chromates in solution act as inhibitors.

Recently, substances of lower **toxicity** than the above-mentioned traditional inhibitors have been introduced. Of the latest compounds, zinc phosphate and calcium molybdate are most often used. The relatively insoluble compounds are added as solid pigments to give a first or primary coat which virtually acts as a glue to attach the pigment. Second or top coats are applied to increase the diffusion paths of oxygen and moisture, and these paths can be extended by the addition of flat particles, *i.e.* **flake**. These may be glass, metal (Al or Zn), mica, or micaceous iron oxide (silicate). The flakes also reduce the penetration of UV light, so extending the life of the paint binder by reducing the embrittlement of the paint polymer.

Zinc metal in a powdered form is often added to paint in order to provide cathodic protection. However, modern research suggests that zinc oxide is formed which acts as a barrier to the penetration of water and air to the metal interface.

7 CONCLUSIONS AND FUTURE TRENDS

In the **electrochemical view** of aqueous corrosion there are **two half cells** (at least one of which is a metal), separated by an electrically-conducting environment. Corrosion inhibition methods can therefore be focused on either electrode – *e.g.* by coating it in a more resistant metal or paint, or by building into the vulnerable metal a corrosion allowance, or by modifying the environment. **Inhibition** seeks to do the latter by adding small quantities of appropriate **chemicals**.

Inhibition is currently a very active subject but the theory of inhibition is far from complete. Many factors contribute to inhibition and a satisfactory interpretation in many cases can be provided only by invoking several of them in a composite mechanism. On the practical side, many substances have been found to possess inhibitive properties. It is difficult to suggest one which in suitable conditions could not act as an inhibitor. Even fluoride, one of the most reactive substances in general, can inhibit the corrosion of aluminium in strong nitric acid. Inhibition must be seen as an overall process, involving the **nature and condition of the electrodes**, their geometry, the composition, temperature, pressure, degree of movement of the environment, as well as the design, construction, and operational factors.

In this overall context, it is not surprising that a substance may restrain corrosion in one case but aggravate it in another. In many real situations the business of corrosion control is complicated by problems of scale deposition, biological fouling, *etc.* An overall strategy must aim to deal with all of these if it is to be effective and in many cases there is **no single correct inhibitor**. Information is available to interpret a situation and then to formulate an approach using either single chemicals or some of the many mixtures which are on the market.

In both theory and practice, inhibition is by no means a closed subject and research remains to be done both in the area of explaining why a substance inhibits corrosion, and in identifying a use for an old inhibitor in a new situation, or finding a new inhibitor. It may be that better metal protection in a number of situations is to be achieved more easily by finding **new synergistic systems** than by developing novel inhibitors. Changing circumstances can mean that the use of a technically efficient substance may become questioned on other grounds. Chromate is a particular case in point, when after many decades of predominantly satisfactory use **toxicity** considerations led to its withdrawal from many applications. However, other inhibitor systems soon appeared to replace it.

8 REFERENCES

1. L. I. Shrier, 'Corrosion', Butterworth-Heinemann, 3rd edn, 1994.
2. I. N. Putilova, S. A. Balezin, and V. P. Barrannik, 'Metallic Corrosion Inhibitors', Pergamon Press, Oxford, 1960.
3. M. Pourbaix, 'Atlas of Electrochemical Equilibria in Aqueous Solutions', Pergamon Press, Oxford, 1966; (also from NACE).
4. I. L. Rozenfeld, 'Corrosion Inhibitors', McGraw–Hill, New York, 1981.
5. British Standard BSS 2486, 'Treatment of water for land boilers', 1978.
6. T. Swan, D. Bradbury, M. G. Segal, R. M. Sellers, and C. G. Wood, *CEGB Research*, 1982, No. 13, June, Central Electricity Generating Board.
7. W. F. Langelier, 'Chemical equilibria in water', *J. Am. Water Works Assoc.*, 1946, **38**, 169.
8. J. S. Ryznar, 'A new index for determining the amount of calcium carbonate formed by a water', *J. Am. Water Works Assoc.*, 1944, **36**, 472.

9 RECOMMENDED READING: CORROSION JOURNALS

Corrosion Science
British Corrosion Journal
Industrial Corrosion
Materials Performance
Corrosion
Corrosion Abstracts
Journal of the Electrochemical Society
Journal of Applied Electrochemistry
Electrochemical Acta

CHAPTER 16

Inorganic Chemicals for Water Purification

EDWARD W. HOOPER

1 INTRODUCTION

In recent years the authorized **maximum limits** for the discharge of toxic or hazardous materials into the environment have been **reduced** considerably. There has also been increased interest in the **recovery** of process materials and intermediate products from waste streams for economic reasons. The recycle and reuse of water can also be of economic interest and be environmentally desirable.

In order to meet the above requirements, considerable effort is being expended to improve existing treatment processes and to develop new processes. This work has led to the use of more specific precipitants and ion exchange materials. Additionally the use of combinations of some existing processes or of an existing process with a new technique such as membrane filtration is becoming current practice. Photocatalytic water purification to remove organic pollutants is covered in Chapter 18, Section 3.4.

In the nuclear industry, the need to reduce the concentration of radionuclides to extremely low levels in order to meet discharge authorizations has led to the development of **highly efficient decontamination processes** which concentrate the radionuclides into small volumes of wet solids suitable for long-term storage and compatible with immobilization processes such as cementation. The need for long-term radiation stability of the secondary waste arising from the treatment processes has favoured the use of inorganic sorbents.

This chapter considers briefly the state of the art in the treatment of aqueous waste streams using **precipitation** and **sorption** processes. A number of the examples presented are from studies on the treatment of radioactive wastes where much greater clean-up is required than is at present needed for non-radioactive industrial wastes. Since many of the toxic metals in industrial wastes also occur as fission or activation products in radioactive wastes, the treatment processes described are applicable generally. These advanced processes can be considered either as alternatives to existing techniques or as an addition to existing facilities. With increasing maturity it can be envisaged that some of these could become routine operations in the near future.

2 PRECIPITATION PROCESSES

2.1 General Principles

The **objective** of a chemical precipitation process is to **use an insoluble finely divided solid material to remove contaminants from a liquid waste**. The insoluble material or floc is generally, but not necessarily, formed **in situ** in the waste stream as a result of a chemical reaction. A typical chemical precipitation method involves four main stages:

(i) the addition of reagents and/or adjustment of pH to form the precipitate
(ii) flocculation
(iii) sedimentation
(iv) solid–liquid separation

Two parameters are commonly used to define the performance of a precipitation process:

(i) the volume reduction factor (VRF), which is the ratio of the volume of waste before treatment to the volume of secondary waste arisings
(ii) the decontamination factor (DF) which can be calculated in several ways. For precipitation processes in which volume changes occur during treatment it is usually defined as:

$$\mathrm{DF} = \frac{\text{total contaminant in feed}}{\text{total contaminant in effluent}} = \frac{C_f V_f}{C_e V_e} \quad (1)$$

where C = concentration per unit volume and V = total volume

The volume reduction factor and the decontamination factors achieved with a precipitation process strongly depend on the method of solid–liquid separation used. **Gravitational settling** is usually rather slow so that the resultant volume of floc, and hence the overall volume reduction factor, depend on settling time. The physical nature of some flocs (*e.g.* gelatinous metal hydroxide flocs) limit the extent to which they can settle under gravity. In these cases, secondary processes may be necesary to dewater the floc and make it suitable for subsequent treatment or storage/disposal. Two recently developed methods for floc dewatering involve **cross-flow filtration** or an **electrokinetic technique**. Both processes have been shown to be very effective; for example, a sedimented iron(III) (ferric) floc containing approximately 5 wt% solids can be dewatered to 40 wt% solids.

The decontamination achieved by a precipitation process will depend on the particular precipitate, the chemistry of the contaminant concerned and the degree of separation of the precipitate from the liquid. It may also be affected by the presence of other components of the waste stream such as complexants, trace organics, or particulates. Generally, **contaminants are removed from solution** by one or more of the following mechanisms:

(a) coprecipitation or isomorphous precipitation with the carrier, where the contaminant itself precipitates under the conditions of the process and is subsequently swept out of solution by the bulk (or scavenging) precipitate, or where the contaminant is incorporated into the crystal structure of an analogous precipitate, *e.g.* radiostrontium removal by barium sulfate precipitate
(b) removal of contaminant already sorbed onto particulates present in the waste effluent, which will be scavenged from solution
(c) adsorption onto the precipitate or on added absorbers, *e.g.* by ion exchange, chemisorption, physical adsorption, *etc.*

The majority of precipitation methods use metal hydroxide precipitates under neutral or alkaline conditions to remove the contaminant. In these processes, a number of the contaminants will be extensively **hydrolysed** and are likely to be either coprecipitated or sorbed onto the precipitate.

2.2 Pretreatment

With some aqueous wastes, **conditioning** may be necessary before the formation of a precipitate in order to increase the decontamination provided by the precipitation stage. These processes may be carried out to oxidize organic contaminants, decompose complexed species or residual complexing agents, alter the valency state of elements or adjust the ionic species in solution to those with a greater affinity for the precipitate.

In considering pH adjustment, oxidation or reduction processes for pretreatment of a radioactive waste stream, it must be appreciated that a particular treatment may produce both desirable and undesirable effects. For example, the use of a reducing agent can often improve ruthenium decontamination in the precipitation stage but may have an adverse effect on the removal of other contaminants by converting them to a lower, more soluble valency state. The use of **computer modelling** can greatly assist in defining the overall effect of a particular pretreatment.

2.2.1 pH Adjustment

The adjustment of the solution pH value can sometimes be advantageously employed in the treatment of wastes containing metal ion complexes in order to form an undissociated acid or base. By pH depression or elevation, the **undissociated acid or base species is formed** and the complex is broken, thus providing opportunities for further treatment procedures.

The tendency of EDTA metal complexes to hydrolyse follows the sequence

$$\mathrm{Ni} < \mathrm{Co} < \mathrm{Cd} < \mathrm{Mn} < \mathrm{Zn} < \mathrm{Cu} < \mathrm{Fe(III)}$$

but in all these cases it is difficult to precipitate the hydroxide unless a very high pH value is used. The precipitative removal of metals from complexes using calcium hydroxide appears in some cases to be improved by first lowering the pH to a value of *e.g.* 1–2, where the complex dissociates and the free ligand acid is formed, and then raising the pH to a high value such as 12 to gain maximum advantage from the competitive mass actión effects of OH^- and Ca^{2+}. However, this treatment does not remove the complexing agent, and so there is the possibility of further problems downstream if other waste streams are blended in.

The complexing action of the neutral ammonium molecule with metals is significant. The metal ammines are cationic, dissociating at low pH, and this property can be used, for example, for the precipitation of silver from its ammine by addition of excess chloride; but they are less susceptible to calcium hydroxide than the anionic EDTA complexes. It is not always possible to achieve a satisfactory degree of metal removal by simple pH adjustment of nickel ammine solution. However, zinc ammines can be treated relatively satisfactorily with calcium hydroxide.

Complex formation is much more of a problem in waste treatment than is often realized. Inadequate stream segregation before treatment can sometimes create problems in plant performance when the possible effect of complexing agents is not appreciated. There are two ways of circumventing the problem (apart from avoiding the use of complexing agents): one is to **isolate the stream** for special treatment, for example with techniques to recover the complexing agent for reuse; the other is chemical destruction of the complexing reagent which generally involves oxidative or reductive attack.

Adjustment of the pH value can be used to modify the ionic species present in the waste stream. This may affect the choice of precipitates and the operating conditions used in the treatment process. Computer modelling is being increasingly used to predict the ionic speciation in solution and the effect of constituents of the waste stream on the solubility of other components. Figures 1 and 2 are examples of such modelling studies. Figure 1 shows how the chemical speciation of iron varies with pH for a particular redox potential (Eh)–pH relationship determined experimentally[1] as

$$\mathrm{Eh} = 0.965 - 0.059\ \mathrm{pH} \tag{2}$$

Figure 2 shows how the presence of EDTA affects the solubility of iron.

2.2.2 *Chemical Oxidation*

Chemical oxidation is used in liquid waste treatment to reduce odour, decolorize, destroy organic matter to improve precipitation and flocculation, and to oxidize ions such as iron and manganese to higher valency states and thereby improve the removal of these elements by the precipitation treatment.

Common oxidants and their application are:

Oxygen. Whilst, in principle, oxygen gas is an excellent source of oxidizing power, in practice the **low solubility** of oxygen in water (10^{-3} mol l^{-1} at 1 bar) makes its use as a general oxidant for aqueous waste treatment **unattractive**.

The practicability of oxidation of Fe^{2+} and Mn^{2+} followed by precipitation of

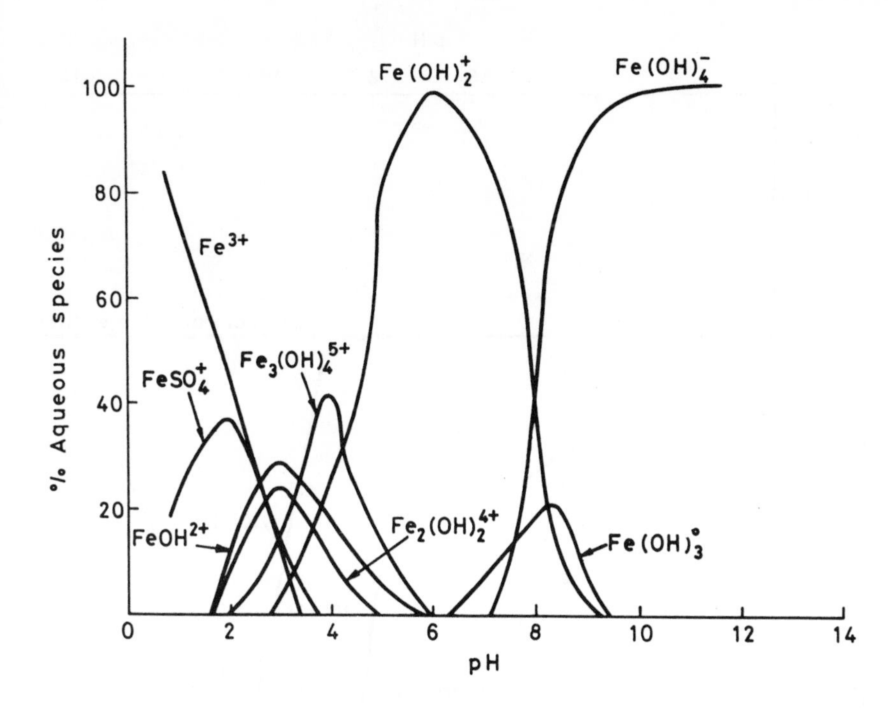

Figure 1 *Variation of chemical speciation of iron with pH*

Fe(III) and Mn(IV) hydrous oxides has been well established as a treatment process. The oxidation reaction is strongly pH dependent, occurring rapidly in neutral or alkaline solution but much more slowly in acidic conditions.

Air oxidation of some anions can be achieved. The anions that can be oxidized are nitrite, sulfide, and sulfur oxy-anions such as SO_3^{2-}, $S_2O_3^{2-}$, and $S_4O_6^{2-}$.

Ozone is a very efficient sterilizing and oxidizing agent featuring several advantages over chlorination. In particular, it does not leave prejudicial decomposition products and allows destruction of complexing and chelating agents. Ozone features the same coagulating, bleaching, and deodorizing effects as chlorine and its by-products. Unfortunately, most of the advantages of ozone are to some extent outweighed by the relatively **high cost** of ozone generation equipment and the inefficiencies associated with the low solubility of the gas in water.

Chlorine is a powerful oxidant and, like oxygen, can support combustion, thus presenting a potential fire risk. **Sodium hypochlorite** is a more acceptable form in which to store and dose chlorine.

Chlorine dioxide is a powerful oxidizing agent used in water treatment mainly for disinfection and the removal of tastes and odour. It is a yellow, explosive gas, very soluble in water. It is spontaneously explosive at partial pressures in excess of 70 mmHg but its solutions in water are safe and reasonably stable provided they are not stronger than 1–2 g l^{-1}. Solutions containing up to 8 g l^{-1} decompose only

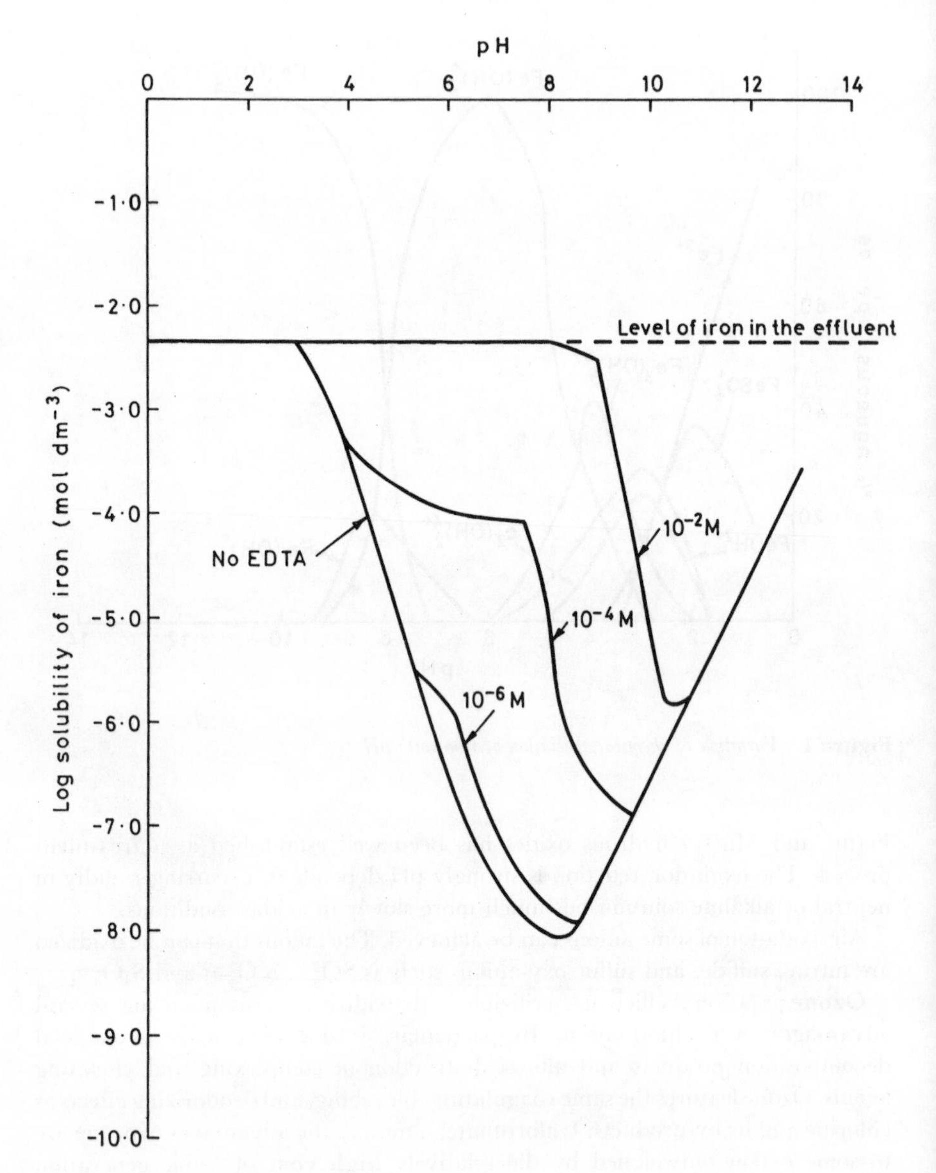

Figure 2 *Iron solubility as a function of EDTA presence*

slowly to give a mixture of HCl and $HClO_3$, but in strong light or in alkaline solution the decomposition is more rapid.

Hydrogen peroxide is a powerful oxidant and a somewhat less powerful reductant. It can, for example, reduce permanganate to Mn^{2+} and chromate to Cr^{3+} at a pH value <8.5.

Hydrogen peroxide will oxidize many organic substances, particularly when there are unsaturated carbon bonds where attack can take place. The **oxidation** is often **enhanced** by the addition of a **transition metal** ion as **catalyst**. This catalytic oxidation process has been shown to oxidize EDTA present in solution as metal–EDTA but further work is necessary to optimize the process.

Hydrogen peroxide can be used to oxidize most sulfur containing compounds, nitrites, and hydrazine and has the advantage of not adding to the dissolved solid concentration.

The use of hydrogen peroxide in the presence of **ultraviolet irradiation** has been shown to increase the rate of oxidation.

Potassium permanganate is a powerful oxidizing agent and will rapidly oxidize Fe^{2+}, Mn^{2+}, sulfides, and many organic substances. In the pH range commonly encountered in waste treatment (pH = 3–11) MnO_2 is formed, and this is known to be a cation absorber. Even in more acid solutions, MnO_2 will form if there is an excess of permanganate ion.

2.2.3 *Chemical Reduction*

Reduction reactions are employed in waste treatment with the objective of converting the pollutant to a solid form as in the reductive recovery of metals or through precipitation of an insoluble material.

Common reductants and their applications are:

Sulfur dioxide is a physiologically harmful gas with a pungent odour and is also corrosive to metals when moist. Aqueous solutions of **sulfites** are stronger **reducing agents** than aqueous solutions of sulfur dioxide. Sulfur dioxide will reduce selenates and tellurates in acid solution (pH = 2) to the elemental form, but reductions are not complete.

Sulfur(IV) compounds are used for reducing anionic hexavalent chromium ions to the trivalent cations to permit precipitation of the hydrous oxide. One application that has been examined for treatment of radioactive wastes is the use of sulfite for the partial reduction of Fe^{3+} in solution in order to precipitate magnetite by increasing the pH value of the solution.

Sodium dithionite ($Na_2S_2O_4$). The potential for dithionite reduction in alkaline solution (+1.12 V) suggests that the reagent should be capable of reducing several types of metal ions to the metal, notably silver, nickel, and copper, and this does occur. In practice, the metal when formed may be in a colloidal condition and so requires further treatment for separation. The reduction procedure can be useful when metal ions are bound with complexing agents which render normal pH adjustment for precipitation of the hydrous oxide inadequate for full treatment. In alkaline systems the rate of reduction may be slow when metal ions are complexed. **Lowering the pH** value to dissociate the complexes may be necessary but at lower pH values the reducing power of dithionite is less; so a compromise has to be adopted.

The pertechnetate ion TcO_4^- is reduced to the insoluble TcO_2 by dithionite but reoxidation occurs readily and it is necessary to carry out operations in an inert atmosphere or maintain a level of reducing agent in the treated waste solution.

The **iron(II) ion** is a well known reducing agent and is an intermediate reductant in terms of the more widely used reagents. It will reduce Cr(VI) to Cr(III) but in solutions with pH > 2 the iron(III) ion resulting from the reduction stage will precipitate and this may be an undesirable addition to the solids produced in the precipitation stage of the process.

Metals. Static or fluidized beds of Al, Mg, or Zn powder are capable of reducing other metals such as Ag, Cd, Cr, Ni, and U to an extent depending on operating conditions. One problem can be the formation of colloidal products.

Hydrazine (N_2H_4) is a powerful reductant but its use does present certain hazards. The residues of hydrazine can be regarded as polluting and the substance is potentially carcinogenic. The possible formation of the **unstable azides** of heavy metals makes it unattractive for use in the treatment of some wastes.

Hydroxylamine (NH_2OH) is a less powerful reductant than hydrazine and its use avoids the possibility of azide formation. Its use is very limited because better, less expensive general reductants are available.

Sodium borohydride ($NaBH_4$) is an extremely powerful reducing agent which will reduce many metal ions, either simple aquo-ions or complexes, to a lower oxidation state or even to the elemental metal. Solutions in water are stable at high pH value but undergo rapid hydrolysis under neutral or acid conditions.

The principal application of sodium borohydride in wastewater treatment is in the **recovery of metals**. Metals whose compounds are reducible include cobalt, nickel, and silver. Although in many cases the metal is precipitated as the element, in the case of cobalt and nickel the corresponding borides Co_2B and Ni_2B may be formed.

Hydrogen peroxide (H_2O_2), although normally considered an oxidizing agent, is also a reducing agent as mentioned in Section 2.2.2 on chemical oxidation.

Hydrogen peroxide will reduce Cr(VI) compounds provided that the pH value is slightly alkaline. Silver complexed with thiosulfate, as in photographic fixing solutions, can be precipitated as a silver–silver oxide sludge.

2.3 Simple Processes

2.3.1 General Processes

In this section, the various processes which operate on the basis of a single or specific chemical reaction to form the precipitates are discussed.

The most commonly used of these simple treatments are:

- the lime–soda process
- phosphate precipitation
- hydroxide processes
- oxalate precipitation

(a) The Lime–Soda Process. This process removes 'hardness' from water and produces a precipitate of calcium carbonate:

$$Ca(HCO_3)_2 + Ca(OH)_2 \rightarrow 2\ CaCO_3 + 2\ H_2O \qquad (3)$$

$$Ca(OH)_2 + Na_2CO_3 \rightarrow CaCO_3 + 2\ NaOH \qquad (4)$$

This method was used in the past, primarily to remove radioactive strontium as an analogue of calcium, but has now been superseded by more efficient processes.

(b) Phosphate Precipitation. Soluble phosphates, especially tri-sodium phosphate, are added to the waste to form insoluble compounds with other ions:

$$3\ M^{n+} + n\ PO_4^{3-} \rightarrow M_3(PO_4)_n \quad (5)$$

where M^{n+} is Fe^{3+}, Al^{3+}, Ca^{2+}, *etc.*, including radioactive strontium. Calcium ion is usually added as a bulk coprecipitant to enhance the removal of other cations.

The reaction is complex and some kind of hydroxyapatite is probably formed, whose structure can include strontium ions. The method also achieves good removal of plutonium and uranium which have insoluble phosphates. In general, the higher the pH value of the operation, the better the DF achieved.

(c) Hydroxide Processes. Many metal ions can be hydrolysed to form insoluble compounds, a number of which exhibit an affinity for sorption of other ions:

$$M^{n+} + n\ OH^- \rightarrow M(OH)_n \quad (6)$$

where M^{n+} is Fe^{3+}, Al^{3+}, Ti^{4+}, *etc.*

Although, in conventional water treatment, aluminium hydroxide precipitation is widely used, in the practice of radioactive waste management the use of **iron(III) hydroxide precipitation** is more common, partly because iron(III) ions may already be present in some waste streams. In general, iron(III) hydroxide floc particles are larger and settle more easily than those of aluminium hydroxide.

Iron(III) floc processes essentially involve the precipitation of iron(III) hydroxide either alone or in conjunction with other precipitates (*e.g.* sulfates, phosphates). Since some plant waste streams already contain iron at levels of 10–200 mg l^{-1}, iron(III) hydroxide can be readily precipitated by the addition of alkali (*e.g.* NaOH, $Ca(OH)_2$, or NH_4OH). Iron(III) hydroxide is stable over a wide pH range 5–14 but generally removes contaminants most effectively at higher pH values.

Iron(III) hydroxide forms as a voluminous, gelatinous precipitate, which may be difficult to handle. Conventional filtration is not very effective, so gravity settling is usually favoured for the initial separation, requiring a relatively large and expensive plant. There is also the possibility of carrying over fine particles of floc suspended in the supernatant liquor. Generally, iron(III) flocs will require further dewatering before immobilization and the supernates often need to be 'polished' in order to achieve the required decontamination. The physical properties of the floc may be significantly improved by the presence of other precipitates (for example, calcium salts).

(d) Oxalate Precipitation. Oxalate precipitation processes are being investigated for the treatment of both highly active and intermediate-level radioactive waste streams. Their main advantage is that they provide **precipitation** of the

actinides and of the **lanthanides** at **low pH** value, leaving most of the fission products and iron in solution. The oxalate precipitate is fairly crystalline and settles quickly to small floc volumes.

Whilst providing separation of actinides from some fission products, the DFs obtained for plutonium are generally lower than those achieved by iron(III) floc and related processes, and oxalate precipitants have not been investigated for the removal of low level activity in waste streams. Finally, the chemicals used are not particularly attractive since they are relatively expensive and could also create a toxicity hazard in the effluent.

(e) Other Precipitates. **Diuranate**. Ammonium diuranate is, at present, used to precipitate the actinides from waste effluents at Dounreay, UK. Limited information is available concerning the process but plutonium DFs in the range 100–500 and a total β–γ DF of approximately 7 have been reported. A potential advantage in using a uranium(VI) floc process is that the floc could be dissolved and recycled to the main reprocessing plant to recover the actinides. Diuranate floc can be more crystalline than metal hydroxide flocs, which makes it easier to handle and results in smaller floc volumes.

Magnetite or iron ferrite floc. Another iron treatment currently under investigation uses magnetite ($FeO.Fe_2O_3$) precipitation. The process was designed to enable magnetic separation of the floc from the supernate. Magnetite is the simplest iron ferrite with a spinel type structure. It was proposed that by forming ferrites *in situ* in radioactive effluents, radionuclides would be incorporated in the ferrite lattice.

Magnetite formation *in situ* in real waste streams has proved rather less satisfactory, because of interference from several ions commonly present in the wastes.

As an alternative process the addition of preformed magnetite has been investigated, and was generally found to be more successful since it was less sensitive to other ions. Up to now, any advantage in using preformed magnetite instead of conventional iron(III) floc processes has yet to be demonstrated.

2.3.2 *Specific Processes*

The treatments initially applied to radioactive wastes were those used to purify municipal and industrial waste water; they were described in the preceding section.

In some cases, depending on the specific chemical and radiochemical compositions of the waste, these treatments will not provide sufficient decontamination from certain nuclides. Special treatments have, therefore, been developed; those for two particular elements, ruthenium and caesium, are described below.

(a) Ruthenium Removal. Ruthenium is a transition metal with several valence states. It is one of the most troublesome nuclides to remove from liquid wastes since it can be present in **cationic, anionic, or non-ionic forms**. Ruthenium is easily complexed; for example, in a nitric acid media it exists as nitroso- and nitrosyl-ruthenium complexes.

The efficiency of ruthenium removal will be variable, depending on the species in solution. It seems that the cationic form can be readily removed using hydroxide precipitation and also using ion exchange. Many metal hydroxides have been tried; the decontamination obtained depends on the sorptive power of the various hydroxides. Decontamination factors are in the range of 2–10 for all ruthenium species. Neutral ruthenium complexes in solution are difficult to remove.

The observation that iron(II) and titanium(II) hydroxides are good absorbers suggests that ruthenium removal is improved in the presence of a reducing agent. Treatments involving **reduction** include:

- precipitation of lead paraperiodate
- coprecipitation of copper and iron(II) hydroxides
- precipitation of metal sulfide
- use of sodium borohydride

In a nitric acid medium, metal sulfide precipitation is the most effective of these four methods. Many metal ions have been tried; iron and cobalt provide the best decontamination. Coprecipitation with cobalt sulfide in acidic medium yields DFs of about 100. Since it is not always possible or desirable (because of corrosion problems) to work at low pH values, the precipitation can be carried out at higher pH values, but the efficiency of the treatment will be decreased. For example, the cobalt sulfide treatment operated at the La Hague reprocessing plant at pH values greater than 4 achieves DFs of 30–40.

The use of a reducing agent such as sodium borohydride in conjunction with copper precipitation seems to be promising for ruthenium removal.

(b) Caesium Removal. A number of precipitation processes exist for the removal of caesium from aqueous waste streams:

(i) use of transition metal hexacyanoferrates (Cu, Ni, Co)
(ii) use of phosphotungstates or phosphomolybdates
(iii) precipitation of the tetraphenylborate
(iv) adsorption on inorganic phosphates (Zr, Ti)

The transition metal **hexacyanoferrates** can be precipitated *in situ* or added as a preformed slurry. The product is generally non-stoichiometric because particular transition metal to $Fe^{II}(CN)_6{}^{4-}$ ratios are used for different metals. Possibly, a mixture of

$$M^{II}_2[Fe^{II}(CN)_6] \text{ and } M^{I}_2M^{II}_3[Fe^{II}(CN)_6]_2$$

is formed where M^I is an alkali metal or the ammonium ion and M^{II} is a transition metal, typically cobalt, nickel, or copper. Caesium adsorption appears to occur by **ion exchange**. Decontamination factors higher than 100 are frequently observed. Different metal hexacyanoferrates provide maximum decontamination at specific pH values. The most commonly used are copper and nickel hexacyanoferrates which are effective over the pH range of 2 to 10.5 and in the presence of high salt loadings but DFs decrease with increasing salt content.

At pH values greater than 11, decomposition of the transition metal hexacyanoferrates occurs, yielding the hexacyanoferrate ion and a precipitate of the transition metal hydroxide.

In a nitric acid medium, granular **phosphotungstic acid** has provided a caesium decontamination factor greater than 100 at room temperature; this material can be prepared as a finely divided precipitate which should show improved kinetics of sorption.

Ammonium phosphomolybdate can also be used to remove caesium from acid solutions, and its use at pH values of up to 9.5 is claimed.

In both alkaline and acid media, caesium can be precipitated as its **tetraphenylborate** by addition of sodium tetraphenylborate (NaTPB). Caesium removal is independent of pH in the range of 10–13 and a 30 minute contact time is sufficient to obtain a caesium DF greater than 10^3. However, there is a strong competition among the cations in the solution, and the caesium DF will be affected by the sodium and potassium ion concentration. Some authors have reported problems occurring with sludges containing NaTPB: the precipitate is highly viscous and gelatinous and is difficult to separate.

Radiolysis of the tetraphenylborate ion produces benzene and benzene derivatives which are toxic and flammable.

Zirconium phosphate and **titanium phosphate** both have an affinity for caesium. Very recent work using precipitates of these materials is achieving very low levels of caesium in effluents.

2.4 Combined Processes

Since the DFs and VRFs obtained by chemical precipitation processes are generally rather low, processing by chemical precipitation is used only to treat high volume waste steams or if more efficient treatments, such as concentration by thermal evaporation or ion exchange, are not possible (for example, when the salt load of the waste or the suspended solids content is high). Alternatively, a **precipitation process** can **precede** another treatment technique such as **ion exchange** or **evaporation**.

When the waste stream composition is variable in nature, either in radioactive or non-radioactive content, a single chemical precipitation process may be inadequate.

To provide good decontamination of the liquid waste a combination of the general or specific treatments described previously is frequently necessary. Often the final combined process is a compromise between the optimal conditions of each single process so that the best overall decontamination factor for the liquid waste is achieved.

Combination of the single processes described above can be used as **multi-stage batch processes** or as a **continuous precipitation process**. For example, phosphate treatment for strontium removal could follow a hexacyanoferrate(II) treatment for caesium removal, the only requirement after the precipitation being to raise the pH value before the next stage. Multistage batch processes will produce several sludges for disposal and require extra equipment, thus increasing the capital outlay. A continuous precipitation process needs only one facility and, of course, produces only one sludge.

The use of these various techniques is strongly dependent on the requirements of each site, on the local discharge authorizations for radiochemical and chemical components, and on the intended solid waste treatment.

3 SORPTION PROCESSES

Sorption is a separation process involving two phases between which certain components can become differentially distributed. There are three types of sorption, classified according to the type of bonding involved:

(a) **Physical Sorption.** This results from the action of relatively weak **van der Waals forces**. The strength of these forces increases with increased molecular weight. There is no exchange of electrons; rather intermolecular attractions occur between 'valency happy' sites and are therefore independent of the electronic properties of the molecules involved. This explains why physical adsorption is a non-specific phenomenon and is generally exhibited between non-polar compounds. The heat of adsorption, or activation energy, is low and therefore this type of adsorption is stable only at temperatures below about 150 °C. Multi-layer build-up of adsorbate is possible.

(b) **Chemical Sorption.** Chemical adsorption or chemisorption involves an exchange of electrons between specific surface sites and solute molecules resulting in the **formation of a chemical bond**. Chemisorption is typified by a much stonger adsorption energy than physical adsorption. Such a bond is therefore more stable at higher temperatures.

(c) **Electrostatic Sorption (Ion Exchange).** This is a term reserved for Coulombic attractive forces between ions and charged functional groups and is more commonly classified as **ion exchange**. In this process, mobile ions from an external solution are exchanged for ions which are electrostatically bound to functional groups contained within a solid inert matrix. The major selectivity here is based on polarity, *i.e.* whether the material is designed for exchanging anions or cations.

For sorption from liquid solutions, consideration must also be given to the nature of the solution, *i.e.* the extent to which the solvent is capable of accommodating the solute. Any chemical incompatibility (*e.g.* a significant thermodynamic gradient) can result in what are termed solvophobic forces, which may drive the solute out of solution. This effect substantially increases the adsorption bond energy beyond that which would result from the surface reaction alone and in extreme cases can prompt adsorption to any available solid surface whether intended as an adsorbent or not. This phenomenon is called **hydrophobic bonding**.

Another consideration worth mentioning arises from the fact that a solution contains at least two components. In general there will be competitive adsorption of both solvent and solute molecules. In practice however, material selection ensures that the solute is by far the most dominant adsorbate.

3.1 Inorganic Sorbents and Ion Exchangers

Inorganic sorbents may be either naturally occurring minerals or synthetic materials. They perform the same basic exchange operations as those involved for organic exchangers.

The naturally occurring materials comprise aluminosilicate materials such as zeolites, clays, and feldspars. Synthetic ion exchangers generally fall into the following categories:

(1) hydrated metal oxides, *e.g.* hydrous titanium oxide, polyantimonic acid
(2) insoluble salts of polyvalent metals, *e.g.* titanium phosphate
(3) insoluble salts of heteropoly acids, *e.g.* ammonium molybdophosphate
(4) complex salts based on insoluble hexacyanoferrates
(5) synthetic zeolites

3.1.1 Natural Inorganic Materials

The clay minerals and naturally occurring zeolites, though replaced to a large extent by synthetic exchangers, continue to be used and have been applied extensively to the treatment of radioactive waste solutions. Limitations to their use include:

(i) relatively low ion exchange capacities
(ii) low abrasion resistance of the zeolites
(iii) clay minerals tend to peptize, *i.e.* to form colloids in some solutions
(iv) these materials are partially decomposed by acids and alkalis

The advantage of naturally occurring materials is their **low cost** but their industrial use is often difficult (impurities, bad physical properties for packed bed operations) and they sometimes need a chemical or thermal pretreatment.

Synthetic Zeolites. These materials are commercially available; they act as molecular sieves by excluding ions larger than the opening in the crystalline matrix. Limitations to the use of synthetic zeolites are similar to those listed above for naturally occurring zeolites but the synthetic zeolites possess **improved stability** and a **higher cation exchange capacity**.

3.1.2 Oxides and Hydrous Oxides

Only a few of these materials are available in commercial quantities as sorbents. Many show high exchange capacity, low solubility, and good radiation and thermal stability.

Cation or anion exchange sorption by oxides and hydrous oxides is known to occur predominantly by displacement of hydrogen or hydroxide ions from the sorbents provided that these have previously been washed free from impurity ions. The predominant functional groups are probably metal–hydroxide bonds which can exhibit amphoteric properties favouring **anion exchange at low pH** and **cation exchange at high pH**.

Although hydrous oxides are at their most active when freshly precipitated, they are at their least useful as regards ion exchange column applications. Their relatively high ion exchange capacities are offset by their ease of dissolution in acids or bases and the poor filtration characteristics of the precipitates in general render them unsuitable for use as column sorbents.

If gelatinous precipitates of the hydrous oxides are dried they undergo considerable shrinkage to give glassy gels which break down in water to produce granular particles suitable for **column applications**. The structure and ion exchange properties of these dried gels depend very much on the conditions of preparation of the precipitates and the drying procedure used to produce the granular product. During the drying stage it is probable that, apart from the removal of free water, elimination of water from –OH groups occurs to produce –M–O–M– bonds (M = metal) giving a stronger absorber material but reducing the ion exchange capacity.

The quantitative effect of heat treatment of different oxides and hydrous oxides on their ion exchange properties varies considerably. Generally, **ion exchange capacities show a steady decrease with heating** and at about 200 °C have fallen to 20–30% of their values for low temperature dried material. Heat treatment can, however, **improve** both the **selectivity** and the **chemical stability** of these adsorbers.

The use of an oxide or hydrous oxide adsorber as a column material is obviously governed to a large extent by its resistance towards breakdown or dissolution when treated with various reagents. In general, the insoluble acidic oxides and hydrous oxides are stable in acid solutions but dissolve readily at $pH > 7$. In contrast, the more basic hydrous oxides such as TiO_2 and ThO_2 dissolve readily in mineral acids above about 0.1 M but are stable in less acid solutions.

3.1.3 Acidic Salts of Multivalent Metals

A wide range of compounds of this type has been described as ion exchangers. These salts, acting mostly as cation exchangers, are gel-like or micro-crystalline materials with the composition and properties depending on the method of preparation.

In general they are not available in commercial quantities and many are likely to be **expensive** to produce because of the limited availability and high cost of the materials required in their preparation.

3.1.4 Hexacyanoferrates

The ion exchange properties of a large number of insoluble hexacyanoferrates of various metals have been studied. These absorbers act as **cation exchangers** with an especially high affinity for caesium. The exchange capacity depends strongly on the method of preparation.

The hexacyanoferrate ion exchangers are stable in mineral acids; in contact with strong HNO_3 a slight oxidation of Fe(II) to Fe(III) can occur. In alkaline solutions a tendency to peptize has been observed but this can be greatly reduced by addition of neutral salts.

Attempts to elute caesium absorbed on hexacyanoferrates have been only partially successful and in general is accompanied by decomposition of the absorber.

Hexacyanoferrates are currently available in research quantities but because of the relatively simple method of preparation it is thought that larger scale manufacture would be feasible and economic.

Inorganic exchangers can be prepared as **granular** materials but they can also be used as **finely divided** materials in chemical precipitation processes to enhance the decontamination factor for some contaminants.

3.2 Process Applications

3.2.1 Packed Beds

The use of inorganic sorbents in packed beds has been limited mainly to zeolites because the preparation of the absorber in a suitable form is frequently a problem since many granular products have insufficient mechanical strength. The use of zeolites for water softening has been superseded by organic resins but zeolites still find applications at nuclear plants because of their superior radiation stability.

An example of a large scale application is the site ion exchange effluent plant (SIXEP) at the Sellafield Site of British Nuclear Fuels plc.[2] The plant is used to treat the cooling water from fuel storage ponds. Typical decontamination factors achieved are 2000 for caesium and 500 for strontium and the number of bed volumes of waste treated between bed discharges is 20 000.

In Finland, IVO International Ltd have introduced the IVO-Cs Treat System[3] based on the use of an inorganic ion exchange material (hexacyanoferrate) which is very selective for caesium.

Inorganic sorbents are used in portable renal dialysis machines to adsorb ammonium ions produced by enzymatic decomposition of urea.[4] The Redy dialysis regeneration filter consists of four purification layers:

(1) hydrated zirconium oxide – for removal of heavy metal cations and oxidants
(2) urease immobilized on aluminium oxide – for the decomposition of urea by the reactions (7) and (8):

$$(NH)_2CO + H_2O \rightarrow NH_3 + NH_2COOH \qquad (7)$$

$$NH_2COOH + H_2O \rightarrow NH_4^+ + HCO_3^- \qquad (8)$$

(3) hydrous zirconium phosphate – the inorganic cation exchanger, for the removal of ammonium cations

(4) active charcoal – removal of creatinine, uric acid, phenols, indols, organic acids, polypeptides, and other organic impurities

Since divalent cations such as Ca^{2+} and Mg^{2+} are more favourably exchanged than the ammonium ion on the zirconium phosphate layer, the regeneration system needs supplementation of these cations to maintain a physiological range of electrolyte concentration, and an additional amount of ion exchangers is required to remove ammonium ions completely.

Activated carbons and impregnated activated carbons can be considered as inorganic sorbents although they are derived from organic material. Literature available from Sutcliffe Speakman Carbon Ltd provides the following information. Sutcliffe Speakman manufacture activated carbons primarily based on coal and coconut shells as raw materials. The types available are

1. **Base activated carbons** – these are carbons activated to a range of activity levels and are available in a variety of forms and sizes, and function by the process of physical adsorption. These carbons are used mainly for air treatment, odour control, water treatment, solvent recovery, and decolorization.
2. **Specialized impregnated activated carbons** – these are carbons activated to a high activity level and then impregnated with a range of chemicals in order to enhance their activity for specific applications. These carbons function by the process of chemical adsorption. They are used for specialized air and water treatment, respirators, and nuclear and military applications.

In an attempt to overcome the poor mechanical strength exhibited by granular forms of inorganic sorbents, **pelletizing** with use of a binder and incorporation into a suitable matrix have been examined. Incorporation into silica gel or a polymeric organic resin has been studied and has been partially successful. These composite materials usually have slower sorption kinetics than the sorbent alone and also loss of sorbent from the support matrix can occur in the absence of any chemical bonding. Recent studies in the Czech Republic,[5] incorporating finely divided inorganic sorbents into polyacrylonitrile at the polymerization stage, are indicating that these difficulties can be resolved.

3.2.2 Filter Precoats

The use of materials such as diatomaceous earth to form a **removable layer** on a screen so as to act as a filter is well known. **Ion exchange** materials can be used in the same way and **powdered organic resins** are used in a number of applications.

There has been some use made of finely divided inorganic sorbents in this type of process but most recent work is examining the use of finely divided sorbents in combination with efficient filtration systems.

3.2.3 Combined Processes

'Seeded cross-flow filtration' is the term commonly used to describe a process in which a finely divided sorbent or seed is used in combination with a membrane filter of small pore size (0.1–0.001 μm). The slurry to be filtered is pumped at high velocity tangential to the membrane surface and this prevents build-up of a fouling layer which would reduce the flux through the membrane. Cross-flow filtration can also be used to separate precipitates from the supernatant liquid, can be used to dewater sludges to solids contents of 30–40 wt%, and, by using a suitable mixture of precipitating agents and sorbents, can enable a treatment process to be tailored to a specific waste stream.

Seeded cross-flow filtration offers a number of advantages over conventional ion exchange processes:

(a) good solid–liquid separation including colloidal material
(b) smaller volumes of sorbent are required
(c) faster kinetics of sorption
(d) no need to granulate sorbents
(e) no need to remove particulate material that would blind a packed bed of sorbent.
(f) relatively unaffected by foaming agents *etc.*

This process is now being employed at nuclear sites in the United Kingdom and in France. The French site uses a hexacyanoferrate sorbent and activated carbon in combination with a membrane filter to remove caesium, cobalt, and other radioelements from laundry wastes.

At the British Nuclear Fuels reprocessing site at Sellafield, UK, a new waste treatment plant is under construction. Known as the enhanced actinide removal plant (EARP), the plant will process bulked effluents from reprocessing activities and concentrates from evaporation plants. The total plant throughput will exceed 60 000 m^3/annum. The treatment involves **precipitation of iron(III) hydroxide** using the iron component of the waste and sodium hydroxide to increase the pH value to 9–10.5. Addition of nickel hexacyanoferrate as a performed slurry, at levels of 10–200 ppm, depending on the caesium loading of the waste, may be made after pH adjustment. Primary and secondary cross-flow ultrafiltration stages respectively provide an effluent for subsequent discharge after analysis and a dewatered sludge for cementation. Figure 3 shows a simplified process flow diagram of the plant.[6]

Electrochemical processes combined with **sorption** processes are being studied. Electrodes impregnated with sorbent material, organic or inorganic, are used with a counter electrode. The applied voltage induces migration of ions to the sorbent and the loaded sorbent can be eluted by reversal of polarity without the need to add any reagents.

4 FUTURE TRENDS

It appears that **chemical precipitation** processes and **solid–liquid separation** techniques will continue to be widely used in waste treatment. Current research and development is showing that combining different processes in one treatment plant can provide higher decontamination factors and smaller secondary waste

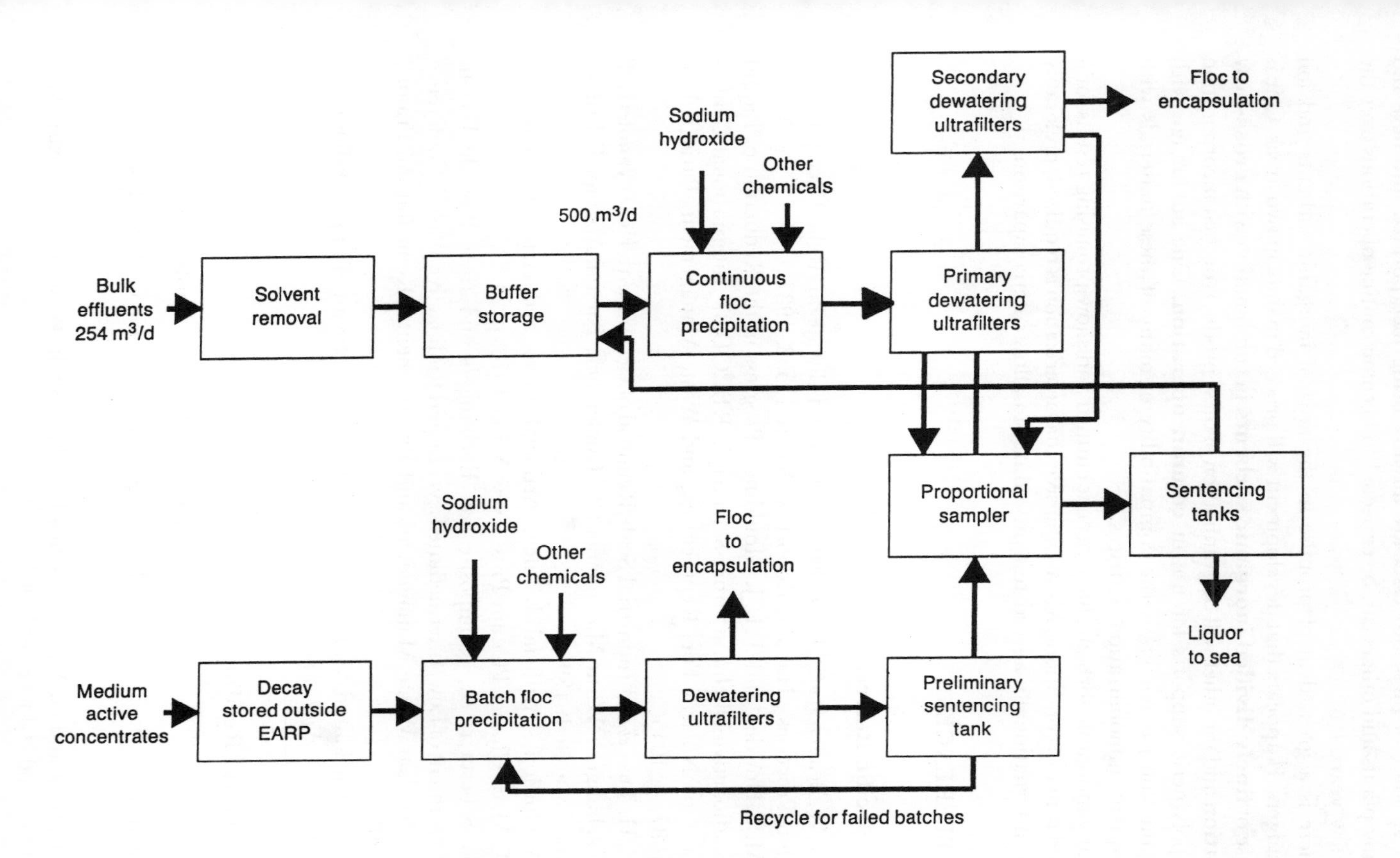

Figure 3 *Simplified process flow diagram for the enhanced actinide removal plant (EARP)*

arisings. Some of these processes are already being incorporated into new and existing plants and others can be expected to become routine operations over the next few years.

There is a great deal of promise in the area of inorganic sorbents and ion exchangers. It appears that development will proceed mainly in two areas. One is the use of **finely divided inorganic sorbents** in combination with **cross-flow filtration** and the other will be granulation of these products or their incorporation into polymeric supports for use in **column operation**. One advantage with inorganic ion exchangers is the compatibility of many of these materials with subsequent immobilization matrices.

Many inorganic sorbents have been identified and show promising results for a limited range of conditions. More extensive characterization is required, preferably by an internationally agreed test procedure, to allow intercomparison.

5 REFERENCES

5.1 Specific References

1. G. W. Beaven *et al.*, 'The effect of EDTA and Citrate on Alpha Decontamination Processes', Harwell Report AERE R 12558, 1987.
2. M. J. Howden and T. L. J. Moulding, 'Progress in the Reduction of Liquid Radioactive Discharges from Sellafield', 'RECOD 87' International Conference on Nuclear Fuel Reprocessing and Waste Management, Paris, vol. 3, 1987, p. 1045.
3. E. H. Tusa *et al.*, 'Industrial Scale Removal of Cesium with Hexacyanoferrate Exchanger', 'Waste Management 93' Conference, Tucson, Arizona, February 28 – March 4, 1993.
4. A Gordon, 'Artificial Kidney, Artificial Liver and Artificial Cells', ed. T. M. S. Chang, Plenum Press, New York, 1978, p. 23.
5. F. Sebesta *et al.*, 'Composite Ion Exchangers and their Possible Use in Treatment of Low/Intermediate Level Liquid Radioactive Waste', Conference on Nuclear Waste Management and Environmental Remediation, Prague, 1993.
6. W. Heafield and M. Howden, 'The Future Treatment of Liquid Effluents at Sellafield', *Nucl. Eng. Int.*, 1988, 149.

5.2 General Reading

C. B. Amphlett, 'Inorganic Ion Exchangers', Elsevier, Amsterdam, 1964.

A. Clearfield, 'Inorganic Ion Exchange Materials', CRC Press, Boca Raton, Florida, 1982.

W. J. Eilbeck and G. Mattock, 'Chemical Processes in Waste Water Treatment', Ellis Horwood, Chichester, 1987.

'Chemical Precipitation Processes for the Treatment of Aqueous Radioactive

Wastes', Technical Report Series No. 337, International Atomic Energy Agency, Vienna, 1992.

5.3 Core Journals

Journal of Inorganic and Nuclear Chemistry
Talanta
Solvent Extraction and Ion Exchange
Separation Science and Technology
Hydrometallurgy
Journal of Physical Chemistry
Journal of the Chemical Society
Journal of the American Chemical Society
Journal of Chromatography
Effluent and Water Treatment Journal

CHAPTER 17

Inorganic Chemistry in the Nuclear Industry

ALEXANDER HARPER

1 INTRODUCTION

It is said that in the early days of the nuclear industry a senior physicist remarked that he encountered 'chemists, chemists, and yet more chemists'. Whatever the truth of this anecdote it serves to illustrate the central role played by **chemistry** in the safe and efficient **production of electrical energy from nuclear fission**, an enterprise often thought of as the domain solely of the physicist and engineer. The aims of this Chapter are to introduce the reader to some of the ways in which chemistry plays its part in this technology, and to indicate the links between industrial practice and underlying science. Before defining the scope of this Chapter in detail, it is useful to review the background to the production of nuclear energy.

The basic **fuel** for the nuclear industry, **uranium**, belongs to the actinide series of elements, and is the heaviest element occurring in significant quantity in nature. It was first identified as a constituent of the mineral **pitchblende** by Klaproth in 1789, but was not isolated in its metallic form until 1841, when Peligot succeeded in reducing uranium tetrachloride with potassium. Three **isotopes** of uranium occur naturally: ^{238}U (99.28%), ^{235}U (0.72%), and ^{234}U (0.005%). These isotopes are all radioactive (as indeed are all known isotopes of the actinide elements), and decay by emission of **alpha particles**. An account of radioactive decay and related topics can be found in many texts including some of those in the Bibliography; see, for example, Chapter 2 of Benedict *et al.*[5]

Although uranium found some small scale uses (for example as a colourant in the glass industry) it was of little industrial interest before the discovery of the fission process by Hahn and Strassmann in 1939. The discovery of the **neutron** by Chadwick in 1932 had suggested the possibility of synthesizing new elements heavier than uranium (**transuranic elements**) by **beta decay** of the products of bombarding uranium with neutrons. Attempts to synthesize element 93 by this route led to the discovery that not only could the synthesis of heavier elements be achieved, but also the splitting (**fission**) of the nucleus of the ^{235}U isotope into two

Table 1 *Characteristics of commercial nuclear reactors*

Reactor	*Fuel*	*Enrichment*	*Moderator*	*Coolant*
MAGNOX (Mg/Al alloy used to encase fuel)	U metal	Natural (0.72%)	Graphite	CO_2
AGR (Advanced gas-cooled reactor)	UO_2	About 2%	Graphite	CO_2
BWR (Boiling water reactor)	UO_2	About 2.2%	H_2O	H_2O
CANDU (Canadian deuterium–uranium reactor)	UO_2	Natural (0.72%)	D_2O	D_2O
PWR (Pressurized water reactor)	UO_2	About 3%	H_2O	H_2O

smaller fragments (fission fragments or fission products). This fission was accompanied by the release of further neutrons and energy, largely in the form of kinetic energy imparted to the fission fragments as shown in Equation 1. Although the amount of energy per fission released is small, fission of all the ^{235}U atoms in 1 tonne of natural uranium would release about 6×10^{14} J, making uranium an energy source of vast potential.

$$^{235}_{92}U + ^{1}_{0}n \rightarrow 2 \text{ fission fragments} + (2\text{–}3)\, ^{1}_{0}n + 3 \times 10^{-11}\ J \quad (1)$$

The release of more neutrons in the fission process than were involved in its initiation raised the possibility of creating a self-sustaining chain reaction. This was first artificially achieved on 9 December 1942 in an experimental reactor built in a squash court at the University of Chicago. Fourteen years later, in 1956, the **first power station** to use the energy released in nuclear fission for the production of electricity was opened at Calder Hall in the UK. Since that time the industry has grown worldwide, and a variety of different reactor types have been developed.

The main reactor designs important in commercial operation are described in Table 1. Here they are characterized in terms of the medium used to transport heat from the core (the **coolant**), the material used to slow down the neutrons released by the fission process so that they can more effectively bring about the fission of further ^{235}U nuclei (the **moderator**), and the chemical form of the fuel. Since fission in these reactors is brought about by slow (thermal) neutrons, they are often referred to as thermal reactors.

One line of reactor development has been to adopt carbon dioxide as a coolant, with graphite as the moderator; commercial use of reactors of this type is confined to the UK. In these reactors the gaseous coolant is circulated over the fuel and passed through a heat exchanger, where steam is raised in a secondary water circuit to drive turbines and produce electricity. The earliest design is represented by the **MAGNOX** reactors, so-called because of the magnesium–aluminium alloy used to clad the metallic uranium fuel. The gas-cooled reactor concept has been further developed to produce the **Advanced Gas-cooled Reactor** or **AGR**. This design uses uranium dioxide clad in stainless steel as a fuel. Uranium dioxide melts at a substantially higher temperature (2865 °C) than uranium metal (1133 °C), thus allowing the fuel to operate at higher temperatures and enhancing thermal efficiency. One disadvantage of the use of UO_2 is the lower density of uranium atoms and hence of fissile material in the fuel. To compensate for this effect the proportion of fissile ^{235}U to non-fissile ^{238}U is increased, a process known as **enrichment**.

The other line of development is the use of water, either as 'light water' (H_2O) or 'heavy water' (D_2O), as both moderator and coolant. **Water-cooled reactors** may be conveniently subdivided into reactors of the **Boiling Water** type (**BWR**), which includes CANDU, in which the moderator/coolant is boiled directly to produce steam to drive the turbines, and reactors of the **Pressurized Water** (**PWR**) type, in which the moderator/coolant is passed through a heat exchanger to raise steam indirectly. Reactors of the BWR and PWR type are the most commonly used outside Eastern Europe.

An understanding of inorganic chemistry is vital to every stage of nuclear energy production, from the extraction of uranium from its ores to the final disposal of waste material produced by reprocessing spent reactor fuel. In addition to the issues associated directly with nuclear fuel, a sound understanding of inorganic chemistry is crucial to a comprehension of the interactions between reactor coolants and the materials with which they come into contact. Furthermore the behaviour of radioactive materials in the environment depends crucially on their chemical form, and this can only be predicted if their chemistry is understood in detail.

From this wide field of endeavour we have selected for discussion the chemistry of the nuclear fuel cycle, which includes the processes of uranium extraction, fuel fabrication, fuel reprocessing, and waste disposal. Such a selection of topics is made on the basis that the information is largely independent of the type of reactor system discussed, and allows the discussion of some elegant chemistry. For those readers interested in pursuing other aspects of chemistry in the nuclear industry the Bibliography contains a number of references of general interest.

The Chapter begins with a discussion of some of the chemical consequences of irradiating nuclear fuel, and continues by describing the concept of the nuclear fuel cycle. After a discussion of relevant aspects of the chemistry of uranium and some of the transuranic elements, the practical applications of chemistry in the nuclear fuel cycle are outlined. Throughout these discussions, the emphasis is placed on the principles of those processes which find application in practice.

2 THE CHEMICAL CONSEQUENCES OF IRRADIATING NUCLEAR FUEL

Irradiated nuclear fuel is an extremely complex chemical system. This degree of complexity is brought about by the twin processes of **fission**, which produces lighter nuclei from heavier ones, and the process of **neutron capture**, which initiates the formation of the transuranic elements.

The splitting of the nucleus of a fissile isotope by collision with a neutron of suitable energy results in two fragments of unequal mass; two different isotopes are therefore formed in each fission event. A wide variety of pairs of isotopes may be produced; this is exemplified in Figure 1 which shows the yield of each mass number as a fraction of total fissions (the primary fission **yield curve**). It is clear from the figure that the fractional yields of each isotope vary widely, from values as low as 10^{-7} to as high as 0.1 (expressed as 1E-07 and 1E-01 respectively in the Figure). The exact form of the curve will depend not only on the fissile nucleus in question, but also upon the energy of the neutron causing the fission: the example shown here refers to the fission of ^{235}U by thermal neutrons.

Not only does the fission process itself give rise to a wide variety of new materials, but the subsequent behaviour of these isotopes adds a further layer of complexity. The primary products of the fission process are all unstable, and undergo radioactive decay. As an illustration of this, Equation 2 shows the decay chain associated with mass number 135 together with associated half-lives.

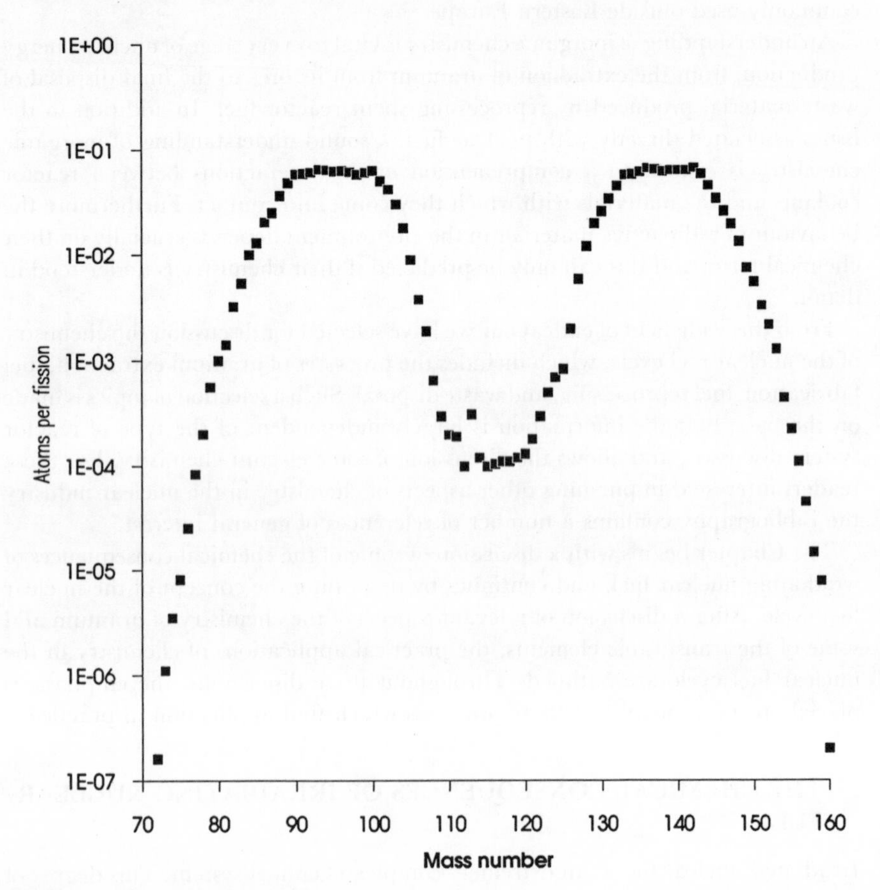

Figure 1 *Fission yields for* ^{235}U

$$^{135}_{52}\text{Te} \xrightarrow{29\,\text{s}} {}^{135}_{53}\text{I} \xrightarrow{6.7\,\text{h}} {}^{135}_{54}\text{Xe} \xrightarrow{9.2\,\text{h}} {}^{135}_{55}\text{Cs} \xrightarrow{3\times10^{6}\,\text{y}} {}^{135}_{56}\text{Ba} \quad (2)$$

It is clear, therefore, that not only is a complex spectrum of materials produced by the fission process, but that the composition of this mixture changes with time.

The formation of the **transuranic elements** in irradiated nuclear fuel proceeds from the capture of a neutron followed by **β-decay** of the resulting isotope. In this β-decay, a neutron is converted within the unstable nucleus into a proton with the emission of an electron (β-particle) (Equation 3), so that a different element with a higher atomic number is formed (*e.g.* Equation 4):

$$^{1}_{0}n \longrightarrow {}^{1}_{1}p + {}^{0}_{-1}e \quad (3)$$

$$e.g.\ {}^{238}_{92}\text{U} + {}^{1}_{0}n \xrightarrow[\text{capture}]{\text{neutron}} {}^{239}_{92}\text{U} \xrightarrow[-e^-]{\beta\text{-decay}} {}^{239}_{93}\text{Np} \xrightarrow[-e^-]{\beta\text{-decay}} {}^{239}_{94}\text{Pu} \quad (4)$$

Both ^{235}U and ^{238}U undergo this process, forming the transuranic elements neptunium Np (atomic number, Z, 93), plutonium Pu ($Z=94$), and americium Am ($Z=95$). The principal routes by which this occurs are summarized in Figure 2. As indicated in the Figure, the ^{238}U nucleus, although not itself fissile, can give rise to fissile isotopes of Pu, which can themselves be used as a reactor fuel, thus substantially enhancing the amount of energy to be obtained from a given amount of uranium.

It is clear from this brief discussion that the composition of irradiated fuel depends upon the detailed conditions under which it is irradiated. As an

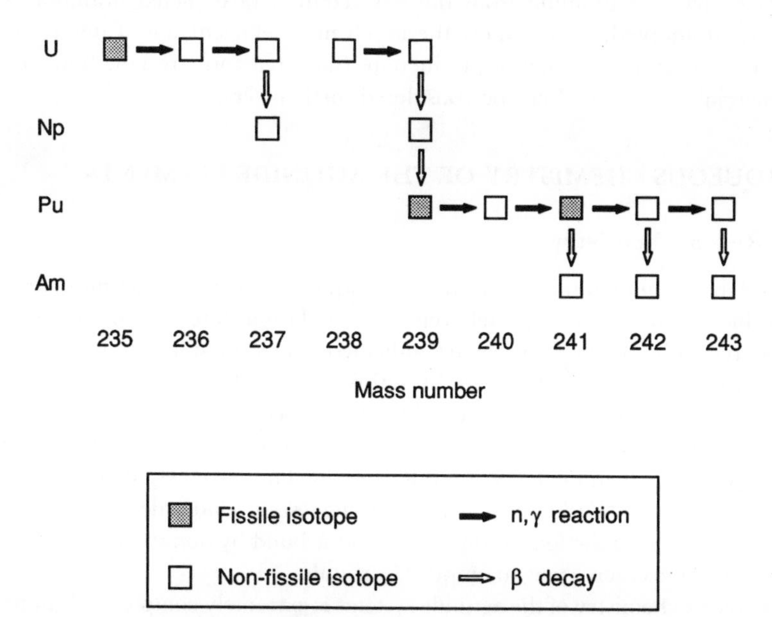

Figure 2 *The formation of transuranic elements from ^{235}U and ^{238}U*

illustration of the complexity of the resulting system, Figure 3 shows the elemental composition of fuel, expressed as parts in 10^6 (1E6) by weight, irradiated in a pressurized water reactor, 150 days after discharge from the reactor.

3 THE NUCLEAR FUEL CYCLE

The objectives of the nuclear fuel cycle are to produce uranium fuel for use in nuclear reactors, and to reprocess used fuel, thus recovering unused uranium and the plutonium formed on irradiation; these materials may then be recycled. The detailed aspects of a nuclear fuel cycle will depend upon the types of the nuclear power station in which the fuel is used, but Figure 4 shows the main features, based on the use of reactors fuelled with enriched UO_2.

After extraction from its ores and subsequent purification, the uranium is passed to an enrichment plant. Here the ^{235}U content is enhanced in that part of the uranium destined for conversion to UO_2 fuel. The remainder of the uranium, depleted in ^{235}U, and therefore known as **'depleted uranium'** is either stored, or can be used to generate **plutonium** in fast reactors as described below.

Following its conversion to fuel and use in nuclear power stations, the fuel passes to a reprocessing plant where the fission products are removed. Unused uranium and plutonium formed during irradiation in the reactor are separated, and the recovered uranium recycled.

Fast reactors, not yet used for generating electricity on a significant scale but in an advanced state of development, can make use of the depleted uranium from the enrichment process, and plutonium recovered from reprocessing. The fast reactor core is designed to make optimum use of the fissile ^{239}Pu isotope, and to produce further plutonium from the ^{238}U content of depleted uranium. This strategy, if adopted, would allow the much more efficient use of the available reserves of uranium in energy production. Such reactors are not, however, in commercial use and will not be considered further here.

4 AQUEOUS CHEMISTRY OF THE ACTINIDE ELEMENTS

4.1 Redox Chemistry

Table 2 shows the oxidation states known in aqueous solution for actinide elements of significance in the nuclear fuel cycle. The oxidation states which are normally considered to be the most stable are indicated. In states III and IV the species in solution is the metal ion M^{n+}, whilst actinides in oxidation states V and VI are present as the oxo-ions $[MO_2]^+$ and $[MO_2]^{2+}$ respectively. The dioxo groups are linear both in the solid state and in solution, and exhibit remarkably strong M–O bonds, particularly in the U, Np, and Pu species. The extremely short U–O bond distance in $[UO_2]^{2+}$ (180 picometres) even suggests a bond order greater than 2. This may be due to the formation of a second π-bond by donation of an electron pair from the oxygen atom to empty U orbitals.

The redox chemistry of the actinide systems is extremely complex; solutions are often comprised of equilibria between a number of oxidation states. A full

Figure 3 *Typical composition of irradiated fuel*

treatment of the systems is to be found in references given in the Bibliography (notably Ahrland *et al.*[7]); the present discussion will be confined to a general description of redox chemistry relevant to the nuclear fuel cycle. Formal oxidation potentials for actinide couples in 1 M $HClO_4$ are shown in Table 3.

As may be seen from Table 3, U^{3+} is unstable in aqueous solution, oxidizing to U^{4+}; the process is much accelerated if the solution is aerated. U^{4+} is relatively

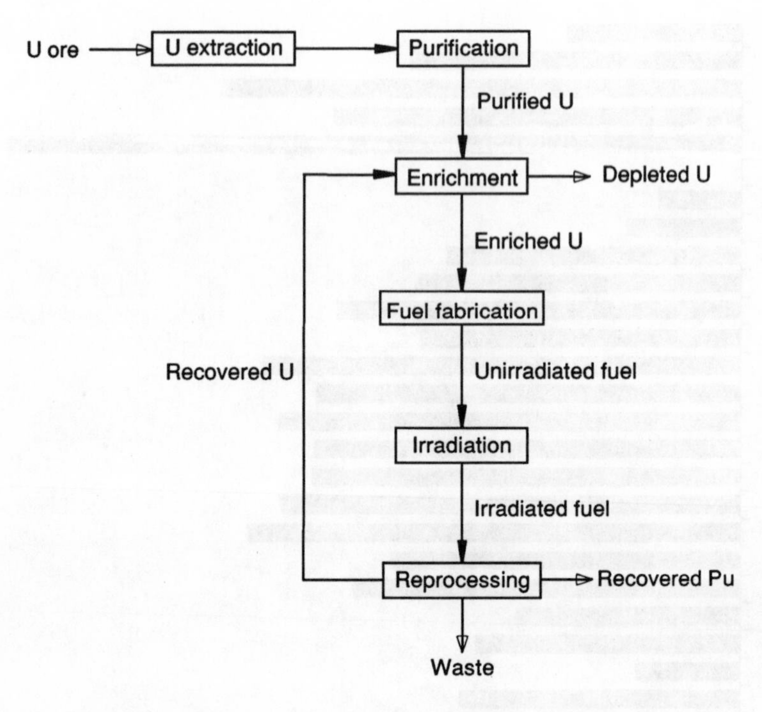

Figure 4 *The nuclear fuel cycle*

Table 2 *Oxidation states of some actinide elements in aqueous solution*

U	*Np*	*Pu*	*Am*	*Cm*
3	3	3	$\underline{3}$	$\underline{3}$
4	4	$\underline{4}$	4	4
5	$\underline{5}$	5	5	5
$\underline{6}$	6	6	6	6
	7	7		

Note: The oxidation state underlined is that generally considered the most stable

stable, oxidizing to $[UO_2]^{2+}$ slowly in the presence of oxygen. $[UO_2]^{+}$, which can be produced only with difficulty and under acidic conditions, undergoes rapid disproportionation to the more stable species $[UO_2]^{2+}$ and U^{4+}. It follows from the above discussion that the only soluble U species of technological significance are U^{4+} and $[UO_2]^{2+}$.

The Np^{V} species $[NpO_2]^{+}$ is generally considered the most stable Np species in solution. Np^{4+} and Np^{3+} also possess significant stability, but both undergo slow oxidation under aerated conditions. Although both $[NpO_2]^{2+}$ and the Np^{VII} species $[NpO_5]^{3-}$ possess some stability, both are readily reduced to $[NpO_2]^{+}$.

Table 3 *Formal electrode potentials (volts) for some actinide couples in 1 M $HClO_4$*

	U	*Np*	*Pu*	*Am*	*Cm*
O/III	1.85	1.83	2.08	2.42	2.31
III/IV	0.63	−0.15	−0.98	−2.34	−3.24
IV/V	−0.61	−0.74	−1.17	−1.16	
V/VI	−0.06	−1.14	−0.92	−1.60	
IV/VI	−0.34	−0.94	−1.04	−1.38	
VI/VII		< −2.07	−0.85		

All the oxidation states of Pu between III and VII are known in solution. Pu^{4+} is generally regarded as the most stable. Both the dioxo anions decompose: $[PuO_2]^{2+}$ is relatively easily reduced to Pu^{4+}, whilst $[PuO_2]^+$ disproportionates to Pu^{4+} and $[PuO_2]^{2+}$. Pu^{3+} is stable in aqueous solutions, even when aerated, but can be oxidized relatively easily to Pu^{4+}.

Americium exists in aqueous solution in oxidation states III to VI. Only Am^{3+}, however, is stable in solution as all the other oxidation states are strong oxidizing agents. Similarly although Cm^{3+} and Cm^{4+} are known in solution, Cm^{4+} acts as an extremely strong oxidizing agent, and only Cm^{3+} can be regarded as stable in aqueous solution.

4.2 Coordination Chemistry

The actinide elements possess a varied coordination chemistry, which, as will be seen later in this Chapter, is of great technological importance in the extraction of uranium and the reprocessing of spent nuclear fuel. The discussion here centres on a basic outline of the behaviour of uranium and plutonium with ligands of industrial significance.

Strong uranium and plutonium complexes are formed in oxidation states IV and VI (M^{4+} and $[MO_2]^{2+}$), with oxidation state IV forming the strongest complexes. In general those ligands coordinating via an oxygen atom form particularly stable complexes. The simple anions NO_3^-, SO_4^{2-}, and CO_3^{2-} all form relatively strong complexes with uranium; as described below the sulfate and carbonate complexes are of particular importance in uranium extraction.

The first stepwise stability constants for some uranium complexes, given in Table 4, illustrate the relative stabilities of different complexes as a function of the

Table 4 *First stepwise stability constant for some uranium complexes*

	SO_4^{2-}	NO_3^-
U^{4+}	4500	1.1
$[UO_2]^{2+}$	65	0.5

The first stepwise stability constant is the equilibrium constant for the reaction $M + L \rightleftharpoons ML$

oxidation state of uranium and ligand charge. In the case of the uranyl ion ($[UO_2]^{2+}$), complexes containing up to three carbonate or sulfate ligands are known. Compounds containing the tricarbonato complex are known in the solid phase, *e.g.* $Ca_2[UO_2(CO_3)_3]$. Structural studies of this material have revealed that the carbonate ions are bidentate, with the coordinating oxygen atoms at the vertices of a planar hexagon normal to the O=U=O structure. It seems likely that this structure is maintained in solution, and is typical for $[UO_2]^{2+}$ complexes where the uranium atom is eight-coordinate.

Technologically important complexes of uranium and plutonium are formed with phosphoryl compounds, notably tributyl phosphate [TBP, $(n\text{-}C_4H_9O)_3PO$]. The importance of these complexes lies in their solubility in organic solvents, thus allowing the purification of uranium and plutonium by the well-established technology of **solvent extraction** (Chapter 2). In nitric acid solution, Pu^{4+} forms a mixed complex of formula $Pu(NO_3)_4.2TBP$, whilst U^{VI} and Pu^{VI} form complexes with formulae of the type $[MO_2(NO_3)_2].2TBP$. In these complexes the nitrate and TBP ligands coordinate to the central uranium atom via oxygen atoms, as shown in Figure 5; the structural similarities between these complexes and the simpler complexes discussed above are obvious.

5 URANIUM MILLING

5.1 Introduction

Commercially important ores of uranium may be broadly classified into three types; ores containing U^{IV} (*e.g.* pitchblende, U_3O_8), hydrated ores containing U^{VI} (*e.g.* Gummite, $UO_3.nH_2O$), and refractory materials containing U^{IV} (*e.g.* Davidite, $UFe_5Ti_8O_{24}$). Although some ore bodies, notably those containing pitchblende, contain relatively high concentrations of uranium (occasionally of the order of 1%), most other commercially exploited ores contain much less, with uranium contents an order of magnitude lower. Although preconcentration by physical means can be used as part of the extraction process, particularly for richer ore bodies, chemical processes are often used at an early stage.

Two main processes are used for the chemical extraction of uranium: **leaching** with sulfuric acid and leaching with sodium carbonate. The choice of leaching agent depends on the nature of the ore being processed. An overview of the milling process is given in Figure 6.

5.2 Acid Leaching

The process of sulfuric acid leaching is preferred when the gangue (non-uraniferous portion of the ore) is acid insoluble, as the process is cheaper and faster than the alternative process based on sodium carbonate. Addition of sulfuric acid to the crushed ore results in the dissolution of uranium from the more readily leached components. In order to convert U^{IV} to the more easily dissolved U^{VI}, and thus complete the leaching process, sodium chlorate is added to the reaction mixture

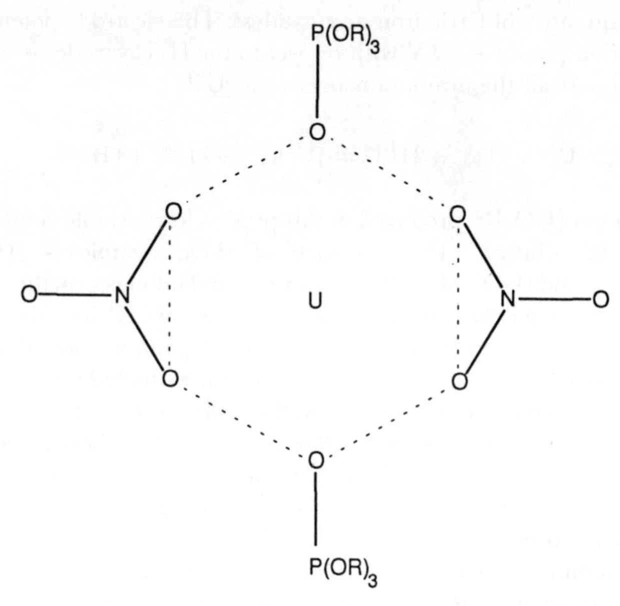

Figure 5 *The structure of* [$UO_2(NO_3)_2$].2TBP (TBP = tributyl phosphate)

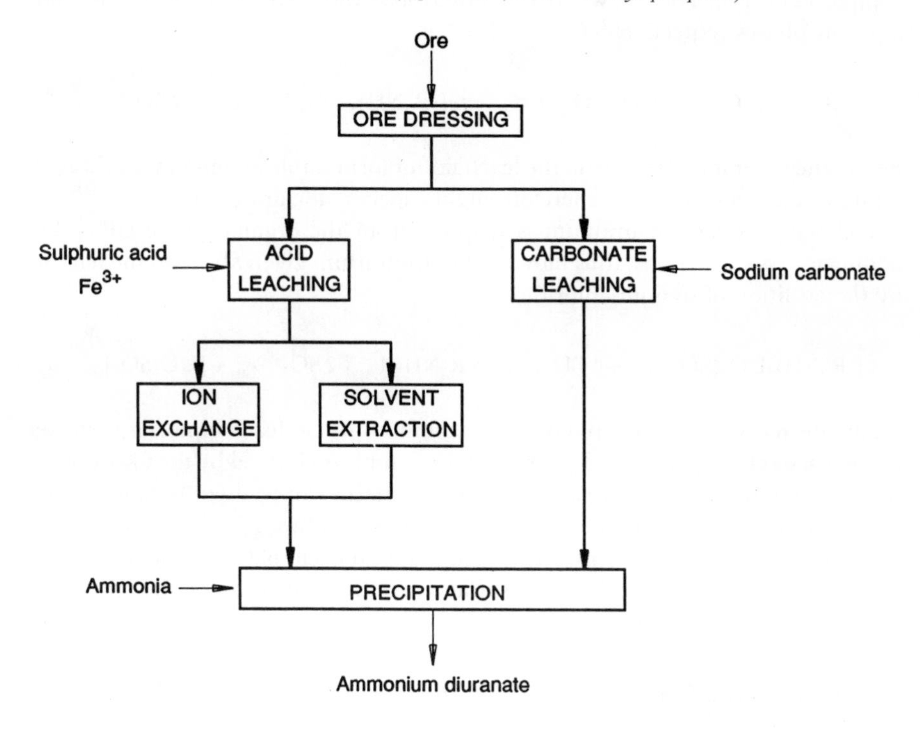

Figure 6 *Uranium milling*

along with a quantity of ferric iron as a catalyst. The electrode potential of the resulting solution (about -0.7 V with respect to the H_2 electrode) is sufficient to ensure that almost all the uranium is present as U^{VI}:

$$U^{4+} + 2\,Fe^{3+} + 2\,H_2O \rightleftharpoons [UO_2]^{2+} + 2\,Fe^{2+} + 4\,H^+ \quad (5)$$

The uranyl ion $[UO_2]^{2+}$ produced in this process forms stable complexes with sulfate ions in solution. The presence of these complexes ($[UO_2SO_4]$, $[UO_2(SO_4)_2]^{2-}$, and $[UO_2(SO_4)_3]^{4-}$) has a profound influence on the subsequent processing of the leach liquors. In addition to the uranyl ion and its sulfate complexes, the liquors resulting from acid leaching of uranium ores also contain a large number of other metallic imputities. These may include iron, vanadium, titanium, calcium, and lead, depending on the type of ore, and it is necessary to remove these materials from solution before conversion of the extracted uranium to a solid concentrate prior to final purification. The principal methods used for this process are **solvent extraction** and **ion exchange**.

Solvent extraction of leach liquors is usually effected with long chain organic amines in solution in an inert organic diluent. The amines act as a liquid anion exchanger by removing the uranyl ion in the form of its di- and tri-sulfate complexes as shown below, where the subscripts o and aq refer to the organic and aqueous phases respectively:

$$2\,(R_3NH)_2SO_{4\,(o)} + [UO_2(SO_4)_3]^{4-}{}_{(aq)} \rightleftharpoons [(R_3NH)_4UO_2(SO_4)_3]_{(o)} + 2\,SO_4{}^{2-}{}_{(aq)} \quad (6)$$

Few of the other metal species in the leach liquor form stable anions in sulfuric acid solution, and this process is therefore highly specific for uranium.

Following extraction uranium is stripped from the organic phase either by adding excess sulfate ions, thus moving the equilibrium given above to the left, or by the addition of sodium chloride:

$$[(R_3NH)_4UO_2(SO_4)_3]_{(o)} + 4\,Cl^-{}_{(aq)} \rightleftharpoons 4\,R_3NHCl_{(o)} + 2\,SO_4{}^{2-}{}_{(aq)} + [UO_2SO_4]_{(aq)} \quad (7)$$

An alternative method of removing uranium from acid leach liquors is to make use of **ion exchange resins**. By virtue of the complexes formed by the uranyl and sulfate ions either anionic or cationic exchange resins can be used; because of the greater selectivity they provide, anion exchangers are always preferred. The most commonly used systems are quaternary ammonium resins of the strong base type. The principal reaction at the resin surface is that given below, although some $[UO_2(SO_4)_2]^{2-}$ is also bound:

$$4\,RCH_2(CH_3)_3NX + [UO_2(SO_4)_3]^{4-} \rightleftharpoons [RCH_2(CH_3)_3N]_4[UO_2(SO_4)_3] + 4\,X^- \quad (8)$$

R represents a long-chain alkyl group, and X^- is commonly Cl^-, $NO_3{}^-$, or $\frac{1}{2}\,SO_4{}^{2-}$. When loaded the resin is stripped with Cl^-, $NO_3{}^-$, or $SO_4{}^{2-}$.

Whichever method is adopted for removing U from the acid leach liquors, the extracted uranium is precipitated from aqueous solution with ammonium hydroxide to form ammonium diuranate, $(NH_4)_2U_2O_7$, which forms the feedstock for uranium purification and fuel fabrication facilities.

5.3 Carbonate Leaching

In the case of those ores where the gangue interacts strongly with acid, *e.g.* limestone, it is usual to adopt a process of leaching with a mixture of sodium carbonate and sodium hydrogen carbonate. Whilst this reduces the quantity of reagent consumed, and indeed results in a liquor containing substantially lower amounts of impurities than the acid leaching process, the extraction of uranium is less efficient and thus requires the ore to be more finely ground and to be in contact with the leaching agent for longer periods at higher temperatures.

In the case of carbonate leaching the less soluble U^{IV} compounds are oxidized to U^{VI}, in the form of the diuranate ion, using atmospheric oxygen as the oxidizing agent. These ions then interact with carbonate ions to form a stable adduct as shown below:

$$2\,UO_2 + O_2 + 2\,CO_3^{2-} + 4\,HCO_3^- \rightarrow 2\,[UO_2(CO_3)_3]^{4-} + 2\,H_2O \quad (9)$$

The hydrogen carbonate ions are necessary to prevent the formation of hydroxide ions which would attack the uranate/carbonate complex anion.

Unlike the acid leaching process, carbonate leaching produces a liquor relatively low in metallic impurities. It is not, therefore, customary to purify the liquors prior to precipitation of the uranium with ammonium hydroxide to form sodium diuranate.

6 URANIUM ENRICHMENT AND FUEL FABRICATION

This stage in the fuel cycle addresses three objectives: to complete purification of the uranium, to enrich the ^{235}U content of the uranium to the level required and to prepare metallic uranium or UO_2 for use as reactor fuel. The main steps are shown schematically in Figure 7.

The input to the process is ammonium diuranate prepared during milling. The final purification of the uranium is effected by dissolving the ammonium diuranate in nitric acid followed by solvent extraction (usually using tributyl phosphate in a hydrocarbon solvent, a process described in more detail in Section 7). Stripping the uranium from the organic phase leaves an aqueous solution of pure uranyl nitrate.

The purified material is converted to UO_3 either by evaporation and direct calcining of the $[UO_2(NO_3)_2].2H_2O$, or by precipitating uranium as ammonium diuranate with ammonia, followed by calcining the solid. The UO_3 produced is converted to UO_2 by reduction with hydrogen at about 600 °C. Higher calcining temperatures yield a denser product, suitable for use when unenriched UO_2 fuel is required (*e.g.* in CANDU reactors).

The usual method for increasing the proportion of the ^{235}U isotope in uranium is based upon the effusion of the volatile compound UF_6 through a porous medium under a pressure gradient; a process is usually loosely known as **gaseous**

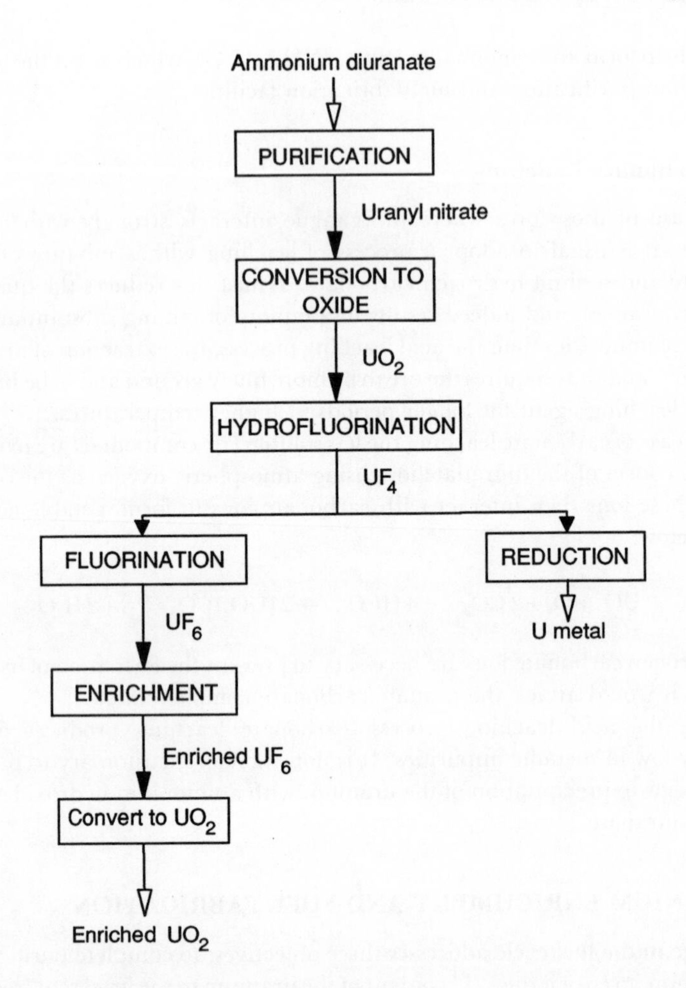

Figure 7 *Uranium enrichment and fuel fabrication*

diffusion. UF_6 is selected because of its high vapour pressure (115 mmHg at 25 °C). Since the molecular weight of $^{235}UF_6$ is marginally lower than that of $^{238}UF_6$, its rate of passage through a porous barrier under a pressure drop is slightly greater. The UF_6 on the low pressure side is therefore slightly richer in the ^{235}U compound than the UF_6 on the high pressure side. The maximum enhancement factor is less than 1.005 per diffusion step, and a large number of separate steps are therefore necessary to produce a substantial increase in the ^{235}U content.

The UF_6 for use in gaseous diffusion plants is produced from purified UO_2 in a two-step process. In the first step, UO_2 is converted to UF_4 by hydrofluorination:

$$UO_2 + 4\,HF \rightarrow UF_4 + 4\,H_2O \tag{10}$$

UF_6 is then manufactured by the direct reaction of solid UF_4 with gaseous fluorine.

Following the enrichment step it is necessary to convert UF_6 to UO_2 for use as a reactor fuel. A number of processes are used to achieve this objective. The first method is a straightforward reversal of the steps by which the UF_6 was formed: reduction of UF_6 to UF_4 followed by the hydrolysis of UF_4 to UO_2. An alternative is direct hydrolysis of UF_6 to form UO_2F_2, followed by the precipitation of the uranium as ammonium diuranate. UO_2 is then prepared by the reduction of the diuranate with hydrogen at temperatures in excess of 800 °C.

Uranium metal, used as a fuel in the UK MAGNOX reactors, is prepared by the reduction of UF_4. The reduction takes place at high temperatures so that the uranium melts and consolidates rather than remaining dispersed among the reactants. At the temperatures necessary for this operation the use of hydrogen as a reductant is not thermodynamically favoured, and it is necessary to turn to **metallothermic methods**. Magnesium and calcium are both suitable reductants in terms of the thermodynamics of the reduction process and their physical properties (*e.g.* melting and boiling points). Of the two materials, magnesium is usually preferred on the basis that it is cheaper and easier to obtain and maintain in a pure state. The reduction may be represented by the equation (11):

$$UF_4 + 2\,Mg \rightarrow U + 2\,MgF_2 \tag{11}$$

Molten uranium separates into a graphite mould, whilst the MgF_2 slag floats, and can be removed easily. The solidified uranium billet can be converted to fuel by casting and machining.

7 REPROCESSING

7.1 Introduction

The main objectives in reprocessing spent nuclear fuel are twofold. Uranium and plutonium are separated from fission products and other materials formed during irradiation which act as poisons, and are thus available for recycling as fuel in thermal and fast reactors. The unwanted, often highly radioactive, waste products are converted into a form suitable for safe long term storage or ultimate disposal.

A number of different processes have been developed for fuel reprocessing. Early methods developed beyond laboratory scale studies include coprecipitation of plutonium as Pu^{IV} with bismuth phosphate and a process based on ion exchange. Neither of these methods proved satisfactory for use on a large scale, and all commercial operations use a process based on **solvent extraction** from a nitrate medium. The principal advantages of such processes are that fabrication of stainless steel plant compatible with nitric acid is relatively straightforward, many of the reagents can be easily recycled, and actinide nitrites are readily extracted into organic solvents whilst fission product nitrates are not. Of those procedures based on solvent extraction the most widely used is the **Purex process**, and it is this which is now described. It should be emphasized that the description below refers to the generic features of the Purex process, and is not intended to be representative of any particular plant.

A schematic outline of the Purex process is shown in Figure 8. The fuel is

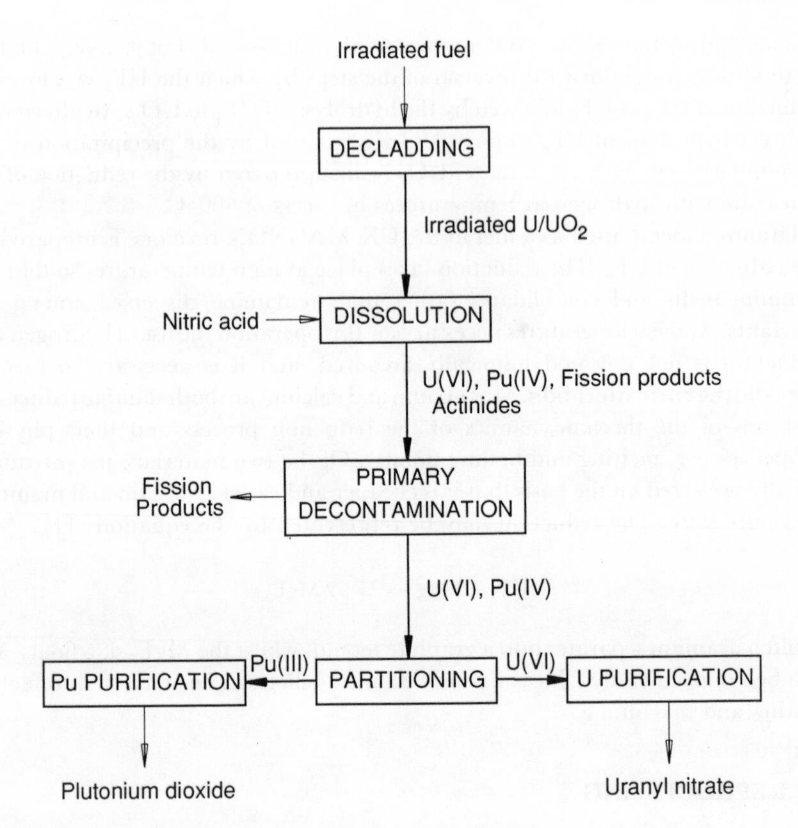

Figure 8 *The Purex process for reprocessing spent nuclear fuel*

removed from its cladding, dissolved in nitric acid, and uranium and plutonium removed from the resulting solution by extraction with tributyl phosphate (TBP) in a hydrocarbon solvent. Uranium and plutonium are partitioned by converting plutonium to Pu^{III}, which cannot be extracted into organic solvents, whilst maintaining the uranium in the form of U^{VI}, which can be so extracted. Separated uranium and plutonium are then purified and available for conversion into fresh nuclear fuel. Each of the steps in this process is now described in more detail.

7.2 Decladding and Dissolution

The function of this stage in the process is to remove the spent fuel from its cladding and to convert the uranium, plutonium, fission products, and other materials formed on irradiation into a form favouring their subsequent separation. Before dissolution can be achieved, however, the cladding which surrounds the fuel during its residence in the nuclear reactor must be removed, usually by mechanical means. For some specialized applications chemical processes have been developed for removing fuel clad (for example, selective dissolution of the aluminium cladding used in some early reactors in the USA), but these methods are relatively slow and complex, and have only very restricted use.

Fuel stripped of its clad is dissolved in nitric acid. The ease of dissolution depends on a number of factors, principally the form of the **fuel** (**oxide** or **metal**) and the proportion of ^{235}U which has undergone fission (the **burn-up**). In general oxide fuels dissolve much more slowly than metal fuels, and the greater concentration of fission products in high burn-up fuel makes dissolution of the final fraction of the fuel more difficult as discussed below.

The reaction between uranium metal and nitric acid is described by the reaction

$$U + 4\,HNO_3 \rightarrow [UO_2(NO_3)_2] + 2\,NO + 2\,H_2O \tag{12}$$

which is exothermic by 1025 kJ (mol UO_2)$^{-1}$. It is therefore necessary to provide for the removal of heat when processing large quantities of metal fuel.

The dissolution of oxide fuel proceeds according to the two reactions below, the first of which is dominant at acid concentrations below about 10 M:

$$3\,UO_2 + 8\,HNO_3 \rightarrow 3\,[UO_2(NO_3)_2] + 2\,NO + 4\,H_2O \tag{13}$$

$$UO_2 + 4\,HNO_3 \rightarrow [UO_2(NO_3)_2] + 2\,NO_2 + 2\,H_2O \tag{14}$$

The first of the reactions above is endothermic by about 100 J (mol UO_2)$^{-1}$, and heating is therefore required when dissolving oxide fuel on a commercial scale. The production of oxides of nitrogen as off-gases may be avoided by carrying out the dissolution of oxide fuels in the presence of oxygen, a process known as fumeless dissolving, which has been used at some installations, particularly in Europe:

$$2\,UO_2 + 4\,HNO_3 + O_2 \rightarrow 2\,[UO_2(NO_3)_2] + 2\,H_2O \tag{15}$$

Plutonium present in oxide fuels dissolves to form a mixture of Pu^{IV} and Pu^{VI}. The rate of Pu dissolution varies, depending on the U/Pu ratio and the conditions of irradiation, but is in general below that of uranium.

Dissolution of metal fuels normally proceeds cleanly, and leaves little if any solid residue. In contrast, oxide fuels may, after the initial stages of the process, leave substantial quantities of solid material, perhaps as much as 0.3% of the original fuel by weight. The problem is particularly acute for those fuels irradiated to high burn-ups at high temperatures. The solid material includes, in addition to a quantity of plutonium, transition metal **fission products**, particularly radioactive isotopes of Ru, Mo, Rh, Te, Zr, and Pd. At the higher temperatures associated with the irradiation of oxide fuels these species can migrate within the structure of the fuel to form metallic aggregates which are relatively inert towards nitric acid.

If it is desired to bring these materials into solution, and thus enable the recovery of the plutonium with which they are associated, the solid residues may be treated with a mixture of nitric and hydrofluoric acids or alternatively with a mixture of nitric acid and Ce^{IV} nitrite. The latter reagent is generally favoured for oxidizing the solid residues as Ce isotopes are already present in solution as fission products, and the addition of further Ce does not complicate subsequent extraction steps. Ce^{IV} is not added to the nitric acid in the first stage of dissolution, as it would oxidize the Pu present to the less easily dissolved Pu^{VI} state.

7.3 Primary Decontamination

The output from the dissolver step is a mixture of U^{VI} and Pu^{IV} in nitric acid solution. This solution also contains small quantities of Np^{V}, Np^{VI}, Am^{III}, and Cm^{III}, and almost all the non-volatile fission products present in the fuel fed to the process. The first stage in the further processing of these liquors is to remove the fission products, which are responsible for much of the radioactivity in the solution. This is done at an early stage in the process primarily to reduce the burden of radioactivity in the process liquors and thus make them easier to handle. An additional advantage is that by reducing the amount of radioactivity in solution, degradation of solvents by radiolysis later in the process is minimized.

Primary decontamination of the process liquors is achieved by extraction with **tributyl phosphate** (TBP) in a hydrocarbon solvent at about 25 °C. Pu^{IV} and U^{VI} are highly extractable under these conditions, and pass into the organic phase, as does most of the Np^{VI}. Np^{V}, Am^{III}, and Cm^{III}, together with the vast bulk of the fission products, remain in the aqueous raffinate. Final cleaning of the organic phase is achieved by scrubbing the organic layer with nitric acid (about 3 M). This is performed at slightly elevated temperature (about 60 °C) to ensure complete extraction of Ru.

7.4 Partitioning of Uranium and Plutonium

The objective of this phase of the reprocessing operation is to separate the U and Pu present in the organic output stream from the primary decontamination phase. The technique is to reduce the Pu^{IV} present to Pu^{III}, which is inextractable into the organic phase, whilst maintaining U as U^{VI}. As a result of this process U remains in the organic phase, whilst Pu passes into the aqueous raffinate. A number of reductants have been proposed for this process: Fe^{II}, U^{IV}, hydroxylamine, and a process based on direct cathodic reduction. Of these methods the use of Fe^{II} and U^{IV} as reductants have seen the most extensive use on a process scale and are therefore selected for discussion.

Fe^{II} in the form of **iron(II) sulfamate**, $Fe(SO_3NH_2)_2$, found early and extensive employment as the reductant for the partitioning process, and is still widely used. This particular iron(II) salt has the advantage that the Fe^{2+} ion is stabilized against oxidation by the acid medium by the sulfamate ion. Fe^{II} acts as a rapid and effective reductant for Pu^{IV} and from this point of view iron(II) sulfamate is a most satisfactory reagent. However its use introduces significant quantities of non-volatile impurities, principally Fe^{III} and sulfamate ions, into the waste streams from the reprocessing plant. This can limit the degree to which these highly active waste streams can be concentrated to assist in their safe storage and ultimate disposal.

An alternative method, which avoids this difficulty, is the use of U^{IV} as a reductant:

$$2\,Pu^{4+} + U^{4+} + 2\,H_2O \rightleftharpoons 2\,Pu^{3+} + [UO_2]^{2+} + 4\,H^+ \tag{16}$$

The oxidized U^{IV} simply passes into the organic phase as $[UO_2]^{2+}$ where it is processed with the rest of the uranium in the system. Clearly it is necessary to incorporate facilities within the reprocessing plant to provide a supply of U^{IV}. In addition, control of the hydrogen ion concentration in the aqueous phase is critical to the efficiency of the partitioning process. The advantages offered in simplifying the handling of highly active waste from the reprocessing operation, however, are extremely attractive, and the use of U^{IV} as a reductant is finding increasing favour.

7.5 Purification of Uranium and Plutonium

Further purification of the uranium and plutonium streams from the partitioning step is necessary before these materials can be fed back into the fuel cycle. The single partitioning step effects U and Pu separation at about 99% efficiency, but this still leaves unacceptably large quantities of cross contamination which must be dealt with. In addition both streams contain significant amounts of fission products and neptunium which must be removed to achieve the objective of U and Pu free of highly radioactive material (principally the fission products) and unwanted neutron absorbers (of which Np is one) which would affect the performance of fuel fabricated from recycled uranium.

Final purification of the uranium stream is achieved by back-extraction of uranium to the aqueous phase followed by at least one further cycle of TBP extraction with plutonium held as inextractable Pu^{III}. If it is necessary to remove further traces of fission products this may be achieved either by passage through silica gel (as in early plants) or by further solvent extraction using different extractants.

The aqueous plutonium stream leaving the partitioning process still contains about 1% of the feed uranium by weight. One technique for removal of this material involves two further TBP extraction cycles. In the first U and Pu are extracted together, leaving the bulk of the residual fission products behind. The oxidation potential of the aqueous solution is carefully controlled to maintain the Np as inextractable Np^{V}. In the second stage, Pu is back extracted into the aqueous phase leaving U behind: this is achieved either by reducing Pu to Pu^{III} or by careful control of the acid concentration of the aqueous phase.

Further clean-up of the Pu stream can be accomplished either by precipitation of the plutonium as the oxalate or by anion exchange from nitrate solution. Of these two processes, oxalate precipitation is often preferred because of difficulties associated with radiolysis of ion exchange resins. Following these final purification steps the plutonium nitrate or oxalate is normally calcined to PuO_2 before being fed back into the fuel cycle.

8 MANAGEMENT OF WASTES FROM THE NUCLEAR FUEL CYCLE

Whilst considerable efforts are expended to ensure that the volume of waste materials from the nuclear fuel cycle is minimized, some unwanted material is inevitably generated. **Radioactive** waste materials are classified as high, intermediate, or low level waste on the basis of the amount of radioactivity they contain. Some of the more interesting chemical issues associated with waste

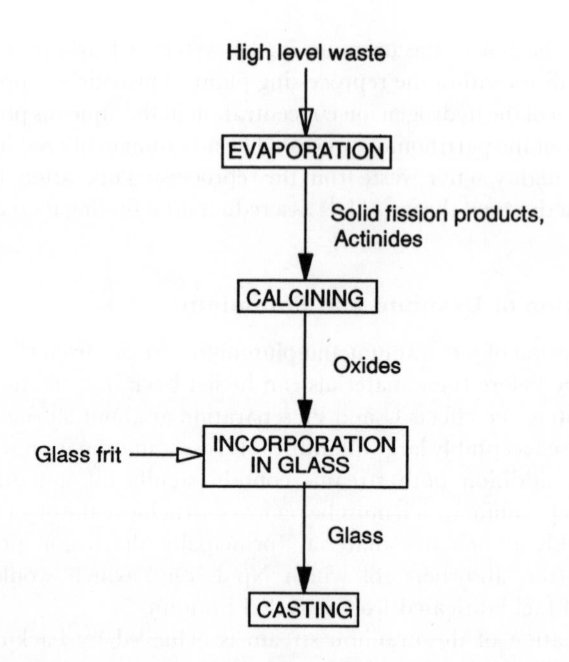

Figure 9 *Vitrification of high level waste*

management arise from dealing with **high level wastes**, and it is these which will be discussed here.

High level waste arises in both solid and liquid forms. The liquid form is the concentrated aqueous raffinates from fuel reprocessing. High level waste in its solid form consists largely of the cladding removed from fuel prior to reprocessing. It is clear, therefore, that high level waste, and in particular liquid waste, has the potential to represent a considerable hazard, and needs to be safely isolated from the environment.

Liquid high level waste consists of a nitric acid solution containing the bulk of the fission products and of the transuranic elements, except of course plutonium, present in the fuel after irradiation. After discharge from the reprocessing plant the liquid waste is held in stainless steel storage tanks after adjusting the nitric acid concentration to optimize compatibility between the solution and the tank. The radioactive nature of the waste has two principal consequences for waste management: the chemical composition of the waste is slowly changing, and the decay of the radionuclides generates heat. The latter phenomenon requires that the storage tanks be continually cooled.

Whilst storage as a liquid is technically satisfactory on a timescale measured in decades, long-term disposal requires that the long-lived radionuclides be fixed in some stable matrix before final disposal in a permanent repository. The matrix chosen must be able to resist both heat and radiation. In addition it must possess considerable chemical stability; in particular, resistance to leaching by any groundwaters finding their way into the repository is of great importance. One widely researched option is the use of some form of glass for radionuclide fixation. Whilst there are a number of detailed differences in the detailed composition of proposed glass matrices, the principles of the **vitrification** process are straightforward and illustrated in Figure 9.

The liquid wastes must first be evaporated to dryness, and the solid residues calcined at temperatures in the range 350–800 °C to convert metallic species to an oxide form. The calcined waste is then mixed with a glass frit (see Chapter 12), and the mixture heated to a temperature in the range 1000–1300 °C to incorporate the oxides into a glass matrix, which may then be cast into blocks suitable for final disposal.

9 CONCLUSIONS AND FUTURE TRENDS

In this chapter we have discussed the chemistry of the **nuclear fuel cycle**, a technology developed to help harness nuclear fission for the **safe and efficient production of electricity**. We have seen how, from the extraction of uranium from its ores to the recycling of irradiated nuclear fuel, the careful application of chemical principles ensures the safe and efficient use of raw materials. Although the technology is well developed, improvements in all aspects of its performance are continually sought, and in this endeavour chemists have their part to play. In the design of advanced thermal reactors and the production of fuel with improved performance, the application of sound inorganic chemistry will be important.

The more **efficient use of raw materials** is a subject in which the nuclear industry, like all others, takes an interest. One way in which this has been approached is to develop **'fast reactors'** which, as described in Section 3, are designed to make optimum use of ^{238}U, the non-fissile uranium isotope, and the plutonium produced in thermal reactors. The production of fuel for these systems, its reprocessing, and the chemistry of the molten sodium used to cool the reactor core present substantial challenges. Fast reactors are in an advanced state of development and a number of prototypes have operated successfully. Although the economics of electricity generation are unlikely to favour the widespread commercial use of fast reactors in the near future, they remain an important strategic resource for the electricity industry in the longer term.

Although this Chapter has concentrated on the nuclear fuel cycle, the vital role played by inorganic chemistry in other activities within the nuclear industry must not be forgotten. Areas as diverse as the behaviour of **reactor coolants** and the analysis of the behaviour of **fission products** in the environment rely heavily on a sound knowledge of chemical sciences. The quest for improved economics and ever increasing standards of safety will continue to provide important challenges for the chemist.

10 BIBLIOGRAPHY

10.1 General Background

1. J. R. Findlay, K. M. Glover, I. L. Jenkins, N. R. Large, J. A. C. Marples, P. E. Potter, and P. W. Sutcliffe, 'The Inorganic Chemistry of Nuclear Fuel Cycles', in 'The Modern Inorganic Chemicals Industry', ed. R. Thompson, Special Publication No. 31, The Chemical Society (now The Royal Society of

Chemistry), London, 1977, p. 419–466.
2. D. J. Littler, 'Chemistry in the Service of Electricity Generation', in 'Energy and Chemistry', ed. R. Thompson, Special Publication No. 41, The Royal Society of Chemistry, London, 1981, p. 187–220.
3. J. A. C. Marples, R. L. Nelson, P. E. Potter, and L. E. J. Roberts, 'Chemistry in the Development of Nuclear Power', in 'Energy and Chemistry', ed. R. Thompson, Special Publication No. 41, The Royal Society of Chemistry, London, 1981, p. 131–163.
4. W. L. Wilkinson, 'Chemistry of the Nuclear Fuel Cycle', in 'Energy and Chemistry', ed. R. Thompson, Special Publication No. 41, The Royal Society of Chemistry, London, 1981, p. 164–186.

10.2 Nuclear Technology

5. M. Benedict, T. Pigford, and H. Levi, 'Nuclear Chemical Engineering', McGraw–Hill, New York, 1981.
6. A. R. Foster and R. L. Wright, 'Basic Nuclear Engineering', Allyn and Bacon, Massachusetts, 1983.

10.3 Actinide Chemistry and Physics

7. S. Ahrland, K. W. Bagnall, D. Brown, R. M. Dell, S. H. Eberle, C. Keller, J. A. Lee, J. O. Liljenzin, P. G. Mardon, J. A. C. Marples, G. W. C. Milner, P. E. Potter, and J. Rydberg, 'The Chemistry of the Actinides', Pergamon Press, Oxford, 1973.
8. A. J. Freeman and G. H. Lander, 'Handbook on the Physics and Chemistry of the Actinides', North–Holland, Amsterdam, 1984.
9. J. J. Katz, G. T. Seaborg, and L. R. Morss, 'The Chemistry of the Actinide Elements', Chapman and Hall, London, 1986.

10.4 Journal Literature

Papers relevant to this field appear in 'standard' journals such as the Journal of Physical Chemistry and the Journal of the Chemical Society, Faraday Transactions. The following two publications provide good background material in this topic:

Nuclear Energy (British Nuclear Energy Society)
Nuclear Engineering International (Reed Business Publishing)

The following journals cover chemistry relevant to the nuclear industry, the first being a review journal:

Progress in Nuclear Energy
Journals in Nuclear Materials
Radiochimica Acta
Journal of Radioanalytical and Nuclear Chemistry

CHAPTER 18

Catalysts and Photocatalysts for Solar Energy Conversion

ANDREW MILLS

1 INTRODUCTION

The purpose of this chapter is to highlight the role of speciality inorganic materials as **catalysts** and **photocatalysts** in the conversion of **solar** energy into **chemical** or **electrical** energy. The growing interest in such systems arises because solar energy, *i.e.* sunlight, represents a major, largely untapped energy source which could easily satisfy the energy requirements of humankind if successfully harnessed, and which will last indefinitely.

At present our modern world requires a continual supply of a vast amount of energy (*ca.* 3×10^{20} J year^{-1}) which is provided mainly by the fossil fuels, *i.e.* coal, oil, and gas, as indicated by the data in Table 1. These fossil fuels originated from the anearobic decomposition of organic matter, such as dead plants and animals, over a timescale of millions of years and, as a consequence, fossil fuels are not renewable on any timescale meaningful to humankind. The time is fast approaching when **major alternatives to the fossil fuels** will have to be found and brought on-line. From a comparison of the major energy resources available in the world, illustrated by the data in Figure 1, it is clear that the future alternative energy to fossil fuels will be derived from a nuclear process, but will it be **nuclear fission** or **fusion** in reactors on Earth, or **nuclear fusion** in a reactor, 93 million miles away, *i.e.* the sun? Only further research will tell.

2 FEATURES OF SOLAR ENERGY

Although using the sun as a source of energy has many attractive features, there are a number of inherent problems which must be overcome in order to capture and utilize it fully.

2.1 The Polychromic Nature of Sunlight

As Isaac Newton discovered in 1665, sunlight consists of a rainbow of colours, *i.e.* it is **polychromatic**, as illustrated by the solar spectrum shown in Figure 2. The

Table 1 *Total annual world energy consumption (1990)*

Energy type	Amount (10^{19} *J*)
oil	13.30
natural gas	7.47
coal	9.43
nuclear	1.98
hydro	2.32
Total	*34.50*

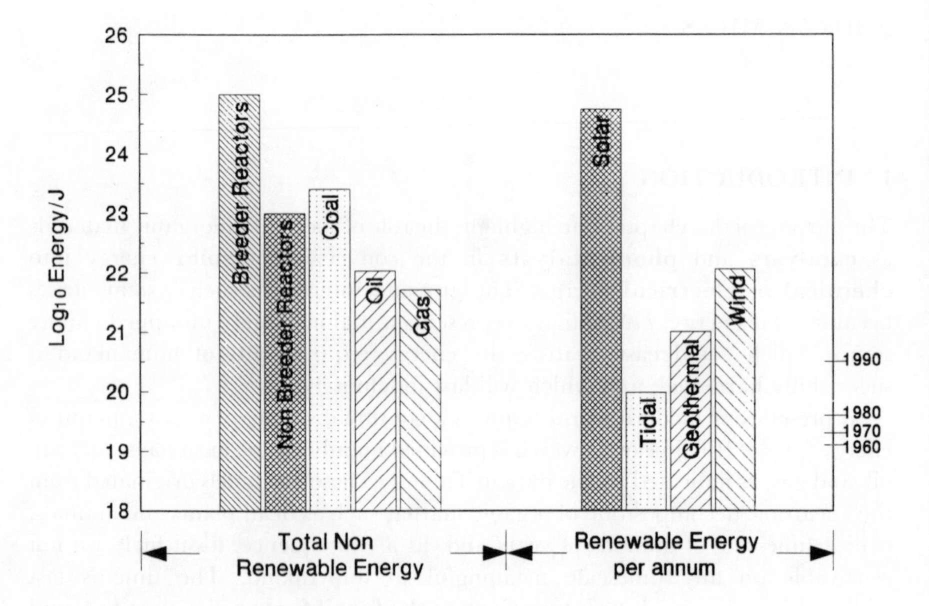

Figure 1 *World energy consumption in 1960, 1970, 1980, and 1990; and world energy resource estimates in 1990*

dips in the spectrum are due to absorption by various atmospheric species like CO_2 and H_2O.

Initially, it might seem reasonable to design a solar energy conversion device which absorbs all the wavelengths of sunlight. However, it should be recognized that whatever absorbs the light is likely to be excited electronically, *i.e.* the light will promote electrons in the absorbing species into a higher electronic energy level to create an **electronically excited state**. Electronic excitation has two distinctive features:

(i) the absorbing species will have a **threshold wavelength** (λ_{thres}) corresponding to an excitation energy ΔE, above which it will not absorb
(ii) although the absorbing species may absorb photons of wavelengths $<\lambda_{thres}$,

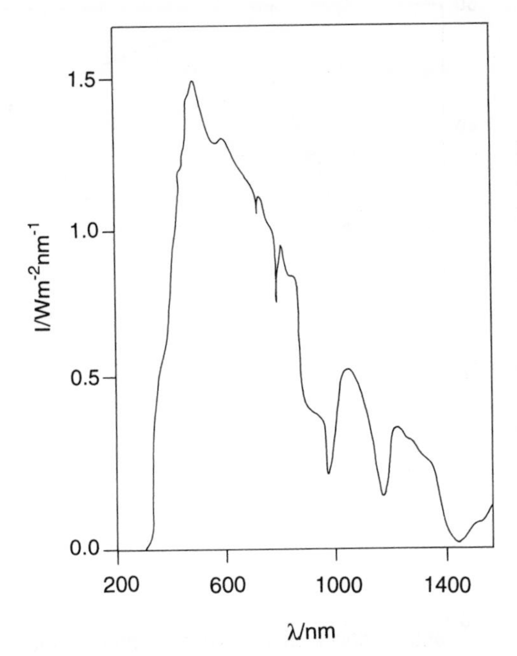

Figure 2 *Typical light intensity versus wavelength distribution for sunlight on a clear day*

only part of the energy of the photon, equivalent to $hc/\lambda_{thres} = \Delta E$, will be used, and the rest of the energy will be converted into thermal energy

These two features are illustrated by the simple energy level diagram in Figure 3. Thus, a solar energy conversion system with a threshold energy small enough to ensure all the wavelengths of the solar spectrum are absorbed would not be very efficient, since most of the energy of the absorbed photons of different wavelengths below λ_{thres} would be converted into thermal energy.

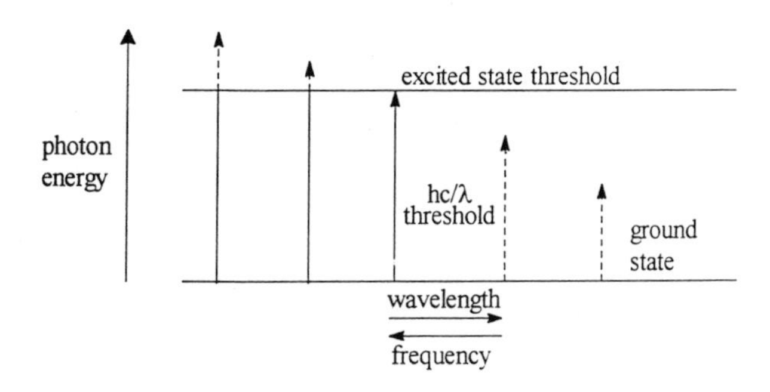

Figure 3 *Utilization of a broad-band spectrum by an absorber which absorbs all light of* $\lambda < \lambda_{thres}$, i.e. *a threshold conversion device. For each photon absorbed, a threshold device is only able to utilize* hc/λ_{thres} *of the photon's energy and cannot utilize photons of energy less than this value*

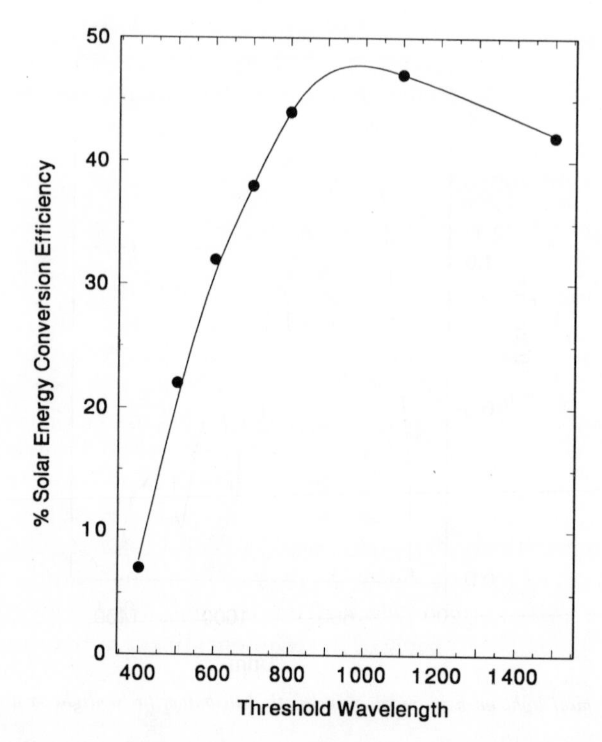

Figure 4 *Plot of % solar energy conversion efficiency* η *vs. the threshold wavelength,* λ_{thres}*. The data was generated using the spectral profile for sunlight illustrated in Figure 2 and Equation (1).*

For a single absorbing species which absorbs all wavelengths below a threshold value, λ_{thres}, the **% efficiency of solar energy conversion** (η) can be expressed quantitatively by the following mathematical expression:

$$\eta = \frac{100(hc/\lambda_{\text{thres}}) \int_0^{\lambda_{\text{thres}}} I(\lambda)\mathrm{d}\lambda}{\int_0^{\infty} I(\lambda)E(\lambda)\mathrm{d}\lambda} \tag{1}$$

where $I(\lambda)$ is the intensity of the solar radiation in incident quanta per wavelength interval and $E(\lambda)$ is given by hc/λ. Figure 4 illustrates the variation in η as a function of threshold wavelength for a clear sky with the sun directly overhead; under these conditions, the optimum threshold wavelength is at *ca.* 1100 nm. It can be shown that for a system containing two absorbers with different threshold wavelengths, η will be higher, and that for three absorbers η will be even higher, and so on.

2.2 Collecting Sunlight

Solar energy falls over a large area of the earth and, therefore, it must be collected before it can be used. Any solar energy collector, which may also incorporate the converter, must be cheap per m^2 of surface area. In order to compete with conventional methods of electricity production it has been estimated[1] that a **collection/conversion device should cost less than ca. \$100 m^{-2}**.

2.3 Converting Sunlight

A whole range of different types of energy are used in everyday life, for example:

(i) **low temperature heat** for water and space heating
(ii) **electricity** for light and communications
(iii) **storable chemical fuels** for transport and heating

If we are going to use solar energy as a major energy source we must be able to convert it efficiently into any one of these three, more useful, forms of energy. The conversion of solar energy into **low temperature heat** energy is now a well established technology and devices for this purpose can be seen on rooftops throughout the Mediterranean. However, such devices are typically only 40% efficient, and this efficiency is drastically reduced to uneconomical levels if the heat energy is then used to produce electricity or a chemical fuel. Thus, there is a real need to develop devices capable of the **efficient direct conversion** of solar to **chemical** or **electrical** energy and, in the following pages, the main approaches to this goal are described and the vital role played by speciality inorganic chemicals is highlighted.

3 CONVERSION OF SOLAR ENERGY TO CHEMICAL ENERGY

It is possible to select any number of chemical reactions which are thermodynamically uphill and in which a fuel is one of the products; by a fuel we mean a material which can be used to provide a source of heat or power. In Table 2 we have collected together several fuel-forming reactions involving reactants which are cheap and readily available. These reactions are of particular interest because the technology already exists for using the fuels generated.

In all the fuel-forming reactions given in Table 2, $\Delta G^{\ominus}$ is positive and therefore energy must be put into the system if substantial amounts of products are to be formed. This energy is usually in the form of heat but light could be used instead. Indeed, it should be remembered that we owe our continued existence on this planet to a fuel-forming reaction which is driven photochemically, *i.e.*

$$6\,CO_2(g) + 6\,H_2O(l) \rightarrow C_6H_{12}O_6(s) + 6\,O_2(g) \qquad (2)$$

ΔG = 1.25 eV/electron transferred,
Energy density = 15.1 kJ g^{-1} of fuel generated

which is the basis of **green plant photosynthesis**. Indeed, for most of the reactions listed in Table 2 there exists natural photosynthetic systems capable of effecting them, albeit often inefficiently.

In all these biological systems there are materials which promote the reaction

Table 2 *Some fuel-forming reactions*

Reaction	$\Delta G^{\ominus}$	*Energy*
(eV/molecule)	*Density**	
$H_2O(l) \rightarrow H_2(g) + \frac{1}{2}O_2(g)$	2.46	118.6
$N_2(g) + 3H_2O(l) \rightarrow 2NH_3(g) + \frac{3}{2}O_2(g)$	3.51	20.0
$CO_2(g) + 2H_2O_2(l) \rightarrow CH_3OH(l) + \frac{3}{2}O_2(g)$	7.28	20.7
$CO_2(g) + 2H_2O(l) \rightarrow CH_4(g) + 2O_2(g)$	8.47	51.1

* in units of kJ/gram of fuel generated

without undergoing any overall change themselves and such materials are called 'catalysts'. In green plant photosynthesis, for example, there is a catalyst, believed to be a manganese tetramer,[2] which allows an oxidant, generated photochemically, to carry out the very difficult process of oxidizing water to oxygen. Fundamental to green plant photosynthesis are two light absorbing systems, P_{690} and P_{700}, coupled together in series, which, upon electronic excitation, help to promote Reaction (2) without undergoing any overall change themselves. Such materials are often referred to as **'photocatalysts'**.

Although there already exists in nature many solar to chemical energy conversion systems, the current demand for energy is much greater than could ever be supplied by natural photosynthesis. It appears that even the best laboratory photosystems have a solar efficiency of only 5% and in the field this is reduced to an average of *ca.* 1%, even under the best conditions. In addition, the plants need fertilizer, irrigation, and harvesting. If a high energy density fuel is required like alcohol then the agricultural product must be fermented. All this work represents energy input before energy can be produced and, therefore, the overall efficiency of converting solar energy to chemical energy by this method is lowered even further.

In nature there isn't a solar to chemical energy conversion system which is efficient enough for our needs. Thus, a highly efficient, **artifical alternative photosynthetic system** must be devised which preferably generates a fuel with a **high energy density**. The photocleavage of water into hydrogen and oxygen has attracted particular attention in this regard, and the different photochemical systems devised so far for this purpose, which employ either dye or semiconductor photocatalysts, are discussed in the following sections.

3.1 Water Splitting using Dye Photocatalysts

When a dye, D, is **promoted to an electronically excited state** by the **absorption of a photon of light** of the appropriate energy, $h\nu$, *i.e.*

$$D + h\nu \rightarrow D^* \quad (3)$$

it becomes a stronger reductant, so $E^{\ominus}(D^+/D^*) < E^{\ominus}(D^+/D)$. In addition, D^* is a stronger oxidizing agent than D, thus $E^{\ominus}(D^*/D^-) > E^{\ominus}(D/D^-)$. These shifts in redox potential are illustrated in Figure 5.

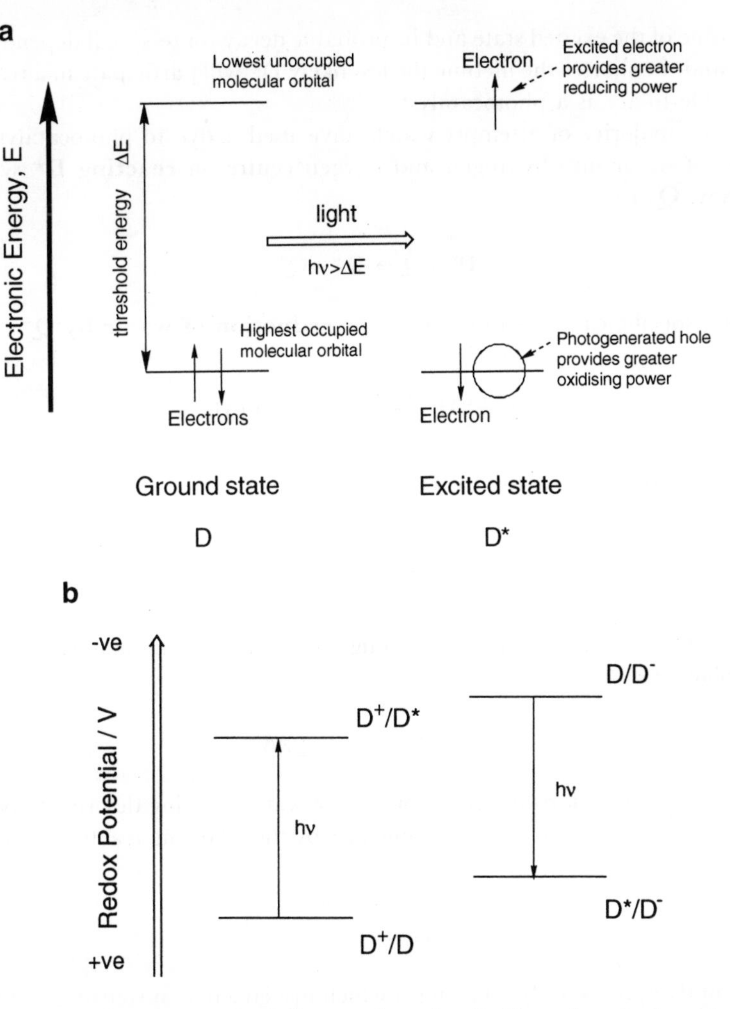

Figure 5 *(a) One electron orbital diagrams for a dye, D, illustrating why its excited state is simultaneously a more powerful reductant and oxidant than the ground state. In the electronically excited state of the dye, D*, the excited electron is in a higher energy level than it was in the ground state of the dye, D, and, therefore, it is more easily removed. In addition, in D* an electron can be more easily added to it than in D, since in the former species the vacancy lies at a lower energy level.*
(b) The net result of electronic excitation of D is to change (i) its oxidation potential, $E(D^+|D)$, *to* $E(D^+|D^*)$, *which is more negative because D* is a better reducing agent than D; and (ii) its reduction potential,* $E(D|D^-)$, *to* $E(D^*|D^-)$, *which is more positive because D* is also a better oxidizing agent than D*

In the absence of an added quencher, the excited state will usually decay via either a non-radiative process, liberating heat, Δ, or a radiative intramolecular process, liberating light, *hv*′:

$$D^* \rightarrow D + \Delta \text{ or } h\nu' \tag{4}$$

The lifetime of the excited state and its probable decay route(s) will depend upon the dye and the shorter the lifetime the less likely D* will participate in a reaction and be able to act as a photocatalyst.

The vast majority of attempts which have used a dye to photocatalyse the cleavage of water into hydrogen and oxygen centre on **reacting D* with a quencher, Q**, *i.e.*

$$D^* + Q \rightarrow D^+ + Q^- \tag{5}$$

and using specific catalysts to promote the **reduction of water by Q^-**:

$$2\,Q^- + 2\,H^+ \xrightarrow{H_2 \text{ catalyst}} 2\,Q + H_2 \tag{6}$$

and the **oxidation of water by D^+**:

$$4\,D^+ + 2\,H_2O \xrightarrow{O_2 \text{ catalyst}} 4\,D + 4\,H^+ + O_2 \tag{7}$$

The H_2 and O_2 catalysts must act fast enough to prevent the thermal back reaction taking place, *i.e.*

$$D^+ + Q^- \rightarrow D + Q + \Delta \tag{8}$$

A reaction scheme for the overall process of water splitting described above is illustrated in Figure 6 and can be summed up by the following reaction equation:

$$2\,H_2O \xrightarrow[D + h\nu]{H_2 \text{ catalyst} + O_2 \text{ catalyst}} 2\,H_2 + O_2 \tag{9}$$

A list of dye photocatalysed systems which are effective in splitting water are given in Table 3.

The list of dye-based water splitting photosystems contained in Table 3, is dominated by one particular dye, namely the dication of ruthenium(II) tris(2,2′-bipyridine), *i.e.* $[Ru(bpy)_3]^{2+}$ and some of the photophysical and redox properties[3] of this very important dye are given in Table 4. From this data it can be seen that $[Ru(bpy)_3]^{2+}$ absorbs visible light, a definite 'plus' for solar energy conversion, and has an excited state which is highly reactive and quite long lived. In addition, the oxidized form of $[Ru(bpy)_3]^{2+}$, $[Ru(bpy)_3]^{3+}$, is quite stable and does not readily decompose, which happens with most other dyes.

Although water splitting, photocatalysed by dyes such as $[Ru(bpy)_3]^{2+}$, is possible in theory and has been claimed to be achieved on numerous occasions, as evidenced by the examples in Table 3, it is worth noting that the photosystems reported so far are notoriously irreproducible and many question the initial claims.[4]

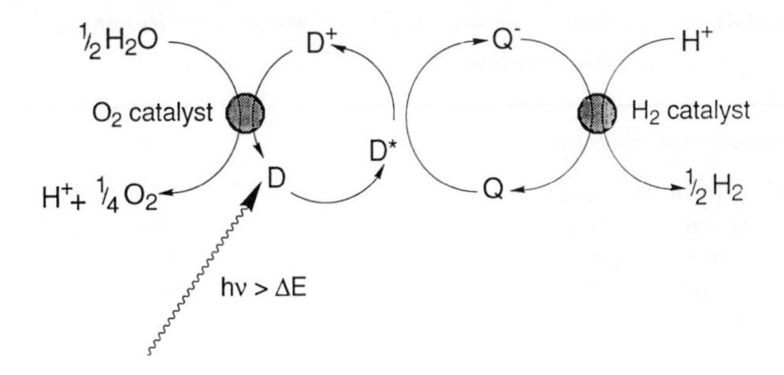

Figure 6 *Schematic illustration of the mode of action of the principal components in a typical dye-sensitized water splitting photosystem. Upon electronic excitation of the dye, D, the excited state, D*, is formed which can then undergo oxidative quenching with Q to form D^+ and Q^-. D^+ is a strong oxidizing agent and can oxidize water to O_2 in the presence of an O_2 catalyst, such as RuO_2. Q^- is a strong reducing agent and can reduce water to H_2 in the presence of an H_2 catalyst, such as colloidal Pt*

Table 3 *Cleavage of water into hydrogen and oxygen using dye photocatalysts*

Sensitizer	*Quencher*	*O_2 catalyst*	*H_2 catalyst*	*$\Phi(H_2)$*
$[Ru(bpy)_3]^{2+}$	$[MV]^{2+}$	$RuO_2.nH_2O$ colloid	Pt colloid	0.0015
$[Ru(bpy)_3]^2$/NafioN/ $[Ru(bpz)_3]^{2+}$	$[MV]^{2+}$	$RuO_2.nH_2O$	Pt colloid	—
$[Ru(bpy)_3]^{2+}$ absorbed on sepolite	$[Al_xEu_{1-x}(OH)_3]$	$RuO_2.nH_2O$	Pt	—
$[Ru(bpy)_3]^{2+}$	$[MV]^{2+}$	$RuO_2.nH_2O/TiO_2Pt$ colloid (a bifunctional catalyst)		—
$[Ru(bpy)_2(X)]^{2+}$ X = C_{12}bpy, C_{18}bpy or 2C_{12}bpy	none	$RuO_2.nH_2O/TiO_2/Pt$ colloid (a bifunctional catalyst and quencher)		up to 0.05
$[Ru(bpy)_3]^{2+}$	none	prussian blue (a bifunctional catalyst and quencher)		10^{-4}

Abbreviations: $C_{14}MV^{2+}$ = *N*-tetradecyl-*N*'-methyl viologen; PVP-C_{16} = a cationic polysoap in which 50% of the pyridine groups in poly(4-vinylpyridine) are quaternized by hexadecyl chains; C_{12}bpy = 4,4'-(Me)(n-dodecyl)bpy; C_{18}bpy = 4,4'-(Me)(n-octadecyl)bpy; 2C_{16}bpy = 4,4'-(n-hexadecyl)$_2$bpy

Table 4 *Photophysical and redox properties of $[Ru(bpy)_3]^{2+}$ in aqueous solution*

Photophysical Properties	
Absorption λ_{max} (nm)	452, 345, 1285
(Molar absorbtivity)	(14 600, 6500, 87 000)
Emission λ_{max} (nm)	610
(Quantum yield)	(0.042)
Excited state lifetime (μs)	0.62
Redox Properties	
(half-wave potentials vs NHE)	
$E_{\frac{1}{2}}(3+/2+)$	1.26
$E_{\frac{1}{2}}(2+/1+)$	−1.28
$E_{\frac{1}{2}}(3+/2+*)$	−0.86
$E_{\frac{1}{2}}(2+*/1+)$	0.84

* indicates that the dye is in its electronically excited state

3.2 The Role of H_2 and O_2 Catalysts in Water Splitting Photosystems

Two essential components of any photosystem for water splitting are a hydrogen catalyst and an oxygen catalyst to mediate reactions (6) and (7), respectively (Figure 6). Hydrogen catalysts are well established and **colloidal Pt** can now be used to mediate the **reduction of water** with **near 100% efficiency** and on a microsecond timescale.[5] In contrast, **oxygen catalysis** has proved much **more difficult** to effect. Many heterogeneous materials when used for O_2 catalysis are either inert (and therefore inactive) or rapidly rendered so through the formation of a passivating layer of surface oxide upon exposure to the oxidant.

The redox potential for the oxidation of water to oxygen in aqueous solution is large and positive and given by the following expression: $(1.23-0.0591 \times pH)$ V vs NHE. In order to oxidize water, the oxidant must have a redox potential which is more positive than that of water. Unfortunately, many materials which might appear possible O_2 catalysts are unstable towards anodic corrosion under the strong oxidizing conditions necessary for the oxidation of water, *i.e.* their oxidation potentials are less than or similar to that of water. A classic example[6] of this is highly hydrated ruthenium dioxide hydrate, $RuO_2.xH_2O$ where $x \geq 2.3$, which is completely oxidized to RuO_4 by a variety of different oxidants with oxidation potentials suitable for water oxidation, such as Ce^{IV}, MnO_4^- or BrO_3^-. In contrast, by annealing $RuO_2.xH_2O$ at *ca.* 144 °C in air for 5 h, a highly active, stable O_2 catalyst is produced, namely thermally activated ruthenium dioxide hydrate, $RuO_2.yH_2O*$ for short (where $y \simeq 1$).[6]

Early work into the development of an **O_2 catalyst** concentrated on **heterogeneous** systems, the most active of which appears to be $RuO_2.yH_2O*$. In this area, current research is directed towards finding ways in which the specific activity (*i.e.* activity per g) of the heterogeneous O_2 catalyst materials obtained so far can be increased, *e.g.* by depositing the heterogenous O_2 catalysts onto high surface area, inert materials (such as zeolites, alumina, or TiO_2) or preparing them in a more finely divided form, such as a colloid.[6] A list of the heterogeneous O_2 catalysts which have been used to date is given in Table 5. There have been attempts to develop stable and efficient **homogeneous** O_2 catalysts, although most appear prone to oxidative decomposition.[7] Table 6 lists some of the homogeneous O_2 catalysts developed to date.

Although the materials listed in Tables 5 and 6 have been formally identified as O_2 catalysts, in most cases this has not been established with any rigour. Thus, many groups involved in this work monitor the disappearance of the strong oxidant, used to test the activity of the potential O_2 catalyst, via Reaction (10):

$$4\,OX + 2\,H_2O \xrightarrow{O_2 \text{ catalyst}} 4\,RED + 4\,H^+ + O_2 \qquad (10)$$

where OX is an oxidant, such as the Ce^{IV} ion, which is readily available and has a redox potential which is greater than that of the O_2/H_2O couple. However, the concomitant oxidation of water to oxygen is often assumed and this assumption may not be valid if there are other chemical species, including the catalyst itself, in the reaction solution which are more readily oxidized than water. For example, $RuO_2.x\,H_2O$ has been reported as being a good O_2 catalyst,[8] but this was before it was found to readily undergo anodic corrosion to form RuO_4.

Table 5 *Heterogeneous oxygen catalysts*

Catalyst	*Oxidant*
Pt foil	Ce^{IV} ions
$RuO_2.xH_2O$ (powder or colloid)	Ce^{IV} ions, $[Ru(bpy)_3]^{3+}$, $[Fe(bpy)_3]^{3+}$, Tl^{3+}, BrO_3^- and MnO_4^-
$RuO_2.xH_2O$ on various supports including TiO_2, Al_2O_3, clay, and zeolite	Ce^{IV} ions, $[Ru(bpy)_3]^{3+}$, and other oxidants
PtO_2	Ce^{IV} ions
IrO_2, $IrO_2.xH_2O$	Ce^{IV} ions
MnO_2	$[Ru(bpy)_3]^{3+}$
Rh_2O_3	$[Ru(bpy)_3]^{3+}$

Table 6 *Homogeneous O_2 catalysts*

Catalyst	*Oxidant*
$[(bpy)_2(H_2O)RuORu(H_2O)(bpy)_2]^{4+}$	Ce^{IV} ions, $[Ru(bpy)_3]^{3+}$
$[(NH_3)_5RuORu(NH_3)_4ORu(NH_3)_5]^{6+}$ (ruthenium red: $Ru^{III}Ru^{IV}Ru^{III}$)	Ce^{IV} ions
$[(NH_3)_4Ru(NH_2)_2Ru(NH_3)_4]^{4+}$	Ce^{IV} ions
$[Ru(NH_3)_5Cl]^{2+}$	Ce^{IV} ions
$[Ru(NH_3)_5(H_2O)]^{3+}$	Ce^{IV} ions
trans-$[(bpy)_2Ru(OH_2)_2]^{2+}$	$[Ru(bpy)_3]^{3+}$

3.3 Water Splitting using Semiconductor Photocatalysts

Electronic excitation of a dye molecule can be considered as the creation of a strong reducing agent (the **promoted electron**) and a strong oxidizing agent (the **vacancy created** by electronic excitation) within the same molecule, *i.e.* the creation of an **electron–hole pair** (e^-h^+). We have seen that one way in which this e^-h^+ pair may be separated is via electron transfer to a suitable electron donor, such as Q in Reaction (5). In this way the absorbed light energy is converted into redox chemical energy, but unfortunately this redox chemical energy can be rapidly converted into heat energy via a thermal back electron transfer reaction, *i.e.* Reaction (8). In addition, the e^-h^+ pairs themselves often recombine very quickly with the captured light degraded to heat, or sometimes with the emission of a photon (fluorescence or phosphorescence), see Reaction (4).

In order to **utilize the absorbed light** in a form other than heat we must **achieve separation of the e^-h^+ pairs** before their recombination. In dye-sensitized systems this separation is promoted by a difference in redox potential between the dye excited state and an added quencher and extended through the use of H_2 and O_2 catalysts. In contrast to dye-based photosystems, when a **semiconductor** is used as a photocatalyst e^-h^+ separation is achieved by means of an electric field, *i.e.* a **difference in electrical potential**.

In a solid, the electrons occupy energy bands as a consequence of the extended bonding network. In a semiconductor the highest occupied and lowest unoccupied energy bands (conductance and valence bands, respectively) are separated by a band gap, ΔE, a region devoid of energy bands. When a semiconductor is immersed in a solution, **charge transfer** occurs at the interface because of a difference in the tendency of the two phases to gain or lose electrons, *i.e.* there is a **difference in electrochemical potential of the two phases**. The net result is the formation of an electrical field at the surface of the semiconductor to a depth of 5–200 nm, the so-called **depletion zone** or **space charge region**, and as a consequence the conductance and valence bands are bent. The direction of this electrical field depends upon the relative electrochemical potentials of the semiconductor and solution.

In most of the work associated with water splitting using semiconductor photocatalysts, the **semiconductors used are n-type**, *i.e.* doped with a donor species so that some electrons are in the conduction band. For an n-type semiconductor the electrical field frequently forms in the direction from the bulk of the semiconductor toward the interface and causes bending of the conduction and valence bands of the semiconductor in the manner illustrated in Figure 7. Figure 8 illustrates the charge distribution under these conditions. The bending of the bands is such that any excess holes created in the space charge region move toward the surface, and any excess electrons move toward the bulk of the semiconductor.

Electron–hole pairs can be generated in the space charge region of the semiconductor by the absorption of light of energy $> \Delta E$. Separation of these e^-h^+ pairs, corresponding to the electron transfer step Reaction (4) in a homogeneous dye-based photoredox reaction, is often highly efficient due to the electric field which forms spontaneously at the semiconductor–electrolyte interface. The e^-h^+

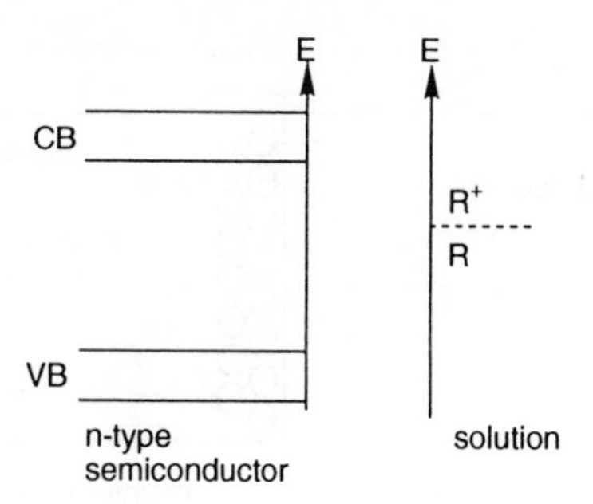

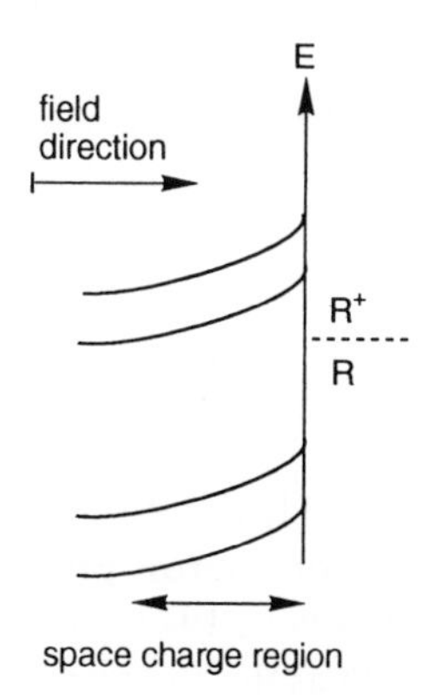

Figure 7 *Electronic energy levels before and after immersion of an n-type semiconductor in solution with a redox couple R^+/R. The bent bands of the semiconductor after immersion is due to the formation of an electric field between the semiconductor and the solution which extends over the space charge region*

pairs generated beyond the space charge region will suffer recombination, except for those that diffuse into that region before recombination occurs.

If the solution contains a species (R) which has an electronic energy level, *i.e.* redox potential, which is more negative than that of the photogenerated hole at the surface, then the electron transfer Reaction (11) is thermodynamically feasible:

$$R + h^+ \rightarrow R^+ \tag{11}$$

The excited electron, which can have an energy approaching that of the conduction band edge, can be transferred through a wire connected to the semiconductor to a second, non-photoactive electrode, such as carbon or a metal, where some electron acceptor (A) can be reduced:

$$A + e^- \rightarrow A^- \tag{12}$$

The combination of a semiconductor photoelectrode and dark counter electrode connected together to form an electrochemical cell is called a

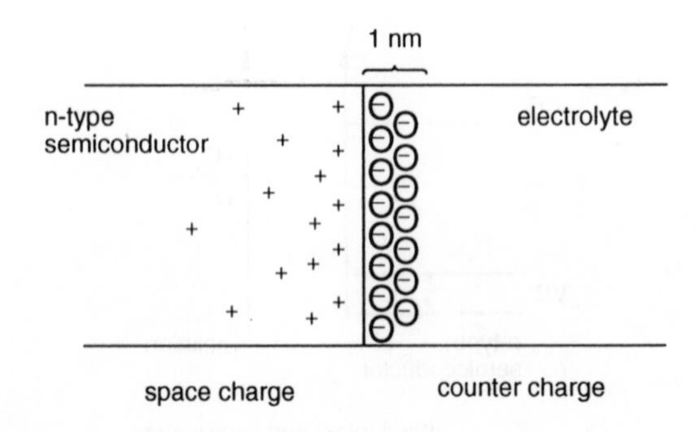

Figure 8 *Charge distribution at the semiconductor|electrolyte junction illustrated in Figure 7. This distribution is responsible for the electric field across the space charge region*

photoelectrochemical cell, or **PEC**. The energetics associated with the overall semiconductor photocatalysed electrochemical reaction is illustrated in Figure 9 and summarized by the following reaction equation:

$$A + R \xrightarrow[h\nu \geq \Delta E]{\text{n-type semiconductor PEC}} A^- + R^+ \quad (13)$$

If both A and R are H_2O and A^- is H_2 and R^+ is O_2, then Reaction (13) represents the splitting of water by **semiconductor photocatalysis** and this was first achieved[9] by Fujishima and Honda in 1972 using UV light, titanium dioxide as the n-type semiconductor, and Pt as the counter electrode, as illustrated in Figure 10.

In order to split water into hydrogen and oxygen using a PEC, the photogenerated holes in the valence band must be thermodynamically able to oxidize water to oxygen, *i.e.* $E_{VB} > E^\ominus(\frac{1}{2}O_2/H_2O)$, and the photogenerated electrons in the conductance band must be thermodynamically able to reduce water to hydrogen, *i.e.* $E_{CB} < E^\ominus(\frac{1}{2}H_2/H^+)$; Figure 11 illustrates the band positions for a variety of different, common semiconductors. In addition, the semiconductor itself must, ideally, be able to support the oxidation of water to O_2 on its surface without undergoing anodic corrosion. Unfortunately, extensive research has shown that, so far, few materials meet these energetic and stability requirements and those that do invariably absorb only ultraviolet radiation, *i.e.* ΔE_g is usually < 400 nm or > 3.1 eV, which is of little use for solar energy con- version, since the fraction of UV light in sunlight is very small.

Micro-Fujishima/Honda cells for water splitting can be made by putting the Pt counter electrode on the surface of TiO_2 particles, as illustrated in Figure 12. Compared with PECs which use macro-semiconductor electrodes, such particulate systems have the advantage of being much simpler and much less expensive to construct and use. Moreover, the efficiency of light absorption in the suspensions or slurries which form can be very high. Table 7 lists some of the different particulate PECs which have been shown to be capable of sensitizing the photocleavage of water.

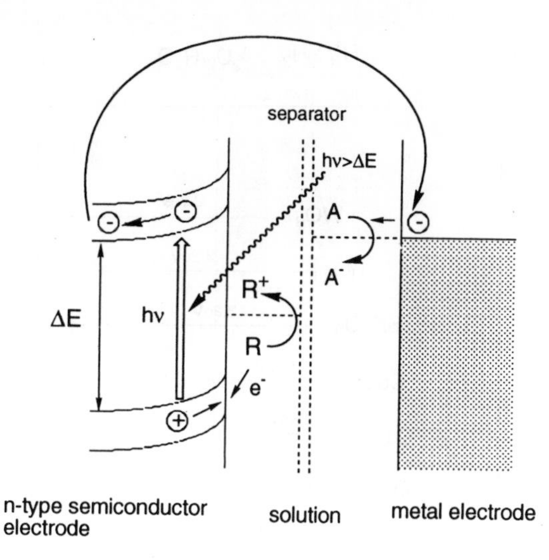

Figure 9 *Illustration of the energetics, circuit, and redox chemistry behind an n-type photoelectrochemical cell (PEC) for driving the photosynthetic redox reaction: $R + A \rightarrow R^+ + A^-$. In this cell, absorption of a photon of* ultra-*band gap energy in the space charge region of the semiconductor creates an electron–hole pair. The hole, h^+, migrates to the surface of the semiconductor where it can oxidize R to R^+, and the photogenerated electron migrates to the bulk of the semiconductor and then onto the metal counter electrode where it can reduce A to A^-*

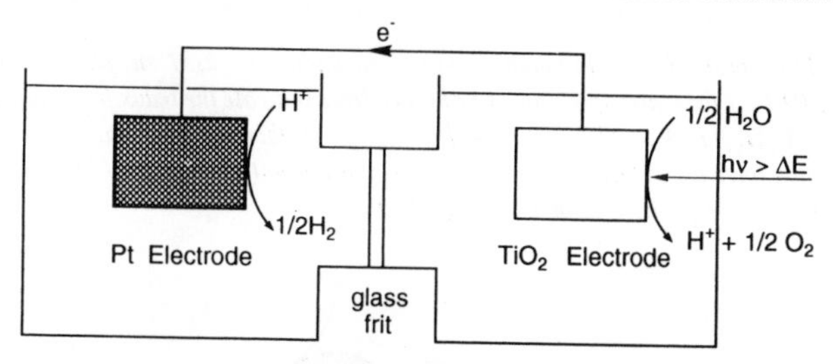

Figure 10 *Schematic illustration of the Fujishima/Honda TiO_2 photoelectrolysis cell. In this cell, absorption of a photon of* ultra-*band gap energy in the space charge region of the n-type TiO_2 creates an electron–hole pair. The hole, h^+, migrates to the surface of semiconductor where it can oxidize H_2O to O_2, and the photogenerated electron migrates to the bulk of the semiconductor and then onto the metal counter electrode where it can reduce H^+ to H_2*

As with the macro-PECs, the **micro-PECs** for water splitting usually utilize oxide semiconductors such as TiO_2 and $SrTiO_3$ which only absorb UV radiation, and, therefore, are of limited application to solar energy conversion. Thus, the use of CdS as a semiconductor for water splitting in a particulate PEC is of particular interest since it absorbs visible light, see Table 7. However, CdS also undergoes photoanodic dissolution, *i.e.*

$$CdS + 2\,h^+ \rightarrow Cd^{2+} + S \qquad (14)$$

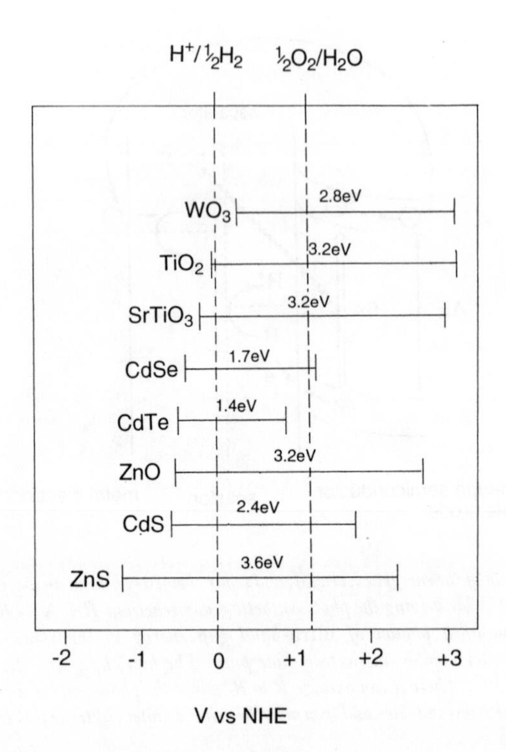

Figure 11 *Band positions of common n-type semiconductors used in photoelectrochemical cells (PECs) for water splitting. The broken lines indicate the redox potentials for the $H^+/\frac{1}{2}H_2$ and $\frac{1}{2}O_2/H_2O$ couples at pH 0. In order to be able to photocleave water a semiconductor must have a conductance band which is more negative than $E^\ominus(H^+/\frac{1}{2}H_2)$ and a valence band more positive than $E^\ominus(\frac{1}{2}O_2/H_2O)$*

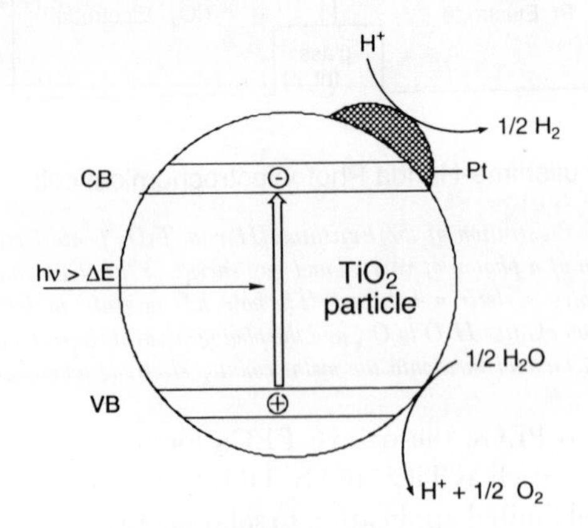

Figure 12 *Schematic illustration of a micro-Fujishima/Honda TiO_2 particulate photoelectrolysis cell. The principles of operation of this cell are the same as those for the macro-Fujishima/Honda cell illustrated in Figure 10*

Table 7 *Particulate photoelectrochemical cell (PEC) systems for the photocleavage of water*

Semiconductor	*H_2 catalyst*	*O_2 catalyst*
TiO_2 powder	Pt	—
$SrTiO_3$ powder	Rh, Pt	—
CdS powder	Pt	RuO_2
TiO_2 colloid	Pt	RuO_2
CdS colloid	Pt	RuO_2

and, therefore, although an O_2 catalyst, RuO_2, has been added to promote water oxidation over Reaction (14), in practice the CdS PEC system for water splitting in Table 7 doesn't function for very long, if at all.

3.4 Other Novel Applications of Semiconductor Photocatalysis

There is a growing interest in using semiconductors, usually in particulate form, to sensitize a wide variety of reactions,[10–15] some of which are listed in Table 8. The use of semiconductors as photosensitizers for the **complete oxidative mineralization of pollutants by oxygen** has attracted increasing attention over recent years. The overall process can be summarized as follows:

$$\begin{matrix}\text{organic}\\ \text{pollutant}\end{matrix} + O_2 \xrightarrow[\textit{ultra}\text{-bandgap light}]{\text{semiconductor, } O_2} CO_2 + H_2O + \text{mineral acids} \tag{15}$$

Photoexcitation of the semiconductor photocatalyst generates an electron–hole pair, the electron of which is trapped through a reaction with adsorbed oxygen, leading to the eventual generation of H_2O_2 and/or H_2O, whereas the photogenerated hole is believed to react with surface OH groups to form adsorbed $OH^{\cdot}$ radicals which then go on to attack the organic pollutant. This approach to **water purification** offers a number of advantages over existing methods (described in

Table 8 *Other novel reactions photocatalysed by TiO_2*

Application	*Reference*
Oxidative mineralization of organic pollutants	10
Photoinduced cytotoxic action towards cancer cells	11
Photodeodorizer for kitchens and bathrooms	12
Photooxidation of oil slicks using TiO_2 coated hollow glass microbeads	13
Recovery of platinum group metals from industrial wastes or dilute solution	14
Synthesis of amino acids (including: glycine, alanine, aspartic acid, and glutamic acid)	15

Chapter 16). These include: the mineral 'effluent' it produces is harmless to the environment, the process can be turned on or off at the flick of a switch, and there exists a real possiblity of incorporating it into current UV-only water purification systems.[10]

In this and the other types of semiconductor photocatalysis listed in Table 8, the semiconductor TiO_2 has attracted particular attention as a photocatalyst because it is biologically and chemically inert, very photostable, and cheap. A comparison of the relative photoactivities for Reaction (15), where the organic pollutant is pentachlorophenol, is illustrated in Figure 13; and a list of some of the organic pollutants which have been destroyed by Reaction (15) sensitized by TiO_2 is given in Table 9.

As indicated in Table 8, TiO_2 has been used as a photocatalyst for mediating the destructing of cancer cells.[11] It is believed that the photoinduced death of the cancer cells is due to attack from photogenerated hydroxyl radicals and hydrogen peroxide. The recovery of platinum group metals[14] from industrial wastes or dilute solution by semiconductor photocatalysis, is rather nicely illustrated by the data in Figure 14 and the overall reaction can be summarized as follows:

$$M^{n+} + H_2O \xrightarrow[h\nu \geq \Delta E]{TiO_2} M + 2H^+ + \tfrac{1}{2}O_2 \tag{16}$$

Of the other novel applications of semiconductor photocatalysis listed in Table 8, the work of Bard *et al.*[15] on the photosynthesis of amino acids from H_2O and CO_2

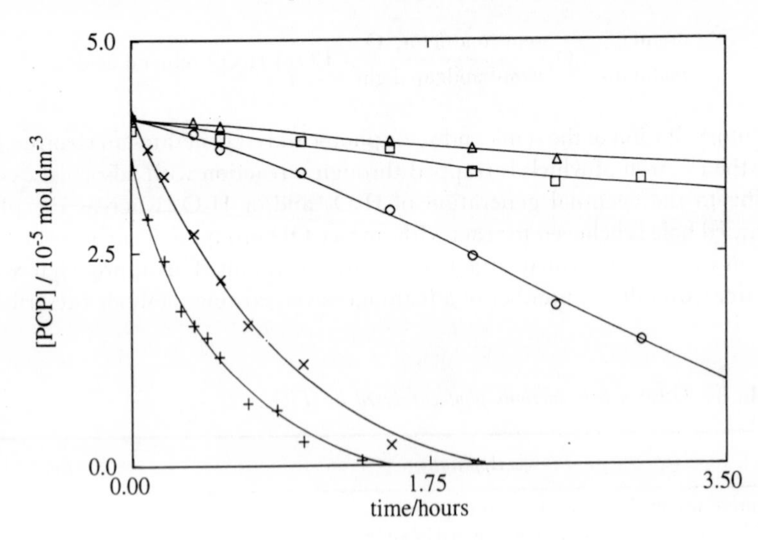

Figure 13 *Photodegradation of pentachlorophenol (PCP) sensitized by dispersions of the following semiconductors:* + TiO_2, × *ZnO*, ○ *CdS*, □ WO_3, *and* △ SnO_2. *In each case the solution was air-saturated, and the semiconductor and initial PCP concentrations were* 2 g cm^{-3} *and* 45 μmol dm^{-3}, *respectively*

(Data taken from M. Barbenie, E. Pramauro, E. Pelizzetti, E. Borgarello, M. Grätzel, and N. Serpone, *Chemosphere*, 1985, **14**, 195)

Table 9 *Photomineralization of organic pollutants sensitized by by TiO_2: examples of compounds studied*

Class	*Examples*
Alkanes	methane; *n*-dodecane; cyclohexane; paraffin
Haloalkanes	mono-, di-, tri- and tetra-chloromethane; fluorotrichloromethane; tribromoethane
Aliphatic alcohols	methanol; iso-propanol; benzyl alcohol; glucose; sucrose
Aliphatic carboxylic acids	formic; ethanoic; oxalic
Alkenes	propene; cyclohexene
Haloalkenes	chloroethene; hexafluoropropene
Aromatics	benzene; naphthalene
Haloaromatics	chlorobenzene; bromobenzene; 2,4-dinitrophenol; 1,2-dichlorobenzene
Aromatic alcohols	phenol; hydroquinone; resorcinol; *p*-cresol
Aromatic carboxylic acids	benzoic; phthalic; salicylic
Polymers	polyethylene; polyvinyl chloride
Surfactants	sodium dodecyl sulfate; trimethyl phosphate; trimethyl phosphite; tetrabutylammonium phosphate
Herbicides	methyl viologen; atrazine; simazine; prometon; prometryne; bentazon
Pesticides	DDT; parathion; lindane
Dyes	methylene blue; rhodamine B; methyl orange; fluorescein; umbelliferone

DDT = 1,1-bis(*p*-chlorophenyl)-2,2,2-trichloroethane

sensitized using TiO_2/Pt particles is most intriguing and does make one wonder if some of the building blocks of life were generated from a primordial chemical 'soup' by semiconductor photocatalysis.

4 CONVERSION OF SOLAR ENERGY TO ELECTRICAL ENERGY

4.1 Thin-layer Photogalvanic Cells

One of the earliest devices for converting solar energy to electrical energy was the thin-layer photogalvanic cell, more often referred to simply as a photogalvanic cell, and was developed by Rabinowitch at MIT some 50 years ago.[16] In most types of **photogalvanic cell** the light is **absorbed by a dye in a thin electrolyte solution sandwiched between two electrodes**, as illustrated in

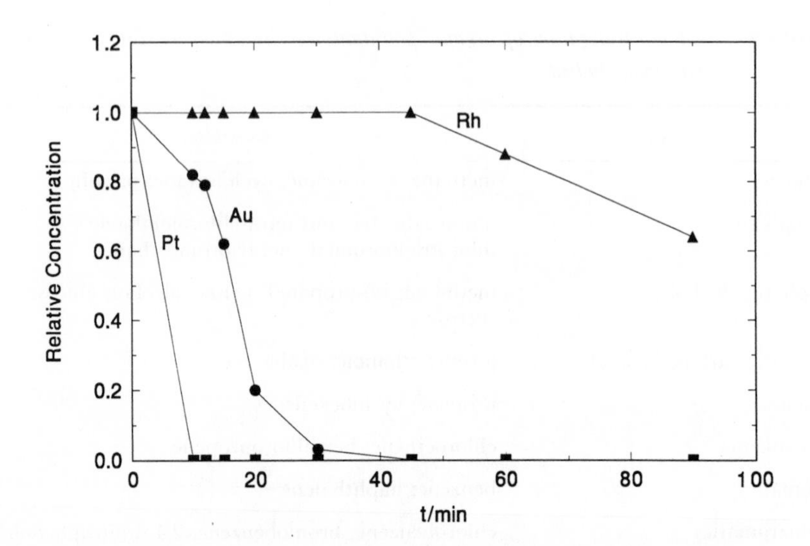

Figure 14 *Photoreduction of a mixture of platinum group metals (PGMs) {Rh(III) (13 ppm); Au(III) (30 ppm); Pt(IV) (45 ppm)}, sensitized by* TiO_2 (2 mg cm^{-3}), *using AM1 simulated sunlight. The initial solution was air-equilibrated and at pH 7.39*
(Data taken from E. Borgarello, G. Emo, R. Harris, E. Pelizzetti, and N. Serpone, *Inorg. Chem.*, 1986, **25**, 4499)

Figure 15. This electrolyte solution also contains a species, B, which will undergo an electron transfer reaction with the excited state of D, *i.e.*

$$D^* + B \rightarrow E + C \tag{17}$$

The products E and C will usually undergo a rapid thermal back reaction. However, it is possible in some cases to prevent this in part if E is able to react at the illuminated, transparent electrode and regenerate the dye, *i.e.*

$$E \pm e^- \xrightarrow{\text{illuminated electrode}} D \tag{18}$$

For the photogalvanic cell to work efficiently it is essential that the illuminated electrode is selective, so that the reaction

$$C \mp e^- \rightarrow B \tag{19}$$

is blocked and C must diffuse across the cell to the dark electrode before it can react via Reaction (19). Provided the electrode kinetics are rapid for the D,E and C,B redox couples at the illuminated and dark electrodes, respectively, then a photopotential (ΔE) will be developed by the cell where

$$\Delta E = E_{D,E} - E_{C,B} \tag{20}$$

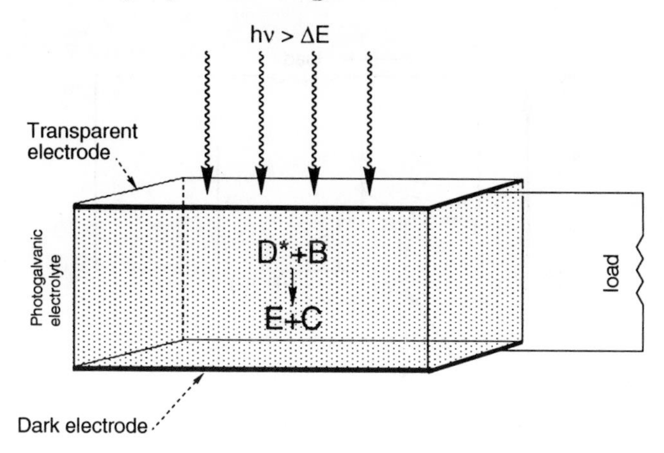

Figure 15 *Schematic illustration of the basic components of a photogalvanic cell. Absorption of a photon of light by the dye, D, creates an electronically excited state, D*, which is quenched by B to form E, an oxidized or reduced form of the dye, and C, a reduced or oxidized form of the quencher. The D/E and B/C redox couples pose the illuminated and dark electrodes, respectively, at their Nernst potentials, thereby creating a potential difference from which electrical power can be drawn via a load resistance in an external circuit*

where $E_{D,E}$ and $E_{C,B}$ are the Nernst potentials at the illuminated and dark electrodes, respectively. The Nernst potential, E, for a general one electron redox couple OX/RED is related to the concentrations of OX and RED, *i.e.* [OX] and [RED] respectively, by the **Nernst equation**, *i.e.*

$$E = E^{\ominus'} + RT/F\ln\{[OX]/RED]\} \tag{21}$$

where $E^{\ominus'}$ is the formal redox potential for the couple. If a load resistance is placed across the terminals of the two electrodes, then during irradiation electrical power can be drawn from the cell. An example of a photogalvanic cell is one in which D is $[Ru(bpy)_3]^{2+}$ and B is Fe^{2+}.

Unfortunately, there are numerous problems associated such photogalvanic cells, not least of which is that of achieving high electrode selectivity. As a result of these problems few photogalvanic devices can be made to work efficiently[17] and even the best single dye photogalvanic cell, *i.e.* the iron–thionine cell, has a theoretical optimum solar energy conversion efficiency (η) of only *ca.* 1%. Thus, although photogalvanic cells were, at one time, a popular focus of solar energy research, interest has slackened considerably over recent years.

4.2 Photoelectrochemical Cells (PECs)

Not only can photoelectrochemical cells be used to generate chemical fuels, such as hydrogen in the Fujishima/Honda cell, they can also be used to **generate electricity**, as illustrated in Figure 16. In this n-type PEC, photogenerated e^-h^+ pairs are produced in the depletion zone of the semiconductor by the absorption of a photon of light of energy $\geq \Delta E$ and separated by the electric field within this

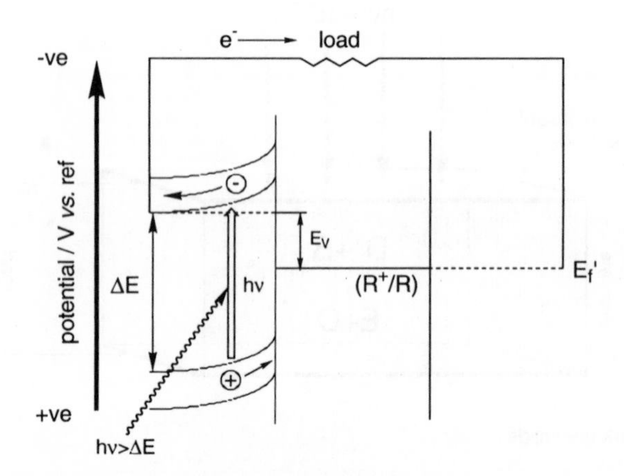

Figure 16 *Illustration of the energetics, circuit, and redox chemistry behind an n-type photoelectrochemical cell for electricity generation. In this cell, absorption of a photon of* ultra-*band gap energy in the space charge region of the semiconductor creates an electron–hole pair. The hole, h^+, migrates to the surface of the semiconductor where it can oxidize R to R^+ and the photogenerated electron migrates to the bulk of the semiconductor and then onto the metal counter electrode where it can reduce R^+ back to R. If a load resistance is incorporated in the external circuit, a photopotential, E_V, is developed to drive the current through the circuit*

region of the semiconductor. The **photogenerated holes migrate to the surface of the semiconductor** where they are able to oxidize some redox species, R, to R^+. For the PEC to operate efficiently the **photogenerated electrons must move through an external circuit to the other electrode** and there reduce R^+ to R.

In PECs for electricity generation, the electrochemical reactions which take place in the electrolyte serve only to carry current. However, one of the major problems with the development of an efficient PEC for solar energy conversion is that all semiconductors which absorb visible light are thermodynamically unstable to oxidation by the photogenerated holes. This is not surprising given that the concentration of holes may be more than 10^6 times that in the bulk of the semiconductor. What is required in order to make an efficient PEC is a **method for making the semiconductor kinetically stable**.

Kinetic stabilization of PECs for electricity generation has been achieved through the **careful choice of the R^+/R redox couple**, such that the rate of oxidation of R by the holes is much greater than that for the semiconductor. Thus, it has been shown[18] that n-type CdS and CdSe are stabilized against anodic corrosion when used in a PEC if Na_2S_2/NaS_2 in 1 mol dm^{-3} NaOH is used as the redox couple R^+/R.

Table 10 lists a selection of different PECs which have been developed to date, in which **the semiconductor used was in the form of a single crystal**. When

Table 10 *Single crystal photoelectrochemical cells (PECs)*

Semiconductor	*Redox electrolyte*	*% solar energy efficiency (η)*
n-GaAs	1 ml dm^{-3} K_2Se, 0.01 mol dm^{-3} K_2Se_2, 1 mol dm^{-3} KOH, Ru^{3+}	12[19]
n-GaAs	1 mol dm^{-3} K_2Se, 0.01 mol dm^{-3} K_2Se_2, 1 mol dm^{-3} KOH, Os^{3+}	15
p-InP	0.3 mol dm^{-3} V^{3+}, 0.05 mol dm^{-3} V^{2+}, 5 mol dm^{-3} HCl	11.5
n-WSe	1 mol dm^{-3} KI, 0.01 mol dm^{-3} KI_3	10.2
n-CdSe	1 mol dm^{-3} Na_2S_2, 1 mol dm^{-3} NaOH	7.5
n-WS_2	1 mol dm^{-3} NaBr, 0.01 mol dm^{-3} Br_2	6

the semiconductor is in this form the major efficiency-lowering problem of recombination of electrons and holes at grain boundaries is minimized.

The first entry in Table 10, the highly efficient n-type GaAs/K_2Se–K_2Se_2–KOH/C cell developed by Heller *et al.*[19] at Bell Laboratories, is worthy of some discussion. In this PEC the n-type GaAs single crystal semiconductor had Ru^{3+} ions chemisorbed onto its surface, a process which is believed to decrease the rate of e^-h^+ recombination, and its surface was etched to provide a textured structure which reduced light reflection losses. If the GaAs in this PEC was not treated in this manner η was only 7–10%.

Although the single crystal PECs listed in Table 10 work with high efficiencies, the cost of producing the semiconductors in single crystal form is also high. If, instead of a single crystal, a polycrystalline form of the semiconductor was used in any of these PECs then their respective solar energy conversion efficiencies would usually be expected to be drastically reduced. This is due to the large number of grain boundaries, and therefore the increased probability of electron–hole recombination, associated with polycrystalline materials. This state of affairs is unfortunate since polycrystalline semiconductors are much easier, and therefore cheaper, to fabricate. Recently, it has been shown that there are methods of preparation and/or subsequent chemical modifications of polycrystalline semiconductors, *e.g.* diffusion of Ru^{3+} ions into the grain boundaries, which can produce polycrystalline semiconductors with low grain boundary and surface state electron–hole recombination rates. As a result of this work a **number of polycrystalline PECs** have now been developed which work with **high solar efficiencies** ($\eta > 5\%$) and a list of some of these is given in Table 11.

4.3 Dye-sensitized PECs

There have been many attempts to make the established, efficient (typically up to 36%), and chemically robust UV-absorbing PECs for electricity generation, *e.g.* a TiO_2/Pt PEC, work using visible light. One major approach has been to dope the semiconductor with transition metals, such as Fe or Cr, to produce a semiconductor which absorbs visible light and is still chemically very stable. Unfortunately, the

Table 11 *Polycrystalline photoelectrochemical cells (PECs)*

Semiconductor	*Redox electrolyte*	*% solar energy efficiency* (η)
n-$CdSe_{0.65}Te_{0.35}$	1 mol dm^{-3} Na_2S_2, 1 mol dm^{-3} KOH	7.9
n-GaAs	1 mol dm^{-3} K_2Se, 0.01 mol dm^{-3} K_2Se_2, 1 mol dm^{-3} KOH	7.8
n-CdSe	1 mol dm^{-3} Na_2S_2, 1 mol dm^{-3} NaOH	6.5

PECs which utilize such transition metal doped semiconductors do not, however, appear to work very well, if at all, when exposed to visible light.

Another approach to extend the operating wavelength range of PECs based on UV-absorbing semiconductors is to absorb dyes onto the surface of the semiconductor. A schematic representation of the energetics associated with a **dye-sensitized PEC** is given in Figure 17. In a dye-sensitized PEC, the electronically excited dye, D*, injects an electron into the conduction band of the semiconductor to produce a photocurrent which flows through an external circuit to the dark counter electrode where R^+ is converted to R. The oxidized form of the dye, D^+, is then converted back to D by the following reaction:

$$D^+ + R \rightarrow D + R^+ \tag{22}$$

This approach has not met with much success, mainly because of two conflicting problems, namely: (i) if a very little amount of dye is adsorbed onto the surface of the semiconductor, *i.e.* a monolayer, then the amount of light absorbed by such a small amount of dye will be very small and the system will be very inefficient, and (ii) if a large amount of dye is absorbed onto the surface of the semiconductor, *i.e.* many monolayers, then self-quenching of the dye occurs very efficiently, *i.e.*

$$D^* + D \rightarrow 2D + \Delta \tag{23}$$

The overall result is that most dye-sensititized PECs are inefficient at harvesting the sunlight, with most having solar to electrical energy conversion efficiencies of $<1\%$. In addition, many of the dyes used are not very stable chemically or photochemically.

A major breakthrough in the area of dye-sensitized PECs has been made recently,[20] based on: a 10 μm thick, optically transparent film of TiO_2 particles, coated with a monolayer of trimeric ruthenium complex dye. The energetics for the cell are the same as illustrated for a general dye-sensitized PEC in Figure 17, with, in this case: semiconductor electrode = colloidal TiO_2 deposited onto a conducting glass electrode, $D = RuL_2(\mu\text{-}(CN)Ru(CN)L'_2)_2$, where L is 2,2′-bipyridine-4,4′-dicarboxylic acid and L′ is 2,2′-bipyridine, $R^+/R = I_3^-/I^-$, and counter electrode = conducting glass electrode. The use of a dye-coated colloidal TiO_2 electrode ensures that the effective surface area of the electrode will be very high and much greater than that for an electrode with a smooth surface. The result

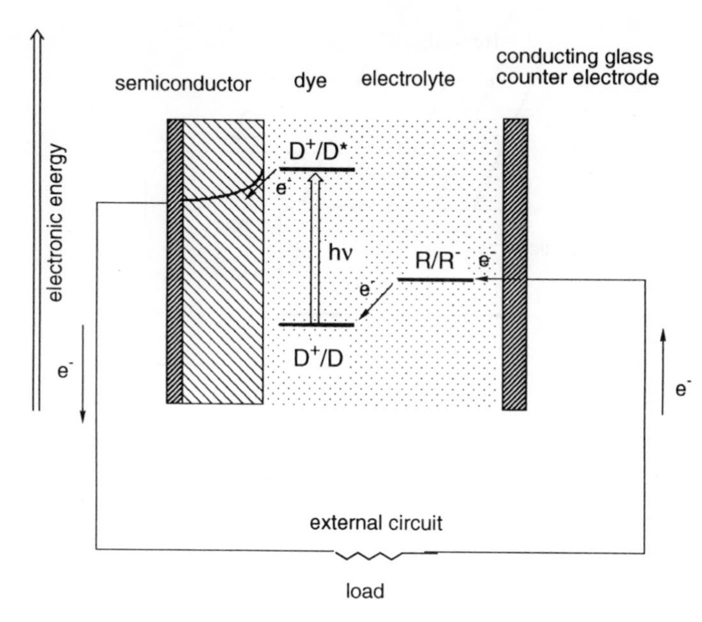

Figure 17 *Illustration of the energetics, circuit, and redox chemistry behind a dye-sensitized n-type photoelectrochemical cell for electricity generation. In this cell, absorption of a photon of light by the dye, D, creates an electronically excited state, D*, which is able to inject an electron into the n-type semiconductor, which is typically TiO_2. The photoinjected electron migrates to the bulk of the semiconductor, through the influence of the intrinsic electrical field set up between the semiconductor and electrolyte. The electron then passes through an electric circuit where it loses energy by passing through a load resistance before reaching the counter electrode. Here it can reduce an electron relay, R, to R^-, which then, in turn, can reduce D^+ back to D, allowing the process to start all over again. In the O'Regan/Grätzel cell of this type,*[20] *$D = RuL_2(\mu\text{-}(CN)Ru(CN)L'_2)_2$, where L is 2,2′-bipyridine-4,4′-dicarboxylic acid and L′ is 2,2′-bipyridine), and $R^+/R = I_3^-/I^-$*

is that **over 80% of the light which can be absorbed by the dye is used to produce electricity** and that this device works with a solar to electrical energy conversion efficiency of 7.1–7.9% in direct sunlight and 12% in a diffuse daylight (the spectral distribution of diffuse daylight overlaps more favourably with the absorption spectrum of the PEC than that of direct daylight). The dye is also very chemically and photochemically stable and can undergo at least 5 million turnovers without decomposition. As a result, the amount of light absorbed by the electrode will be much greater than that for a smooth single crystal.[20]

4.4 Solar Cells

A solar cell, or **photovoltaic cell**, is a solid state semiconductor device containing a region of varying chemical composition which is associated with an electric field gradient which is used to separate any photogenerated electron–hole pairs in the region. The **gradation in chemical composition** can be achieved either by **layering two dissimilar materials** together or by **doping a single semiconductor asymmetrically**. The most well known and advanced of the

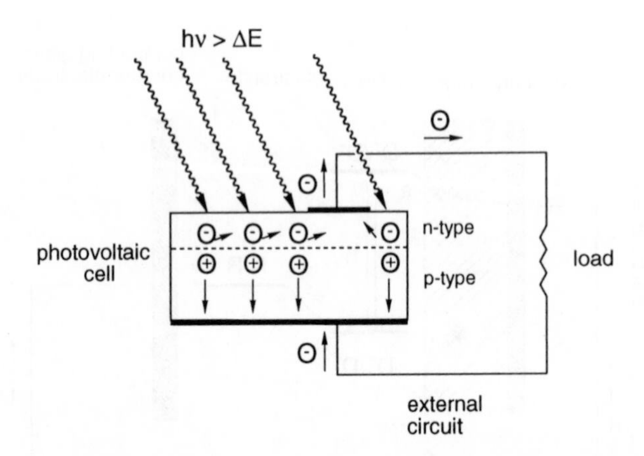

Figure 18 *Schematic representation of current generation by a solar cell. In this cell, absorption of a photon of* ultra-*band gap energy in the space charge region of the n-type|p-type semiconductor junction creates an electron–hole pair. Under the influence of the electric field present in the space charge region, the hole,* h^+*, migrates into the bulk of the p-type semiconductor and to the back plate metal connector. The electric field in the space charge region also drives the photogenerated electon into the bulk of the n-semiconductor and then onto one of the collector 'finger' metal connectors, through an external circuit, and finally to the p-type semiconductor where it fills the photogenerated hole. If a load resistance is incorporated in the external circuit, a photopotential,* E_V*, is developed to drive the current through the circuit and electrical power can be drawn from the cell*

solar cells is the **silicon solar cell** and we shall limit our discussion to this device. A typical silicon solar cell consists of 1 μm layer of n-type silicon (silicon doped with phosphorus) in intimate contact with a 100–300 μm layer of p-type silicon (silicon doped with boron). A schematic representation of the way in which a silicon solar cell generates current is given in Figure 18.

The first silicon solar cell was developed at Bell Telephone laboratories in 1954 by Chapin, Fuller, and Pearson,[21] and worked with a 4% solar efficiency. This efficiency was soon bettered by the same laboratories and cells with efficiencies of 11% were available by the end of the decade. However, research into silicon solar cells really took off in the late fifties after they had proved enormously successful at providing power for the radio in the US Vanguard space satellite. Until recently most silicon solar cells were made from large single crystal silicon to avoid problems of low efficiency due to recombination of the photogenerated electrons and holes at grain boundaries and other trapping states. The band gap of **single crystal silicon** is 1.1 eV (1127 nm) and the efficiencies of single crystal silicon solar cells used in satellites are **typically 15%** under solar space flux and cells with **efficiencies in excess of 20% can be produced on a laboratory scale**. However, single crystal silicon is expensive to manufacture and, consequently, this also applies to the single crystal solar cells. As a result, single crystal solar cells are much too expensive to compete with the conventional forms of energy currently available, such as fossil fuels and nuclear power.

In 1976 Carlson and Wronski[22] at the RCA Laboratories in Princeton made a major breakthrough in solar cell technology by producing an **amorphous**

silicon solar cell with a solar conversion efficiency of 5.5%. Until that time, amorphous silicon was thought to be totally inappropriate for use in solar cells because its structural and electrical properties are closer to those of glass, an insulator, than to those of single crystal silicon or similar semiconducting materials. It is now known that amorphous silicon can be used in solar cells through careful control of the conditions under which it is deposited and modification of its composition. Amorphous silicon has a larger band gap (1.65 eV or 752 nm) than single crystal silicon and absorbs light about 40 times more strongly than does single crystal silicon and this is the key to the success of amorphous silicon. Since carriers in amorphous silicon have a low mobility and recombine rapidly, the only way they can be collected efficiently, *i.e.* separated by the electric field, during illumination is if the time needed for their collection is short. Although this requires a thin cell, the strongly absorbing properties of amorphous silicon make it possible for most of the sunlight to be absorbed in such a cell. Thus amorphous silicon solar cells have only a thin layer of silicon and this has additional economic advantages since less material is required and deposition tends to be easier.

The amorphous silicon solar cell is not without its problems, the main one being a slow photoelectronic degradation at the high solar flux levels found in space, the reason for which is not as yet clearly understood. However, despite such problems, amorphous silicon solar cells are now available which work with a **10% solar energy conversion efficiency** and are in direct competition with single crystal silicon solar cells for terrestrial applications. They are now to be found providing power to **calculators**, **miniature televisions**, and **portable battery chargers**. Japanese companies have been particularly good at marketing these cells in such items and the sales profits have allowed them to support ongoing research in amorphous silicon solar cells.

5 CONCLUSIONS AND NEW POTENTIAL APPLICATIONS

There already exists a large number of photochemical and photoelectrochemical systems capable of converting solar energy into chemical or electrical energy, most of which utilize specialist inorganic catalysts and photocatalysts and are described in this chapter. At present, only **solar cells** represent anything near a viable **challenge** to **nuclear energy** in terms of cost, and both of these alternative energy approaches are, at present, appreciably more expensive than fossil fuels. It is difficult to say when, or even if, solar energy will supply an appreciable fraction of the world's energy needs through an artificial solar energy conversion device. It is clear that further research is necessary in order to generate relevant innovative ideas and yet, at present, research funding is largely directed towards other projects; projects with specific, short-term goals.

Photoelectrochemical systems hold out the most promise in the area of new potential applications. For example, the recent success story of an efficient PEC based on dye sensitization of a thin layer of colloidal TiO_2 on a transparent conducting electrode has reawakened hopes of a **cheap solar to electrical energy conversion system**, since such cells are much easier to fabricate than solar cells. Such is the promise of the initial results in this area that further research is likely to be intense over the next few years.

The use of TiO_2 as a photocatalyst for **purifying water** of organic pollutants has proved to be of real interest to the water purification companies and is a growing area of research of great potential. One area of promise which is still in its infancy and deserves greater attention is the **cytotoxic action** of TiO_2 photocatalysts. It appears quite likely that this area of research will be extended from killing cancer cells into the general research area of photosterilization, where semiconductor photocatalysts are used to destroy bacteria and viruses.

6 REFERENCES

6.1 Specific

1. Solar Energy Research Institute, 'Basic Photovoltaic Principles and Methods', Van Nostrand Reinhold, New York, 1984.
2. J. C. de Paula, W. F. Beck, and G. W. Beck, *Nouv. J. Chim.*, 1987, **11**, 103.
3. K. Kalyanasundaram, *Coord. Chem. Rev.*, 1982, **46**, 159.
4. A Mills, 'Chemistry of the Platinum Metals; Recent Developments', ed. F. R. Hartley, Studies in Inorganic Chemistry, vol. 11, Elsevier, Amsterdam, 1991, ch. 11.
5. A Demortier, M. de. Backer, and G. Lepoute, *Nouv. J. Chim.*, 1983, **7**, 421.
6. A. Mills, *Chem. Soc. Rev.*, 1989, **18**, 285.
7. A. Mills and T. Russell, *J. Chem. Soc., Faraday Trans.*, 1991, **87**, 313.
8. J. Kiwi and M. Grätzel, *Chimia*, 1979, **33**, 289.
9. A. Fujishima and K. Honda, *Nature (London)*, 1972, **238**, 37.
10. A. Mills, R. H. Davies, and D. Worsley, *Chem. Soc. Rev.*, 1993, **22**, 417.
11. R. Cai, Y. Kubota, T. Shuin, H. Sakai, K. Hashimoto, and A. Fujishima, *Cancer Res.*, 1992, **52**, 2346.
12. T. Ogawa, T. Saito, T. Hasegawa, H. Shinozaki, K. Hashimoto, and A. Fujishima, in The First Interntional Conference on TiO_2 Photocatalytic Purification and Treatment of Water and Air: Book of Abstracts', London, Ontario, Canada, November, 1992, p. 192.
13. I. Rosenberg, J. R. Brock, and A. Heller, *J. Phys. Chem.*, 1992, **96**, 3423.
14. E. Borgarello, R. Harris, and N. Serpone, *Nouv. J. Chim.*, 1985, **9**, 743.
15. A. J. Bard, W. W. Dunn, and Y. Aikawa, *J. Am. Chem. Soc.*, 1981, **103**, 6893.
16. E. Rabinowitch, *J. Phys. Chem.*, 1940, **8**, 551.
17. W. J. Albery, *Acc. Chem. Res.*, 1982, **15**, 142.
18. A. J. Nozik, *Appl. Phys. Lett.*, 1977, **29**, 1019.
19. A. Heller, B. A. Parkinson, and B. Miller, *Appl. Phys. Lett.*, 1978, **36**, 521.
20. B. O'Regan and M. Grätzel, *Nature (London)*, 1991, **353**, 737.
21. D. M. Chapin, C. S. Fuller, and G. L. Pearson, *Appl. Phys. Lett.*, 1954, **25**, 676.
22. D. E. Carlson and C. R. Wronski, *Appl. Phys. Lett.*, 1976, **28**, 671.

6.2 General

1. 'Energy Resources through Photochemistry and Catalysis', ed. M. Grätzel, Academic Press, London, 1993.
2. 'Photocatalysis: Fundamentals and Applications', ed. N. Serpone and E. Pelizzetti, Wiley–Interscience, New York, 1989.
3. G. Foley, 'The Energy Question', Penguin Press, Harmondsworth, UK, 4th edn, 1992.
4. 'Light, Chemical Change and Life: A Source Book in Photochemistry', ed. J. D. Coyle, R. R. Hill, and D. R. Roberts, Open University Press, Milton Keynes, 1982.
5. Y. V. Pleskov, 'Solar Energy Conversion: A Photoelectrochemical Approach', Springer-Verlag, Heidelberg, 1990.
6. Solar Energy Research Institute, 'Basic Photovoltaic Principles and Methods', Van Nostrand Reinhold, New York, 1984.

6.3 Journal Literature

For articles relevant to the topic of catalysts and photocatalysts for solar energy conversion the following core journals are recommended:

Journal of the American Chemical Society
Journal of Physical Chemistry
Inorganic Chemistry
Applied Physics Letters
Nature
Solar Energy Materials
Journal of the Chemical Society, Faraday Transactions

Subject Index

CAMBRIDGE UNIVERSITY PRESS

THE SYNDICS OF THE PRESS

have pleasure in sending for review

a copy of

LEAKEY: Olduvai Gorge 1960-1963
Volume III

Price: £10.00 net.

The date of publication is 30th December 1971

It is requested that no review should appear before this date.

THE SYNDICS *will be obliged if the* EDITOR *will send a copy of any review he may publish to*

BENTLEY HOUSE, 200 EUSTON ROAD
LONDON, N.W.1